Student Solutions Manual

for Stewart's

CALCULUS

Concepts AND Contexts

SINGLE VARIABLE

Jeffery A. Cole

Anoka-Ramsey Community College

BROOKS/COLE PUBLISHING COMPANY

I(T)P® An International Thomson Publishing Company

Pacific Grove • Albany • Belmont • Bonn • Boston • Cincinnati • Detroit • Johannesburg • London
Madrid• Melbourne • Mexico City • New York • Paris • Singapore • Tokyo • Toronto• Washington

 A GARY W. OSTEDT BOOK

Sponsoring Editor: *Beth Wilbur*
Marketing Representative: *Bob Johnson*
Editorial Associate: *Nancy Conti*
Production: *Mary Vezilich*
Cover Design: *Vernon T. Boes*

Cover Photo: *Ian Sabell, Linear Photographs Ltd.; violin created by David Bruce Johnson*
Typesetting/Production Coordination: *Andy Bulman-Fleming*
Printing and Binding: *Malloy Lithographing*

 The ITP logo is a registered trademark under license.

For more information, contact:

BROOKS/COLE PUBLISHING COMPANY
511 Forest Lodge Road
Pacific Grove, CA 93950
USA

International Thomson Editores
Seneca 53
Col. Polanco
11560 México, D. F., México

International Thomson Publishing Europe
Berkshire House 168-173
High Holborn
London WC1V 7AA
England

International Thomson Publishing Japan
Hirakawacho Kyowa Building, 3F
2-2-1 Hirakawacho
Chiyoda-ku, Tokyo 102
Japan

Thomas Nelson Australia
102 Dodds Street
South Melbourne, 3205
Victoria, Australia

International Thomson Publishing Asia
221 Henderson Road
#05-10 Henderson Building
Singapore 0315

Nelson Canada
1120 Birchmount Road
Scarborough, Ontario
Canada M1K 5G4

International Thomson Publishing GmbH
Königswinterer Strasse 418
53227 Bonn
Germany

Printed in the United States of America

10 9 8 7 6 5 4 3

ISBN 0-534-34435-6

Preface

This *Student Solutions Manual* contains strategies for solving and solutions to selected exercises in the text *Calculus: Concepts and Contexts, Single Variable* by James Stewart. It contains solutions to the odd-numbered exercises in each section, the review sections, the True-False Quizzes, and the Problem Solving sections, as well as solutions to all of the exercises in the Concept Checks.

This manual is a text supplement and should be read along *with* the text. You should read all exercise solutions in this manual because many concept explanations are given and then used in subsequent solutions. All concepts necessary to solve a particular problem are not reviewed for every exercise. If you are having difficulty with a previously covered concept, refer back to the section where it was covered for more complete help.

A significant number of today's students are involved in various outside activities, and find it difficult, if not impossible, to attend all class sessions; this manual should help meet the needs of these students. In addition, it is my hope that this manual's solutions will enhance the understanding of all readers of the material and provide insights to solving other exercises.

I appreciate feedback concerning errors, solution correctness or style, and manual style. Any comments may be sent directly to me at the address below or in care of the publisher: Brooks/Cole Publishing Company, 511 Forest Lodge Road, Pacific Grove, CA 93950.

I would like to thank Jim Stewart, for his keen insights to solutions and for the opportunity to work on several aspects of the text; Andy Bulman-Fleming, for typesetting the manuscript and providing invaluable help in many areas; Dan Clegg, of Palomar College, for his careful assistance with most of the new solutions; Brian Betsill and Kathi Townes, of TECH-arts, for rendering and assembling the art package; and Gary Ostedt and Beth Wilbur, of Brooks/Cole Publishing Company, for entrusting me with this project as well as for their patience and support.

I dedicate this book to my wife, Joan.

Jeffery A. Cole
Anoka-Ramsey Community College
11200 Mississippi Blvd. NW
Coon Rapids, MN 55433

Contents

1 Functions and Models 1

1.1 Four Ways to Represent a Function 1
1.2 New Functions from Old Functions 5
1.3 Graphing Calculators and Computers 11
1.4 Parametric Curves 15
1.5 Exponential Functions 19
1.6 Inverse Functions and Logarithms 21
1.7 Models and Curve Fitting 25
      Review 27

Principles of Problem Solving 32

2 Limits and Derivatives 35

2.1 The Tangent and Velocity Problems 35
2.2 The Limit of a Function 37
2.3 Calculating Limits Using the Limit Laws 39
2.4 Continuity 42
2.5 Limits Involving Infinity 45
2.6 Tangents, Velocities, and Other Rates of Change 49
2.7 Derivatives 52
2.8 The Derivative as a Function 55
2.9 Linear Approximations 61
2.10 What Does f' Say about f? 63
      Review 66

Focus on Problem Solving 72

3 Differentiation Rules 74

3.1 Derivatives of Polynomials and Exponential Functions 74
3.2 The Product and Quotient Rules 78
3.3 Rates of Change in the Natural and Social Sciences 81
3.4 Derivatives of Trigonometric Functions 85
3.5 The Chain Rule 88
3.6 Implicit Differentiation 93
3.7 Derivatives of Logarithmic Functions 98

Contents

3.8 Linear Approximations and Differentials 100
         Review 102

Focus on Problem Solving 108

4 Applications of Differentiation 116

4.1 Related Rates 116
4.2 Maximum and Minimum Values 120
4.3 Derivatives and the Shapes of Curves 124
4.4 Graphing with Calculus and Calculators 132
4.5 Indeterminate Forms and L'Hospital's Rule 140
4.6 Optimization Problems 145
4.7 Applications to Economics 151
4.8 Newton's Method 153
4.9 Antiderivatives 157
         Review 161

Focus on Problem Solving 172

5 Integrals 176

5.1 Areas and Distances 176
5.2 The Definite Integral 181
5.3 Evaluating Definite Integrals 184
5.4 The Fundamental Theorem of Calculus 187
5.5 The Substitution Rule 190
5.6 Integration by Parts 194
5.7 Integration Using Tables and Computer Algebra Systems 198
5.8 Approximate Integration 201
5.9 Improper Integrals 207
         Review 211

Focus on Problem Solving 216

6 Applications of Integration 218

6.1 More about Areas 218
6.2 Volumes 222
6.3 Arc Length 228
6.4 Average Value of a Function 231
6.5 Applications to Physics and Engineering 232
6.6 Applications to Economics and Biology 237

Contents

6.7 Probability   238
Review   240

Focus on Problem Solving   244

7 **Differential Equations**   247
7.1 Modeling with Differential Equations   247
7.2 Direction Fields   249
7.3 Euler's Method   252
7.4 Separable Equations   254
7.5 Exponential Growth and Decay   258
7.6 The Logistic Equation   260
7.7 Predator-Prey Systems   265
Review   267

Focus on Problem Solving   271

8 **Infinite Sequences and Series**   273
8.1 Sequences   273
8.2 Series   276
8.3 The Integral and Comparison Tests; Estimating Sums   282
8.4 Other Convergence Tests   285
8.5 Power Series   288
8.6 Representations of Functions as Power Series   292
8.7 Taylor and Maclaurin Series   296
8.8 The Binomial Series   302
8.9 Applications of Taylor Polynomials   305
8.10 Using Series to Solve Differential Equations   311
Review   314

Focus on Problem Solving   321

Appendixes   325
A Intervals, Inequalities, and Absolute Values   325
B Coordinate Geometry   327
C Trigonometry   332
D Precise Definitions of Limits   335
F Integration of Rational Functions by Partial Fractions   338
G Polar Coordinates   342
H Complex Numbers   353

Chapter 1   Functions and Models

Section 1.1   Four Ways to Represent a Function

Note: In exercises requiring estimations or approximations, your answers may vary slightly from the answers listed.

1. (a) The point $(-1, -2)$ is on the graph of f, so $f(-1) = -2$.

(b) When $x = 2$, y is about 2.8, so $f(2) \approx 2.8$.

(c) $f(x) = 2$ is equivalent to $y = 2$. When $y = 2$, we have $x = -3$ and $x = 1$.

(d) Reasonable estimates for x when $y = 0$ are $x = -2.5$ and $x = 0.3$.

(e) The domain of f consists of all x-values on the graph of f. For this function, the domain is $-3 \le x \le 3$. The range of f consists of all y-values on the graph of f. For this function, the range is $-2 \le y \le 3$.

(f) As x increases from -1 to 3, y increases from -2 to 3. Thus, f is increasing on the interval $[-1, 3]$.

3. From Figure 1 in the text, the lowest point occurs at about $(t, a) = (12, -85)$. The highest point occurs at about $(17, 115)$. Thus, the range of the vertical ground acceleration is $-85 \le a \le 115$. In Figure 11, the range of the north-south acceleration is approximately $-325 \le a \le 485$. In Figure 12, the range of the east-west acceleration is approximately $-210 \le a \le 200$.

5. Yes, the curve is the graph of a function because it passes the Vertical Line Test. The domain is $[-3, 2]$ and the range is $[-2, 2]$.

7. No, the curve is not the graph of a function since for $x = -1$ there are infinitely many points on the curve.

9. The person's weight increased to about 160 pounds at age 20 and stayed fairly steady for 10 years. The person's weight dropped to about 120 pounds for the next 5 years, then increased rapidly to about 170 pounds. The next 30 years saw a gradual increase to 190 pounds. Possible reasons for the drop in weight at 30 years of age: diet, exercise, health problems.

11. The water will cool down almost to freezing as the ice melts. Then, when the ice has melted, the water will slowly warm up to room temperature.

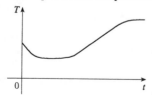

13. Of course, this graph depends strongly on the geographical location!

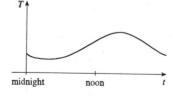

1

15.

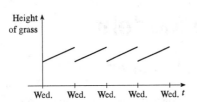

17. (a)

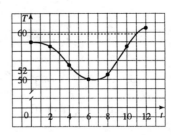

(b) $T(11) \approx 59°F$

19. $f(x) = 2x^2 + 3x - 4$, so $f(0) = 2(0)^2 + 3(0) - 4 = -4$,
$f(2) = 2(2)^2 + 3(2) - 4 = 10$, $f(\sqrt{2}) = 2(\sqrt{2})^2 + 3(\sqrt{2}) - 4 = 3\sqrt{2}$,
$f(1 + \sqrt{2}) = 2(1 + \sqrt{2})^2 + 3(1 + \sqrt{2}) - 4 = 2(1 + 2 + 2\sqrt{2}) + 3 + 3\sqrt{2} - 4 = 5 + 7\sqrt{2}$,
$f(-x) = 2(-x)^2 + 3(-x) - 4 = 2x^2 - 3x - 4$, $f(x + 1) = 2(x + 1)^2 + 3(x + 1) - 4 =$
$2(x^2 + 2x + 1) + 3x + 3 - 4 = 2x^2 + 7x + 1$, $2f(x) = 2(2x^2 + 3x - 4) = 4x^2 + 6x - 8$, and
$f(2x) = 2(2x)^2 + 3(2x) - 4 = 2(4x^2) + 6x - 4 = 8x^2 + 6x - 4$.

21. $f(x) = x - x^2$, so $f(2 + h) = 2 + h - (2 + h)^2 = 2 + h - 4 - 4h - h^2 = -(h^2 + 3h + 2)$,
$f(x + h) = x + h - (x + h)^2 = x + h - x^2 - 2xh - h^2$, and
$$\frac{f(x + h) - f(x)}{h} = \frac{x + h - x^2 - 2xh - h^2 - x + x^2}{h} = \frac{h - 2xh - h^2}{h} = 1 - 2x - h.$$

23. $f(x) = x^4/(x^2 + x - 6)$ is defined for all x except when $0 = x^2 + x - 6 = (x + 3)(x - 2) \iff$
$x = -3$ or 2, so the domain is $\{x \in \mathbb{R} \mid x \neq -3, 2\}$.

25. $f(t) = \sqrt[3]{t - 1}$ is defined for every t, since every real number has a cube root. The domain is the set of
all real numbers, $\mathbb{R}$.

27. $f(x) = 3 - 2x$. Domain is $\mathbb{R}$.

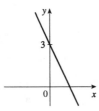

29. $G(x) = |x| + x$. Since $|x| = \begin{cases} x & \text{if } x \geq 0 \\ -x & \text{if } x < 0 \end{cases}$ we have

$G(x) = \begin{cases} 2x & \text{if } x \geq 0 \\ 0 & \text{if } x < 0 \end{cases}$ Domain is $\mathbb{R}$. Note that the negative

x-axis is part of the graph of G.

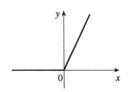

31. $f(x) = \dfrac{x}{|x|} = \begin{cases} x/x & \text{if } x > 0 \\ -x/x & \text{if } x < 0 \end{cases} = \begin{cases} 1 & \text{if } x > 0 \\ -1 & \text{if } x < 0 \end{cases}$

Note that we did not use $x \geq 0$, because $x \neq 0$. Hence, the domain of f is $\{x \mid x \neq 0\}$.

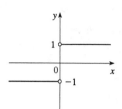

33. $f(x) = \begin{cases} x & \text{if } x \leq 0 \\ x+1 & \text{if } x > 0 \end{cases}$

Domain is $\mathbb{R}$.

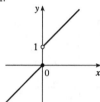

35. $f(x) = \begin{cases} x+2 & \text{if } x \leq -1 \\ x^2 & \text{if } x > -1 \end{cases}$

Domain is $\mathbb{R}$.

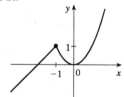

37. Recall that the slope m of a line between the two points (x_1, y_1) and (x_2, y_2) is $m = \dfrac{y_2 - y_1}{x_2 - x_1}$ and an equation of the line connecting those two points is $y - y_1 = m(x - x_1)$. The slope of this line segment is

$\dfrac{-6 - 1}{4 - (-2)} = -\dfrac{7}{6}$, so an equation is $y - 1 = -\dfrac{7}{6}(x + 2)$. The function is $f(x) = -\dfrac{7}{6}x - \dfrac{4}{3}, -2 \leq x \leq 4$.

39. We need to solve the given equation for y. $x + (y-1)^2 = 0 \implies (y-1)^2 = -x \implies$
$y - 1 = \pm\sqrt{-x} \implies y = 1 \pm \sqrt{-x}$. The expression with the positive radical represents the top half of the parabola, and the one with the negative radical represents the bottom half. Hence, we want $f(x) = 1 - \sqrt{-x}, x \leq 0$.

41. For $-1 \leq x \leq 2$, the graph is the line with slope 1 and y-intercept 1, that is, the line $y = x + 1$. For $2 < x \leq 4$, the graph is the line with slope $-\dfrac{3}{2}$ and x-intercept 4, so $y = -\dfrac{3}{2}(x - 4) = -\dfrac{3}{2}x + 6$. So

the function is $f(x) = \begin{cases} x+1 & \text{if } -1 \leq x \leq 2 \\ -\dfrac{3}{2}x + 6 & \text{if } 2 < x \leq 4 \end{cases}$

43. Let the length and width of the rectangle be L and W. Then the perimeter is $2L + 2W = 20$ and the

area is $A = LW$. Solving the first equation for W in terms of L gives $W = \dfrac{20 - 2L}{2} = 10 - L$. Thus,

$A(L) = L(10 - L) = 10L - L^2$. Since lengths are positive, the domain of A is $0 < L < 10$. If we further restrict L to be larger than W, then $5 < L < 10$ would be the domain.

45. Let the length of a side of the equilateral triangle be x. Then by the Pythagorean Theorem, the height y of the triangle satisfies $y^2 + \left(\dfrac{1}{2}x\right)^2 = x^2$, so that $y = \dfrac{\sqrt{3}}{2}x$. Using the formula for the area A of a

triangle, $A = \dfrac{1}{2}$ (base) (height), we obtain $A(x) = \dfrac{1}{2}(x)\left(\dfrac{\sqrt{3}}{2}x\right) = \dfrac{\sqrt{3}}{4}x^2$, with domain $x > 0$.

47. Let each side of the base of the box have length x, and let the height of the box be h. Since the volume is 2, we know that $2 = hx^2$, so that $h = 2/x^2$, and the surface area is $S = x^2 + 4xh$. Thus, $S(x) = x^2 + 4x(2/x^2) = x^2 + 8/x$, with domain $x > 0$.

49. The height of the box is x and the length and width are $L = 20 - 2x$, $W = 12 - 2x$. Then $V = LWx$ and so

$$V(x) = (20 - 2x)(12 - 2x)(x) = 4(10 - x)(6 - x)(x) = 4x(60 - 16x + x^2)$$
$$= 4x^3 - 64x^2 + 240x$$

The sides L, W, and x must be positive. Thus, $L > 0 \iff 20 - 2x > 0 \iff x < 10$; $w > 0 \iff 12 - 2x > 0 \iff x < 6$; and $x > 0$. Combining these restrictions gives us the domain $0 < x < 6$.

51. (a)

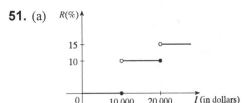

(c)

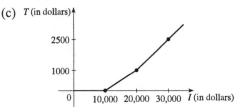

(b) On \$14,000, tax is assessed on \$4000, and 10% (\$4000) = \$400.

On \$26,000, tax is assessed on \$16,000, and

10% (\$10,000) + 15% (\$6000) = \$1000 + \$900 = \$1900.

53. (a) Because an even function is symmetric with respect to the y-axis, and the point $(5, 3)$ is on the graph of this even function, the point $(-5, 3)$ must also be on its graph.

(b) Because an odd function is symmetric with respect to the origin, and the point $(5, 3)$ is on the graph of this odd function, the point $(-5, -3)$ must also be on its graph.

55. $f(-x) = (-x)^{-2} = \dfrac{1}{(-x)^2} = \dfrac{1}{x^2} = x^{-2} = f(x)$, so f is an even function.

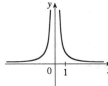

57. $f(-x) = (-x)^2 + (-x) = x^2 - x$. Since this is neither $f(x)$ nor $-f(x)$, the function f is neither even nor odd.

59. $f(-x) = (-x)^3 - (-x) = -x^3 + x = -(x^3 - x) = -f(x)$, so f is odd.

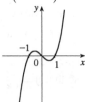

Section 1.2 New Functions from Old Functions

1. (a) $f(x) = \sqrt[5]{x}$ is a root function.

(b) $g(x) = \sqrt{1 - x^2}$ is an algebraic function because it is a root of a polynomial.

(c) $h(x) = x^9 + x^4$ is a polynomial of degree 9.

(d) $r(x) = \dfrac{x^2 + 1}{x^3 + x}$ is a rational function because it is a ratio of polynomials.

(e) $s(x) = \tan 2x$ is a trigonometric function.

(f) $t(x) = \log_{10} x$ is a logarithmic function.

3. (a) The graph of $y = x^8$ must be the graph labelled g, because g is the graph of a power function of even degree, as shown in Figure 3.

(b) The graph of $y = \log_8 x$ must be the graph labelled h, because h is a graph similar to the graphs of logarithmic functions shown in Figure 12.

(c) The graph of $y = 2 + \sin 2x$ must be the graph labelled f, because f is the graph of a periodic function.

5. (a) If the graph of f is shifted 3 units upward, its equation becomes $y = f(x) + 3$.

(b) If the graph of f is shifted 3 units downward, its equation becomes $y = f(x) - 3$.

(c) If the graph of f is shifted 3 units to the right, its equation becomes $y = f(x - 3)$.

(d) If the graph of f is shifted 3 units to the left, its equation becomes $y = f(x + 3)$.

(e) If the graph of f is reflected about the x-axis, its equation becomes $y = -f(x)$.

(f) If the graph of f is reflected about the y-axis, its equation becomes $y = f(-x)$.

(g) If the graph of f is stretched vertically by a factor of 3, its equation becomes $y = 3f(x)$.

(h) If the graph of f is shrunk vertically by a factor of 3, its equation becomes $y = \frac{1}{3}f(x)$.

7. (a) To graph $y = f(2x)$ we shrink the graph of f horizontally by a factor of 2.

(b) To graph $y = f\left(\frac{1}{2}x\right)$ we stretch the graph of f horizontally by a factor of 2.

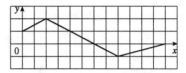

(c) To graph $y = f(-x)$ we reflect the graph of f about the y-axis.

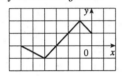

(d) To graph $y = -f(-x)$ we reflect the graph of f about the y-axis, then about the x-axis.

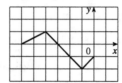

9. The graph of $y = 2\sin x$ can be obtained from the graph of $y = \sin x$ by stretching it vertically by a factor of 2.

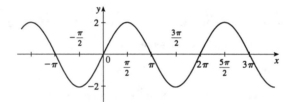

11. $y = -1/x$: Start with the graph of $y = 1/x$ (Figure 4) and reflect about the x-axis.

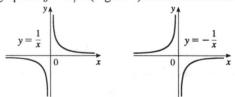

13. $y = \tan 2x$: Start with the graph of $y = \tan x$ (Figure 11) and compress horizontally by a factor of 2.

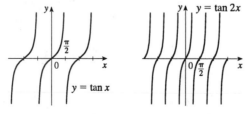

15. $y = \cos(x/2)$: Start with the graph of $y = \cos x$ [Figure 10(b)] and stretch horizontally by a factor of 2.

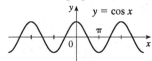 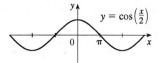

17. $y = \dfrac{1}{x-3}$: Start with the graph of $y = 1/x$ (Figure 4) and shift 3 units to the right.

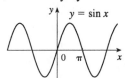 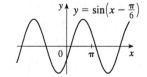 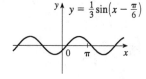

19. $y = \frac{1}{3}\sin\left(x - \frac{\pi}{6}\right)$: Start with the graph of $y = \sin x$ (Figure 19), shift $\frac{\pi}{6}$ units to the right, and then compress vertically by a factor of 3.

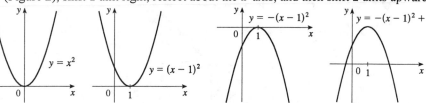

21. $y = 1 + 2x - x^2 = -x^2 + 2x + 1 = -\left(x^2 - 2x + 1\right) + 1 + 1 = -(x-1)^2 + 2$: Start with the graph of $y = x^2$ (Figure 2), shift 1 unit right, reflect about the x-axis, and then shift 2 units upward.

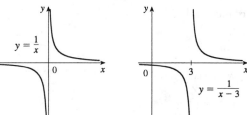 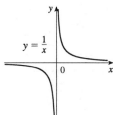

23. $y = 2 - \sqrt{x+1}$: Start with the graph of $y = \sqrt{x}$ [Figure 5(a)], reflect about the x-axis, shift 1 unit to the left, and then shift 2 units upward.

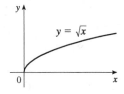 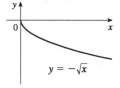 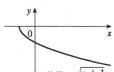 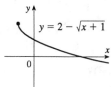

25. $y = |\cos x|$: Start with the graph of $y = \cos x$ [Figure 10(b)] and reflect the parts of the graph that lie below the x-axis about the x-axis.

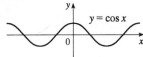
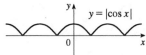

27. (a) To obtain $y = f(|x|)$, the portion of
   $y = f(x)$ to the right of the y-axis is
   reflected about the y-axis.
   (b) $y = \sin|x|$

(c) $y = \sqrt{|x|}$

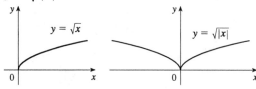

Note: *For the rest of this section, " $D =$ " stands for "The domain of the function is".*

29. $f(x) = x^3 + 2x^2$; $g(x) = 3x^2 - 1$. $D = \mathbb{R}$ for both f and g.
   $(f + g)(x) = x^3 + 2x^2 + 3x^2 - 1 = x^3 + 5x^2 - 1$, $D = \mathbb{R}$.
   $(f - g)(x) = x^3 + 2x^2 - (3x^2 - 1) = x^3 - x^2 + 1$, $D = \mathbb{R}$.
   $(fg)(x) = (x^3 + 2x^2)(3x^2 - 1) = 3x^5 + 6x^4 - x^3 - 2x^2$, $D = \mathbb{R}$.
   $\left(\dfrac{f}{g}\right)(x) = \dfrac{x^3 + 2x^2}{3x^2 - 1}$, $D = \left\{x \mid x \neq \pm\frac{1}{\sqrt{3}}\right\}$ since $3x^2 - 1 \neq 0$.

31. $f(x) = x$, $g(x) = 1/x$

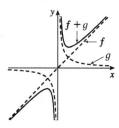

33. $f(x) = 2x^2 - x$; $g(x) = 3x + 2$. $D = \mathbb{R}$ for both f and g, and hence for their composites.
   $(f \circ g)(x) = f(g(x)) = f(3x + 2) = 2(3x + 2)^2 - (3x + 2) = 18x^2 + 21x + 6$.
   $(g \circ f)(x) = g(f(x)) = g(2x^2 - x) = 3(2x^2 - x) + 2 = 6x^2 - 3x + 2$.
   $(f \circ f)(x) = f(f(x)) = f(2x^2 - x) = 2(2x^2 - x)^2 - (2x^2 - x) = 8x^4 - 8x^3 + x$.
   $(g \circ g)(x) = g(g(x)) = g(3x + 2) = 3(3x + 2) + 2 = 9x + 8$.

35. $f(x) = \sqrt{x^2 - 1}$, $D = (-\infty, -1] \cup [1, \infty)$; $g(x) = \sqrt{1-x}$, $D = (-\infty, 1]$.

$(f \circ g)(x) = f(g(x)) = f(\sqrt{1-x}) = \sqrt{(\sqrt{1-x})^2 - 1} = \sqrt{-x}$. To find the domain

of $(f \circ g)(x)$, we must find the values of x that are in the domain of g such that $g(x)$ is in

the domain of f. In symbols, we have $D = \{x \in (-\infty, 1] \mid \sqrt{1-x} \in (-\infty, -1] \cup [1, \infty)\}$.

First, we concentrate on the requirement that $\sqrt{1-x} \in (-\infty, -1] \cup [1, \infty)$. Because

$\sqrt{1-x} \geq 0$, $\sqrt{1-x}$ is not in $(-\infty, -1]$. If $\sqrt{1-x}$ is in $[1, \infty)$, then we must have

$\sqrt{1-x} \geq 1 \Longrightarrow 1 - x \geq 1 \Longrightarrow x \leq 0$. Combining the restrictions $x \leq 0$ and $x \in (-\infty, 1]$,

we obtain $D = (-\infty, 0]$. $(g \circ f)(x) = g(f(x)) = g(\sqrt{x^2 - 1}) = \sqrt{1 - \sqrt{x^2 - 1}}$,

$D = \{x \in (-\infty, -1] \cup [1, \infty) \mid \sqrt{x^2 - 1} \in (-\infty, 1]\}$. Now $\sqrt{x^2 - 1} \leq 1 \Longrightarrow x^2 - 1 \leq 1 \Longrightarrow$

$x^2 \leq 2 \Longrightarrow |x| \leq \sqrt{2} \Longrightarrow -\sqrt{2} \leq x \leq \sqrt{2}$. Combining this restriction with $x \in (-\infty, -1] \cup [1, \infty)$,

we obtain $D = [-\sqrt{2}, -1] \cup [1, \sqrt{2}]$.

$(f \circ f)(x) = f(f(x)) = f(\sqrt{x^2 - 1}) = \sqrt{(\sqrt{x^2 - 1})^2 - 1} = \sqrt{x^2 - 2}$,

$D = \{x \in (-\infty, -1] \cup [1, \infty) \mid \sqrt{x^2 - 1} \in (-\infty, -1] \cup [1, \infty)\}$. Now

$\sqrt{x^2 - 1} \geq 1 \Longrightarrow x^2 - 1 \geq 1 \Longrightarrow x^2 \geq 2 \Longrightarrow |x| \geq \sqrt{2} \Longrightarrow x \geq \sqrt{2}$ or

$x \leq -\sqrt{2}$. Combining this restriction with $x \in (-\infty, -1] \cup [1, \infty)$, we obtain

$D = (-\infty, -\sqrt{2}] \cup [\sqrt{2}, \infty)$. $(g \circ g)(x) = g(g(x)) = g(\sqrt{1-x}) = \sqrt{1 - \sqrt{1-x}}$,

$D = \{x \in (-\infty, 1] \mid \sqrt{1-x} \in (-\infty, 1]\}$. Now $\sqrt{1-x} \leq 1 \Longrightarrow 1 - x \leq 1 \Longrightarrow x \geq 0$. Combining

this restriction with $x \in (-\infty, 1]$, we obtain $D = [0, 1]$.

37. $(f \circ g \circ h)(x) = f(g(h(x))) = f(g(x^2 + 2)) = f((x^2 + 2)^3) = \dfrac{1}{(x^2 + 2)^3}$

39. Let $g(x) = x - 9$ and $f(x) = x^5$. Then $(f \circ g)(x) = (x-9)^5 = F(x)$.

41. Let $g(x) = x^2$ and $f(x) = \dfrac{x}{x+4}$. Then $(f \circ g)(x) = \dfrac{x^2}{x^2 + 4} = G(x)$.

43. Let $h(x) = x^2$, $g(x) = 3^x$, and $f(x) = 1 - x$. Then $(f \circ g \circ h)(x) = 1 - 3^{x^2} = H(x)$.

45. (a) $g(2) = 5$, because the point $(2, 5)$ is on the graph of g. Thus, $f(g(2)) = f(5) = 4$, because the
point $(5, 4)$ is on the graph of f.

(b) $g(f(0)) = g(0) = 3$ 　　　　　　　　(c) $(f \circ g)(0) = f(g(0)) = f(3) = 0$

(d) $(g \circ f)(6) = g(f(6)) = g(6)$. This value is not defined, because there is no point on the graph of
g that has x-coordinate 6.

(e) $(g \circ g)(-2) = g(g(-2)) = g(1) = 4$ 　　　　(f) $(f \circ f)(4) = f(f(4)) = f(2) = -2$

47. (a) Using the relationship $distance = rate \cdot time$ with the radius r as the distance, we have $r(t) = 60t$.

(b) $A = \pi r^2 \Longrightarrow (A \circ r)(t) = A(r(t)) = \pi (60t)^2 = 3600\pi t^2$. This formula gives us the number of cm^2 encircled by the rippled area at any time t.

49. (a)

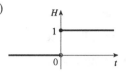

(b)

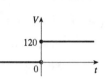

$$V(t) = \begin{cases} 0 & \text{if } t < 0 \\ 120 & \text{if } t \geq 0 \end{cases}$$

so $V(t) = 120H(t)$.

(c)

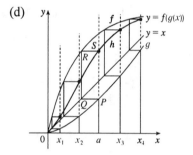

Starting with the formula in part (b), we replace 120 with 240 to reflect the different voltage. Also, because we are starting 5 units to the right of $t = 0$, we replace t with $t - 5$. Thus, the formula is $V(t) = 240H(t - 5)$.

51. We need to examine $h(-x)$.

$$h(-x) = (f \circ g)(-x) = f(g(-x)) = f(g(x)) \quad \text{[because } g \text{ is even]} \quad = h(x).$$

Because $h(-x) = h(x)$, h is an even function.

53. (a) $P = (a, g(a))$ and $Q = (g(a), g(a))$ because Q has the same y-value as P and it is on the line $y = x$.

(b) The x-value of Q is $g(a)$; this is also the x-value of R. The y-value of R is therefore $f(x$-value$)$, that is, $f(g(a))$. Hence, $R = (g(a), f(g(a)))$.

(c) The coordinates of S are $(a, f(g(a)))$ or, equivalently, $(a, h(a))$.

(d)

Section 1.3 Graphing Calculators and Computers

1. $f(x) = 10 + 25x - x^3$

(a) $[-4, 4]$ by $[-4, 4]$

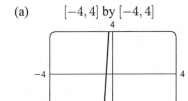

(b) $[-10, 10]$ by $[-10, 10]$

(c) $[-20, 20]$ by $[-100, 100]$

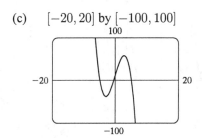

(d) $[-100, 100]$ by $[-200, 200]$

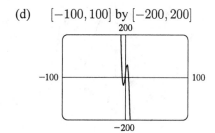

The most appropriate graph is produced in viewing rectangle (c) because the maximum and minimum points are fairly easy to see and estimate..

3. $f(x) = \sqrt[4]{256 - x^2}$. To find an appropriate viewing rectangle, we calculate f's domain and range: $256 - x^2 \geq 0 \iff$ $x^2 \leq 256 \iff |x| \leq 16 \iff -16 \leq x \leq 16$, so the domain is $[-16, 16]$. Also, $0 \leq \sqrt[4]{256 - x^2} \leq \sqrt[4]{256} = 4$, so the range is $[0, 4]$. Thus, we choose the viewing rectangle to be $[-20, 20]$ by $[-2, 6]$.

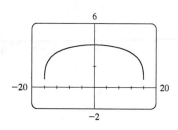

5. $y = \dfrac{1}{x^2 + 25}$

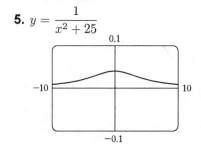

7. $y = x^4 - 4x^3$

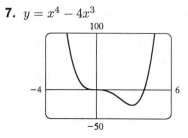

9. $f(x) = \cos(100x)$

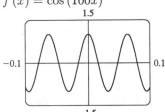

11. $f(x) = \sin(x/40)$

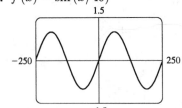

13. $y = 3^{\cos(x^2)}$

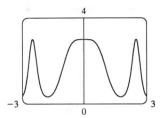

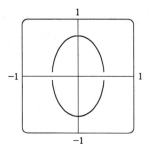

15. We must solve the given equation for y to obtain equations for the upper and lower halves of the ellipse.

$$4x^2 + 2y^2 = 1 \Longrightarrow 2y^2 = 1 - 4x^2 \Longrightarrow$$

$$y^2 = \frac{1 - 4x^2}{2} \Longrightarrow y = \pm\sqrt{\frac{1 - 4x^2}{2}}$$

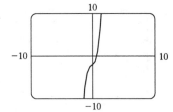

17. Graphing $f(x) = 3x^3 + x^2 + x - 2$ in a standard viewing rectangle, $[-10, 10]$ by $[-10, 10]$, reveals one real root between 0 and 1. The second figure shows a close-up of this region. By using a root finder or by zooming in, we find the value of the root to be approximately 0.67.

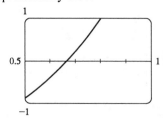

19. From the graph of $f(x) = 2\sin x$ and $g(x) = x$, we see that there are three points of intersection. The intersection point $(0,0)$ is obvious and due to the symmetry of the graphs (both functions are odd), we only need to find one of the other two points of intersection. Using an intersection finder or zooming in, we find the x-value of the intersection to be approximately 1.90. Hence, the solutions are $x = 0$ and $x \approx \pm 1.90$.

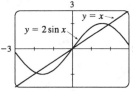 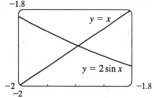

21. $g(x) = x^3/10$ is larger than $f(x) = 10x^2$ whenever $x > 100$.

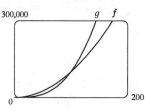

23.

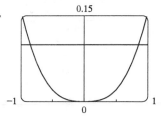

We see from the graph of $y = |\sin x - x|$ that there are two solutions to the equation $|\sin x - x| = 0.1$: $x \approx -0.85$ and $x \approx 0.85$. The condition $|\sin x - x| < 0.1$ holds for any x lying between these two values.

25. (a) The root functions $y = \sqrt{x}$, $y = \sqrt[4]{x}$ and $y = \sqrt[6]{x}$

(b) The root functions $y = x$, $y = \sqrt[3]{x}$ and $y = \sqrt[5]{x}$

(c) The root functions $y = \sqrt{x}$, $y = \sqrt[3]{x}$, $y = \sqrt[4]{x}$ and $y = \sqrt[5]{x}$

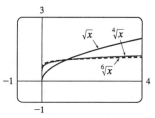 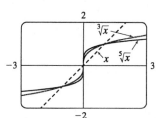 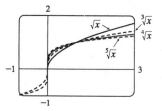

(d) • For any n, the nth root of 0 is 0 and the nth root of 1 is 1; that is, all nth root functions pass through the points $(0,0)$ and $(1,1)$.

• For odd n, the domain of the nth root function is $\mathbb{R}$, while for even n, it is $\{x \in \mathbb{R} \mid x \geq 0\}$.

• Graphs of even root functions look similar to that of $\sqrt{x}$, while those of odd root functions resemble that of $\sqrt[3]{x}$.

• As n increases, the graph of $\sqrt[n]{x}$ becomes steeper near 0 and flatter for $x > 1$.

13

27. $f(x) = x^4 + cx^2 + x$. If $c < 0$, there are three humps: two minimum points and a maximum point. These humps get flatter as c increases, until at $c = 0$ two of the humps disappear and there is only one minimum point. This single hump then moves to the right and approaches the origin as c increases.

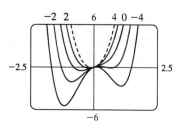

29. $y = x^n 2^{-x}$. As n increases, the maximum of the function moves further from the origin, and gets larger. Note, however, that regardless of n, the function approaches 0 as $x \to \infty$.

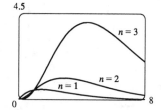

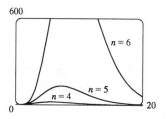

31. $y^2 = cx^3 + x^2$

If $c < 0$, the loop is to the right of the origin, and if c is positive, it is to the left. In both cases, the closer c is to 0, the larger the loop is. (In the limiting case, $c = 0$, the loop is "infinite," that is, it doesn't close.) Also, the larger $|c|$ is, the steeper the slope is on the loopless side of the origin.

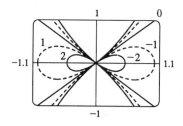

Section 1.4 Parametric Curves

1. (a) $x = 2t - 1,\ y = 2 - t,\ -3 \le t \le 3$

t	x	y
-3	-7	5
-2	-5	4
-1	-3	3
0	-1	2
1	1	1
2	3	0
3	5	-1

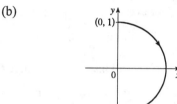

(b) $x = 2(2 - y) - 1 = 3 - 2y$, so
$x + 2y = 3$, with $-7 \le x \le 5$.

5. (a) $x = \sin\theta,\ y = \cos\theta,\ 0 \le \theta \le \pi$.
$x^2 + y^2 = \sin^2\theta + \cos^2\theta = 1,\ 0 \le x \le 1$.

(b)

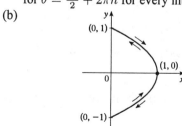

3. (a) $x = \sqrt{t},\ y = 1 - t$

t	x	y
0	0	1
1	1	0
2	1.414	-1
3	1.732	-2
4	2	-3

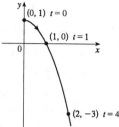

(b) $x = \sqrt{t} \implies t = x^2.\ y = 1 - t = 1 - x^2$.
Since $t \ge 0,\ x \ge 0$.

7. (a) $x = \sin^2\theta,\ y = \cos^2\theta$.

$x + y = \sin^2\theta + \cos^2\theta = 1,\ 0 \le x \le 1$.

Note that the curve is at $(0, 1)$ whenever
$\theta = \pi n$ and is at $(1, 0)$ whenever $\theta = \frac{\pi}{2}n$
for every integer n.

(b)

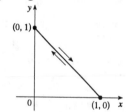

9. (a) $x = \cos^2\theta,\ y = \sin\theta$.

$x + y^2 = \cos^2\theta + \sin^2\theta = 1,\ -1 \le y \le 1$.

Note that the curve is at $(1, 0)$ for $\theta = \pi n$,
at $(0, 1)$ for $\theta = \frac{\pi}{2} + 2\pi n$, and at $(0, -1)$
for $\theta = \frac{3\pi}{2} + 2\pi n$ for every integer n.

(b)

11. $x^2 + y^2 = \cos^2\pi t + \sin^2\pi t = 1,\ 1 \le t \le 2$,
so the particle moves counterclockwise
along the circle $x^2 + y^2 = 1$ from $(-1, 0)$ to
$(1, 0)$, along the lower half of the circle.

13. $\left(\frac{1}{2}x\right)^2 + \left(\frac{1}{3}y\right)^2 = \sin^2 t + \cos^2 t = 1$, so the particle moves once clockwise along the ellipse $\frac{1}{4}x^2 + \frac{1}{9}y^2 = 1$, starting and ending at $(0, 3)$.

15. $y = \csc t = 1/\sin t = 1/x$. The particle slides down the first quadrant branch of the hyperbola $xy = 1$ from $\left(\frac{1}{2}, 2\right)$ to $(\sin 1, \csc 1) \approx (0.84147, 1.1884)$ as t goes from $\frac{\pi}{6}$ to 1.

17. As $t \to -\infty$, $x \to \infty$ and $y \to -\infty$. The graph passes through the origin at $t = -1$, and then goes through the second quadrant (x negative, y positive), passing through the point $(-1, 1)$ at $t = 0$. As t increases, the graph passes through the point $(0, 2)$ at $t = 1$, and then as $t \to \infty$,

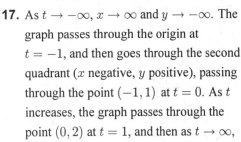 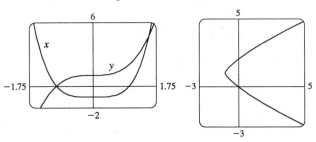

both x and y approach ∞. The first figure was obtained using $x_1 = t$, $y_1 = t^4 - 1$; $x_2 = t$, $y_2 = t^3 + 1$; and $-2\pi \le t \le 2\pi$.

19. Clearly the curve passes through (x_1, y_1) when $t = 0$ and through (x_2, y_2) when $t = 1$. For $0 < t < 1$, x is strictly between x_1 and x_2 and y is strictly between y_1 and y_2. For every value of t, x and y satisfy the relation $y - y_1 = \dfrac{y_2 - y_1}{x_2 - x_1}(x - x_1)$, which is the equation of the line through (x_1, y_1) and (x_2, y_2).

Finally, any point (x, y) on that line satisfies $\dfrac{y - y_1}{y_2 - y_1} = \dfrac{x - x_1}{x_2 - x_1}$; if we call that common value t, then the given parametric equations yield the point (x, y); and any (x, y) on the line between (x_1, y_1) and (x_2, y_2) yields a value of t in $[0, 1]$. So the given parametric equations exactly specify the line segment from (x_1, y_1) to (x_2, y_2).

21. As in Example 4, we let $y = t$ and $x = t - 3t^3 + t^5$ and use a t-interval of $[-2\pi, 2\pi]$.

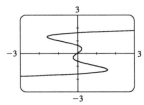

23. The circle $x^2 + y^2 = 4$ can be represented parametrically by $x = 2\cos t$, $y = 2\sin t$; $0 \le t \le 2\pi$. The circle $x^2 + (y - 1)^2 = 4$ can be represented by $x = 2\cos t$, $y = 1 + 2\sin t$; $0 \le t \le 2\pi$. This representation gives us the circle with a counterclockwise orientation starting at $(2, 1)$.

(a) To get a clockwise orientation, we could change the equations to $x = 2\cos t$, $y = 1 - 2\sin t$.

(b) To get three times around in the counterclockwise direction, we use the original equations $x = 2\cos t$, $y = 1 + 2\sin t$ with the domain expanded to $0 \le t \le 6\pi$.

(c) To start at $(0, 3)$ using the original equations, we must have $x_1 = 0$; that is, $2\cos t = 0$. Hence, $t = \frac{\pi}{2}$. So we use $x = 2\cos t$, $y = 1 + 2\sin t$; $\frac{\pi}{2} \le t \le \frac{3\pi}{2}$.

Alternatively, if we want t to start at 0, we could change the equations of the curve. For example, we could use $x = -2\sin t$, $y = 1 + 2\cos t$, $0 \le t \le \pi$.

25. (a) Let $x^2/a^2 = \sin^2 t$ and $y^2/b^2 = \cos^2 t$ to obtain $x = a \sin t$ and $y = b \cos t$ with $0 \le t \le 2\pi$ as possible parametric equations for the ellipse $x^2/a^2 + y^2/b^2 = 1$.

(c) As b increases, the ellipse is stretched vertically.

(b) The equations are $x = 3 \sin t$ and $y = b \cos t$ for $b \in \{1, 2, 4, 8\}$.

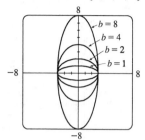

27. The case $\frac{\pi}{2} < \theta < \pi$ is illustrated. C has coordinates $(r\theta, r)$ as before, and Q has coordinates $(r\theta, r + r \cos(\pi - \theta)) = (r\theta, r(1 - \cos\theta))$ [since $\cos(\pi - \alpha) = \cos\pi\cos\alpha + \sin\pi\sin\alpha = -\cos\alpha$], so P has coordinates

$$(r\theta - r\sin(\pi - \theta), r(1 - \cos\theta)) = (r(\theta - \sin\theta), r(1 - \cos\theta))$$

[since $\sin(\pi - \alpha) = \sin\pi\cos\alpha - \cos\pi\sin\alpha = \sin\alpha$]. Again we have the parametric equations $x = r(\theta - \sin\theta)$, $y = r(1 - \cos\theta)$.

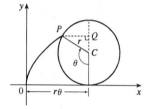

29. It is apparent that $x = |OQ|$ and $y = |QP| = |ST|$. From the diagram, $x = |OQ| = a\cos\theta$ and $y = |ST| = b\sin\theta$. Thus, the parametric equations are $x = a\cos\theta$ and $y = b\sin\theta$. To eliminate θ we rearrange: $\sin\theta = y/b \implies \sin^2\theta = (y/b)^2$ and $\cos\theta = x/a \implies \cos^2\theta = (x/a)^2$. Adding the two equations: $\sin^2\theta + \cos^2\theta = 1 = x^2/a^2 + y^2/b^2$. Thus, we have an ellipse.

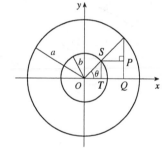

31. $C = (2a\cot\theta, 2a)$, so the x-coordinate of P is $x = 2a\cot\theta$. Let $B = (0, 2a)$. Then $\angle OAB$ is a right angle and $\angle OBA = \theta$, so $|OA| = 2a\sin\theta$ and $A = (2a\sin\theta\cos\theta, 2a\sin^2\theta)$. Thus, the y-coordinate of P is $y = 2a\sin^2\theta$.

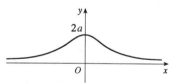

33. $x = t^2, y = t^3 - ct$. We use a graphing device to produce the graphs for various values of c with $-\pi \le t \le \pi$. Note that all the members of the family are symmetric about the x-axis. For $c < 0$, the graph does not cross itself, but for $c = 0$ it has a cusp at $(0,0)$ and for $c > 0$ the graph crosses itself at $x = c$, so the loop grows larger as c increases.

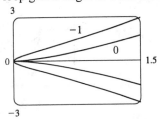 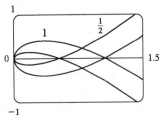

35. Note that all the Lissajous figures are symmetric about the x-axis. The parameters a and b simply stretch the graph in the x- and y-directions respectively. For $a = b = n = 1$ the graph is simply a circle with radius 1. For $n = 2$ the graph crosses itself at the origin and there are loops above and below the x-axis. In general, the figures have $n - 1$ points of intersection, all of which are on the y-axis, and a total of n closed loops.

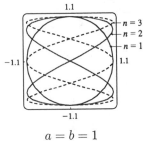

$a = b = 1$

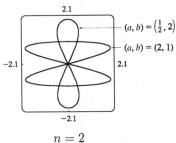

$n = 2$

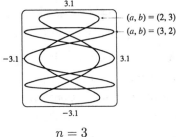

$n = 3$

Section 1.5 Exponential Functions

1. (a) $f(x) = a^x$, $a > 0$ (b) $\mathbb{R}$

(c) $(0, \infty)$ (d) See Figures 4(c), 4(b), and 4(a), respectively.

3. All of these graphs approach 0 as $x \to -\infty$, all of them pass through the point $(0, 1)$, and all of them are increasing and approach ∞ as $x \to \infty$. The larger the base, the faster the function increases for $x > 0$, and the faster it approaches 0 as $x \to -\infty$.

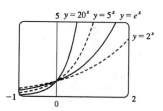

5. The functions with bases greater than 1 (3^x and 10^x) are increasing, while those with bases less than 1 $\left[\left(\frac{1}{3}\right)^x \text{ and } \left(\frac{1}{10}\right)^x\right]$ are decreasing. The graph of $\left(\frac{1}{3}\right)^x$ is the reflection of that of 3^x about the y-axis, and the graph of $\left(\frac{1}{10}\right)^x$ is the reflection of that of 10^x about the y-axis. The graph of 10^x increases more quickly than that of 3^x for $x > 0$, and approaches 0 faster as $x \to -\infty$.

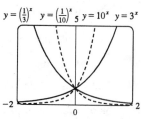

7. We start with the graph of $y = 2^x$ and shift 1 unit upward.

9. We start with the graph of $y = 3^x$ and reflect it about the y-axis.

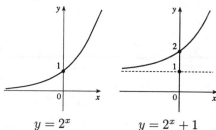

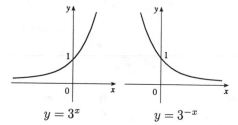

11. We start with the graph of $y = 3^{-x}$ (from Exercise 9) and reflect it about the x-axis.

13. We start with the graph of $y = e^x$ (Figure 12), reflect it about the x-axis, and then shift 3 units upward. Note the horizontal asymptote at $y = 3$.

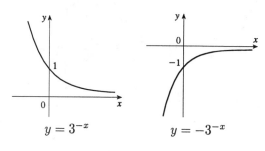

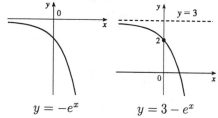

19

15. (a) To find the equation of the graph that results from shifting the graph of $y = e^x$ 2 units downward, we subtract 2 from the original function to get $y = e^x - 2$.

(b) To find the equation of the graph that results from shifting the graph of $y = e^x$ 2 units to the right, we replace x with $x - 2$ in the original function to get $y = e^{(x-2)}$.

(c) To find the equation of the graph that results from reflecting the graph of $y = e^x$ about the x-axis, we multiply the original function by -1 to get $y = -e^x$.

(d) To find the equation of the graph that results from reflecting the graph of $y = e^x$ about the y-axis, we replace x with $-x$ in the original function to get $y = e^{-x}$.

(e) To find the equation of the graph that results from reflecting the graph of $y = e^x$ about the x-axis and then about the y-axis, we first multiply the original function by -1 (to get $y = -e^x$) and then replace x with $-x$ in this equation to get $y = -e^{-x}$.

17. Use $y = Ca^x$ with the points $(1, 6)$ and $(3, 24)$. $6 = Ca^1$ and $24 = Ca^3 \implies 24 = \left(\dfrac{6}{a}\right)a^3 \implies$ $4 = a^2 \implies a = 2$ (since $a > 0$) and $C = 3$. The function is $f(x) = 3 \cdot 2^x$.

19. 2 ft = 24 in, $f(24) = 24^2$ in = 576 in = 48 ft. $g(24) = 2^{24}$ in = $2^{24} / (12 \cdot 5280)$ mi ≈ 265 mi

21. The graph of g finally surpasses that of f at $x \approx 35.8$.

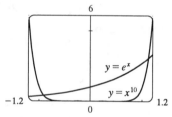

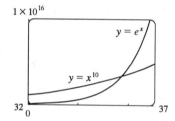

23. (a) Fifteen hours represents 5 doubling periods. $100 \cdot 2^5 = 3200$

(b) In t hours, there will be $t/3$ doubling periods. The initial population is 100, so the population y at time t is $y = 100 \cdot 2^{t/3}$.

(c) $t = 20 \implies y = 100 \cdot 2^{20/3} \approx 10{,}159$

(d) We graph $y_1 = 100 \cdot 2^{x/3}$ and $y_2 = 50{,}000$. The two curves intersect at $x \approx 26.9$.

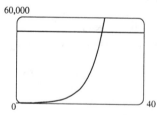

Section 1.6   Inverse Functions and Logarithms

1. (a) See Definition 1.

(b) It must pass the Horizontal Line Test.

3. f is not one-to-one because $2 \neq 6$, but $f(2) = f(6)$.

5. The diagram shows that there is a horizontal line which intersects the graph more than once, so the function is not one-to-one.

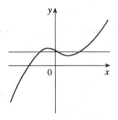

7. The function is one-to-one because no horizontal line intersects the graph more than once.

9. $x_1 \neq x_2 \Longrightarrow 7x_1 \neq 7x_2 \Longrightarrow 7x_1 - 3 \neq 7x_2 - 3 \Longrightarrow f(x_1) \neq f(x_2)$, so f is 1-1.

11. $g(x) = |x| \Longrightarrow g(-1) = 1 = g(1)$, so g is not one-to-one.

13. A football will attain every height h up to its maximum height twice: once on the way up, and again on the way down. Thus, even if t_1 does not equal t_2, $f(t_1)$ may equal $f(t_2)$, so f is not 1-1.

15. f does not pass the Horizontal Line Test, so f is not 1-1.

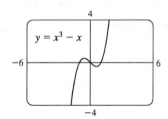

17. Since $f(2) = 9$ and f is 1-1, we know that $f^{-1}(9) = 2$. Remember, if the point $(2, 9)$ is on the graph of f, then the point $(9, 2)$ is on the graph of f^{-1}.

19. First, we must determine x such that $g(x) = 4$. By inspection, we see that if $x = 0$, then $g(x) = 4$. Since g is 1-1, it has an inverse, and $g^{-1}(4) = 0$.

21. We solve $C = \frac{5}{9}(F = 32)$ for F: $\frac{9}{5}C = F - 32 \Longrightarrow F = \frac{9}{5}C + 32$. This gives us the Fahrenheit temperature F as a function of the Celsius temperature C. $F \geq -459.67 \Longrightarrow \frac{9}{5}C + 32 \geq -459.67 \Longrightarrow \frac{9}{5}C \geq -491.67 \Longrightarrow C \geq -273.15$, the domain of the inverse function.

23. $y = \dfrac{1+3x}{5-2x} \implies 5y - 2xy = 1 + 3x \implies 5y - 1 = 3x + 2xy \implies x(3+2y) = 5y - 1 \implies$

$x = \dfrac{5y-1}{2y+3}$. Interchange x and y: $y = \dfrac{5x-1}{2x+3}$. So $f^{-1}(x) = \dfrac{5x-1}{2x+3}$.

25. $y = \sqrt{2+5x} \implies y^2 = 2 + 5x$ and $y \geq 0 \implies 5x = y^2 - 2 \implies x = \dfrac{y^2 - 2}{5}$, $y \geq 0$. Interchange x and

y: $y = \dfrac{x^2 - 2}{5}$, $x \geq 0$. So $f^{-1}(x) = \dfrac{x^2 - 2}{5}$, $x \geq 0$.

27. $y = \ln(x+3) \implies x + 3 = e^y \implies x = e^y - 3$. Interchange x and y: $y = e^x - 3$. So $f^{-1}(x) = e^x - 3$.

29. $y = 1 - \dfrac{2}{x^2} \implies 1 - y = \dfrac{2}{x^2} \implies x^2 = \dfrac{2}{1-y} \implies$

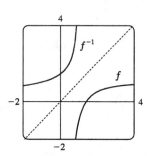

$x = \sqrt{\dfrac{2}{1-y}}$, since $x > 0$. Interchange x and y:

$y = \sqrt{\dfrac{2}{1-x}}$. So $f^{-1}(x) = \sqrt{\dfrac{2}{1-x}}$.

31. The function f is one-to-one, so its inverse exists and the graph of its inverse can be obtained by reflecting the graph of f through the line $y = x$.

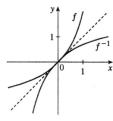

33. (a) It is defined as the inverse of the exponential function with base a, that is, $\log_a x = y \iff a^y = x$.

(b) $(0, \infty)$ (c) $\mathbb{R}$ (d) See Figure 13.

35. (a) $\log_2 64 = 6$ since $2^6 = 64$.   (b) $\log_6 \frac{1}{36} = -2$ since $6^{-2} = \frac{1}{36}$.

37. (a) $\log_{10} 1.25 + \log_{10} 80 = \log_{10}(1.25 \cdot 80) = \log_{10} 100 = \log_{10} 10^2 = 2$

(b) $\log_5 10 + \log_5 20 - 3\log_5 2 = \log_5(10 \cdot 20) - \log_5 2^3 = \log_5 \frac{200}{8} = \log_5 25 = \log_5 5^2 = 2$

39. $2\ln 4 - \ln 2 = \ln 4^2 - \ln 2 = \ln 16 - \ln 2 = \ln \frac{16}{2} = \ln 8$

41. (a) $\log_2 5 = \dfrac{\ln 5}{\ln 2} \approx 2.321928$ (b) $\log_5 26.05 = \dfrac{\ln 26.05}{\ln 5} \approx 2.025563$

43. To graph these functions, we use $\log_{1.5} x = \dfrac{\ln x}{\ln 1.5}$ and

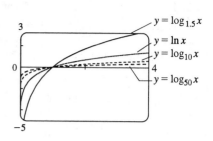

$\log_{50} x = \dfrac{\ln x}{\ln 50}$. These graphs all approach $-\infty$ as $x \to 0^+$, and they all pass through the point $(1, 0)$. Also, they are all increasing, and all approach ∞ as $x \to \infty$. The functions with larger bases increase extremely slowly, and the ones with smaller bases do so somewhat more quickly. The functions with large bases approach the y-axis more closely as $x \to 0^+$.

45. 3 ft $= 36$ in, so we need x such that $\log_2 x = 36 \Longleftrightarrow x = 2^{36} = 68{,}719{,}476{,}736$. In miles, this is

$$68{,}719{,}476{,}736 \text{ in} \cdot \frac{1 \text{ ft}}{12 \text{ in}} \cdot \frac{1 \text{ mi}}{5280 \text{ ft}} \approx 1{,}084{,}587.7 \text{ mi}.$$

47. (a)   (b)

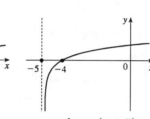   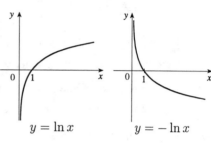

$$y = \log_{10} x \qquad\qquad y = \log_{10}(x+5) \qquad\qquad y = \ln x \qquad\qquad y = -\ln x$$

49. (a) $e^x = 16 \Longleftrightarrow \ln e^x = \ln 16 \Longleftrightarrow x = \ln 16 = \ln 2^4 = 4 \ln 2$

(b) $\ln x = -1 \Longleftrightarrow e^{\ln x} = e^{-1} \Longleftrightarrow x = 1/e$

51. (a) $2^{x-5} = 3 \Longleftrightarrow \log_2 3 = x - 5 \Longleftrightarrow x = 5 + \log_2 3$. Or: $2^{x-5} = 3 \Longleftrightarrow \ln\left(2^{x-5}\right) = \ln 3 \Longleftrightarrow$

$$(x-5) \ln 2 = \ln 3 \Longleftrightarrow x - 5 = \frac{\ln 3}{\ln 2} \Longleftrightarrow x = 5 + \frac{\ln 3}{\ln 2}$$

(b) $\ln x + \ln(x-1) = \ln(x(x-1)) = 1 \Longleftrightarrow x(x-1) = e^1 \Longleftrightarrow x^2 - x - e = 0$. The quadratic formula gives $x = \frac{1}{2}\left(1 \pm \sqrt{1+4e}\right)$, but we reject the negative root since the natural logarithm is not defined for $x < 0$. So $x = \frac{1}{2}\left(1 + \sqrt{1+4e}\right)$.

53. We graph f in several viewing rectangles until we are convinced that f satisfies the Horizontal Line Test. Now we sketch the following sets of parametric equations for f, f^{-1}, and $y = x$ with $-2\pi \le t \le 2\pi$:

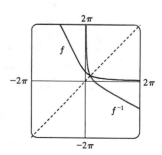

$$\begin{aligned} x_1 &= t &\text{and}&& y_1 &= \sqrt{t^2+1} - t \\ x_2 &= \sqrt{t^2-1} - t &\text{and}&& y_2 &= t \\ x_3 &= t &\text{and}&& y_3 &= t \end{aligned}$$

55. Enter $x = \sqrt{y^3 + y^2 + y + 1}$ and use your CAS to solve the equation for y. Using Derive, we get two (irrelevant) solutions involving imaginary expressions, as well as one which can be simplified to the following: $y = f^{-1}(x) = -\frac{\sqrt[3]{4}}{6}\left(\sqrt[3]{D - 27x^2 + 20} - \sqrt[3]{D + 27x^2 - 20} + \sqrt[3]{2}\right)$, where $D = 3\sqrt{3}\sqrt{27x^4 - 40x^2 + 16}$. Maple and Mathematica each give two complex expressions and one real expression, and the real expression is equivalent to that given by Derive. For example, Maple's expression simplifies to $\frac{1}{6}\frac{M^{2/3} - 8 - 2M^{1/3}}{2M^{1/3}}$, where $M = 108x^2 + 12\sqrt{48 - 120x^2 + 81x^4} - 80$.

57. (a) $n = 100 \cdot 2^{t/3} \implies \frac{n}{100} = 2^{t/3} \implies \log_2\left(\frac{n}{100}\right) = \frac{t}{3} \implies t = 3\log_2\left(\frac{n}{100}\right)$. This function tells us how long it will take to obtain n bacteria (given the number n).

(b) $n = 50{,}000 \implies t = 3\log_2\frac{50{,}000}{100} = 3\log_2 500 = 3\left(\frac{\ln 500}{\ln 2}\right) \approx 26.9$ hours

59. (a) To find the equation of the graph that results from shifting the graph of $y = \ln x$ 3 units upward, we add 3 to the original function to get $y = \ln x + 3$.

(b) To find the equation of the graph that results from shifting the graph of $y = \ln x$ 3 units to the left, we replace x with $x + 3$ in the original function to get $y = \ln(x + 3)$.

(c) To find the equation of the graph that results from reflecting the graph of $y = \ln x$ about the x-axis, we multiply the original equation by -1 to get $y = -\ln x$.

(d) To find the equation of the graph that results from reflecting the graph of $y = \ln x$ about the y-axis, we replace x with $-x$ in the original equation to get $y = \ln(-x)$.

(e) To find the equation of the graph that results from reflecting the graph of $y = \ln x$ about the line $y = x$, we interchange x and y in the original equation to get $x = \ln y \iff y = e^x$.

(f) To find the equation of the graph that results from reflecting the graph of $y = \ln x$ about the x-axis and then about the line $y = x$, we first multiply the original equation by -1 [to get $y = -\ln x$] and then interchange x and y in this equation to get $x = -\ln y \iff \ln y = -x \iff y = e^{-x}$.

(g) To find the equation of the graph that results from reflecting the graph of $y = \ln x$ about the y-axis and then about the line $y = x$, we first replace x with $-x$ in the original equation [to get $y = \ln(-x)$] and then interchange x and y to get $x = \ln(-y) \iff -y = e^x \iff y = -e^x$.

(h) To find the equation of the graph that results from shifting the graph of $y = \ln x$ 3 units to the left and then reflecting it about the line $y = x$, we first replace x with $x + 3$ in the original equation [to get $y = \ln(x + 3)$] and then interchange x and y in this equation to get $x = \ln(y + 3) \iff y + 3 = e^x \iff y = e^x - 3$.

Section 1.7 Models and Curve Fitting

Note: Some values are given to many decimal places. These are the results given by several computer algebra systems — rounding is left to the reader.

1. (a)

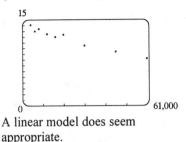

A linear model does seem appropriate.

(b) Using the points $(4000, 14.1)$ and $(60{,}000, 8.2)$, we obtain $y - 14.1 = \dfrac{8.2 - 14.1}{60{,}000 - 4000}(x - 4000)$ or, equivalently, $y \approx -0.000105357x + 14.521429$.

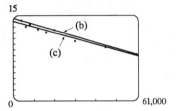

(c) Using a computing device, we obtain the following least squares regression line:
$$y = -0.0000997855x + 13.950764.$$

(d) When $x = 25{,}000$, $y \approx 11.456$; or about 11.5 per 100 population.

(e) When $x = 80{,}000$, $y \approx 5.968$; or about a 6% chance.

3. (a) An exponential model is $y = 301.8130539\,(0.8197450332)^x$ or $y = 301.8130539e^{-0.1987619222x}$.
If your calculator gives the first expression, you can convert to the second by finding $\ln(0.8197450332) \approx -0.1987619222$. A fourth-degree polynomial that fits these data is
$$y = -0.0024304185x^4 + 0.1351589312x^3 - 2.014321592x^2 - 4.055294453x + 199.0922273.$$

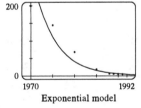

Exponential model

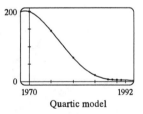

Quartic model

(b) Let $x = 2$ and $x = 12$ for 1972 and 1982, respectively.

	1972	1982
Exponential	202.8	27.8
Fourth-degree	184.0	43.5

5. By looking at the scatter plot of the data, we rule out a logarithmic model.

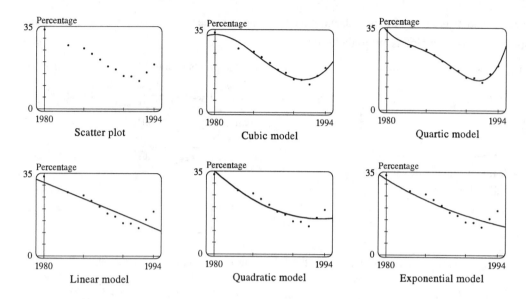

The cubic and quartic polynomial models seem to be the best. The linear, quadratic and exponential models seem reasonable for a 1982 estimate (interpolation), but not very good for a 1995 estimate (extrapolation). Using the cubic polynomial model (with $x = 0$ corresponding to 1980), $y = 0.0272238976x^3 - 0.4626415472x^2 + 0.0685010555x + 32.891372364$, we obtain estimates of 31.4% for 1982 and 21.7% for 1995. The estimate for 1996 would be 27.1%, which seems too high. It's risky to use the model to make predictions beyond 1995.

Chapter 1 Review

Concept Check

1. (a) Function — see Section 1.1 (page 12).

   (b) Domain and range of a function — see the discussion immediately following the definition of function in Section 1.1 (page 12).

   (c) Graph of a function — see the definition and Figure 4 in Section 1.1 (page 13).

   (d) Increasing function — see Section 1.1 (page 23).

   (e) Composition of two functions — see the definition in Section 1.2 (page 36).

   (f) Parametric Curve — see Section 1.4 (page 48).

2. (a) An even function f satisfies $f(-x) = f(x)$ for every number x in its domain. It is symmetric with respect to the y-axis.

   (b) An odd function g satisfies $g(-x) = -g(x)$ for every number x in its domain. It is symmetric with respect to the origin.

3. (a) See the definition of a one-to-one function in Section 1.6 (page 64). If no horizontal line intersects the graph of a function more than once, then the function is one-to-one.

   (b) See the definition of inverse function in Section 1.6 (page 64). The graph of f^{-1} can be obtained from the graph of f by reflecting it about the line $y = x$.

4. (a) See Section 1.1 (page 18).

   (b) See Section 1.6 (page 64).

5. (a) Constant function: $f(x) = 2$, $f(x) = c$

   (b) Power function: $f(x) = x^2$, $f(x) = x^a$

   (c) Exponential function: $f(x) = 2^x$, $f(x) = a^x$

   (d) Linear function: $f(x) = 2x + 1$, $f(x) = ax + b$

   (e) Quadratic function: $f(x) = x^2 + x + 1$, $f(x) = ax^2 + bx + c$

   (f) Polynomial of degree 5: $f(x) = x^5 + 2$

   (g) Rational function: $f(x) = \dfrac{x}{x+2}$, $f(x) = \dfrac{P(x)}{Q(x)}$ where $P(x)$ and $Q(x)$ are polynomials

6. (a)

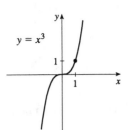

$y = x^3$

(b)

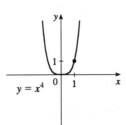

$y = x^4$

(c)

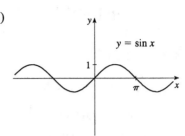

$y = \sin x$

(d)

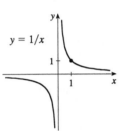

$y = \tan x$

(e)

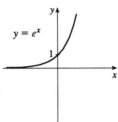

$y = e^x$

(f)

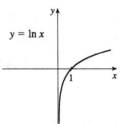

$y = \ln x$

(g)

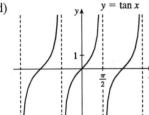

$y = 1/x$

(h)

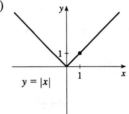

$y = |x|$

(i)

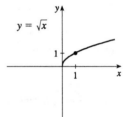

$y = \sqrt{x}$

True-False Quiz

1. False. Let $f(x) = x^3$. Then f is one-to-one and $f^{-1}(x) = \sqrt[3]{x}$. But $1/f(x) = 1/x^3$, which is not equal to $f^{-1}(x)$.

3. False. Let $f(x) = x^2$, $s = -1$, and $t = 1$. Then $f(s + t) = (-1 + 1)^2 = 0^2 = 0$, but $f(s) + f(t) = (-1)^2 + 1^2 = 2 \neq 0 = f(s + t)$.

5. True. See the Vertical Line Test.

7. True. The function $\ln x$ is an increasing function on $(0, \infty)$.

Exercises

1. (a) When $x = 2$, $y \approx 2.7$. Thus, $f(2) \approx 2.7$.   (b) $f(x) = 3 \Longrightarrow x \approx 2.3, 5.6$

(c) The domain of f is $-6 \le x \le 6$, or $[-6, 6]$.   (d) The range of f is $-4 \le y \le 4$, or $[-4, 4]$.

(e) f is increasing on $[-4, 4]$.

(f) f is not one-to-one since it fails the Horizontal Line Test.

(g) f is odd since its graph is symmetric with respect to the origin.

3. (a)

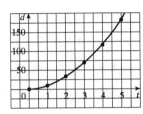

(b) From the graph, we see that the distance is slightly less than 150 feet.

5. $f(x) = \sqrt{4 - 3x^2}$. Domain: $4 - 3x^2 \ge 0 \Longrightarrow 3x^2 \le 4 \Longrightarrow x^2 \le \frac{4}{3} \Longrightarrow |x| \le \frac{2}{\sqrt{3}}$. Range: $y \ge 0$ and $y \le \sqrt{4} \Longrightarrow 0 \le y \le 2$.

7. $h(t) = e^{-t^2}$. Domain: $t \in \mathbb{R}$. Range: $e^{t^2} \ge 1 \Longrightarrow 0 < \dfrac{1}{e^{t^2}} \le 1$, so $0 < t \le 1$.

9. (a) To obtain the graph of $y = f(x) + 8$, we shift the graph of $y = f(x)$ up 8 units .

(b) To obtain the graph of $y = f(x + 8)$, we shift the graph of $y = f(x)$ left 8 units.

(c) To obtain the graph of $y = 1 + 2f(x)$, we stretch the graph of $y = f(x)$ vertically by a factor of 2, and then shift the resulting graph 1 unit upward.

(d) To obtain the graph of $y = f(x - 2) - 2$, we shift the graph of $y = f(x)$ right 2 units, and then shift the resulting graph 2 units downward.

(e) To obtain the graph of $y = -f(x)$, we reflect the graph of $y = f(x)$ about the x-axis.

(f) To obtain the graph of $y = f^{-1}(x)$, we reflect the graph of $y = f(x)$ about the line $y = x$ (assuming f is one–to–one).

11. To sketch the graph of $y = 1 + \sqrt{x + 2}$, we shift the graph of $y = \sqrt{x}$ left 2 units and up 1 unit.

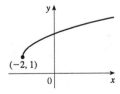

13. To sketch the graph of $y = \cos 3x$, we compress the graph of $y = \cos x$ horizontally by a factor of 3.

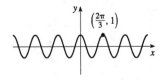

15. To sketch the graph of $y = -e^x$, we reflect the graph of $y = e^x$ about the x-axis.

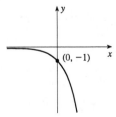

17. (a) The terms of f are a mixture of odd and even powers of x, so f is neither even nor odd.

(b) The terms of f are all odd powers of x, so f is odd.

(c) $f(-x) = e^{-(-x)^2} = e^{-x^2} = f(x)$, so f is even.

(d) $f(-x) = 1 + \sin(-x) = 1 - \sin x$. Now $f(-x) \neq f(x)$ and $f(-x) \neq -f(x)$, so f is neither even nor odd.

19. $(f \circ g)(x) = f(g(x)) = f(x^2 - 9) = \ln(x^2 - 9)$.
Domain: $x^2 - 9 > 0 \Longrightarrow x^2 > 9 \Longrightarrow |x| > 3 \Longrightarrow x \in (-\infty, -3) \cup (3, \infty)$

$(g \circ f)(x) = g(f(x)) = g(\ln x) = (\ln x)^2 - 9$. Domain: $x > 0$, or $(0, \infty)$

$(f \circ f)(x) = f(f(x)) = f(\ln x) = \ln(\ln x)$. Domain: $\ln x > 0 \Longrightarrow x > e^0 = 1$, or $(1, \infty)$

$(g \circ g)(x) = g(g(x)) = g(x^2 - 9) = (x^2 - 9)^2 - 9$. Domain: $x \in \mathbb{R}$, or $(-\infty, \infty)$

21. We need to know the value of x such that $f(x) = 2$. Since $x = 1$ gives us $y = 2$, $f^{-1}(2) = 1$.

23. (a) $e^{2\ln 3} = (e^{\ln 3})^2 = 3^2 = 9$

(b) $\log_{10} 25 + \log_{10} 4 = \log_{10}(25 \cdot 4) = \log_{10} 100 = \log_{10} 10^2 = 2$

25. (a) After 4 days, $\frac{1}{2}$ gram remains; after 8 days, $\frac{1}{4}$ g; after 12 days, $\frac{1}{8}$ g; after 16 days, $\frac{1}{16}$ g.

(b) $m(4) = \dfrac{1}{2}$, $m(8) = \dfrac{1}{2^2}$, $m(12) = \dfrac{1}{2^3}$, $m(16) = \dfrac{1}{2^4}$. From the pattern, we see that $m(t) = \dfrac{1}{2^{t/4}}$, or $2^{-t/4}$.

(c) $m = 2^{-t/4} \Longrightarrow \log_2 m = -t/4 \Longrightarrow t = -4\log_2 m$; this is the time elapsed when there are m grams of ^{100}Pd.

(d) $m = 0.01 \Longrightarrow t = -4\log_2 0.01 = -4 \left(\dfrac{\ln 0.01}{\ln 2} \right) \approx 26.6$ days

27. $f(x) = \ln(x^2 - c)$. If $c < 0$, the domain of f is $\mathbb{R}$. If $c = 0$, the domain of f is $(-\infty, 0) \cup (0, \infty)$. If $c > 0$, the domain of f is $(-\infty, -\sqrt{c}) \cup (\sqrt{c}, \infty)$. As c increases, the dip at $x = 0$ becomes deeper. For $c \geq 0$, the graph has asymptotes at $x = \pm\sqrt{c}$.

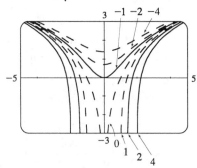

29. (a)

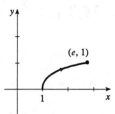

(b) $x = e^t \Longrightarrow t = \ln x$; $y = \sqrt{t}$ so $y = \sqrt{\ln x}$.

31. We sketch $x = t$, $y = 2t + \ln t$ (the function) and $x = 2t + \ln t$, $y = t$ (its inverse) for $t > 0$.

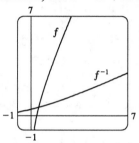

33.

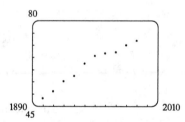

Many models appear to be plausible. Your choice depends on whether you think medical advances will keep increasing life expectancy, or if there is bound to be a natural leveling-off of life expectancy. A linear model, $y = 0.263x - 450.034$, gives us an estimate of 76.0 years for the year 2000.

Principles of Problem Solving

1.

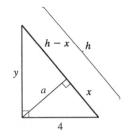

By using the area formula for a triangle, $\frac{1}{2}$ (base) (height), in

two ways, we see that $\frac{1}{2}\,(4)\,(y) = \frac{1}{2}\,(h)\,(a)$, so $a = \dfrac{4y}{h}$.

Since $4^2 + y^2 = h^2$, $y = \sqrt{h^2 - 16}$, and $a = \dfrac{4\sqrt{h^2 - 16}}{h}$.

3. $|2x - 1| = \begin{cases} 1 - 2x & \text{if } x < \frac{1}{2} \\ 2x - 1 & \text{if } x \geq \frac{1}{2} \end{cases}$ and $|x + 5| = \begin{cases} -x - 5 & \text{if } x < -5 \\ x + 5 & \text{if } x \geq -5 \end{cases}$

Therefore, we consider the three cases $x < -5$, $-5 \leq x < \frac{1}{2}$, and $x \geq \frac{1}{2}$.

If $x < -5$, we must have $1 - 2x - (-x - 5) = 3 \iff x = 3$, which is false, since we are considering $x < -5$.

If $-5 \leq x < \frac{1}{2}$, we must have $1 - 2x - (x + 5) = 3 \iff x = -\frac{7}{3}$.

If $x \geq \frac{1}{2}$, we must have $2x - 1 - (x + 5) = 3 \iff x = 9$.

So the two solutions of the equation are $x = -\frac{7}{3}$ and $x = 9$.

5. $f(x) = |x^2 - 4|x| + 3|$. If $x \geq 0$, then $f(x) = |x^2 - 4x + 3| = |(x - 1)(x - 3)|$.

Case (i):   If $0 < x \leq 1$, then $f(x) = x^2 - 4x + 3$.

Case (ii):   If $1 < x \leq 3$, then $f(x) = -(x^2 - 4x + 3) = -x^2 + 4x - 3$.

Case (iii):   If $x > 3$, then $f(x) = x^2 - 4x + 3$.

This enables us to sketch the graph for $x \geq 0$. Then we use the fact that f is an even function to reflect this part of the graph about the y-axis to obtain the entire graph. Or, we could consider also the cases $x < -3$, $-3 \leq x < -1$, and $-1 \leq x < 0$.

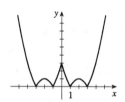

7. $|x| + |y| \leq 1$. The boundary of the region has equation $|x| + |y| = 1$.

In quadrants I, II, III, and IV, this becomes the lines $x + y = 1$, $-x + y = 1$, $-x - y = 1$, and $x - y = 1$ respectively.

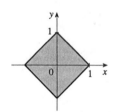

9. $(\log_2 3)(\log_3 4)(\log_4 5)\cdots(\log_{31} 32) = \left(\dfrac{\ln 3}{\ln 2}\right)\left(\dfrac{\ln 4}{\ln 3}\right)\left(\dfrac{\ln 5}{\ln 4}\right)\cdots\left(\dfrac{\ln 32}{\ln 31}\right)$

$$= \dfrac{\ln 32}{\ln 2} = \dfrac{\ln 2^5}{\ln 2} = \dfrac{5\ln 2}{\ln 2} = 5$$

11. $\ln\left(x^2 - 2x - 2\right) \le 0 \Longrightarrow x^2 - 2x - 2 \le e^0 = 1 \Longrightarrow x^2 - 2x - 3 \le 0 \Longrightarrow$

$(x - 3)(x + 1) \le 0 \Longrightarrow x \in [-1, 3]$. Since the argument must be positive, $x^2 - 2x - 2 > 0 \Longrightarrow$

$\left[x - \left(1 - \sqrt{3}\right)\right]\left[x - \left(1 + \sqrt{3}\right)\right] > 0 \Longrightarrow x \in \left(-\infty, 1 - \sqrt{3}\right) \cup \left(1 + \sqrt{3}, \infty\right)$. The intersection of

these intervals is $\left[-1, 1 - \sqrt{3}\right) \cup \left(1 + \sqrt{3}, 3\right]$.

13. Let $\alpha = \arcsin x$ and $\beta = \arcsin y$ (see the figures.) Using $\sin(\alpha + \beta) = \sin\alpha\cos\beta + \cos\alpha\sin\beta$, we

take the sine of the left-hand side, obtaining $x\sqrt{1 - y^2} + \sqrt{1 - x^2}\,(y)$. We obtain the same expression

if we take the sine of the right-hand side. Since both sides are between $-\frac{\pi}{2}$ and $\frac{\pi}{2}$ and their sines are

equal, the sides are equal.

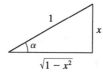
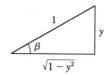

15. Let S_n be the statement that $7^n - 1$ is divisible by 6.

- S_1 is true because $7^1 - 1 = 6$ is divisible by 6.
- Assume S_k is true, that is, $7^k - 1$ is divisible by 6. In other words, $7^k - 1 = 6m$ for some positive
  integer m. Then $7^{k+1} - 1 = 7^k \cdot 7 - 1 = (6m + 1) \cdot 7 - 1 = 42m + 6 = 6(7m + 1)$, which is
  divisible by 6, so S_{k+1} is true.
- Therefore, by mathematical induction, $7^n - 1$ is divisible by 6 for every positive integer n.

17. $f_0(x) = x^2$ and $f_{n+1}(x) = f_0(f_n(x))$ for $n = 0, 1, 2, \ldots$.

$f_1(x) = f_0(f_0(x)) = f_0(x^2) = (x^2)^2 = x^4$, $f_2(x) = f_0(f_1(x)) = f_0(x^4) = (x^4)^2 = x^8$,

$f_3(x) = f_0(f_2(x)) = f_0(x^8) = (x^8)^2 = x^{16}, \ldots$. Thus, a general formula is $f_n(x) = x^{2^{n+1}}$.

Chapter 2   Limits and Derivatives

Section 2.1   The Tangent and Velocity Problems

1. (a) Slopes of the secant lines:

x	m_{PQ}
0	$\frac{2.6-1.3}{0-3} \approx -0.43$
1	$\frac{2.0-1.3}{1-3} = -0.35$
2	$\frac{1.1-1.3}{2-3} = 0.2$
4	$\frac{2.1-1.3}{4-3} = 0.8$
5	$\frac{3.5-1.3}{5-3} = 1.1$

(b) We average the slopes of the two closest secant lines from part (a): $\frac{1}{2}(0.2+0.8) = 0.5$.

(c) Using the points $(0.6, 0)$ and $(5, 2.5)$ from the graph, the slope of the tangent line at P is about $\dfrac{2.5-0}{5-0.6} \approx 0.57$.

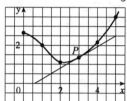

3. For the curve $y = \sqrt{x}$ and the point $P(4, 2)$:

(a)

	x	Q	m_{PQ}
(i)	5	$(5, 2.236068)$	0.236068
(ii)	4.5	$(4.5, 2.121320)$	0.242641
(iii)	4.1	$(4.1, 2.024846)$	0.248457
(iv)	4.01	$(4.01, 2.002498)$	0.249844
(v)	4.001	$(4.001, 2.000250)$	0.249984
(vi)	3	$(3, 1.732051)$	0.267949
(vii)	3.5	$(3.5, 1.870829)$	0.258343
(viii)	3.9	$(3.9, 1.974842)$	0.251582
(ix)	3.99	$(3.99, 1.997498)$	0.250156
(x)	3.999	$(3.999, 1.999750)$	0.250016

(b) The slope appears to be $\frac{1}{4}$.

(c) $y - 2 = \frac{1}{4}(x - 4)$ or $y = \frac{1}{4}x + 1$.

5. (a) At $t = 2$, $y = 40(2) - 16(2)^2 = 16$. The average velocity between times 2 and $2 + h$ is
$$\frac{40(2+h) - 16(2+h)^2 - 16}{h} = \frac{-24h - 16h^2}{h} = -24 - 16h, \text{ if } h \neq 0.$$

(i) $h = 0.5$, -32 ft/s (ii) $h = 0.1$, -25.6 ft/s

(iii) $h = 0.05$, -24.8 ft/s (iv) $h = 0.01$, -24.16 ft/s

(b) The instantaneous velocity when $t = 2$ is -24 ft/s.

7. Average velocity between times 1 and $1 + h$ is

$$\frac{s\,(1+h) - s\,(1)}{h} = \frac{(1+h)^3\,/6 - 1/6}{h} = \frac{h^3 + 3h^2 + 3h}{6h} = \frac{h^2 + 3h + 3}{6} \text{ if } h \neq 0.$$

(a)   (i) $[1,3]$: $h = 2$, $\frac{13}{6}$ ft/s   (ii) $[1,2]$: $h = 1$, $\frac{7}{6}$ ft/s

    (iii) $[1,1.5]$: $h = 0.5$, $\frac{19}{24}$ ft/s   (iv) $[1,1.1]$: $h = 0.1$, $\frac{331}{600}$ ft/s

(b) As h approaches 0, the velocity approaches $\frac{1}{2}$ ft/s.

(c) 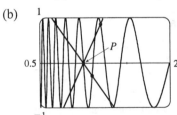   (d)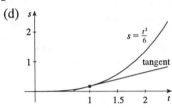

9. For the curve $y = \sin\left(10\pi/x\right)$ and the point $P\,(1,0)$:

(a)

x	Q	m_{PQ}	x	Q	m_{PQ}
2	$(2,0)$	0	0.5	$(0.5,0)$	0
1.5	$(1.5, 0.8660)$	1.7321	0.6	$(0.6, 0.8660)$	-2.1651
1.4	$(1.4, -0.4339)$	-1.0847	0.7	$(0.7, 0.7818)$	-2.6061
1.3	$(1.3, -0.8230)$	-2.7433	0.8	$(0.8, 1)$	-5
1.2	$(1.2, 0.8660)$	4.3301	0.9	$(0.9, -0.3420)$	3.4202
1.1	$(1.1, -0.2817)$	-2.8173	0.99	$(0.99, 0.3120)$	-31.2033

As x approaches 1, the slopes do not appear to be approaching any particular value.

(b) 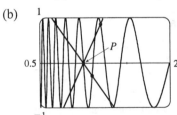   We see that problems with estimation are caused by the frequent oscillations of the graph. The tangent is so steep at P that we need to take x-values much closer to 1 in order to get accurate estimates of its slope.

(c) If we choose $x = 1.001$, then the point Q is $(1.001, -0.0314)$ and $m_{PQ} \approx -31.3794$. If $x = 0.999$, then Q is $(0.999, 0.0314)$ and $m_{PQ} = -31.4422$. Averaging these two slopes gives us the estimate -31.4108.

Section 2.2   The Limit of a Function

1. As x approaches 2, $f(x)$ approaches 5. [Or, the values of $f(x)$ can be made as close to 5 as we like by taking x sufficiently close to 2 (but $x \neq 2$).] Yes, the graph could have a hole at $(2, 5)$ and be defined such that $f(2) = 3$.

3. (a) $\lim\limits_{x \to 1} f(x) = 3$

(b) $\lim\limits_{x \to 3^-} f(x) = 2$

(c) $\lim\limits_{x \to 3^+} f(x) = -2$

(d) $\lim\limits_{x \to 3} f(x)$ doesn't exist because the limits

   in part (b) and part (c) are not equal.

(e) $f(3) = 1$

(f) $\lim\limits_{x \to -2^-} f(x) = -1$

(g) $\lim\limits_{x \to -2^+} f(x) = -1$

(h) $\lim\limits_{x \to -2} f(x) = -1$

(i) $f(-2) = -3$

5.

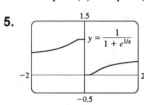

(a) $\lim\limits_{x \to 0^-} f(x) = 1$

(b) $\lim\limits_{x \to 0^+} f(x) = 0$

(c) $\lim\limits_{x \to 0} f(x) = 0$ does not exist because the limits in part (a) and part (b) are not equal.

7.

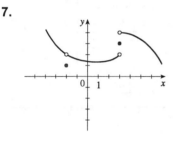

9. For $g(x) = \dfrac{x - 1}{x^3 - 1}$:

x	$g(x)$		x	$g(x)$
0.2	0.806452		1.8	0.165563
0.4	0.641026		1.6	0.193798
0.6	0.510204		1.4	0.229358
0.8	0.409836		1.2	0.274725
0.9	0.369004		1.1	0.302115
0.99	0.336689		1.01	0.330022

It appears that $\lim\limits_{x \to 1} \dfrac{x - 1}{x^3 - 1} = 0.\overline{3} = \dfrac{1}{3}$.

11. For $f(x) = \dfrac{1 - \cos x}{x^2}$:

x	$f(x)$
1	0.459698
0.5	0.489670
0.4	0.493369
0.3	0.496261
0.2	0.498336
0.1	0.499583
0.05	0.499896
0.01	0.499996

It appears that

$$\lim\limits_{x \to 0} \dfrac{1 - \cos x}{x^2} = 0.5.$$

13. (a) From the following graphs, it seems that $\lim\limits_{x \to 0} \dfrac{\tan(4x)}{x} = 4$.

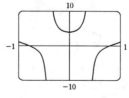

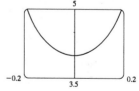

(b)

x	$f(x)$
± 0.1	4.227932
± 0.01	4.002135
± 0.001	4.000021
± 0.0001	4.000000

15. (a) Let $h(x) = (1+x)^{1/x}$

(b)

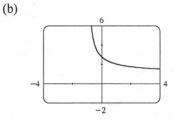

x	$h(x)$
-0.001	2.71964
-0.0001	2.71842
-0.00001	2.71830
-0.000001	2.71828
0.000001	2.71828
0.00001	2.71827
0.0001	2.71815
0.001	2.71692

It appears that $\lim\limits_{x \to 0} (1+x)^{1/x} \approx 2.71828$, which is approximately e. In Section 3.7 we'll see that the value of the limit is exactly e.

17. For $f(x) = x^2 - (2^x/1000)$:

(a)

x	$f(x)$
1	0.998000
0.8	0.638259
0.6	0.358484
0.4	0.158680
0.2	0.038851
0.1	0.008928
0.05	0.001465

It appears that $\lim\limits_{x \to 0} f(x) = 0$.

(b)

x	$f(x)$
0.04	0.000572
0.02	-0.000614
0.01	-0.000907
0.005	-0.000978
0.003	-0.000993
0.001	-0.001000

It appears that $\lim\limits_{x \to 0} f(x) = -0.001$.

19.

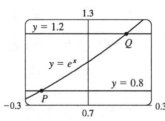

We need to have $0.8 < e^x < 1.2$. From the graph we obtain the approximate points of intersection $P(-0.2231436, 0.8)$ and $Q(0.18232156, 1.2)$. So if x is within 0.182 of 0, then y will be within 0.2 of 1. If we must have e^x within 0.1 of 1, we get $P(-0.1053605, 0.9)$ and $Q(0.09531018, 1.1)$. We would then need x to be within 0.095 of 0.

Section 2.3   Calculating Limits Using the Limit Laws

1. (a) $\lim\limits_{x \to a} [f(x) + h(x)] = \lim\limits_{x \to a} f(x) + \lim\limits_{x \to a} h(x) = -3 + 8 = 5$

(b) $\lim\limits_{x \to a} [f(x)]^2 = \left[\lim\limits_{x \to a} f(x)\right]^2 = (-3)^2 = 9$

(c) $\lim\limits_{x \to a} \sqrt[3]{h(x)} = \sqrt[3]{\lim\limits_{x \to a} h(x)} = \sqrt[3]{8} = 2$ $\qquad$ **(d)** $\lim\limits_{x \to a} \dfrac{1}{f(x)} = \dfrac{1}{\lim\limits_{x \to a} f(x)} = \dfrac{1}{-3} = -\dfrac{1}{3}$

(e) $\lim\limits_{x \to a} \dfrac{f(x)}{h(x)} = \dfrac{\lim\limits_{x \to a} f(x)}{\lim\limits_{x \to a} h(x)} = \dfrac{-3}{8} = -\dfrac{3}{8}$ $\qquad$ **(f)** $\lim\limits_{x \to a} \dfrac{g(x)}{f(x)} = \dfrac{\lim\limits_{x \to a} g(x)}{\lim\limits_{x \to a} f(x)} = \dfrac{0}{-3} = 0$

(g) The limit does not exist, since $\lim\limits_{x \to a} g(x) = 0$ but $\lim\limits_{x \to a} f(x) \neq 0$.

(h) $\lim\limits_{x \to a} \dfrac{2f(x)}{h(x) - f(x)} = \dfrac{2 \lim\limits_{x \to a} f(x)}{\lim\limits_{x \to a} h(x) - \lim\limits_{x \to a} f(x)} = \dfrac{2(-3)}{8 - (-3)} = -\dfrac{6}{11}$

3. $\lim\limits_{x \to 4} \left(5x^2 - 2x + 3\right) = \lim\limits_{x \to 4} 5x^2 - \lim\limits_{x \to 4} 2x + \lim\limits_{x \to 4} 3$ $\qquad$ (Limit Laws 2 & 1)

$\qquad\qquad\qquad\qquad = 5 \lim\limits_{x \to 4} x^2 - 2 \lim\limits_{x \to 4} x + 3$ $\qquad$ (3 & 7)

$\qquad\qquad\qquad\qquad = 5(4)^2 - 2(4) + 3 = 75$ $\qquad$ (9 & 8)

5. $\lim\limits_{t \to -2} (t+1)^9 \left(t^2 - 1\right) = \lim\limits_{t \to -2} (t+1)^9 \lim\limits_{t \to -2} \left(t^2 - 1\right)$ $\qquad\qquad\qquad$ (4)

$\qquad\qquad\qquad\quad = \left[\lim\limits_{t \to -2} (t+1)\right]^9 \lim\limits_{t \to -2} \left(t^2 - 1\right)$ $\qquad\qquad\qquad$ (6)

$\qquad\qquad\qquad\quad = \left[\lim\limits_{t \to -2} t + \lim\limits_{t \to -2} 1\right]^9 \left[\lim\limits_{t \to -2} t^2 - \lim\limits_{t \to -2} 1\right]$ $\qquad$ (1 & 2)

$\qquad\qquad\qquad\quad = [(-2) + 1]^9 \left[(-2)^2 - 1\right] = -3$ $\qquad\qquad\qquad$ (8, 7 & 9)

7. $\lim\limits_{x \to 4^-} \sqrt{16 - x^2} = \sqrt{\lim\limits_{x \to 4^-} \left(16 - x^2\right)}$ $\qquad$ (11)

$\qquad\qquad\qquad = \sqrt{\lim\limits_{x \to 4^-} 16 - \lim\limits_{x \to 4^-} x^2}$ $\qquad$ (2)

$\qquad\qquad\qquad = \sqrt{16 - (4)^2} = 0$ $\qquad$ (7 & 9)

9. $\lim\limits_{x \to -3} \dfrac{x^2 - x + 12}{x + 3}$ does not exist since $x + 3 \to 0$ but $x^2 - x + 12 \to 24$ as $x \to -3$.

11. $\lim\limits_{h \to 0} \dfrac{(h-5)^2 - 25}{h} = \lim\limits_{h \to 0} \dfrac{(h^2 - 10h + 25) - 25}{h} = \lim\limits_{h \to 0} \dfrac{h^2 - 10h}{h} = \lim\limits_{h \to 0} (h - 10) = -10$

13. $\lim\limits_{t \to 9} \dfrac{9-t}{3-\sqrt{t}} = \lim\limits_{t \to 9} \dfrac{(3+\sqrt{t})(3-\sqrt{t})}{3-\sqrt{t}} = \lim\limits_{t \to 9} (3 + \sqrt{t}) = 3 + \sqrt{9} = 6$

15. $\lim\limits_{t \to 0} \dfrac{\sqrt{2-t} - \sqrt{2}}{t} = \lim\limits_{t \to 0} \dfrac{\sqrt{2-t} - \sqrt{2}}{t} \cdot \dfrac{\sqrt{2-t} + \sqrt{2}}{\sqrt{2-t} + \sqrt{2}} = \lim\limits_{t \to 0} \dfrac{-t}{t\left(\sqrt{2-t} + \sqrt{2}\right)} = \lim\limits_{t \to 0} \dfrac{-1}{\sqrt{2-t} + \sqrt{2}}$

$\qquad = -\dfrac{1}{2\sqrt{2}} = -\dfrac{\sqrt{2}}{4}$

17. $\lim\limits_{x \to 1} \left(\dfrac{1}{x-1} - \dfrac{2}{x^2 - 1} \right) = \lim\limits_{x \to 1} \dfrac{(x+1) - 2}{(x-1)(x+1)} = \lim\limits_{x \to 1} \dfrac{x-1}{(x-1)(x+1)} = \lim\limits_{x \to 1} \dfrac{1}{x+1} = \dfrac{1}{2}$

19. Let $f(x) = -x^2$, $g(x) = x^2 \cos 20\pi x$ and $h(x) = x^2$. Then
$-1 \le \cos 20\pi x \le 1 \implies -x^2 \le x^2 \cos 20\pi x \le x^2 \implies$
$f(x) \le g(x) \le h(x)$. So since $\lim\limits_{x \to 0} f(x) = \lim\limits_{x \to 0} h(x) = 0$, by
the Squeeze Theorem we have $\lim\limits_{x \to 0} g(x) = 0$.

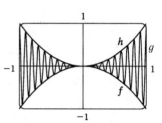

21. $1 \le f(x) \le x^2 + 2x + 2$ for all x. Now $\lim\limits_{x \to -1} 1 = 1$ and

$\lim\limits_{x \to -1} (x^2 + 2x + 2) = \lim\limits_{x \to -1} x^2 + 2 \lim\limits_{x \to -1} x + \lim\limits_{x \to -1} 2 = (-1)^2 + 2(-1) + 2 = 1$. Therefore, by the

Squeeze Theorem, $\lim\limits_{x \to -1} f(x) = 1$.

23. $-1 \le \cos (2/x) \le 1 \implies -x^4 \le x^4 \cos (2/x) \le x^4$. Since $\lim\limits_{x \to 0} (-x^4) = 0$ and $\lim\limits_{x \to 0} x^4 = 0$, we have

$\lim\limits_{x \to 0} \left[x^4 \cos (2/x) \right] = 0$ by the Squeeze Theorem.

25. If $x > -4$, then $|x+4| = x+4$, so $\lim\limits_{x \to -4+} |x+4| = \lim\limits_{x \to -4+} (x+4) = -4 + 4 = 0$.

If $x < -4$, then $|x+4| = -(x+4)$, so $\lim\limits_{x \to -4-} |x+4| = \lim\limits_{x \to -4-} -(x+4) = -(-4+4) = 0$.

Since the right and left limits are equal, $\lim\limits_{x \to -4} |x+4| = 0$.

27. Since $|x| = -x$ for $x < 0$, we have $\lim\limits_{x \to 0-} \left(\dfrac{1}{x} - \dfrac{1}{|x|} \right) = \lim\limits_{x \to 0-} \left(\dfrac{1}{x} - \dfrac{1}{-x} \right) = \lim\limits_{x \to 0-} \dfrac{2}{x}$, which does not

exist since the denominator approaches 0 and the numerator does not.

29. (a) (i) $\lim\limits_{x \to 0^+} h(x) = \lim\limits_{x \to 0^+} x^2 = 0^2 = 0$

(ii) $\lim\limits_{x \to 0^-} h(x) = \lim\limits_{x \to 0^-} x = 0$, so $\lim\limits_{x \to 0} h(x) = 0$.

(iii) $\lim\limits_{x \to 1} h(x) = \lim\limits_{x \to 1} x^2 = 1^2 = 1$

(iv) $\lim\limits_{x \to 2^-} h(x) = \lim\limits_{x \to 2^-} x^2 = 2^2 = 4$

(v) $\lim\limits_{x \to 2^+} h(x) = \lim\limits_{x \to 2^+} (8 - x) = 8 - 2 = 6$

(vi) Since $\lim\limits_{x \to 2^-} h(x) \neq \lim\limits_{x \to 2^+} h(x)$, $\lim\limits_{x \to 2} h(x)$ does not exist.

(b)

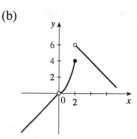

31. (a) (i) $[\![x]\!] = -2$ for $-2 \le x < -1$, so $\lim\limits_{x \to -2^+} [\![x]\!] = \lim\limits_{x \to -2^+} (-2) = -2$

(ii) $[\![x]\!] = -3$ for $-3 \le x < -2$, so $\lim\limits_{x \to -2^-} [\![x]\!] = \lim\limits_{x \to -2^-} (-3) = -3$. The right and left limits are different, so $\lim\limits_{x \to -2} [\![x]\!]$ does not exist.

(iii) $[\![x]\!] = -3$ for $-3 \le x < -2$, so $\lim\limits_{x \to -2.4} [\![x]\!] = \lim\limits_{x \to -2.4} (-3) = -3$.

(b) (i) $[\![x]\!] = n - 1$ for $n - 1 \le x < n$, so $\lim\limits_{x \to n^-} [\![x]\!] = \lim\limits_{x \to n^-} (n - 1) = n - 1$.

(ii) $[\![x]\!] = n$ for $n \le x < n + 1$, so $\lim\limits_{x \to n^+} [\![x]\!] = \lim\limits_{x \to n^+} n = n$.

(c) $\lim\limits_{x \to a} [\![x]\!]$ exists $\Longleftrightarrow$ a is not an integer.

33. The graph of $f(x) = [\![x]\!] + [\![-x]\!]$ is the same as the graph of $g(x) = -1$ with holes at each integer, since $f(a) = 0$ for any integer a. Thus, $\lim\limits_{x \to 2^-} f(x) = -1$ and $\lim\limits_{x \to 2^+} f(x) = -1$, so $\lim\limits_{x \to 2} f(x) = -1$. $f(2) = [\![2]\!] + [\![-2]\!] = 2 + (-2) = 0$.

35. Since $p(x)$ is a polynomial, $p(x) = a_0 + a_1 x + a_2 x^2 + \cdots + a_n x^n$. Thus, by the Limit Laws,

$$\lim\limits_{x \to a} p(x) = \lim\limits_{x \to a} \left(a_0 + a_1 x + a_2 x^2 + \cdots + a_n x^n \right)$$

$$= a_0 + a_1 \lim\limits_{x \to a} x + a_2 \lim\limits_{x \to a} x^2 + \cdots + a_n \lim\limits_{x \to a} x^n$$

$$= a_0 + a_1 a + a_2 a^2 + \cdots + a_n a^n = p(a)$$

Thus, for any polynomial p, $\lim\limits_{x \to a} p(x) = p(a)$.

37. Let $f(x) = [\![x]\!]$ and $g(x) = -[\![x]\!]$. Then $\lim\limits_{x \to 3} f(x)$ and $\lim\limits_{x \to 3} g(x)$ do not exist (Example 9) but

$$\lim\limits_{x \to 3} [f(x) + g(x)] = \lim\limits_{x \to 3} ([\![x]\!] - [\![x]\!]) = \lim\limits_{x \to 3} 0 = 0.$$

39. Since the denominator approaches 0 as $x \to -2$, the limit will exist only if the numerator also approaches 0 as $x \to -2$. In order for this to happen, we need $\lim\limits_{x \to -2} (3x^2 + ax + a + 3) = 0 \Longleftrightarrow$ $3(-2)^2 + a(-2) + a + 3 = 0 \Longleftrightarrow 12 - 2a + a + 3 = 0 \Longleftrightarrow a = 15$. With $a = 15$, the limit becomes

$$\lim\limits_{x \to -2} \frac{3x^2 + 15x + 18}{x^2 + x - 2} = \lim\limits_{x \to -2} \frac{3(x + 2)(x + 3)}{(x - 1)(x + 2)} = \frac{3(-2 + 3)}{-2 - 1} = -1.$$

Section 2.4 Continuity

1. From Equation 1, $\lim\limits_{x \to 4} f(x) = f(4)$.

3. (a) The following are the numbers at which f is discontinuous and the type of discontinuity at that number: -5 (jump), -3 (infinite), -1 (undefined), 3 (removable), 5 (infinite), 8 (jump), 10 (undefined).

(b) f is continuous from the left at -5 and -3, and continuous from the right at 8. It is continuous from neither side at -1, 3, 5, and 10.

5.

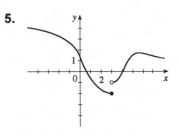

7. (a)

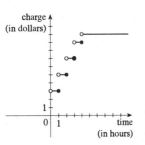

(b) There are discontinuities at $t = 1, 2, 3$, and 4. A person parking in the lot would want to keep in mind that the charge will jump at the beginning of each hour.

9. $\lim\limits_{x \to 4} g(x) = \lim\limits_{x \to 4} \dfrac{x+1}{2x^2 - 1} = \dfrac{\lim\limits_{x \to 4} x + \lim\limits_{x \to 4} 1}{2 \lim\limits_{x \to 4} x^2 - \lim\limits_{x \to 4} 1} = \dfrac{4+1}{2(4)^2 - 1} = \dfrac{5}{31} = g(4)$. So g is continuous at 4.

11. $f(x) = \dfrac{x^2 - 1}{x + 1}$ is discontinuous at -1 because $f(-1)$ is not defined.

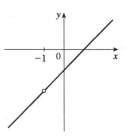

13. Since $f(x) = \dfrac{x^2 - 2x - 8}{x - 4}$ if $x \neq 4$,

$$\lim_{x \to 4} f(x) = \lim_{x \to 4} \frac{x^2 - 2x - 8}{x - 4} = \lim_{x \to 4} \frac{(x - 4)(x + 2)}{x - 4}$$

$$= \lim_{x \to 4} (x + 2) = 4 + 2 = 6.$$

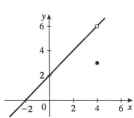

But $f(4) = 3$, so $\lim\limits_{x \to 4} f(x) \neq f(4)$. Therefore, f is discontinuous at 4.

15. $G\left(x\right) = \dfrac{x^4 + 17}{6x^2 + x - 1}$ is a rational function, so by Theorem 5 it is continuous on its domain, which is

$\left\{x \mid \left(3x - 1\right)\left(2x + 1\right) \neq 0\right\} = \left\{x \mid x \neq -\frac{1}{2}, \frac{1}{3}\right\}$.

17. By Theorem 5, the polynomial $5x$ is continuous on $(-\infty, \infty)$. By Theorems 9 and 7, $\sin 5x$ is continuous on $(-\infty, \infty)$. By Theorem 7, e^x is continuous on $(-\infty, \infty)$. By Theorem 4 #4, the product of e^x and $\sin 5x$ is continuous at all numbers which are in both of their domains, that is, on $(-\infty, \infty)$.

19. By Theorem 5, the polynomial $t^4 - 1$ is continuous $(-\infty, \infty)$. By Theorem 7, $\ln x$ is continuous on its domain, $(0, \infty)$. By Theorem 9, $\ln\left(t^4 - 1\right)$ is continuous on its domain, which is

$\left\{t \mid t^4 - 1 > 0\right\} = \left\{t \mid |t| > 1\right\} = (-\infty, 1) \cup (1, \infty)$.

21. The function $y = 1/\left(1 + e^{1/x}\right)$ is discontinuous at $x = 0$

because the left- and right-hand limits at $x = 0$ are different.

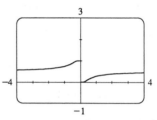

23. Because we are dealing with root functions, $5 + \sqrt{x}$ is continuous on $[0, \infty)$, $\sqrt{x + 5}$ is continuous on $[-5, \infty)$, so the quotient $f\left(x\right) = \dfrac{5 + \sqrt{x}}{\sqrt{5 + x}}$ is continuous on $[0, \infty)$. Since f is continuous at $x = 4$,

$\lim\limits_{x \to 4} f\left(x\right) = f\left(4\right) = \frac{7}{3}$.

25. Because $x^2 - x$ is continuous on $\mathbb{R}$, the composite function $f\left(x\right) = e^{x^2 - x}$ is continuous on $\mathbb{R}$, so

$\lim\limits_{x \to 1} f\left(x\right) = f\left(1\right) = e^{1-1} = e^0 = 1$.

27. f is continuous on $(-\infty, -1)$, $(-1, 1)$ and $(1, \infty)$ since on each of these

intervals it is a polynomial. Now $\lim\limits_{x \to -1^-} f\left(x\right) = \lim\limits_{x \to -1^-} \left(2x + 1\right) = -1$

and $\lim\limits_{x \to -1^+} f\left(x\right) = \lim\limits_{x \to -1^+} 3x = -3$, so f is discontinuous at -1. Since

$f\left(-1\right) = -1$, f is continuous from the left at -1. Also

$\lim\limits_{x \to 1^-} f\left(x\right) = \lim\limits_{x \to 1^-} 3x = 3$ and $\lim\limits_{x \to 1^+} f\left(x\right) = \lim\limits_{x \to 1^+} \left(2x - 1\right) = 1$, so f

is discontinuous at 1. Since $f\left(1\right) = 1$, f is continuous from the right at 1.

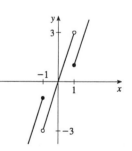

29. f is continuous on $(-\infty, 3)$ and $(3, \infty)$. Now $\lim\limits_{x \to 3^-} f\left(x\right) = \lim\limits_{x \to 3^-} \left(cx + 1\right) = 3c + 1$ and

$\lim\limits_{x \to 3^+} f\left(x\right) = \lim\limits_{x \to 3^+} \left(cx^2 - 1\right) = 9c - 1$. So f is continuous $\iff 3c + 1 = 9c - 1 \iff 6c = 2 \iff$

$c = \frac{1}{3}$. Thus, for f to be continuous on $(-\infty, \infty)$, $c = \frac{1}{3}$.

31. $f\left(x\right) = x^3 - x^2 + x$ is continuous on the interval $[2, 3]$, $f\left(2\right) = 6$, and $f\left(3\right) = 21$. Since $6 < 10 < 21$, there is a number c in $(2, 3)$ such that $f\left(c\right) = 10$ by the Intermediate Value Theorem.

33. $f(x) = x^3 - 3x + 1$ is continuous on the interval $[0, 1]$, $f(0) = 1$, and $f(1) = -1$. Since $-1 < 0 < 1$, there is a number c in $(0, 1)$ such that $f(c) = 0$ by the Intermediate Value Theorem. Thus, there is a root of the equation $x^3 - 3x + 1 = 0$ in the interval $(0, 1)$.

35. $f(x) = \cos x - x$ is continuous on the interval $[0, 1]$, $f(0) = 1$, and $f(1) = \cos 1 - 1 \approx -0.46$. Since $-0.46 < 0 < 1$, there is a number c in $(0, 1)$ such that $f(c) = 0$ by the Intermediate Value Theorem. Thus, there is a root of the equation $\cos x - x = 0$, or $\cos x = x$, in the interval $(0, 1)$.

37. (a) $f(x) = e^x + x - 2$ is continuous on the interval $[0, 1]$, $f(0) = -1$, and $f(1) = e - 1 \approx 1.72$. Since $-1 < 0 < 1.72$, there is a number c in $(0, 1)$ such that $f(c) = 0$ by the Intermediate Value Theorem. Thus, there is a root of the equation $e^x + x - 2 = 0$, or $e^x = 2 - x$, in the interval $(0, 1)$.

(b) $f(0.44) \approx -0.007$ and $f(0.45) \approx 0.018$, so there is a root between 0.44 and 0.45.

39. (a) Let $f(x) = \sqrt{x - 5} - \dfrac{1}{x + 3}$. Then $f(5) = -\frac{1}{8} < 0$ and $f(6) = \frac{8}{9} > 0$, and f is continuous on $[5, \infty)$. So by the Intermediate Value Theorem, there is a number c in $(5, 6)$ such that $f(c) = 0$. This implies that $\dfrac{1}{c + 3} = \sqrt{c - 5}$.

(b) Using the intersect feature of the graphing device, we find that the root of the equation is $x = 5.016$, correct to three decimal places.

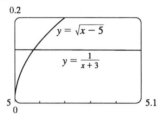

41. $\displaystyle\lim_{h \to 0} \sin(a + h) = \lim_{h \to 0} (\sin a \cos h + \cos a \sin h)$

$\displaystyle = \lim_{h \to 0} (\sin a \cos h) + \lim_{h \to 0} (\cos a \sin h)$

$\displaystyle = \left(\lim_{h \to 0} \sin a\right)\left(\lim_{h \to 0} \cos h\right) + \left(\lim_{h \to 0} \cos a\right)\left(\lim_{h \to 0} \sin h\right)$

$= (\sin a)(1) + (\cos a)(0) = \sin a$

43. If there is such a number, it satisfies the equation $x^3 + 1 = x \iff x^3 - x + 1 = 0$. Let the LHS of this equation be called $f(x)$. Now $f(-2) = -5 < 0$, and $f(-1) = 1 > 0$. Note also that $f(x)$ is a polynomial, and thus continuous. So by the Intermediate Value Theorem, there is a number c between -2 and -1 such that $f(c) = 0$, so that $c = c^3 + 1$.

Section 2.5 Limits Involving Infinity

1. (a) As x approaches 2 (from the right or the left), the values of $f(x)$ become large.

(b) As x approaches 1 from the right, the values of $f(x)$ become large negative.

(c) As x becomes large, the values of $f(x)$ approach 5.

(d) As x becomes large negative, the values of $f(x)$ approach 3.

3. (a) $\lim\limits_{x \to 2} f(x) = \infty$ (d) $\lim\limits_{x \to \infty} f(x) = 1$

(b) $\lim\limits_{x \to -1^{-}} f(x) = \infty$ (e) $\lim\limits_{x \to -\infty} f(x) = 2$

(c) $\lim\limits_{x \to -1^{+}} f(x) = -\infty$ (f) Vertical: $x = -1$, $x = 2$; Horizontal: $y = 1$, $y = 2$

5.

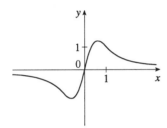

7.

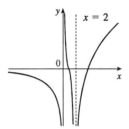

9. If $f(x) = x^2/2^x$, then a calculator gives $f(0) = 0$, $f(1) = 0.5$, $f(2) = 1$, $f(3) = 1.125$, $f(4) = 1$, $f(5) = 0.78125$, $f(6) = 0.5625$, $f(7) = 0.3828125$, $f(8) = 0.25$, $f(9) = 0.158203125$, $f(10) = 0.09765625$, $f(20) \approx 0.00038147$, $f(50) \approx 2.2204 \times 10^{-12}$, $f(100) \approx 7.8886 \times 10^{-27}$. It appears that $\lim\limits_{x \to \infty} \left(x^2/2^x \right) = 0$.

11.

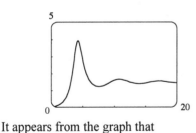

It appears from the graph that
$$\lim_{x \to \infty} \frac{3x^2}{2x^2 + 25 \sin x} \approx 1.5$$

13. Vertical: $x \approx -1.62$, $x \approx 0.62$, $x = 1$;
Horizontal: $y = 1$

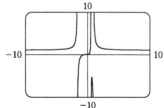

15. $\lim\limits_{x \to 3} \dfrac{1}{(x-3)^8} = \infty$ because $(x-3)^8 \to 0$ as $x \to 3$ and $\dfrac{1}{(x-3)^8} > 0$ whenever $x \neq 3$.

17. $\displaystyle\lim_{x \to -2^+} \frac{x-1}{x^2(x+2)} = -\infty$ since the numerator is negative and the denominator approaches 0 from the positive side as $x \to -2^+$.

19. $\displaystyle\lim_{x \to \infty} \frac{x+4}{x^2-2x+5} = \lim_{x \to \infty} \frac{\dfrac{1}{x}+\dfrac{4}{x^2}}{1-\dfrac{2}{x}+\dfrac{5}{x^2}} = \frac{\displaystyle\lim_{x \to \infty}\left(\dfrac{1}{x}+\dfrac{4}{x^2}\right)}{\displaystyle\lim_{x \to \infty}\left(1-\dfrac{2}{x}+\dfrac{5}{x^2}\right)}$

$$= \frac{\displaystyle\lim_{x \to \infty}\frac{1}{x}+4\lim_{x \to \infty}\frac{1}{x^2}}{\displaystyle\lim_{x \to \infty}1-2\lim_{x \to \infty}\frac{1}{x}+5\lim_{x \to \infty}\frac{1}{x^2}} = \frac{0+4\,(0)}{1-2\,(0)+5\,(0)} = 0$$

21. $\displaystyle\lim_{t \to -\infty} \frac{6t^2+5t}{(1-t)(2t-3)} = \lim_{t \to -\infty} \frac{6t^2+5t}{-2t^2+5t-3} = \lim_{t \to -\infty} \frac{6+5/t}{-2+5/t-3/t^2}$

$$= \frac{\displaystyle\lim_{t \to -\infty}6+5\lim_{t \to -\infty}(1/t)}{\displaystyle\lim_{t \to -\infty}(-2)+5\lim_{t \to -\infty}(1/t)-3\lim_{t \to -\infty}(1/t^2)}$$

$$= \frac{6+5\,(0)}{-2+5\,(0)-3\,(0)} = -3$$

23. $\displaystyle\lim_{x \to \infty}\left(\sqrt{x^2+3x+1}-x\right) = \lim_{x \to \infty}\left(\sqrt{x^2+3x+1}-x\right)\frac{\sqrt{x^2+3x+1}+x}{\sqrt{x^2+3x+1}+x}$

$$= \lim_{x \to \infty} \frac{x^2+3x+1-x^2}{\sqrt{x^2+3x+1}+x} = \lim_{x \to \infty} \frac{3x+1}{\sqrt{x^2+3x+1}+x}$$

$$= \lim_{x \to \infty} \frac{3+1/x}{\sqrt{1+(3/x)+(1/x^2)}+1} = \frac{3+0}{\sqrt{1+3\cdot 0+0}+1} = \frac{3}{2}$$

25. $\displaystyle\lim_{x \to \infty} \cos x$ does not exist because, as x increases, $\cos x$ does not approach any one value, but oscillates between 1 and -1..

27. $\displaystyle\lim_{x \to \infty} \frac{x^7-1}{x^6-1} = \lim_{x \to \infty} \frac{1-1/x^7}{(1/x)-(1/x^7)} = \infty$ since $1 - \dfrac{1}{x^7} \to 1$ while $\dfrac{1}{x}-\dfrac{1}{x^7} \to 0^+$ as $x \to \infty$.

Or: Divide numerator and denominator by x^6 instead of x^7.

29. If $z = 1/x$, then $\displaystyle\lim_{x \to \infty} z = 0$. Thus, $\displaystyle\lim_{x \to \infty} \sin(1/x) = \lim_{z \to 0} \sin z = 0$.

31. Since $x^2 - 1 \to 0$ and $y < 0$ for $-1 < x < 1$ and $y > 0$ for $x < -1$ and $x > 1$, we have

$$\lim_{x \to 1^-} \frac{x^2 + 4}{x^2 - 1} = -\infty, \ \lim_{x \to 1^+} \frac{x^2 + 4}{x^2 - 1} = \infty, \ \lim_{x \to -1^-} \frac{x^2 + 4}{x^2 - 1} = \infty, \text{ and } \lim_{x \to -1^+} \frac{x^2 + 4}{x^2 - 1} = -\infty, \text{ so } x = 1$$

and $x = -1$ are vertical asymptotes. Also $\displaystyle\lim_{x \to \pm\infty} \frac{x^2 + 4}{x^2 - 1} = \lim_{x \to \pm\infty} \frac{1 + 4/x^2}{1 - 1/x^2} = \frac{1 + 0}{1 - 0} = 1$, so $y = 1$ is

a horizontal asymptote. The graph confirms these calculations.

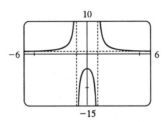

33. (a) This must be graph IV, the only graph that is always negative to the left of $x = 1$ and positive to the right of $x = 1$.

(b) This is graph III, the only graph with a horizontal asymptote of $y = 1$.

(c) This must be graph II, since it's the only graph that is positive everywhere. $[(x - 1)^2 > 0 \text{ for } x \neq 1.]$

(d) $x^2 - 1 < 0$ if $|x| < 1$. Since the function is the reciprocal of $x^2 - 1$, the graph must be negative for $|x| < 1$ and positive elsewhere. The only graph fitting this description is VI.

(e) $(x - 1)^2 > 0$ if $x \neq 1$, so the sign of y is determined by the sign of the numerator, x. Thus, $y < 0$ if $x < 0$ and $y > 0$ if $x > 0$, as is only the case with graph I.

(f) The graph must have vertical asymptotes at $x = \pm 1$ and an x-intercept at $x = 0$. The only graph fitting this description is V.

35. Let's look for a rational function.

(1) $\displaystyle\lim_{x \to \pm\infty} f(x) = 0 \Longrightarrow$ degree of numerator < degree of denominator

(2) $\displaystyle\lim_{x \to 0} f(x) = -\infty \Longrightarrow$ there is a factor of x^2 in the denominator (not just x, since that would produce a sign change at $x = 0$), and the function is negative near $x = 0$.

(3) $\displaystyle\lim_{x \to 3^-} f(x) = \infty$ and $\displaystyle\lim_{x \to 3^+} f(x) = -\infty \Longrightarrow$ vertical asymptote at $x = 3$; there is a factor of $(x - 3)$ in the denominator.

(4) $f(2) = 0 \Longrightarrow 2$ is an x-intercept; there is at least one factor of $(x - 2)$ in the numerator.

Combining all of this information, and putting in a negative sign to give us the desired left- and right-hand limits, gives us $f(x) = \dfrac{2 - x}{x^2(x - 3)}$ as one possibility.

37. Divide numerator and denominator by the highest power of x in $Q(x)$.

(a) If $\deg P < \deg Q$, then numerator $\to 0$ but denominator doesn't. So $\displaystyle\lim_{x\to\infty} \frac{P(x)}{Q(x)} = 0$.

(b) If $\deg P > \deg Q$, then numerator $\to \pm\infty$ but denominator doesn't, so $\displaystyle\lim_{x\to\infty} \frac{P(x)}{Q(x)} = \pm\infty$

(depending on the ratio of the leading coefficients of P and Q).

39. $\displaystyle\lim_{x\to\infty} \frac{4x-1}{x} = \lim_{x\to\infty}\left(4 - \frac{1}{x}\right) = 4$, and $\displaystyle\lim_{x\to\infty} \frac{4x^2+3x}{x^2} = \lim_{x\to\infty}\left(4 + \frac{3}{x}\right) = 4$. Therefore, by the

Squeeze Theorem, $\displaystyle\lim_{x\to\infty} f(x) = 4$.

41. (a) After t minutes, $25t$ liters of brine with 30 g of salt per liter has been pumped into the tank, so
it contains $(5000 + 25t)$ liters of water and $25t \cdot 30 = 750t$ grams of salt. Therefore, the salt
concentration at time t will be $C(t) = \dfrac{750t}{5000 + 25t} = \dfrac{30t}{200 + t} \dfrac{\text{g}}{\text{L}}$.

(b) $\displaystyle\lim_{t\to\infty} C(t) = \lim_{t\to\infty} \frac{30t}{200+t} = \lim_{t\to\infty} \frac{30t/t}{200/t + t/t} = \frac{30}{0+1} = 30$. So the salt concentration

approaches that of the brine being pumped into the tank.

43. (a) If $t = -x/10$, then $x = -10t$ and as $x \to \infty$, $t \to -\infty$. Thus, $\displaystyle\lim_{x\to\infty} e^{-x/10} = \lim_{t\to-\infty} e^t = 0$ by
Equation 7.

(b) $y = e^{-x/10}$ and $y = 0.1$ intersect at $x_1 \approx 23.03$. If $x > x_1$, then $e^{-x/10} < 0.1$.

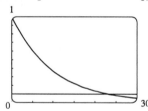

(c) $e^{-x/10} < 0.1 \Longrightarrow -x/10 < \ln 0.1 \Longrightarrow x > -10\ln 0.1 = x_1$

Section 2.6 Tangents, Velocities, and Other Rates of Change

1. (a) This is just the slope of the line through two points:
$$m_{PQ} = \frac{\Delta y}{\Delta x} = \frac{f(x) - f(3)}{x - 3}$$

(b) This is the limit of the slope of the secant line PQ as Q approaches P:
$$m = \lim_{x \to 3} \frac{f(x) - f(3)}{x - 3}$$

3. The slope at D is the largest positive slope, followed by the positive slope at E. The slope at C is zero. The slope at B is steeper than at A (both are negative). In decreasing order, we have the slopes at: D, E, C, A, B.

5. (a) (i) $m = \lim\limits_{x \to -3} \dfrac{f(x) - f(-3)}{x - (-3)} = \lim\limits_{x \to -3} \dfrac{(x^2 + 2x) - (3)}{x - (-3)} = \lim\limits_{x \to -3} \dfrac{(x + 3)(x - 1)}{x + 3} =$
$\lim\limits_{x \to -3} (x - 1) = -4$

(ii) $m = \lim\limits_{h \to 0} \dfrac{f(-3 + h) - f(-3)}{h} = \lim\limits_{h \to 0} \dfrac{\left[(-3 + h)^2 + 2(-3 + h)\right] - (3)}{h}$
$= \lim\limits_{h \to 0} \dfrac{9 - 6h + h^2 - 6 + 2h - 3}{h} = \lim\limits_{h \to 0} \dfrac{h(h - 4)}{h} = \lim\limits_{h \to 0} (h - 4) = -4$

(b) The equation of the tangent line is $y - 3 = -4(x + 3)$ or $y = -4x - 9$.

(c)

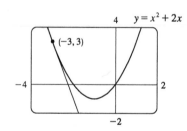

7. Using (1), $m = \lim\limits_{x \to 1} \dfrac{\sqrt{x} - \sqrt{1}}{x - 1} = \lim\limits_{x \to 1} \dfrac{(\sqrt{x} - 1)(\sqrt{x} + 1)}{(x - 1)(\sqrt{x} + 1)} = \lim\limits_{x \to 1} \dfrac{1}{\sqrt{x} + 1} = \dfrac{1}{2}$.

Tangent line: $y - 1 = \frac{1}{2}(x - 1) \Longrightarrow y = \frac{1}{2}x + \frac{1}{2}$

9. Using (1), $m = \lim\limits_{x \to -2} \dfrac{1/x^2 - \frac{1}{4}}{x - (-2)} = \lim\limits_{x \to -2} \dfrac{4 - x^2}{4x^2(x + 2)} = \lim\limits_{x \to -2} \dfrac{(2 - x)(2 + x)}{4x^2(x + 2)} = \lim\limits_{x \to -2} \dfrac{2 - x}{4x^2} = \dfrac{1}{4}$.

Tangent line: $y - \frac{1}{4} = \frac{1}{4}(x + 2) \Longrightarrow y = \frac{1}{4}x + \frac{3}{4}$.

11. (a) Using (1),

$$m = \lim_{x \to a} \frac{\left(x^3 - 4x + 1\right) - \left(a^3 - 4a + 1\right)}{x - a} = \lim_{x \to a} \frac{\left(x^3 - a^3\right) - 4\left(x - a\right)}{x - a}$$

$$= \lim_{x \to a} \frac{(x - a)\left(x^2 + ax + a^2\right) - 4\left(x - a\right)}{x - a} = \lim_{x \to a} \left(x^2 + ax + a^2 - 4\right) = 3a^2 - 4$$

(b) At $(1, -2)$: $m = 3\,(1)^2 - 4 = -1$, so an equation of the
tangent line is $y - (-2) = -1\,(x - 1) \iff y = -x - 1$. At
$(2, 1)$: $m = 3\,(2)^2 - 4 = 8$, so an equation of the tangent line
is $y - 1 = 8\,(x - 2) \iff y = 8x - 15$.

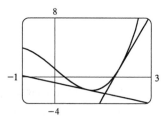

13. (a) Since the slope of the tangent at $s = 0$ is 0, the car's initial velocity was 0.

(b) The slope of the tangent is greater at C than at B, so the car was going faster at C.

(c) Near A, the tangent lines are becoming steeper as x increases, so the velocity was increasing, so the car was speeding up. Near B, the tangent lines are becoming less steep, so the car was slowing down. The steepest tangent near C is the one at C, so at C the car had just finished speeding up, and was about to start slowing down.

(d) Between D and E, the slope of the tangent is 0, so the car did not move during that time.

15. Let $s\,(t) = 40t - 16t^2$.

$$v\,(2) = \lim_{t \to 2} \frac{s\,(t) - s\,(2)}{t - 2} = \lim_{t \to 2} \frac{\left(40t - 16t^2\right) - 16}{t - 2} = \lim_{t \to 2} \frac{-8\,(t - 2)\,(2t - 1)}{t - 2}$$

$$= -8 \lim_{t \to 2} (2t - 1) = -24$$

Thus, the instantaneous velocity when $t = 2$ is -24 ft/s.

17. $v\,(a) = \lim_{h \to 0} \dfrac{s\,(a + h) - s\,(a)}{h} = \lim_{h \to 0} \dfrac{4\,(a + h)^3 + 6\,(a + h) + 2 - \left(4a^3 + 6a + 2\right)}{h}$

$$= \lim_{h \to 0} \frac{4a^3 + 12a^2 h + 12ah^2 + 4h^3 + 6a + 6h + 2 - 4a^3 - 6a - 2}{h}$$

$$= \lim_{h \to 0} \frac{12a^2 h + 12ah^2 + 4h^3 + 6h}{h} = \lim_{h \to 0} \left(12a^2 + 12ah + 4h^2 + 6\right)$$

$$= \left(12a^2 + 6\right) \text{ m/s}$$

So $v\,(1) = 12\,(1)^2 + 6 = 18$ m/s, $v\,(2) = 12\,(2)^2 + 6 = 54$ m/s, and $v\,(3) = 12\,(3)^2 + 6 = 114$ m/s.

19. The sketch shows the graph for a room temperature of $72°$ and a refrigerator temperature of $38°$. The initial rate of change is greater in magnitude than the rate of change after an hour.

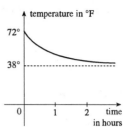

21. (a) (i) $[8, 11]$: $\frac{7.9 - 11.5}{3} = -1.2°/\text{h}$

(ii) $[8, 10]$: $\frac{9.0 - 11.5}{2} = -1.25°/\text{h}$

(iii) $[8, 9]$: $\frac{10.2 - 11.5}{1} = -1.3°/\text{h}$

(b) In the figure, we estimate A to be $(18, 15.5)$ and B as $(23, 6)$. So the slope is $\frac{6 - 15.5}{23 - 18} = -1.9°/\text{h}$ at 8:00 P.M.

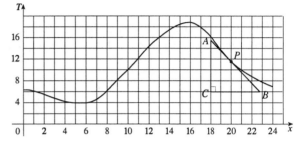

23. (a) (i) $\dfrac{\Delta C}{\Delta x} = \dfrac{C(105) - C(100)}{105 - 100} = \dfrac{6601.25 - 6500}{5} = 20.25/\text{unit.}$

(ii) $\dfrac{\Delta C}{\Delta x} = \dfrac{C(101) - C(100)}{101 - 100} = \dfrac{6520.05 - 6500}{1} = 20.05/\text{unit.}$

(b) $\dfrac{C(100 + h) - C(100)}{h} = \dfrac{\left[5000 + 10(100 + h) + 0.05(100 + h)^2\right] - 6500}{h} = \dfrac{20h + 0.05h^2}{h}$

$$= 20 + 0.05h, \ h \neq 0.$$

So the instantaneous rate of change is $\displaystyle\lim_{h \to 0} \dfrac{C(100 + h) - C(100)}{h} = \lim_{h \to 0} (20 + 0.05h) = \$20/\text{unit.}$

Section 2.7 Derivatives

1.

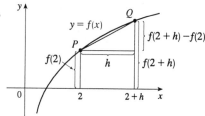

The line from P to Q is the line that has

slope $\dfrac{f(2+h) - f(2)}{h}$.

3. $g'(0)$ is the only negative value. The slope at $x = 4$ is smaller than the slope at $x = 2$ and both are smaller than the slope at $x = -2$. Thus, $g'(0) < 0 < g'(4) < g'(2) < g'(-2)$.

5.

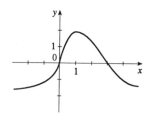

7. $f'(2) = \lim\limits_{h \to 0} \dfrac{f(2+h) - f(2)}{h} = \lim\limits_{h \to 0} \dfrac{\left[3(2+h)^2 - 5(2+h)\right] - \left[3(2)^2 - 5(2)\right]}{h}$

$= \lim\limits_{h \to 0} \dfrac{(12 + 12h + 3h^2 - 10 - 5h) - (2)}{h} = \lim\limits_{h \to 0} \dfrac{3h^2 + 7h}{h} = \lim\limits_{h \to 0} (3h + 7) = 7$

So an equation of the tangent line at $(2, 2)$ is $y - 2 = 7(x - 2)$ or $y = 7x - 12$.

9. (a) $F'(1) = \lim\limits_{x \to 1} \dfrac{F(x) - F(1)}{x - 1} = \lim\limits_{x \to 1} \dfrac{(x^3 - 5x + 1) - (-3)}{x - 1} = \lim\limits_{x \to 1} \dfrac{x^3 - 5x + 4}{x - 1}$

$= \lim\limits_{x \to 1} \dfrac{(x - 1)(x^2 + x - 4)}{x - 1} = \lim\limits_{x \to 1} (x^2 + x - 4) = -2$

So an equation of the tangent line at $(1, -3)$ is $y - (-3) = -2(x - 1) \iff y = -2x - 1$.

Note: Instead of using Equation 3 to compute $F'(1)$, we could have used Equation 1.

(b)

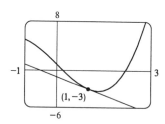

11. (a) $f'(1) = \lim\limits_{h \to 0} \dfrac{f(1+h) - f(1)}{h} = \lim\limits_{h \to 0} \dfrac{3^{1+h} - 3^1}{h}$.

(b)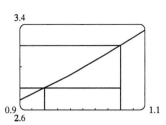

So let $F(h) = \dfrac{3^{1+h} - 3}{h}$. We calculate:

h	$F(h)$
0.1	3.484
0.01	3.314
0.001	3.298
0.0001	3.296
−0.1	3.121
−0.01	3.278
−0.001	3.294
−0.0001	3.296

From the graph, we estimate that the slope of the tangent is about

$$\frac{3.2 - 2.8}{1.06 - 0.94} = \frac{0.4}{0.12} \approx 3.3.$$

We estimate that $f'(1) \approx 3.296$.

13. $f'(a) = \lim\limits_{h \to 0} \dfrac{f(a+h) - f(a)}{h} = \lim\limits_{h \to 0} \dfrac{\left[1 + (a+h) - 2(a+h)^2\right] - \left(1 + a - 2a^2\right)}{h}$

$= \lim\limits_{h \to 0} \dfrac{h - 4ah - 2h^2}{h} = \lim\limits_{h \to 0} (1 - 4a - 2h) = 1 - 4a$

15. $f'(a) = \lim\limits_{h \to 0} \dfrac{f(a+h) - f(a)}{h} = \lim\limits_{h \to 0} \dfrac{\dfrac{a+h}{(a+h)^2 - 1} - \dfrac{a}{a^2 - 1}}{h}$

$= \lim\limits_{h \to 0} \dfrac{(a+h)(a^2 - 1) - a(a^2 + 2ah + h^2 - 1)}{h(a^2 - 1)(a^2 + 2ah + h^2 - 1)} = \lim\limits_{h \to 0} \dfrac{h(-a^2 - 1 - ah)}{h(a^2 - 1)(a^2 + 2ah + h^2 - 1)}$

$= \lim\limits_{h \to 0} \dfrac{-a^2 - 1 - ah}{(a^2 - 1)(a^2 + 2ah + h^2 - 1)} = \dfrac{-a^2 - 1}{(a^2 - 1)(a^2 - 1)} = -\dfrac{a^2 + 1}{(a^2 - 1)^2}$

17. By Equation 1, $\lim\limits_{h \to 0} \dfrac{\sqrt{1+h} - 1}{h} = f'(1)$, where $f(x) = \sqrt{x}$.

[Or $f'(0)$ where $f(x) = \sqrt{1+x}$; the answers to Exercises 17-22 are not unique.]

19. By Equation 3, $\lim\limits_{x \to 1} \dfrac{x^9 - 1}{x - 1} = f'(1)$, where $f(x) = x^9$.

21. By Equation 1, $\lim\limits_{t \to 0} \dfrac{\sin\left(\frac{\pi}{2} + t\right) - 1}{t} = f'\left(\dfrac{\pi}{2}\right)$, where $f(x) = \sin x$.

23. $v(2) = f'(2) = \lim\limits_{h \to 0} \dfrac{f(2+h) - f(2)}{h} = \lim\limits_{h \to 0} \dfrac{\left[(2+h)^2 - 6(2+h) - 5\right] - \left[2^2 - 6(2) - 5\right]}{h}$

$= \lim\limits_{h \to 0} \dfrac{(4 + 4h + h^2 - 12 - 6h - 5) - (-13)}{h} = \lim\limits_{h \to 0} \dfrac{h^2 - 2h}{h} = \lim\limits_{h \to 0} (h - 2) = -2\ \text{m/s}$

25. (a) $f'(x)$ is the rate of change of the production cost with respect to the number of ounces of gold produced. Its units are dollars per ounce.

   (b) After 800 ounces of gold have been produced, the rate at which the production cost is increasing is $17/ounce. So the cost of producing the 800th (or 801st) ounce is about $17.

   (c) In the short term, the values of $f'(x)$ will decrease because more efficient use is made of start-up costs as x increases. But eventually $f'(x)$ might increase due to large-scale operations.

27. (a) $f'(v)$ is the rate at which the fuel consumption is changing with respect to the speed. Its units are $(\text{gal/h}) / (\text{mi/h})$ or gal/mi.

   (b) The fuel consumption is decreasing by 0.05 gal/mi as the car's speed reaches 20 mi/h. So if you increase your speed to 21 mi/h, you could expect to decrease your fuel consumption by about 0.05 gal/mi.

29. For 1983: We will average the difference quotients (slopes of secant lines) obtained using the years 1982 and 1984.

Let $A = \dfrac{C(1982) - C(1983)}{1982 - 1983} = \dfrac{2.44 - 3.05}{-1} = 0.61$ and

$B = \dfrac{C(1984) - C(1983)}{1984 - 1983} = \dfrac{3.52 - 3.05}{1} = 0.47.$ Then

$C'(1983) = \lim\limits_{t \to 1983} \dfrac{C(t) - C(1983)}{t - 1983} \approx \dfrac{A + B}{2} = 0.54.$ This means that the price of coffee beans was rising at about $(54 \text{ cents/kg}) / \text{year}$ in 1983.

For 1990: Using data for 1989 and 1991 in a similar fashion, we obtain

$C'(1990) \approx [-0.04 + (-0.09)] / 2 = -0.065.$ So the price was falling at about $(6.5 \text{ cents/kg}) / \text{year}$ in 1990.

31. Since $f(x) = x \sin(1/x)$ when $x \neq 0$ and $f(0) = 0$, we have

$f'(0) = \lim\limits_{h \to 0} \dfrac{f(0 + h) - f(0)}{h} = \lim\limits_{h \to 0} \dfrac{h \sin(1/h) - 0}{h} = \lim\limits_{h \to 0} \sin(1/h).$ This limit does not exist since $\sin(1/h)$ takes the values -1 and 1 on any interval containing 0. (Compare with Example 4 in Section 2.2.)

Section 2.8 The Derivative as a Function

1. *Note:* Your answers may vary depending on your estimates.

From the graph of f, it appears that

(a) $f'(1) \approx -2$

(b) $f'(2) \approx 0.8$

(c) $f'(3) \approx -1$

(d) $f'(4) \approx -0.5$

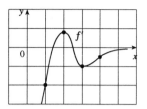

3. (a)$' =$ II, since from left to right, the slopes of the tangents to graph (a) start out negative, become 0, then positive, then 0, then negative again. The actual function values in graph II follow the same pattern.

(b)$' =$ IV, since from left to right, the slopes of the tangents to graph (b) start out at a fixed positive quantity, then suddenly become negative, then positive again. The discontinuities in graph IV indicate sudden changes in the slopes of the tangents.

(c)$' =$ I, since the slopes of the tangents to graph (c) are negative for $x < 0$ and positive for $x > 0$, as are the function values of graph I.

(d)$' =$ III, since from left to right, the slopes of the tangents to graph (d) are positive, then 0, then negative, then 0, then positive, then 0, then negative again, and the function values in graph III follow the same pattern.

5.

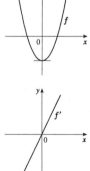

7.

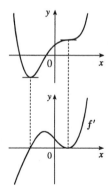

9.

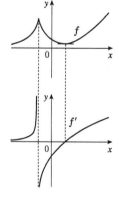

11.

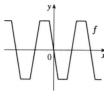

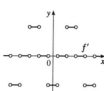

13.

15.

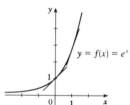

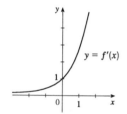

The slope at 0 appears to be 1 and the slope at 1 appears to be 2.7. As x decreases, the slope gets closer to 0. Since the graphs are so similar, we might guess that $f'(x) = e^x$.

17. (a) By zooming in, we estimate that $f'(0) = 0$, $f'\left(\frac{1}{2}\right) = 1$, $f'(1) = 2$, and $f'(2) = 4$.

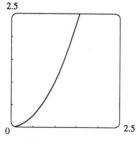

(b) By symmetry, $f'(-x) = -f'(x)$. So $f'\left(-\frac{1}{2}\right) = -1$, $f'(-1) = -2$, and $f'(-2) = -4$.

(c) It appears that $f'(x)$ is twice the value of x, so we guess that $f'(x) = 2x$.

(d) $f'(x) = \lim\limits_{h \to 0} \dfrac{f(x+h) - f(x)}{h} = \lim\limits_{h \to 0} \dfrac{(x+h)^2 - x^2}{h} = \lim\limits_{h \to 0} \dfrac{(x^2 + 2hx + h^2) - x^2}{h}$

$= \lim\limits_{h \to 0} \dfrac{h(2x+h)}{h} = \lim\limits_{h \to 0} (2x + h) = 2x$

19. $f'(x) = \lim\limits_{h \to 0} \dfrac{f(x+h) - f(x)}{h} = \lim\limits_{h \to 0} \dfrac{[5(x+h)+3] - (5x+3)}{h} = \lim\limits_{h \to 0} \dfrac{5h}{h} = \lim\limits_{h \to 0} 5 = 5.$

Domain of f = domain of $f' = \mathbb{R}$.

21. $g'(x) = \lim\limits_{h \to 0} \dfrac{g(x+h) - g(x)}{h} = \lim\limits_{h \to 0} \dfrac{\sqrt{1 + 2(x+h)} - \sqrt{1+2x}}{h} \left[\dfrac{\sqrt{1+2(x+h)} + \sqrt{1+2x}}{\sqrt{1+2(x+h)} + \sqrt{1+2x}} \right]$

$= \lim\limits_{h \to 0} \dfrac{(1 + 2x + 2h) - (1 + 2x)}{h \left[\sqrt{1 + 2(x+h)} + \sqrt{1+2x} \right]} = \lim\limits_{h \to 0} \dfrac{2}{\sqrt{1+2(x+h)} + \sqrt{1+2x}} = \dfrac{1}{\sqrt{1+2x}}$

Domain of $g = \left[-\frac{1}{2}, \infty \right)$, domain of $g' = \left(-\frac{1}{2}, \infty \right)$.

23. $f'(x) = \lim\limits_{h \to 0} \dfrac{f(x+h) - f(x)}{h} = \lim\limits_{h \to 0} \dfrac{\dfrac{x+h+1}{x+h-1} - \dfrac{x+1}{x-1}}{h}$

$= \lim\limits_{h \to 0} \dfrac{(x+h+1)(x-1) - (x+1)(x+h-1)}{h(x+h-1)(x-1)} = \lim\limits_{h \to 0} \dfrac{-2h}{h(x+h-1)(x-1)}$

$= \lim\limits_{h \to 0} \dfrac{-2}{(x+h-1)(x-1)} = \dfrac{-2}{(x-1)^2}$

Domain of f = domain of $f' = \{ x \mid x \neq 1 \}$.

25. (a) $f'(x) = \lim\limits_{h \to 0} \dfrac{f(x+h) - f(x)}{h} = \lim\limits_{h \to 0} \dfrac{\left[x + h - \left(\dfrac{2}{x+h} \right) \right] - \left[x - \left(\dfrac{2}{x} \right) \right]}{h}$

$= \lim\limits_{h \to 0} \dfrac{\left[h + \dfrac{2}{x} - \dfrac{2}{(x+h)} \right]}{h} = \lim\limits_{h \to 0} \left[1 + \dfrac{2(x+h) - 2x}{(h)(x)(x+h)} \right] = \lim\limits_{h \to 0} \left[1 + \dfrac{2}{x(x+h)} \right]$

$= 1 + 2x^{-2}$

(b) Notice that when f has steep tangent lines, $f'(x)$ is very large. When f is flatter, $f'(x)$ is smaller.

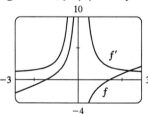

27. (a) $U'(t)$ is the rate at which the unemployment rate is changing with respect to time. Its units are percent per year.

(b) **For 1983:** $U'(t) = \dfrac{U(1984) - U(1983)}{1984 - 1983} = \dfrac{7.4 - 9.5}{1} = -2.1$

For 1984: $U'(t) = \dfrac{U(1985) - U(1983)}{1985 - 1983} = \dfrac{7.1 - 9.5}{2} = -1.2$

t	1983	1984	1985	1986	1987	1988	1989	1990	1991	1992
$U'(t)$	-2.1	-1.2	-0.25	-0.5	-0.75	-0.45	0	0.7	0.95	0.7

29. f is not differentiable at $x = -1$ or at $x = 11$ because the graph has vertical tangents at those points; at $x = 4$, because there is a discontinuity there; and at $x = 8$, because the graph has a corner there.

31. As we zoom in toward $(-1, 0)$, the curve appears more and more like a straight line, so f is differentiable at $x = -1$. But no matter how much we zoom in toward the origin, the curve doesn't straighten out — we can't eliminate the sharp point (a cusp). So f is not differentiable at $x = 0$.

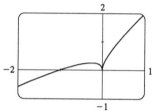

33. $a = f$, $b = f'$, $c = f''$. We can see this because where a has a horizontal tangent, $b = 0$, and where b has a horizontal tangent, $c = 0$. We can immediately see that c can be neither f nor f', since at the points where c has a horizontal tangent, neither a nor b is equal to 0.

35. We can immediately see that a is the graph of the acceleration function, since at the points where a has a horizontal tangent, neither c nor b is equal to 0. Next, we note that $a = 0$ at the point where b has a horizontal tangent, so b must be the graph of the velocity function, so that $b' = a$. We conclude that c is the graph of the position function.

37. $f'(x) = \lim\limits_{h \to 0} \dfrac{f(x+h) - f(x)}{h} = \lim\limits_{h \to 0} \dfrac{\left[1 + 4(x+h) - (x+h)^2\right] - (1 + 4x - x^2)}{h}$

$= \lim\limits_{h \to 0} \dfrac{4h - 2xh - h^2}{h} = \lim\limits_{h \to 0} (4 - 2x - h) = 4 - 2x$

$f''(x) = \lim\limits_{h \to 0} \dfrac{f'(x+h) - f'(x)}{h} = \lim\limits_{h \to 0} \dfrac{[4 - 2(x+h)] - (4 - 2x)}{h} = \lim\limits_{h \to 0} \dfrac{-2h}{h} = \lim\limits_{h \to 0} (-2) = -2$

We see from the graph that our answers are reasonable because the graph of f' is that of a linear function and the graph of f'' is that of a constant function.

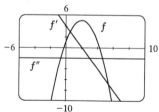

39. $f'(x) = \lim\limits_{h \to 0} \dfrac{f(x+h) - f(x)}{h} = \lim\limits_{h \to 0} \dfrac{\left[2(x+h)^2 - (x+h)^3\right] - (2x^2 - x^3)}{h}$

$= \lim\limits_{h \to 0} \dfrac{h\left(4x + 2h - 3x^2 - 3xh - h^2\right)}{h} = \lim\limits_{h \to 0} \left(4x + 2h - 3x^2 - 3xh - h^2\right) = 4x - 3x^2$

$f''(x) = \lim\limits_{h \to 0} \dfrac{f'(x+h) - f'(x)}{h} = \lim\limits_{h \to 0} \dfrac{\left[4(x+h) - 3(x+h)^2\right] - (4x - 3x^2)}{h}$

$= \lim\limits_{h \to 0} \dfrac{h(4 - 6x - 3h)}{h} = \lim\limits_{h \to 0} (4 - 6x - 3h) = 4 - 6x$

$f'''(x) = \lim\limits_{h \to 0} \dfrac{f''(x+h) - f''(x)}{h} = \lim\limits_{h \to 0} \dfrac{[4 - 6(x+h)] - (4 - 6x)}{h} = \lim\limits_{h \to 0} \dfrac{-6h}{h} = \lim\limits_{h \to 0} (-6)$

$= -6$

$f^{(4)}(x) = \lim\limits_{h \to 0} \dfrac{f'''(x+h) - f'''(x)}{h} = \lim\limits_{h \to 0} \dfrac{-6 - (-6)}{h} = \lim\limits_{h \to 0} \dfrac{0}{h} = \lim\limits_{h \to 0} (0) = 0$

The graphs are consistent with the geometric interpretations of the derivatives because f' has zeros where f has a local minimum and a local maximum, f'' has a zero where f' has a local maximum, and f''' is a constant function equal to the slope of f''.

41. (a) $f'(a) = \lim\limits_{x \to a} \dfrac{f(x) - f(a)}{x - a} = \lim\limits_{x \to a} \dfrac{x^{1/3} - a^{1/3}}{x - a} = \lim\limits_{x \to a} \dfrac{x^{1/3} - a^{1/3}}{\left(x^{1/3} - a^{1/3}\right)\left(x^{2/3} + x^{1/3}a^{1/3} + a^{2/3}\right)}$

$= \lim\limits_{x \to a} \dfrac{1}{x^{2/3} + x^{1/3}a^{1/3} + a^{2/3}} = \lim\limits_{x \to a} \dfrac{1}{3x^{2/3}} = \dfrac{1}{3a^{2/3}}$

(b) $f'(0) = \lim\limits_{h \to 0} \dfrac{f(0+h) - f(0)}{h} = \lim\limits_{h \to 0} \dfrac{\sqrt[3]{h} - 0}{h} = \lim\limits_{h \to 0} \dfrac{1}{h^{2/3}}$. This limit does not exist, and

therefore $f'(0)$ does not exist.

(c) $\lim\limits_{x \to 0} |f'(x)| = \lim\limits_{x \to 0} \dfrac{1}{3x^{2/3}} = \infty$ and f is continuous at $x = 0$ (root function), so f has a vertical tangent at $x = 0$.

43. $f(x) = |x - 6| = \begin{cases} 6 - x & \text{if } x < 6 \\ x - 6 & \text{if } x \geq 6 \end{cases}$

$\displaystyle\lim_{x \to 6^+} \frac{f(x) - f(6)}{x - 6} = \lim_{x \to 6^+} \frac{|x - 6| - 0}{x - 6} = \lim_{x \to 6^+} \frac{x - 6}{x - 6} = \lim_{x \to 6^+} 1 = 1.$

But $\displaystyle\lim_{x \to 6^-} \frac{f(x) - f(6)}{x - 6} = \lim_{x \to 6^-} \frac{|x - 6| - 0}{x - 6} = \lim_{x \to 6^-} \frac{6 - x}{x - 6} = \lim_{x \to 6^-} (-1) = -1.$

So $f'(6) = \displaystyle\lim_{x \to 6} \frac{f(x) - f(6)}{x - 6}$ does not exist. However, $f'(x) = \begin{cases} -1 & \text{if } x < 6 \\ 1 & \text{if } x > 6 \end{cases}$

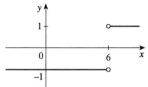

45. (a) If f is even, then

$f'(-x) = \displaystyle\lim_{h \to 0} \frac{f(-x + h) - f(-x)}{h} = \lim_{h \to 0} \frac{f(x - h) - f(x)}{h}$

$= -\displaystyle\lim_{h \to 0} \frac{f(x - h) - f(x)}{-h}$ [let $\Delta x = -h$] $= -\displaystyle\lim_{\Delta x \to 0} \frac{f(x + \Delta x) - f(x)}{\Delta x} = -f'(x)$

Therefore, f' is odd.

(b) If f is odd, then

$f'(-x) = \displaystyle\lim_{h \to 0} \frac{f(-x + h) - f(-x)}{h} = \lim_{h \to 0} \frac{-f(x - h) + f(x)}{h}$

$= \displaystyle\lim_{h \to 0} \frac{f(x - h) - f(x)}{-h}$ [let $\Delta x = -h$] $= \displaystyle\lim_{\Delta x \to 0} \frac{f(x + \Delta x) - f(x)}{\Delta x} = f'(x)$

Therefore, f' is even.

Section 2.9 Linear Approximations

1. (a) $f'(0) = \lim\limits_{h \to 0} \dfrac{f(h) - f(0)}{h}$

$= \lim\limits_{h \to 0} \dfrac{3^h - 1}{h}$. Let $y = \dfrac{3^h - 1}{h}$.

x	y
0.01	1.1047
0.001	1.0992
0.0001	1.0987
-0.01	1.0926
-0.001	1.0980
-0.0001	1.0986

From the table, we conclude that
$f'(0) \approx 1.0986$.

(b) An approximate equation of the tangent line is

$y - 1 = 1.0986\,(x - 0)$, or $y = 1.0986x + 1$.

$L(0.05) = 1.0986\,(0.05) + 1 = 1.05493$ and

$L(0.1) = 1.0986\,(0.1) + 1 = 1.10986$. So we

estimate that $3^{0.05} \approx 1.0549$ and $3^{0.1} \approx 1.1099$.

(c)

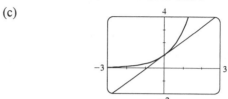

Since the tangent line lies *below* the curve, the
approximations are *less* than the true values.

3. (a)

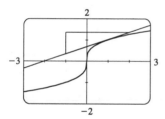

From the triangle in the graph,

$\Delta y / \Delta x = \dfrac{\frac{4}{3} - \frac{1}{3}}{2 - (-1)} = \frac{1}{3}$.

(b) $L(x) = f(1) + f'(1)\,(x - 1)$

$= 1 + \frac{1}{3}\,(x - 1) = \frac{1}{3}x + \frac{2}{3}$.

(c)

x	$L(x)$	Calculator
0.5	$0.83\overline{3}$	0.794
0.9	$0.96\overline{6}$	0.965
0.99	$0.99\overline{6}$	0.997
1.01	$1.00\overline{3}$	1.003
1.1	$1.03\overline{3}$	1.032
1.5	$1.16\overline{6}$	1.145
2	$1.33\overline{3}$	1.260

We see from the table that all of the estimates are
overestimates (ignore the rounding for $x = 0.99$).
The most accurate estimates are for $x = 0.99$ and

$x = 1.01$, which are the closest values to $x = 1$.

(d) The graph in part (a) shows that the tangent line lies above the curve. That explains why the
estimates are overestimates.

5. (a) From Exercise 2.8.17(d), $f'(x) = 2x$, so

$$f'(1) = 2.$$

(c)

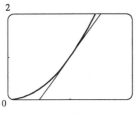

Since the tangent line lies under the graph, our underestimate claim in part (b) is supported.

(b) $L(x) = f(1) + f'(1)(x - 1)$

$$= 1 + 2(x - 1) = 2x - 1$$

x	$L(x) = 2x - 1$	$f(x) = x^2$
0.9	0.8	0.81
0.95	0.9	0.9025
0.99	0.98	0.9801
1.01	1.02	1.0201
1.05	1.1	1.1025
1.1	1.2	1.21

The estimates using L are all underestimates of the actual function values.

7. As in Example 3, $T(0) = 185$, $T(10) = 172$, $T(20) = 160$, and

$$T'(20) \approx \frac{T(10) - T(20)}{10 - 20} = \frac{172 - 160}{-10} = -1.2° \text{ F/min.}$$

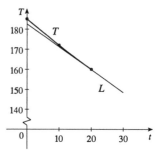

$T(30) \approx T(20) + T'(20)(30 - 20) \approx 160 - 1.2(10) = 148°$ F.
We would expect the temperature of the turkey to get closer to $75°$ F as time increases. Since the temperature decreased $13°$ F in the first 10 minutes and $12°$ F in the second 10 minutes, we can assume that the slopes of the tangent line are increasing through negative values: $-1.3, -1.2, \ldots$. Hence, the tangent lines are under the curve and $148°$ F is an underestimate. From the figure, we estimate the slope of the tangent line at $t = 20$ to be $\frac{184 - 147}{0 - 30} = -\frac{37}{30}$. Then the linear approximation becomes $T(30) \approx T(20) + T'(20) \cdot 10 \approx 160 - \frac{37}{30}(10) = 147\frac{2}{3} \approx 147.7$.

9. $S'(1993) \approx \dfrac{S(1992) - S(1993)}{1992 - 1993} = \dfrac{27.5 - 30}{-1} = 2.5\%/\text{year.}$

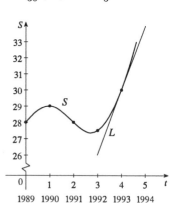

$S(1994) \approx S(1993) + S'(1993)(1994 - 1993)$

$$\approx 30 + 2.5(1) = 32.5\%.$$

Similarly, $S(1995)$ will be 2.5% more than $S(1994)$. So $S(1995) = 35\%$. From the figure, we estimate the slope of the tangent line at $t = 4$ to be $\frac{34 - 26}{5 - 3} = 4$. Then the linear approximation becomes $S(5) = S(4) + S'(4) \cdot 1 \approx 30 + 4 = 34\%$. And in a similar fashion, $S(6) \approx 38\%$.

11. (a) The graph shows that $f'(1) = 2$, so $L(x) = f(1) + f'(1)(x - 1) = 5 + 2(x - 1) = 2x + 3$.

$f(0.9) \approx L(0.9) = 4.8$ and $f(1.1) \approx L(1.1) = 5.2$.

(b) From the graph, we see that $f'(x)$ is positive and decreasing. This means that the slopes of the tangent lines are positive, but the tangents are becoming less steep. So the tangent lines lie *above* the curve. Thus, the estimates in part (a) are too large.

Section 2.10 What Does f' Say about f?

1. (a) Since $f'(x) > 0$ on $(-\infty, 0)$ and $(3, \infty)$, f is increasing on the same intervals. $f'(x) < 0$ and f is decreasing on $(0, 3)$.

(b) Since $f'(x) = 0$ at $x = 0$ and f' changes from positive to negative there, f changes from increasing to decreasing and has a local maximum at $x = 0$. Since $f'(x) = 0$ at $x = 3$ and changes from negative to positive there, f changes from decreasing to increasing and has a local minimum at $x = 3$.

(c)

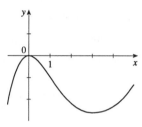

3. The derivative f' is increasing when the slopes of the tangent lines of f are becoming larger as x increases. This seems to be the case on the interval $(2, 5)$. The derivative is decreasing when the slopes of the tangent lines of f are becoming smaller as x increases, and this seems to be the case on $(-\infty, 2)$ and $(5, \infty)$. So f' is increasing on $(2, 5)$ and decreasing on $(-\infty, 2)$ and $(5, \infty)$.

5. If $D(t)$ is the size of the deficit as a function of time, then at the time of the speech $D'(t) > 0$, but $D''(t) < 0$ because $D''(t) = (D')'(t)$ is the rate of change of $D'(t)$.

7. (a) The rate of increase of the population is initially very small, then increases rapidly until about 1932 when it starts decreasing. The rate becomes negative by 1936, peaks in magnitude in 1937, and approaches 0 in 1940.

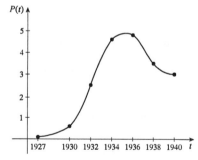

(b) Inflection points (IP) appear to be at $(1932, 2.5)$ and $(1937, 4.3)$. The rates of population increase and decrease have their maximum values at those points.

9. Most students learn more in the third hour of studying than in the eighth hour, so $K(3) - K(2)$ is larger than $K(8) - K(7)$. In other words, as you begin studying for a test, the rate of knowledge gain is large and then starts to taper off, so $K'(t)$ decreases and the graph of K is concave downward.

11. (a) f is increasing where f' is positive, that is, on $(0, 2)$, $(4, 6)$, and $(8, \infty)$; and decreasing where f' is negative, that is, on $(2, 4)$ and $(6, 8)$.

(b) f has local maxima where f' changes from positive to negative, at $x = 2$ and at $x = 6$, and local minima where f' changes from negative to positive, at $x = 4$ and at $x = 8$.

(c) f is concave upward where f' is increasing, that is, on $(3, 6)$ and $(6, \infty)$, and concave downward where f' is decreasing, that is, on $(0, 3)$.

(e)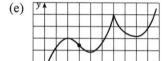

(d) There is a point of inflection where f changes from being CD to being CU, that is, at $x = 3$.

13. The function must be always decreasing and concave downward.

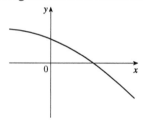

15. $f(-1) = 4$ and $f(1) = 0$ gives us two points to start with. $f'(-1) = f'(1) = 0 \Longrightarrow$ horizontal tangents at $x = \pm 1$. $f'(x) < 0$ if $|x| < 1 \Longrightarrow f$ is decreasing on $(-1, 1)$. $f'(x) > 0$ if $|x| > 1 \Longrightarrow f$ is increasing on $(-\infty, -1)$ and $(1, \infty)$. $f''(x) < 0$ if $x < 0 \Longrightarrow f$ is concave downward on $(-\infty, 0)$. $f''(x) > 0$ if $x > 0 \Longrightarrow f$ is concave upward on $(0, \infty)$ and there is an inflection point at $x = 0$.

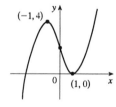

17. First we plot the points which are known to be on the graph: $(2, -1)$ and $(0, 0)$. We can also draw a short line segment of slope 0 at $x = 2$, since we are given that $f'(2) = 0$. Now we know that $f'(x) < 0$ (that is, the function is decreasing) on $(0, 2)$, and that $f''(x) < 0$ on $(0, 1)$ and $f''(x) > 0$ on $(1, 2)$. So we must join the points $(0, 0)$ and $(2, -1)$ in such a way that the curve is concave down on $(0, 1)$ and concave up on $(1, 2)$. The curve must be concave up and increasing on $(2, 4)$ and concave down and increasing on $(4, \infty)$. Now we just need to reflect the curve in the y-axis, since we are given that f is an even function.

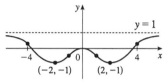

19. (a) Since e^{-x^2} is positive for all x, $f'(x) = xe^{-x^2}$ is positive where $x > 0$ and negative where $x < 0$. Thus, f is increasing on $(0, \infty)$ and decreasing on $(-\infty, 0)$.

(b) Since f changes from decreasing to increasing at $x = 0$, f has a minimum value there.

21. For small x, f is negative, so the graph of its antiderivative must be decreasing. So only b can be f's antiderivative.

23. The graph of F will have a minimum at 0 and a maximum at 2, since the given graph of F' goes from negative to positive at $x = 0$, and from positive to negative at $x = 2$.

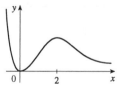

25.

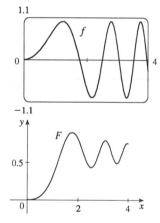

Chapter 2 Review

Concept Check

1. (a) $\lim\limits_{x \to a} f(x) = L$: See Definition 2.2.1.

(b) $\lim\limits_{x \to a^+} f(x) = L$: See Section 2.2, following Definition 2.

(c) $\lim\limits_{x \to a^-} f(x) = L$: See Definition 2.2.2.

(d) $\lim\limits_{x \to a} f(x) = \infty$: See Definition 2.5.1.

(e) $\lim\limits_{x \to \infty} f(x) = L$: See Definition 2.5.3.

(f) Vertical asymptote: See Definition 2.5.2.

(g) Horizontal asymptote: See Definition 2.5.4.

2. See the 11 Limit Laws in Section 2.3.

3. (a) $f'(a) = \lim\limits_{h \to 0} \dfrac{f(a+h) - f(a)}{h}$ and $f'(a) = \lim\limits_{x \to a} \dfrac{f(x) - f(a)}{x - a}$

(b) (i) The derivative of a function f at a number a is the slope of the tangent line to the graph of $y = f(x)$ at the point $(a, f(a))$.

(ii) The derivative $f'(a)$ is the instantaneous rate of change of $y = f(x)$ with respect to x when $x = a$.

4. The second derivative of a function f, denoted f'', is the derivative of the derivative of f: $f'' = (f')'$.

5. (a) The velocity $v(t)$ of the particle at time $t = a$ is the instantaneous rate of change of the displacement s with respect to the time t, that is, $v(a) = f'(a)$.

(b) The speed of the particle is the absolute value of the velocity, that is, $|f'(a)|$.

(c) The acceleration $a(t)$ of the particle at time $t = a$ is the instantaneous rate of change of the velocity v with respect to the time t, that is, $v'(a) = f''(a)$.

6. (a) A function f is continuous at a number a if $f(x)$ gets closer to $f(a)$ as x gets close to a, that is, $\lim\limits_{x \to a} f(x) = f(a)$.

(b) A function f is differentiable at a number a if its derivative f' exists at $x = a$, that is, if $f'(a)$ exists.

(c) See Theorem 2.8.4. This theorem also tells us that if f is *not* continuous at a, then f is *not* differentiable at a.

7. (a) Squeeze Theorem: See Theorem 2.3.3.

(b) Intermediate Value Theorem: See Theorem 2.4.10.

8. (a) $y = x^4$: No asymptote

(b) $y = \sin x$: No asymptote

(c) $y = \tan x$: Vertical asymptotes $x = \frac{\pi}{2} + \pi n$, n an integer

(d) $y = \tan^{-1} x$: Horizontal asymptotes

$y = \pm \frac{\pi}{2}$

(e) $y = e^x$: Horizontal asymptote $y = 0$

($\lim\limits_{x \to -\infty} e^x = 0$)

(f) $y = \ln x$: Vertical asymptote $x = 0$

($\lim\limits_{x \to 0^+} \ln x = -\infty$)

(g) $y = 1/x$: Vertical asymptote $x = 0$, horizontal asymptote $y = 0$

(h) $y = \sqrt{x}$: No asymptote

9. (a) See the first box in Section 2.10.   (b) See the second box in Section 2.10.

10. (a) When x is near a, the linear approximation to f at a is the approximation
$$f(x) \approx f(a) + f'(a)(x - a).$$

(b) An antiderivative of a function f is a function F such that $F' = f$.

True-False Quiz

1. False. Limit Law 2 applies only if the individual limits exist (these don't.)

3. True. Limit Law 5 applies.

5. False. Consider $\lim\limits_{x \to 5} \dfrac{x^2 - 5x}{x - 5}$ or $\lim\limits_{x \to 5} \dfrac{\sin(x - 5)}{x - 5}$. By Example 3 in Section 2.2, we know that the latter limit exists (and it is equal to 1).

7. True. A polynomial is continuous everywhere, so $\lim\limits_{x \to b} p(x)$ exists and is equal to $p(b)$.

9. False. Consider $f(x) = \begin{cases} 1/(x - 1) & \text{if } x \neq 1 \\ 2 & \text{if } x = 1 \end{cases}$

11. True. Use Theorem 2.4.8 with $a = 2$, $b = 5$, and $g(x) = 4x^2 - 11$. Note that $f(4) = 3$ is not needed.

13. False. See the note after Theorem 4 in Section 2.8.

15. False. The equation $y - 4 = 2x(x + 2)$ is not even linear! An equation of the tangent line is
$$y - 4 = -4(x + 2).$$

Exercises

1. (a) (i) $\lim\limits_{x \to 2^+} f(x) = 3$

    (ii) $\lim\limits_{x \to -3^+} f(x) = 0$

    (iii) $\lim\limits_{x \to -3} f(x)$ does not exist since the left and right limits are not equal.

    (iv) $\lim\limits_{x \to 4} f(x) = 2$

    (v) $\lim\limits_{x \to 0} f(x) = \infty$

    (vi) $\lim\limits_{x \to 2^-} f(x) = -\infty$

    (vii) $\lim\limits_{x \to \infty} f(x) = 4$

    (viii) $\lim\limits_{x \to -\infty} f(x) = -1$

(b) The horizontal asymptotes are $y = 4$ and $y = -1$.

(c) The vertical asymptotes are $x = 0$ and $x = 2$.

(d) f is discontinuous at $x = -3, 0, 2,$ and 4.

3. $\lim\limits_{x \to 0} \tan\left(x^2\right) = \tan 0 = 0$ because the tangent function is continuous at $x = 0$.

5. $\lim\limits_{h \to 0} \dfrac{(1+h)^2 - 1}{h} = \lim\limits_{h \to 0} \dfrac{1 + 2h + h^2 - 1}{h} = \lim\limits_{h \to 0} \dfrac{2h + h^2}{h} = \lim\limits_{h \to 0} (2 + h) = 2$

7. $\lim\limits_{x \to -1} \dfrac{x^2 - x - 2}{x^2 + 3x - 2} = \dfrac{\lim\limits_{x \to -1} \left(x^2 - x - 2\right)}{\lim\limits_{x \to -1} \left(x^2 + 3x - 2\right)} = \dfrac{(-1)^2 - (-1) - 2}{(-1)^2 + 3(-1) - 2} = \dfrac{0}{-4} = 0$

9. $\lim\limits_{t \to 6} \dfrac{17}{(t-6)^2} = \infty$ since $(t-6)^2 \to 0$ and $\dfrac{17}{(t-6)^2} > 0$.

11. $\lim\limits_{x \to 8^-} \dfrac{|x - 8|}{x - 8} = \lim\limits_{x \to 8^-} \dfrac{-(x - 8)}{x - 8} = \lim\limits_{x \to 8^-} (-1) = -1$

13. $\lim\limits_{x \to 0} \dfrac{1 - \sqrt{1 - x^2}}{x} \cdot \dfrac{1 + \sqrt{1 - x^2}}{1 + \sqrt{1 - x^2}} = \lim\limits_{x \to 0} \dfrac{1 - \left(1 - x^2\right)}{x\left(1 + \sqrt{1 - x^2}\right)} = \lim\limits_{x \to 0} \dfrac{x^2}{x\left(1 + \sqrt{1 - x^2}\right)}$

$$= \lim\limits_{x \to 0} \dfrac{x}{1 + \sqrt{1 - x^2}} = 0$$

15. $\lim\limits_{x \to \infty} e^{-x^2} = 0$ since $-x^2 \to -\infty$ as $x \to \infty$ and $\lim\limits_{t \to -\infty} e^t = 0$.

17. From the graph of $y = \cos^2 x / x^2$, it appears that $y = 0$ is the horizontal asymptote and $x = 0$ is the vertical asymptote. Now $0 \le (\cos x)^2 \le 1 \implies \dfrac{0}{x^2} \le \dfrac{\cos^2 x}{x^2} \le \dfrac{1}{x^2} \implies 0 \le \dfrac{\cos^2 x}{x^2} \le \dfrac{1}{x^2}$. But $\lim\limits_{x \to \pm\infty} 0 = 0$ and $\lim\limits_{x \to \pm\infty} \dfrac{1}{x^2} = 0$, so by the Squeeze Theorem, $\lim\limits_{x \to \pm\infty} \dfrac{\cos^2 x}{x^2} = 0$.

Thus, $y = 0$ is the horizontal asymptote. $\lim\limits_{x \to 0} \dfrac{\cos^2 x}{x^2} = \infty$

because $\cos^2 x \to 1$ and $x^2 \to 0$ as $x \to 0$, so $x = 0$ is the vertical asymptote.

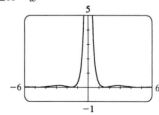

19. Since $2x - 1 \le f(x) \le x^2$ for $0 < x < 3$ and $\lim_{x \to 1} (2x - 1) = 1 = \lim_{x \to 1} x^2$, we have $\lim_{x \to 1} f(x) = 1$ by the Squeeze Theorem.

21. (a) $f(x) = \sqrt{-x}$ if $x < 0$, $f(x) = 3 - x$ if $0 \le x < 3$, $f(x) = (x - 3)^2$ if $x > 3$. So

(i) $\lim_{x \to 0^+} f(x) = \lim_{x \to 0^+} (3 - x) = 3$

(iv) $\lim_{x \to 3^-} f(x) = \lim_{x \to 3^-} (3 - x) = 0$

(ii) $\lim_{x \to 0^-} f(x) = \lim_{x \to 0^-} \sqrt{-x} = 0$

(v) $\lim_{x \to 3^+} f(x) = \lim_{x \to 3^+} (x - 3)^2 = 0$

(iii) Because of (i) and (ii), $\lim_{x \to 0} f(x)$ does not exist.

(vi) Because of (iv) and (v), $\lim_{x \to 3} f(x) = 0$.

(b) f is discontinuous at 0 since $\lim_{x \to 0} f(x)$ does not exist. f is discontinuous at 3 since $f(3)$ does not exist.

(c)

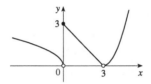

23. $f(x) = 2x^3 + x^2 + 2$ is a polynomial, so it is continuous on $[-2, -1]$ and $f(-2) = -10 < 0 < 1 = f(-1)$. So by the Intermediate Value Theorem there is a number c in $(-2, -1)$ such that $f(c) = 0$, that is, the equation $2x^3 + x^2 + 2 = 0$ has a root in $(-2, -1)$.

25. (a) $s = 1 + 2t + t^2/4$. The average velocity over the time interval $[1, 1 + h]$ is

$$\frac{s(1 + h) - s(1)}{h} = \frac{1 + 2(1 + h) + (1 + h)^2/4 - 13/4}{h} = \frac{10h + h^2}{4h} = \frac{10 + h}{4}. \text{ So for the}$$

following intervals the average velocities are:

(i) $[1, 3]$: $(10 + 2)/4 = 3$ m/s

(iii) $[1, 1.5]$: $(10 + 0.5)/4 = 2.625$ m/s

(ii) $[1, 2]$: $(10 + 1)/4 = 2.75$ m/s

(iv) $[1, 1.1]$: $(10 + 0.1)/4 = 2.525$ m/s

(b) When $t = 1$ the velocity is $\lim_{h \to 0} \frac{s(1 + h) - s(1)}{h} = \lim_{h \to 0} \frac{10 + h}{4} = 2.5$ m/s.

27. Estimating the slopes of the tangent lines at $x = 2$, 3, and 5, we obtain approximate values 0.4, 2, and 0.1. Since the graph is concave downward at $x = 5$, $f''(5)$ is negative. Arranging the numbers in increasing order, we have: $f''(5), 0, f'(5), f'(2), 1, f'(3)$.

29. (a) Estimating $f'(1)$ from the triangle in the graph, we get

$$\frac{\Delta y}{\Delta x} \approx \frac{-0.37}{0.50} = -0.74.$$

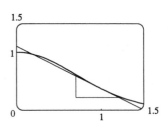

To estimate $f'(1)$ numerically, we have

$$f'(1) = \lim_{h \to 0} \frac{f(1+h) - f(1)}{h} = \lim_{h \to 0} \frac{e^{-(1+h)^2} - e^{-1}}{h} = y.$$

From the table, we have $f'(1) \approx -0.736$.

(b) $y - e^{-1} \approx -0.736(x - 1)$ or $y \approx -0.736x + 1.104$

(c) See the graph in part (a).

h	y
0.01	−0.732
0.001	−0.735
0.0001	−0.736
−0.01	−0.739
−0.001	−0.736
−0.0001	−0.736

31. (a) $f'(r)$ is the rate at which the total cost changes with respect to the interest rate. Its units are dollars/ (percent per year) .

(b) The total cost of paying off the loan is increasing by $1200/ (percent per year) as the interest rate reaches 10%. So if the interest rate goes up from 10% to 11%, the cost goes up approximately $1200.

(c) As r increases, C increases. So $f'(r)$ will always be positive.

33.

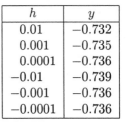
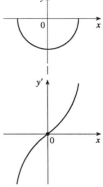

35. (a) $\quad f'(x) = \lim_{h \to 0} \frac{f(x+h) - f(x)}{h}$

$$= \lim_{h \to 0} \frac{\sqrt{3 - 5(x+h)} - \sqrt{3 - 5x}}{h} \frac{\left(\sqrt{3 - 5(x+h)} + \sqrt{3 - 5x}\right)}{\left(\sqrt{3 - 5(x+h)} + \sqrt{3 - 5x}\right)}$$

$$= \lim_{h \to 0} \frac{[3 - 5(x+h)] - (3 - 5x)}{h\left(\sqrt{3 - 5(x+h)} + \sqrt{3 - 5x}\right)} = \lim_{h \to 0} \frac{-5}{\sqrt{3 - 5(x+h)} + \sqrt{3 - 5x}}$$

$$= \frac{-5}{2\sqrt{3 - 5x}}$$

(b) Domain of f: $3 - 5x \geq 0 \Longrightarrow 5x \leq 3 \Longrightarrow x \in \left(-\infty, \frac{3}{5}\right]$

Domain of f': exclude $\frac{3}{5}$; $x \in \left(-\infty, \frac{3}{5}\right)$

(c) Our answer to part (a) is reasonable because $f'(x)$ is always negative and f is always decreasing.

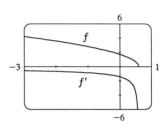

37. f is not differentiable: at $x = -4$ because f is not continuous, at $x = -1$ because f has a corner, at $x = 2$ because f is not continuous, and at $x = 5$ because f has a vertical tangent.

39. (a) Remember that the slope of the tangent for $f(x) = e^x$ at $x = 0$ is 1 (by definition of the number e).

(b) $L(x) = f(0) + f'(0)(x - 0) = 1 + 1(x) = x + 1$

(c)

x	-0.2	-0.1	-0.01	0.01	0.1	0.2
$e^x \approx L(x)$	0.8	0.9	0.99	1.01	1.1	1.2

(d) The graph of $f(x) = e^x$ is concave upward, so its tangent lines are below the graph and thus, the approximations are underestimates. The most accurate estimates are for those closest to $x = 0$, namely, for $e^{-0.01}$ and for $e^{0.01}$.

41. (a) $f'(x) > 0$ on $(-2, 0)$ and $(2, \infty) \implies f$ is increasing on those intervals. $f'(x) < 0$ on $(-\infty, -2)$ and $(0, 2) \implies f$ is decreasing on those intervals.

(b) $f'(x) = 0$ at $x = -2$, 0, and 2, so these are where local maxima or minima will occur. At $x = \pm 2$, f' changes from negative to positive, so f has local minima at those values. At $x = 0$, f' changes from positive to negative, so f has a local maximum there.

(c) f' is increasing on $(-\infty, -1)$ and $(1, \infty) \implies f'' > 0$ and f is concave upward on those intervals. f' is decreasing on $(-1, 1) \implies f'' < 0$ and f is concave downward on this interval.

(d)

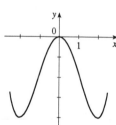

43.

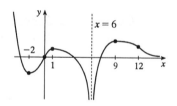

45. (a) Using the data closest to $t = 6$, we have
$$\frac{s(8) - s(6)}{8 - 6} = \frac{180 - 95}{2} = 42.5 \text{ and}$$
$$\frac{s(4) - s(6)}{4 - 6} = \frac{40 - 95}{-2} = 27.5. \text{ Averaging}$$
these two values gives us
$$\frac{42.5 + 27.5}{2} = 35 \text{ ft/s as an estimate for the}$$
speed of the car after 6 seconds.

(b)

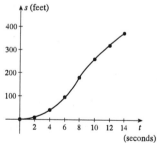

From the graph, it appears that the inflection point is at $(8, 180)$.

(c) The velocity of the car is at a maximum at the inflection point.

Focus on Problem Solving

1. Let $t = \sqrt[6]{x}$, so $x = t^6$. Then $t \to 1$ as $x \to 1$, so

$$\lim_{x \to 1} \frac{\sqrt[3]{x} - 1}{\sqrt{x} - 1} = \lim_{t \to 1} \frac{t^2 - 1}{t^3 - 1} = \lim_{t \to 1} \frac{(t-1)(t+1)}{(t-1)(t^2+t+1)} = \lim_{t \to 1} \frac{t+1}{t^2+t+1} = \frac{1+1}{1^2+1+1} = \frac{2}{3}.$$

Another Method: Multiply both numerator and denominator by $(\sqrt{x} + 1)\left(\sqrt[3]{x^2} + \sqrt[3]{x} + 1\right)$.

3. For $-\frac{1}{2} < x < \frac{1}{2}$, we have $2x - 1 < 0$ and $2x + 1 > 0$, so $|2x - 1| = -(2x - 1)$ and $|2x+1| = 2x+1$.

Therefore, $\displaystyle\lim_{x \to 0} \frac{|2x - 1| - |2x + 1|}{x} = \lim_{x \to 0} \frac{-(2x-1) - (2x+1)}{x} = \lim_{x \to 0} \frac{-4x}{x} = \lim_{x \to 0} (-4) = -4.$

5. Since $[\![x]\!] \le x < [\![x]\!] + 1$, we have $1 \le \dfrac{x}{[\![x]\!]} \le 1 + \dfrac{1}{[\![x]\!]}$ for $x > 0$. As $x \to \infty$, $[\![x]\!] \to \infty$, so $\dfrac{1}{[\![x]\!]} \to 0$

and $1 + \dfrac{1}{[\![x]\!]} \to 1$. Thus, $\displaystyle\lim_{x \to \infty} \frac{x}{[\![x]\!]} = 1$ by the Squeeze Theorem.

7. f is continuous on $(-\infty, a)$ and (a, ∞). To make f continuous on $\mathbb{R}$, we must have continuity

at a. Thus, $\displaystyle\lim_{x \to a^+} f(x) = \lim_{x \to a^-} f(x) \Longrightarrow \lim_{x \to a^+} x^2 = \lim_{x \to a^-} (x+1) \Longrightarrow a^2 = a + 1 \Longrightarrow$

$a = \left(1 \pm \sqrt{5}\right)/2 \approx 1.618$ or -0.618.

9. (a) Consider $G(x) = T(x + 180°) - T(x)$. Fix any number a. If $G(a) = 0$, we

are done: Temperature at a = Temperature at $a + 180°$. If $G(a) > 0$, then

$G(a + 180°) = T(a + 360°) - T(a + 180°) = T(a) - T(a + 180°) = -G(a) < 0$. Also, G is

continuous since temperature varies continuously. So, by the Intermediate Value Theorem, G has a

zero on the interval $[a, a + 180°]$. If $G(a) < 0$, then a similar argument applies.

(b) Yes. The same argument applies.

(c) The same argument applies for quantities that vary continuously, such as barometric pressure. But

one could argue that altitude above sea level is sometimes discontinuous, so the result might not

always hold for that quantity.

11. Let a be the x-coordinate of Q. Since the derivative of $y = 1 - x^2$ is $y' = -2x$, the slope at Q is $-2a$.

But since the triangle is equilateral, $\overline{AO}/\overline{OC} = \sqrt{3}/1$, so the slope at Q is $-\sqrt{3}$. Therefore, we must

have that $-2a = -\sqrt{3} \Longrightarrow a = \frac{\sqrt{3}}{2}$. Thus, the point Q has coordinates $\left(\frac{\sqrt{3}}{2}, 1 - \left(\frac{\sqrt{3}}{2}\right)^2\right) = \left(\frac{\sqrt{3}}{2}, \frac{1}{4}\right)$

and by symmetry, P has coordinates $\left(-\frac{\sqrt{3}}{2}, \frac{1}{4}\right)$.

13. (a) Put $x = 0$ and $y = 0$ in the equation: $f(0) = f(0 + 0) = f(0) + f(0) + 0^2 \cdot 0 + 0 \cdot 0^2 = 2f(0)$.

Subtracting $f(0)$ from each side of this equation gives $f(0) = 0$.

(b) $f'(0) = \lim\limits_{h \to 0} \dfrac{f(0 + h) - f(0)}{h} = \lim\limits_{h \to 0} \dfrac{\left[f(0) + f(h) + 0^2 h + 0h^2\right] - f(0)}{h}$

$= \lim\limits_{h \to 0} \dfrac{f(h)}{h} = \lim\limits_{x \to 0} \dfrac{f(x)}{x} = 1$

(c) $f'(x) = \lim\limits_{h \to 0} \dfrac{f(x + h) - f(x)}{h} = \lim\limits_{h \to 0} \dfrac{\left[f(x) + f(h) + x^2 h + xh^2\right] - f(x)}{h}$

$= \lim\limits_{h \to 0} \dfrac{f(h) + x^2 h + xh^2}{h} = \lim\limits_{h \to 0} \left[\dfrac{f(h)}{h} + x^2 + xh\right] = 1 + x^2$

15. $\lim\limits_{x \to a} f(x) = \lim\limits_{x \to a} \left(\tfrac{1}{2}\left[f(x) + g(x)\right] + \tfrac{1}{2}\left[f(x) - g(x)\right]\right)$

$= \tfrac{1}{2} \lim\limits_{x \to a} \left[f(x) + g(x)\right] + \tfrac{1}{2} \lim\limits_{x \to a} \left[f(x) - g(x)\right]$

$= \tfrac{1}{2} \cdot 2 + \tfrac{1}{2} \cdot 1 = \tfrac{3}{2}$, and

$\lim\limits_{x \to a} g(x) = \lim\limits_{x \to a} \left(\left[f(x) + g(x)\right] - f(x)\right) = \lim\limits_{x \to a} \left[f(x) + g(x)\right] - \lim\limits_{x \to a} f(x) = 2 - \tfrac{3}{2} = \tfrac{1}{2}$.

So $\lim\limits_{x \to a} \left[f(x) g(x)\right] = \left[\lim\limits_{x \to a} f(x)\right] \left[\lim\limits_{x \to a} g(x)\right] = \tfrac{3}{2} \cdot \tfrac{1}{2} = \tfrac{3}{4}$.

Another Solution: Since $\lim\limits_{x \to a} \left[f(x) + g(x)\right]$ and $\lim\limits_{x \to a} \left[f(x) - g(x)\right]$ exist, we must have

$\lim\limits_{x \to a} \left[f(x) + g(x)\right]^2 = \left(\lim\limits_{x \to a} \left[f(x) + g(x)\right]\right)^2$ and $\lim\limits_{x \to a} \left[f(x) - g(x)\right]^2 = \left(\lim\limits_{x \to a} \left[f(x) - g(x)\right]\right)^2$, so

$\lim\limits_{x \to a} \left[f(x) g(x)\right] = \lim\limits_{x \to a} \tfrac{1}{4}\left(\left[f(x) + g(x)\right]^2 - \left[f(x) - g(x)\right]^2\right)$ (because all of the f^2 and g^2 cancel)

$= \tfrac{1}{4}\left(\lim\limits_{x \to a} \left[f(x) + g(x)\right]^2 - \lim\limits_{x \to a} \left[f(x) - g(x)\right]^2\right) = \tfrac{1}{4}\left(2^2 - 1^2\right) = \tfrac{3}{4}$.

17. We are given that $|f(x)| \leq x^2$ for all x. In particular, $|f(0)| \leq 0$, but $|a| \geq 0$ for all a. The only

conclusion is that $f(0) = 0$. Now $\left|\dfrac{f(x) - f(0)}{x - 0}\right| = \left|\dfrac{f(x)}{x}\right| = \dfrac{|f(x)|}{|x|} \leq \dfrac{x^2}{|x|} = \dfrac{|x^2|}{|x|} = |x| \implies$

$-|x| \leq \dfrac{f(x) - f(0)}{x - 0} \leq |x|$. But $\lim\limits_{x \to 0} (-|x|) = 0 = \lim\limits_{x \to 0} |x|$, so by the Squeeze Theorem,

$\lim\limits_{x \to 0} \dfrac{f(x) - f(0)}{x - 0} = 0$. So by the definition of a derivative, f is differentiable at 0 and, furthermore,

$f'(0) = 0$.

Chapter 3   Differentiation Rules

Section 3.1   Derivatives of Polynomials and Exponential Functions

1. (a) e is the number such that $\lim\limits_{h \to 0} \dfrac{e^h - 1}{h} = 1$.

(b)

x	$(2.7^x - 1)/x$
-0.001	0.9928
-0.0001	0.9932
0.001	0.9937
0.0001	0.9933

x	$(2.8^x - 1)/x$
-0.001	1.0291
-0.0001	1.0296
0.001	1.0301
0.0001	1.0297

From the tables (to two decimal places), $\lim\limits_{h \to 0} \dfrac{2.7^h - 1}{h} = 0.99$ and $\lim\limits_{h \to 0} \dfrac{2.8^h - 1}{h} = 1.03$. Since $0.99 < 1 < 1.03$, $2.7 < e < 2.8$.

3. $y = x^8 \implies y' = 8x^{8-1} = 8x^7$

5. $y = x^{-2/5} \implies y' = -\frac{2}{5}x^{(-2/5)-1} = -\frac{2}{5}x^{-7/5} = -2/\left(5x^{7/5}\right)$

7. $f(x) = x^2 - 10x + 100 \implies f'(x) = 2x - 10$

9. $V(r) = \frac{4}{3}\pi r^3 \implies V'(r) = \frac{4}{3}\pi\left(3r^2\right) = 4\pi r^2$

11. $Y(t) = 6t^{-9} \implies Y'(t) = 6(-9)t^{-10} = -54t^{-10}$

13. $F(x) = (16x)^3 = 4096x^3 \implies F'(x) = 4096\left(3x^2\right) = 12{,}288x^2$

15. $g(x) = x^2 + 1/x^2 = x^2 + x^{-2} \implies g'(x) = 2x + (-2)x^{-3} = 2x - 2/x^3$

17. $y = \dfrac{x^2 + 4x + 3}{\sqrt{x}} = x^{3/2} + 4x^{1/2} + 3x^{-1/2} \implies$

$y' = \frac{3}{2}x^{1/2} + 4\left(\frac{1}{2}\right)x^{-1/2} + 3\left(-\frac{1}{2}\right)x^{-3/2} = \frac{3}{2}\sqrt{x} + 2/\sqrt{x} - 3/\left(2x\sqrt{x}\right)$

19. $y = 3x + 2e^x \implies y' = 3 + 2e^{xx+1}$

21. $f(x) = 2x^2 - x^4 \implies f'(x) = 4x - 4x^3$.

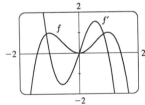

Notice that $f'(x) = 0$ when f has a horizontal tangent and that f' is an odd function while f is an even function.

23. $f(x) = 3x^{15} - 5x^3 + 3 \implies$

$f'(x) = 45x^{14} - 15x^2$.

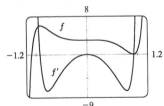

Notice that $f'(x) = 0$ when f has a horizontal tangent.

25. $f(x) = x - 3x^{1/3} \Longrightarrow$

$f'(x) = 1 - x^{-2/3} = 1 - 1/x^{2/3}$.

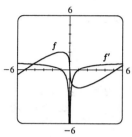

Note that $f'(x) = 0$ when f has a horizontal tangent, f' is positive when f is increasing, and f' is negative when f is decreasing.

27. (a)

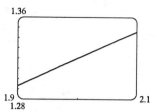

The endpoints of f in this graph are about $(1.9, 1.2927)$ and $(2.1, 1.3455)$. An estimate of $f'(2)$ is

$$\frac{1.3455 - 1.2927}{2.1 - 1.9} = \frac{0.0528}{0.2} = 0.264.$$

(b) $f(x) = x^{2/5} \Longrightarrow$

$f'(x) = \frac{2}{5}x^{-3/5} = 2/(5x^{3/5})$.

$f'(2) = 2/(5 \cdot 2^{3/5}) \approx 0.263902.$

29. $y = f(x) = x + \dfrac{4}{x} \Longrightarrow f'(x) = 1 - \dfrac{4}{x^2}$.

So the slope of the tangent line at $(2, 4)$ is $f'(2) = 0$ and its equation is $y - 4 = 0$ or $y = 4$.

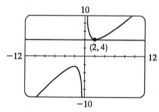

31. $y = f(x) = x + \sqrt{x} \Longrightarrow$

$f'(x) = 1 + \frac{1}{2}x^{-1/2}$. So the slope of the tangent line at $(1, 2)$ is

$f'(1) = 1 + \frac{1}{2}(1) = \frac{3}{2}$ and its equation is

$y - 2 = \frac{3}{2}(x - 1)$ or $y = \frac{3}{2}x + \frac{1}{2}$.

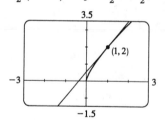

33. $f(x) = x^4 - 3x^3 + 16x \Longrightarrow f'(x) = 4x^3 - 9x^2 + 16 \Longrightarrow f''(x) = 12x^2 - 18x$

35. $f(x) = 2x - 5x^{3/4} \Longrightarrow$

$f'(x) = 2 - \frac{15}{4}x^{-1/4} \Longrightarrow$

$f''(x) = \frac{15}{16}x^{-5/4}$

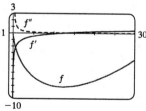

Note that f' is negative when f is decreasing and positive when f is increasing. f'' is always positive since f' is always increasing.

37. (a) $s = t^3 - 3t \Longrightarrow$

$v(t) = s'(t)$

$= 3t^2 - 3 \Longrightarrow$

$a(t) = v'(t) = 6t$

(b) $a(2) = 6(2) = 12 \text{ m/s}^2$

(c) $v(t) = 3t^2 - 3 = 0$ when $t^2 = 1$, that is, $t = 1$ and $a(1) = 6 \text{ m/s}^2$.

39. (a) 1920: $m_1 = \frac{1860 - 1750}{1920 - 1910} = \frac{110}{10} = 11$, $m_2 = \frac{2070 - 1860}{1930 - 1920} = \frac{210}{10} = 21$,

$(m_1 + m_2)/2 = (11 + 21)/2 = 16$ million/year

1980: $m_1 = \frac{4450 - 3700}{1980 - 1970} = \frac{750}{10} = 75$, $m_2 = \frac{5300 - 4450}{1990 - 1980} = \frac{850}{10} = 85$,

$(m_1 + m_2)/2 = (75 + 85)/2 = 80$ million/year

(b) $P(t) = at^3 + bt^2 + ct + d \Longrightarrow P'(t) = 3at^2 + 2bt + c$

(c) $P'(1920) = 3(2325.67)(1920)^2 + 2(-1.306448 \times 10^7)(1920) + 2.44631 \times 10^{10}$
$= 14{,}010{,}464/\text{year}$ [smaller than the answer in part (a), but close to it]
$P'(1980) = 78{,}845{,}204/\text{year}$ (smaller, but close)

41. $f(x) = 1 + 2e^x - 3x \Longrightarrow f'(x) = 2e^x - 3$. $f'(x) > 0 \Longrightarrow 2e^x > 3 \Longrightarrow e^x > 1.5 \Longrightarrow$
$x > \ln 1.5 \approx 0.41$. f is increasing when f' is positive; that is, on $(\ln 1.5, \infty)$.

43. $y = x^3 - x^2 - x + 1$ has a horizontal tangent when $y' = 3x^2 - 2x - 1 = 0$. $(3x + 1)(x - 1) = 0 \Longleftrightarrow$
$x = 1$ or $-\frac{1}{3}$. Therefore, the points are $(1, 0)$ and $\left(-\frac{1}{3}, \frac{32}{27}\right)$.

45. $y = 6x^3 + 5x - 3 \Longrightarrow m = y' = 18x^2 + 5$, but $x^2 \geq 0$ for all x, so $m \geq 5$ for all x.

47.

Let (a, a^2) be a point of intersection. The tangent line has
slope $2a$ and equation $y - (-4) = 2a(x - 0) \Longrightarrow y = 2ax - 4$.
Since (a, a^2) also lies on the line, $a^2 = 2a(a) - 4$, or $a^2 = 4$.
So $a = \pm 2$ and the points are $(2, 4)$ and $(-2, 4)$.

49. $y = f(x) = 1 - x^2 \Longrightarrow f'(x) = -2x$, so the tangent line at
$(2, -3)$ has slope $f'(2) = -4$. The normal line has slope
$-\frac{1}{-4} = \frac{1}{4}$ and equation $y + 3 = \frac{1}{4}(x - 2)$ or $y = \frac{1}{4}x - \frac{7}{2}$.

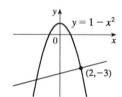

51. $f'(x) = \lim_{h \to 0} \dfrac{f(x + h) - f(x)}{h} = \lim_{h \to 0} \dfrac{\frac{1}{x + h} - \frac{1}{x}}{h} = \lim_{h \to 0} \dfrac{x - (x + h)}{hx(x + h)}$

$\qquad\qquad = \lim_{h \to 0} \dfrac{-h}{hx(x + h)} = \lim_{h \to 0} \dfrac{-1}{x(x + h)} = -\dfrac{1}{x^2}$

53. Let $P(x) = ax^2 + bx + c$. Then $P'(x) = 2ax + b$ and $P''(x) = 2a$. $P''(2) = 2 \Longrightarrow 2a = 2 \Longrightarrow$
$a = 1$. $P'(2) = 3 \Longrightarrow 4a + b = 4 + b = 3 \Longrightarrow b = -1$. $P(2) = 5 \Longrightarrow 2^2 - 2 + c = 5 \Longrightarrow c = 3$. So
$P(x) = x^2 - x + 3$.

55. (a) At this stage, we would guess that an antiderivative of x^2 must have x^3 in it. Differentiating x^3 gives us $3x^2$, so we know that we must divide x^3 by 3. That gives us $F(x) = \frac{1}{3}x^3$. Checking, we have $F'(x) = \frac{1}{3}(3x^2) = x^2 = f(x)$. Because we can add an arbitrary constant C to F without changing its derivative, we have an infinite number of antiderivatives of the form $F(x) = \frac{1}{3}x^3 + C$.

(b) As in part (a), antiderivatives of $f(x) = x^3$ and $f(x) = x^4$ are $F(x) = \frac{1}{4}x^4 + C$ and $F(x) = \frac{1}{5}x^5 + C$.

(c) Similarly, an antiderivative for $f(x) = x^n$ is $F(x) = \dfrac{1}{n+1}x^{n+1} + C$, since then

$$F'(x) = \frac{1}{n+1}[(n+1)x^n] = x^n = f(x) \text{ for } n \neq -1.$$

57. Substituting $x = 1$ and $y = 1$ into $y = ax^2 + bx$ gives us **(1)** $a + b = 1$. The slope of the tangent line $y = 3x - 2$ is 3 and the slope of the tangent to the parabola at (x, y) is $y' = 2ax + b$. At $x = 1$, $y' = 3 \Longrightarrow$ **(2)** $3 = 2a + b$. Subtracting **(1)** from **(2)** gives us $2 = a$ and it follows that $b = -1$. The parabola has equation $y = 2x^2 - x$.

59. *Solution 1:*   Let $f(x) = x^{1000}$. Then, by the definition of derivative,

$$f'(1) = \lim_{x \to 1} \frac{f(x) - f(1)}{x - 1} = \lim_{x \to 1} \frac{x^{1000} - 1}{x - 1}. \text{ But this is just the limit we want to find, and}$$

we know (from the Power Rule) that $f'(x) = 1000x^{999}$, so $f'(1) = 1000(1)^{999} = 1000$. So

$$\lim_{x \to 1} \frac{x^{1000} - 1}{x - 1} = 1000.$$

Solution 2:   Note that $\left(x^{1000} - 1\right) = (x - 1)\left(x^{999} + x^{998} + x^{997} + \cdots + x^2 + x + 1\right)$. So

$$\lim_{x \to 1} \frac{x^{1000} - 1}{x - 1} = \lim_{x \to 1} \frac{(x - 1)\left(x^{999} + x^{998} + x^{997} + \cdots + x^2 + x + 1\right)}{x - 1}$$

$$= \lim_{x \to 1}\left(x^{999} + x^{998} + x^{997} + \cdots + x^2 + x + 1\right) = \underbrace{1 + 1 + 1 + \cdots + 1 + 1 + 1}_{1000 \text{ ones}}$$

$$= 1000, \text{ as above.}$$

Section 3.2 The Product and Quotient Rules

1. Product Rule: $y = \left(x^2 + 1\right)\left(x^3 + 1\right) \implies y' = \left(x^2 + 1\right)\left(3x^2\right) + \left(x^3 + 1\right)\left(2x\right) = 3x^4 + 3x^2 + 2x^4 + 2x = 5x^4 + 3x^2 + 2x.$

Multiplying first: $y = \left(x^2 + 1\right)\left(x^3 + 1\right) = x^5 + x^3 + x^2 + 1 \implies y' = 5x^4 + 3x^2 + 2x$ (equivalent)

3. By the Product Rule, $f\left(x\right) = x^2 e^x \implies$

$$f'\left(x\right) = x^2 \frac{d}{dx}\left(e^x\right) + e^x \frac{d}{dx}\left(x^2\right) = x^2 e^x + e^x\left(2x\right) = xe^x\left(x + 2\right).$$

5. By the Quotient Rule, $y = \dfrac{e^x}{x^2} \implies$

$$y' = \frac{x^2 \dfrac{d}{dx}\left(e^x\right) - e^x \dfrac{d}{dx}\left(x^2\right)}{\left(x^2\right)^2} = \frac{x^2\left(e^x\right) - e^x\left(2x\right)}{x^4} = \frac{xe^x\left(x - 2\right)}{x^4} = \frac{e^x\left(x - 2\right)}{x^3}.$$

7. $h\left(x\right) = \dfrac{x + 2}{x - 1} \implies h'\left(x\right) = \dfrac{\left(x - 1\right)\left(1\right) - \left(x + 2\right)\left(1\right)}{\left(x - 1\right)^2} = \dfrac{x - 1 - x - 2}{\left(x - 1\right)^2} = \dfrac{-3}{\left(x - 1\right)^2}$

9. $G\left(s\right) = \left(s^2 + s + 1\right)\left(s^2 + 2\right) \implies$

$$G'\left(s\right) = \left(s^2 + s + 1\right)\left(2s\right) + \left(s^2 + 2\right)\left(2s + 1\right) = 2s^3 + 2s^2 + 2s + 2s^3 + s^2 + 4s + 2$$
$$= 4s^3 + 3s^2 + 6s + 2$$

11. $y = \dfrac{x^2 + 4x + 3}{\sqrt{x}} = x^{3/2} + 4x^{1/2} + 3x^{-1/2} \implies$

$$y' = \tfrac{3}{2}x^{1/2} + 4\left(\tfrac{1}{2}\right)x^{-1/2} + 3\left(-\tfrac{1}{2}\right)x^{-3/2} = \tfrac{3}{2}\sqrt{x} + \frac{2}{\sqrt{x}} - \frac{3}{2x\sqrt{x}}.$$

Another Method: Use the Quotient Rule.

13. $y = \left(r^2 - 2r\right)e^r \implies y' = \left(r^2 - 2r\right)\left(e^r\right) + e^r\left(2r - 2\right) = e^r\left(r^2 - 2r + 2r - 2\right) = e^r\left(r^2 - 2\right)$

15. $y = \dfrac{1}{x^4 + x^2 + 1} \implies y' = \dfrac{\left(x^4 + x^2 + 1\right)\left(0\right) - 1\left(4x^3 + 2x\right)}{\left(x^4 + x^2 + 1\right)^2} = -\dfrac{2x\left(2x^2 + 1\right)}{\left(x^4 + x^2 + 1\right)^2}$

17. $f\left(x\right) = \dfrac{x}{x + c/x} \implies f'\left(x\right) = \dfrac{\left(x + c/x\right)\left(1\right) - x\left(1 - c/x^2\right)}{\left(x + c/x\right)^2} = \dfrac{x + c/x - x + c/x}{\left(\dfrac{x^2 + c}{x}\right)^2} \cdot \dfrac{x^2}{x^2} = \dfrac{2cx}{\left(x^2 + c\right)^2}$

19. (a) $y = f\left(x\right) = \dfrac{1}{1 + x^2} \implies f'\left(x\right) = \dfrac{-2x}{\left(1 + x^2\right)^2}.$ So the slope of

the tangent line at the point $\left(-1, \tfrac{1}{2}\right)$ is $f'\left(-1\right) = \dfrac{2}{2^2} = \tfrac{1}{2}$ and

its equation is $y - \tfrac{1}{2} = \tfrac{1}{2}\left(x + 1\right)$ or $y = \tfrac{1}{2}x + 1.$

(b)

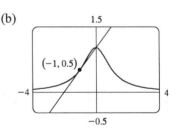

21. (a) $f(x) = \dfrac{e^x}{x^3} \implies$

$$f'(x) = \frac{x^3 (e^x) - e^x (3x^2)}{(x^3)^2}$$
$$= \frac{x^2 e^x (x - 3)}{x^6}$$
$$= \frac{e^x (x - 3)}{x^4}$$

(b)

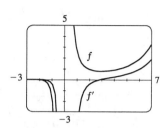

$f' = 0$ when f has a horizontal tangent line, f' is negative when f is decreasing, and f' is positive when f is increasing.

23. (a) $f(x) = (x - 1) e^x \implies f'(x) = (x - 1) e^x + e^x (1) = e^x (x - 1 + 1) = xe^x.$
$f''(x) = x (e^x) + e^x (1) = e^x (x + 1)$

(b)

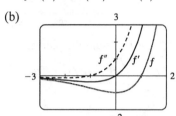

$f' = 0$ when f has a horizontal tangent and $f'' = 0$ when f' has a horizontal tangent. f' is negative when f is decreasing and positive when f is increasing. f'' is negative when f' is decreasing and positive when f' is increasing. f'' is negative when f is concave down and positive when f is concave up.

25. (a) $(fg)'(5) = f(5) g'(5) + g(5) f'(5) = (1)(2) + (-3)(6) = 2 - 18 = -16$

(b) $\left(\dfrac{f}{g}\right)'(5) = \dfrac{g(5) f'(5) - f(5) g'(5)}{[g(5)]^2} = \dfrac{(-3)(6) - (1)(2)}{(-3)^2} = -\dfrac{20}{9}$

(c) $\left(\dfrac{g}{f}\right)'(5) = \dfrac{f(5) g'(5) - g(5) f'(5)}{[f(5)]^2} = \dfrac{(1)(2) - (-3)(6)}{(1)^2} = 20$

27. $f(x) = e^x g(x) \implies f'(x) = e^x g'(x) + g(x) e^x = e^x [g'(x) + g(x)].$
$f'(0) = e^0 [g'(0) + g(0)] = 1 (5 + 2) = 7$

29. (a) $u(x) = f(x) g(x)$, so $u'(1) = f(1) g'(1) + g(1) f'(1) = 2 \cdot (-1) + 1 \cdot 2 = 0$

(b) $v(x) = f(x) / g(x)$, so $v'(5) = \dfrac{g(5) f'(5) - f(5) g'(5)}{[g(5)]^2} = \dfrac{2 \left(-\frac{1}{3}\right) - 3 \cdot \frac{2}{3}}{2^2} = -\dfrac{2}{3}$

31. Let $P(t)$ be the population and let $A(t)$ be the average annual income at time t, where t is measured in years and $t = 0$ corresponds to July 1993. Then the total personal income is given by $T(t) = P(t) A(t)$. We wish to find $T'(0)$. $T'(t) = P(t) A'(t) + A(t) P'(t)$. The term $P(t) A'(t)$ represents the portion of the rate of change of total income due to the existing population's increasing income. The term $A(t) P'(t)$ represents the portion of the rate of change of total income due to the increasing population. $T'(0) = P(0) A'(0) + A(0) P'(0) \approx (3{,}354{,}000) (1900) + (21{,}107) (45{,}000) = 7{,}322{,}415{,}000.$ So the total personal income was rising at a rate of about \$7.322 billion per year.

33. f is increasing when f' is positive. $f(x) = x^3 e^x \implies f'(x) = x^3 e^x + e^x (3x^2) = x^2 e^x (x + 3)$. Now $x^2 \geq 0$ and $e^x > 0$ for all x, so $f'(x) > 0$ when $x \in (-3, 0) \cup (0, \infty)$, so f is increasing on $(-3, \infty)$.

35. If $y = f(x) = \dfrac{x}{x+1}$ then $f'(x) = \dfrac{(x+1)(1) - x(1)}{(x+1)^2} = \dfrac{1}{(x+1)^2}$. When $x = a$, the equation of the

tangent line is $y - \dfrac{a}{a+1} = \dfrac{1}{(a+1)^2}(x - a)$. This line passes through $(1, 2)$ when

$2 - \dfrac{a}{a+1} = \dfrac{1}{(a+1)^2}(1 - a) \iff$

$2(a+1)^2 = a(a+1) + (1-a) = a^2 + 1 \iff a^2 + 4a + 1 = 0$.

The quadratic formula gives the roots of this equation as $-2 \pm \sqrt{3}$,

so there are two such tangent lines, which touch the curve at

$A\left(-2 + \sqrt{3}, \frac{1-\sqrt{3}}{2}\right) \approx (-0.27, -0.37)$ and

$B\left(-2 - \sqrt{3}, \frac{1+\sqrt{3}}{2}\right) \approx (-3.73, 1.37)$.

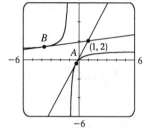

Note: We will sometimes use the form $f'g + fg'$ rather than the form $fg' + gf'$ for the Product Rule.

37. (a) $(fgh)' = [(fg)h]' = (fg)'h + (fg)h' = (f'g + fg')h + (fg)h' = f'gh + fg'h + fgh'$

(b) Putting $f = g = h$ in part (a), we have

$$\frac{d}{dx}[f(x)]^3 = (fff)' = f'ff + ff'f + fff' = 3fff' = 3[f(x)]^2 f'(x).$$

(c) $\dfrac{d}{dx}\left(e^{3x}\right) = \dfrac{d}{dx}\left(e^x\right)^3 = 3\left(e^x\right)^2 e^x = 3e^{2x}e^x = 3e^{3x}$

39. For $f(x) = x^2 e^x$, $f'(x) = x^2 e^x + e^x(2x) = e^x(x^2 + 2x)$. Similarly, we have

$$f''(x) = e^x(x^2 + 4x + 2) \qquad\qquad f^{(4)}(x) = e^x(x^2 + 8x + 12)$$
$$f'''(x) = e^x(x^2 + 6x + 6) \qquad\qquad f^{(5)}(x) = e^x(x^2 + 10x + 20)$$

It appears that the coefficient of x in the quadratic term increases by 2 with each differentiation. The pattern for the constant terms seems to be $0 = 1 \cdot 0$, $2 = 2 \cdot 1$, $6 = 3 \cdot 2$, $12 = 4 \cdot 3$, $20 = 5 \cdot 4$. So a reasonable guess is that $f^{(n)}(x) = e^x[x^2 + 2nx + n(n-1)]$.

Proof: Let S_n be the statement that $f^{(n)}(x) = e^x[x^2 + 2nx + n(n-1)]$.

1. S_1 is true because $f'(x) = e^x(x^2 + 2x)$.

2. Assume that S_k is true, that is, $f^{(k)}(x) = [x^2 + 2kx + k(k-1)]$. Then

$$f^{(k+1)}(x) = \frac{d}{dx}\left[f^{(k)}(x)\right] = e^x(2x + 2k) + [x^2 + 2kx + k(k-1)]e^x$$
$$= e^x[x^2 + (2k+2)x + (k^2 + k)] = e^x[x^2 + 2(k+1)x + (k+1)k]$$

This shows that S_{k+1} is true.

3. Therefore, by mathematical induction, S_n is true for all n; that is,

$f^{(n)}(x) = e^x[x^2 + 2nx + n(n-1)]$ for every positive integer n.

41. $\dfrac{d}{dx}\left(x^{-n}\right) = \dfrac{d}{dx}\left(\dfrac{1}{x^n}\right) = -\dfrac{nx^{n-1}}{(x^n)^2} = -nx^{n-1-2n} = -nx^{-n-1}$

Section 3.3 Rates of Change in the Natural and Social Sciences

1. (a) $s = f(t) = t^3 - 12t^2 + 36t \implies v(t) = f'(t) = 3t^2 - 24t + 36$

(b) $v(3) = 27 - 72 + 36 = -9$ m/s

(c) The particle is at rest when $v(t) = 0$. $3t^2 - 24t + 36 = 0 \implies 3(t-2)(t-6) = 0 \implies t = 2, 6$.

(d) The particle is moving forward when $v(t) > 0$. $3(t-2)(t-6) > 0 \iff 0 \le t < 2$ or $t > 6$.

(e) Since the particle is moving forward and backward, we need to calculate the distance traveled in the intervals $[0, 2]$, $[2, 6]$, and $[6, 8]$ separately. $|f(2) - f(0)| = |32 - 0| = 32$. $|f(6) - f(2)| = |0 - 32| = 32$. $|f(8) - f(6)| = |32 - 0| = 32$. The total distance is $32 + 32 + 32 = 96$ m.

(f)

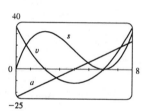

(g) $a(t) = v'(t) = 6t - 24$. $a(3) = -6 \text{ (m/s)} /\text{s or m/s}^2$

(h)

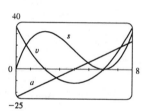

(i) The particle is speeding up when v and a have the same sign. This occurs when $2 < t < 4$ and when $t > 6$. It is slowing down when v and a have opposite signs; that is, when $0 \le t < 2$ and $4 < t < 6$.

3. (a) $s(t) = t^3 - 4.5t^2 - 7t \implies v(t) = s'(t) = 3t^2 - 9t - 7 = 5 \iff 3t^2 - 9t - 12 = 0 \iff 3(t-4)(t+1) = 0 \iff t = 4$ or -1. Since $t \ge 0$, the particle reaches a velocity of 5 m/s at $t = 4$ s.

(b) $a(t) = v'(t) = 6t - 9 = 0 \iff t = 1.5$. The acceleration changes from negative to positive, so the velocity changes from decreasing to increasing. Thus, at $t = 1.5$, the velocity has its minimum value.

5. (a) $A(x) = x^2 \implies A'(x) = 2x$. $A'(15) = 30$ mm^2/mm is the rate at which the area is increasing with respect to the side length as x reaches 15 mm.

(b) The perimeter is $P(x) = 4x$, so $A'(x) = 2x = \frac{1}{2}(4x) = \frac{1}{2}P(x)$. The figure suggests that if Δx is small, then the change in the area of the square is approximately half of its perimeter (2 of the 4 sides) times Δx. From the figure, $\Delta A = 2x(\Delta x) + (\Delta x)^2$. If Δx is small, then $\Delta A \approx 2x(\Delta x)$ and so $\Delta A / \Delta x \approx 2x$.

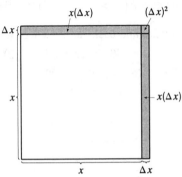

7. (a) $A(r) = \pi r^2$, so the average rate of change is:

(i) $\dfrac{A(3) - A(2)}{3 - 2} = \dfrac{9\pi - 4\pi}{1} = 5\pi$

(ii) $\dfrac{A(2.5) - A(2)}{2.5 - 2} = \dfrac{6.25\pi - 4\pi}{0.5} = 4.5\pi$

(iii) $\dfrac{A(2.1) - A(2)}{2.1 - 2} = \dfrac{4.41\pi - 4\pi}{0.1} = 4.1\pi$

(b) $A'(r) = 2\pi r$, so $A'(2) = 4\pi$.

(c) The circumference is $C(r) = 2\pi r = A'(r)$. The figure suggests that if Δr is small, then the change in the area of the circle (a ring around the outside) is approximately equal to its circumference times Δr. Straightening out this ring gives us a shape that is approximately rectangular with length $2\pi r$ and width Δr, so $\Delta A \approx 2\pi r(\Delta r)$. Algebraically,

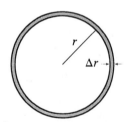

$\Delta A = A(r + \Delta r) - A(r) = \pi(r + \Delta r)^2 - \pi r^2 = 2\pi r(\Delta r) + \pi(\Delta r)^2$. So we see that if Δr is small, then $\Delta A \approx 2\pi r(\Delta r)$ and therefore, $\Delta A / \Delta r \approx 2\pi r$.

9. $S(r) = 4\pi r^2 \implies S'(r) = 8\pi r \implies$

(a) $S'(1) = 8\pi$ ft^2/ft   **(b)** $S'(2) = 16\pi$ ft^2/ft   **(c)** $S'(3) = 24\pi$ ft^2/ft

As the radius increases, the surface area grows at an increasing rate. In fact, the rate of change is linear with respect to the radius.

11. $f(x) = 3x^2$, so the linear density at x is $\rho(x) = f'(x) = 6x$.

(a) $\rho(1) = 6$ kg/m   (b) $\rho(2) = 12$ kg/m   (c) $\rho(3) = 18$ kg/m

Since ρ is an increasing function, the density will be the highest at the right end of the rod and lowest at the left end.

13. $Q(t) = t^3 - 2t^2 + 6t + 2$, so the current is $Q'(t) = 3t^2 - 4t + 6$.

(a) $Q'(0.5) = 3(0.5)^2 - 4(0.5) + 6 = 4.75$ A

(b) $Q'(1) = 3(1)^2 - 4(1) + 6 = 5$ A

The current is lowest when Q' has a minimum. $Q''(t) = 6t - 4 < 0$ when $t < \frac{2}{3}$. So the current decreases when $t < \frac{2}{3}$ and increases when $t > \frac{2}{3}$. Thus, the current is lowest at $\frac{2}{3}$ s.

15. (a) $PV = C \Longrightarrow V = \dfrac{C}{P} \Longrightarrow \dfrac{dV}{dP} = -\dfrac{C}{P^2}$

(b) From the formula for dV/dP in part (a), we see that as P increases, the absolute value of dV/dP decreases. Thus, the volume is decreasing more rapidly at the beginning.

(c) $\beta = -\dfrac{1}{V}\dfrac{dV}{dP} = -\dfrac{1}{V}\left(-\dfrac{C}{P^2}\right) = \dfrac{C}{(PV)P} = \dfrac{C}{CP} = \dfrac{1}{P}$

17. (a) $[C] = \dfrac{a^2 kt}{akt + 1} \Longrightarrow$

rate of reaction $= \dfrac{d[C]}{dt} = \dfrac{(akt + 1)(a^2 k) - (a^2 kt)(ak)}{(akt + 1)^2} = \dfrac{a^2 k(akt + 1 - akt)}{(akt + 1)^2} = \dfrac{a^2 k}{(akt + 1)^2}$

(b) If $x = [C]$, then $a - x = a - \dfrac{a^2 kt}{akt + 1} = \dfrac{a^2 kt + a - a^2 kt}{akt + 1} = \dfrac{a}{akt + 1}$.

So $k(a - x)^2 = k\left(\dfrac{a}{akt + 1}\right)^2 = \dfrac{a^2 k}{(akt + 1)^2} = \dfrac{d[C]}{dt}$ [from part (a)] $= \dfrac{dx}{dt}$.

(c) As $t \to \infty$, $[C] = \dfrac{a^2 kt}{akt + 1} = \dfrac{a^2 k}{ak + (1/t)} \to \dfrac{a^2 k}{ak} = a$ moles/L

(d) As $t \to \infty$, $\dfrac{d[C]}{dt} = \dfrac{a^2 k}{(akt + 1)^2} \to 0$.

(e) As t increases, nearly all of the reactants A and B are converted into product C. In practical terms, the reaction virtually stops.

19. (a) Using $v = \dfrac{P}{4\eta l}\left(R^2 - r^2\right)$ with $R = 0.01$, $l = 3$, $P = 3000$, and $\eta = 0.027$, we have v as a function

of r: $v\left(r\right) = \dfrac{3000}{4\left(0.027\right)3}\left(0.01^2 - r^2\right)$. $v\left(0\right) = 0.\overline{925}$ cm/s, $v\left(0.005\right) = 0.69\overline{4}$ cm/s, $v\left(0.01\right) = 0$.

(b) $v\left(r\right) = \dfrac{P}{4\eta l}\left(R^2 - r^2\right) \Longrightarrow v'\left(r\right) = \dfrac{P}{4\eta l}\left(-2r\right) = -\dfrac{Pr}{2\eta l}$. When $l = 3$, $P = 3000$, and

$\eta = 0.027$, we have $v'\left(r\right) = -\dfrac{3000r}{2\left(0.027\right)3}$. $v'\left(0\right) = 0$, $v'\left(0.005\right) = -92.\overline{592}$ (cm/s) /cm, and

$v'\left(0.01\right) = -185.\overline{185}$ (cm/s) /cm.

(c) The velocity is greatest where $r = 0$ (at the center) and the velocity is changing most where
$r = R = 0.01$ cm (at the edge).

21. (a) $C\left(x\right) = 2000 + 3x + 0.01x^2 + 0.0002x^3 \Longrightarrow C'\left(x\right) = 3 + 0.02x + 0.0006x^2$

(b) $C'\left(100\right) = 3 + 0.02\left(100\right) + 0.0006\left(10{,}000\right) = 3 + 2 + 6 = \$11$/yard. $C'\left(100\right)$ is the rate at
which costs are increasing as the 100th yard is produced. It predicts the cost of the 101st yard.

(c) The cost of manufacturing the 101st yard is
$C\left(101\right) - C\left(100\right) = \left(2000 + 303 + 102.01 + 206.0602\right) - \left(2000 + 300 + 100 + 200\right)$
$$= 11.0702 \approx \$11.07/\text{yard}$$

23. (a) $A\left(x\right) = \dfrac{p\left(x\right)}{x} \Longrightarrow A'\left(x\right) = \dfrac{xp'\left(x\right) - p\left(x\right) \cdot 1}{x^2}$. $A'\left(x\right) > 0 \Longrightarrow A\left(x\right)$ is increasing; that is, the
average productivity increases as the size of the workforce increases.

(b) Suppose $p'\left(x\right) > A\left(x\right)$. Then $p'\left(x\right) > \dfrac{p\left(x\right)}{x} \Longrightarrow xp'\left(x\right) > p\left(x\right) \Longrightarrow xp'\left(x\right) - p\left(x\right) > 0 \Longrightarrow$

$\dfrac{xp'\left(x\right) - p\left(x\right)}{x^2} > 0 \Longrightarrow A'\left(x\right) > 0$.

25. $PV = nRT \Longrightarrow T = \dfrac{PV}{nR} = \dfrac{PV}{\left(10\right)\left(0.0821\right)} = \dfrac{1}{0.821}\left(PV\right)$.

$\dfrac{dT}{dt} = \dfrac{1}{0.821}\left[P\left(t\right)V'\left(t\right) + V\left(t\right)P'\left(t\right)\right] = \dfrac{1}{0.821}\left[\left(8\right)\left(-0.15\right) + \left(10\right)\left(0.10\right)\right] \approx -0.2436$ K/min

27. (a) $\dfrac{dC}{dt} = 0$ and $\dfrac{dW}{dt} = 0$.

(b) The caribou go extinct $\Longleftrightarrow C = 0$.

(c) We have **(1)** $0.05C - 0.001CW = 0$ and **(2)** $-0.05W + 0.0001CW = 0$. Adding 10 times
(2) to **(1)** gives us $0.05C - 0.5W = 0 \Longrightarrow C = 10W$. Substituting $C = 10W$ into **(1)** results in
$W = 0$ or 50 and hence, $C = 0$ or 500. The pairs are $\left(0, 0\right)$ and $\left(500, 50\right)$. So it is possible for the
two species to live in harmony.

Section 3.4 Derivatives of Trigonometric Functions

1. $y = \sin x + \cos x \Longrightarrow dy/dx = \cos x - \sin x$

3. $y = x^2 \cos x \Longrightarrow dy/dx = x^2 (-\sin x) + (\cos x)(2x) = 2x \cos x - x^2 \sin x$ [or $x(2 \cos x - x \sin x)$]

5. $y = 2 \cot x - \sqrt{x} \sec x \Longrightarrow$

$dy/dx = 2 \left(- \csc^2 x \right) - \left[\sqrt{x} \left(\sec x \tan x \right) + (\sec x) \cdot \frac{1}{2} x^{-1/2} \right]$

$\qquad = -2 \csc^2 x - \sqrt{x} \sec x \tan x - \frac{1}{2} x^{-1/2} \sec x$

7. $y = \dfrac{\tan x}{x} \Longrightarrow \dfrac{dy}{dx} = \dfrac{x \sec^2 x - \tan x}{x^2}$

9. $y = \dfrac{x}{\sin x + \cos x} \Longrightarrow$

$\dfrac{dy}{dx} = \dfrac{(\sin x + \cos x) - x (\cos x - \sin x)}{(\sin x + \cos x)^2} = \dfrac{(1+x) \sin x + (1-x) \cos x}{\sin^2 x + \cos^2 x + 2 \sin x \cos x}$

$\qquad\qquad = \dfrac{(1+x) \sin x + (1-x) \cos x}{1 + \sin 2x}$

11. $y = e^x (\tan x - x) \Longrightarrow dy/dx = e^x \left(\sec^2 x - 1 \right) + (\tan x - x) e^x = e^x \left(\tan^2 x + \tan x - x \right)$ since $\sec^2 x - 1 = \tan^2 x$.

13. $\dfrac{d}{dx} (\csc x) = \dfrac{d}{dx} \left(\dfrac{1}{\sin x} \right) = \dfrac{(\sin x)(0) - 1 (\cos x)}{\sin^2 x} = \dfrac{- \cos x}{\sin^2 x} = - \dfrac{1}{\sin x} \cdot \dfrac{\cos x}{\sin x} = - \csc x \cot x$

15. $\dfrac{d}{dx} (\cot x) = \dfrac{d}{dx} \left(\dfrac{\cos x}{\sin x} \right) = \dfrac{(\sin x)(- \sin x) - (\cos x)(\cos x)}{\sin^2 x} = - \dfrac{\sin^2 x + \cos^2 x}{\sin^2 x} = - \dfrac{1}{\sin^2 x}$

$\qquad = - \csc^2 x$

17. $y = \tan x \Longrightarrow y' = \sec^2 x \Longrightarrow$ the slope of the tangent line at $\left(\frac{\pi}{4}, 1 \right)$ is $\sec^2 \frac{\pi}{4} = \left(\sqrt{2} \right)^2 = 2$ and an

equation is $y - 1 = 2 \left(x - \frac{\pi}{4} \right)$ or $y = 2x + 1 - \frac{\pi}{2}$.

19. (a) $y = x \cos x \implies y' = x(-\sin x) + \cos x(1) = \cos x - x \sin x$. So the slope of the tangent at the point $(\pi, -\pi)$ is $\cos \pi - \pi \sin \pi = -1 - \pi(0) = -1$, and its equation is $y + \pi = -(x - \pi) \iff y = -x$.

(b)

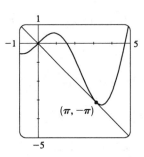

21. (a) $f(x) = 2x + \cot x \implies f'(x) = 2 - \csc^2 x$

(b)

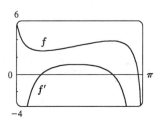

Notice that $f'(x) = 0$ when f has a horizontal tangent. Also $f'(x)$ is large negative when the graph of f is steep.

23. $g(s) = s^2 \cos s \implies g'(s) = s^2(-\sin s) + (\cos s)(2s) = 2s \cos s - s^2 \sin s \implies$
$g''(s) = (2s)(-\sin s) + (\cos s)(2) - \left[s^2(\cos s) + (\sin s)(2s)\right] = (2 - s^2) \cos s - 4s \sin s$

25. $y = x + 2 \sin x$ has a horizontal tangent when $y' = 1 + 2 \cos x = 0 \iff \cos x = -\frac{1}{2} \iff$
$x = \frac{2\pi}{3} + 2\pi n$ or $\frac{4\pi}{3} + 2\pi n$ or, equivalently, $(2n+1)\pi \pm \frac{\pi}{3}$, n an integer.

27. $f(x) = x - 2 \sin x$, $0 \le x \le 2\pi$. $f'(x) = 1 - 2 \cos x$. So $f'(x) > 0 \iff 1 - 2 \cos x > 0 \iff$
$\frac{1}{2} > \cos x \iff \frac{\pi}{3} < x < \frac{5\pi}{3} \implies f$ is increasing on $\left(\frac{\pi}{3}, \frac{5\pi}{3}\right)$.

29. (a) $x(t) = 8 \sin t \implies v(t) = x'(t) = 8 \cos t \implies a(t) = x''(t) = -8 \sin t$

(b) $x\left(\frac{2\pi}{3}\right) = 8\left(\frac{\sqrt{3}}{2}\right) = 4\sqrt{3}$, $v\left(\frac{2\pi}{3}\right) = 8\left(-\frac{1}{2}\right) = -4$, $a\left(\frac{2\pi}{3}\right) = -8\left(\frac{\sqrt{3}}{2}\right) = -4\sqrt{3}$. Since
$v\left(\frac{2\pi}{3}\right) < 0$, the particle is moving to the left. Because v and a have the same sign, the particle is speeding up.

31.

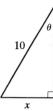

From the diagram we can see that $\sin\theta = x/10 \iff x = 10\sin\theta$. But we want to find the rate of change of x with respect to θ, that is, $dx/d\theta$. Taking the derivative of the above expression,

$dx/d\theta = 10\,(\cos\theta)$. So when $\theta = \frac{\pi}{3}$, $dx/d\theta = 10\cos\frac{\pi}{3} = 10\left(\frac{1}{2}\right) = 5$ ft/rad.

33. $D\sin x = \cos x \Longrightarrow D^2\sin x = -\sin x \Longrightarrow D^3\sin x = -\cos x \Longrightarrow D^4\sin x = \sin x$. The derivatives of $\sin x$ occur in a cycle of four. Since $99 = 4\,(24) + 3$, we have $D^{99}\sin x = D^3\sin x = -\cos x$.

35. $y = A\sin x + B\cos x \Longrightarrow y' = A\cos x - B\sin x \Longrightarrow y'' = -A\sin x - B\cos x$. Substituting into $y'' + y' - 2y = \sin x$ gives us $(-3A - B)\sin x + (A - 3B)\cos x = 1\sin x$, so we must have $-3A - B = 1$ and $A - 3B = 0$. Solving for A and B, we add the first equation to three times the second to get $B = -\frac{1}{10}$ and $A = -\frac{3}{10}$.

37. $\displaystyle\lim_{x\to0}\frac{\tan 4x}{x} = \lim_{x\to0}\left(\frac{\sin 4x}{x}\cdot\frac{1}{\cos 4x}\right) = 4\lim_{x\to0}\frac{\sin 4x}{4x}\cdot\lim_{x\to0}\frac{1}{\cos 4x} = 4\cdot1\cdot1 = 4$

39. $\displaystyle\lim_{\theta\to0}\frac{\sin\theta}{\theta+\tan\theta} = \frac{\displaystyle\lim_{\theta\to0}\frac{\sin\theta}{\theta}}{\displaystyle\lim_{\theta\to0}\frac{\theta+\tan\theta}{\theta}} = \frac{1}{\displaystyle\lim_{\theta\to0}\left(1+\frac{\sin\theta}{\theta}\cdot\frac{1}{\cos\theta}\right)} = \frac{1}{1+1\cdot1} = \frac{1}{2}$

41. By the definition of radian measure, $s = r\theta$, where r is the radius of the circle.

By drawing the bisector of the angle θ, we can see that $\sin\dfrac{\theta}{2} = \dfrac{d/2}{r} \Longrightarrow d = 2r\sin\dfrac{\theta}{2}$.

So $\displaystyle\lim_{\theta\to0^+}\frac{s}{d} = \lim_{\theta\to0^+}\frac{r\theta}{2r\sin(\theta/2)} = \lim_{\theta\to0^+}\frac{2\cdot(\theta/2)}{2\sin(\theta/2)} = \lim_{\theta\to0}\frac{\theta/2}{\sin(\theta/2)} = 1.$ $\Bigg[$This is just the reciprocal

of the limit $\displaystyle\lim_{x\to0}\frac{\sin x}{x} = 1$ combined with the fact that as $\theta\to0$, $\frac{\theta}{2}\to0$ also.$\Bigg]$

Section 3.5 The Chain Rule

1. Let $u = g(x) = x^2 + 4x + 6$ and $y = f(u) = u^5$.

Then $\dfrac{dy}{dx} = \dfrac{dy}{du}\dfrac{du}{dx} = (5u^4)(2x+4) = 5(x^2+4x+6)^4(2x+4) = 10(x^2+4x+6)^4(x+2)$.

3. Let $u = g(x) = \tan x$ and $y = f(u) = \cos u$.

Then $\dfrac{dy}{dx} = \dfrac{dy}{du}\dfrac{du}{dx} = (-\sin u)(\sec^2 x) = -\sin(\tan x)\sec^2 x$.

5. Let $u = g(x) = \sqrt{x}$ and $y = f(u) = e^u$.

Then $\dfrac{dy}{dx} = \dfrac{dy}{du}\dfrac{du}{dx} = \dfrac{1}{2}x^{-1/2}e^u = \dfrac{1}{2}x^{-1/2}e^{\sqrt{x}} = \dfrac{e^{\sqrt{x}}}{2\sqrt{x}}$.

7. $g(x) = \sqrt{x^2 - 7x} = (x^2 - 7x)^{1/2} \Longrightarrow g'(x) = \frac{1}{2}(x^2 - 7x)^{-1/2}(2x-7) = \dfrac{2x-7}{2\sqrt{x^2-7x}}$

9. $y = \cos(x^3) \Longrightarrow y' = -\sin(x^3)(3x^2) = -3x^2\sin(x^3)$

11. Using Formula 5 and the Chain Rule, $y = 5^{-1/x} \Longrightarrow$

$y' = 5^{-1/x}(\ln 5)\left[-1 \cdot (-x^{-2})\right] = 5^{-1/x}(\ln 5)/x^2$

13. $f(x) = xe^{-x^2} \Longrightarrow f'(x) = e^{-x^2} + xe^{-x^2}(-2x) = e^{-x^2}(1 - 2x^2)$

15. $G(x) = (3x-2)^{10}(5x^2 - x + 1)^{12} \Longrightarrow$

$G'(x) = (3x-2)^{10}(12)(5x^2 - x + 1)^{11}(10x-1) + 10(3x-2)^9(3)(5x^2 - x + 1)^{12}$

$\qquad = 6(3x-2)^9(5x^2 - x + 1)^{11}\left[2(3x-2)(10x-1) + 5(5x^2 - x + 1)\right]$

$\qquad = 6(3x-2)^9(5x^2 - x + 1)^{11}(85x^2 - 51x + 9)$

17. $y = e^{x\cos x} \Longrightarrow y' = e^{x\cos x}(\cos x - x\sin x)$

19. $F(y) = \left(\dfrac{y-6}{y+7}\right)^3 \Longrightarrow$

$F'(y) = 3\left(\dfrac{y-6}{y+7}\right)^2\dfrac{(y+7)(1)-(y-6)(1)}{(y+7)^2} = 3\left(\dfrac{y-6}{y+7}\right)^2\dfrac{13}{(y+7)^2} = \dfrac{39(y-6)^2}{(y+7)^4}$

21. $f(z) = (2z-1)^{-1/5} \Longrightarrow f'(z) = -\frac{1}{5}(2z-1)^{-6/5}(2) = -\frac{2}{5}(2z-1)^{-6/5}$

23. $y = x\sin\dfrac{1}{x} \Longrightarrow y' = \sin\dfrac{1}{x} + x\cos\dfrac{1}{x}\left(-\dfrac{1}{x^2}\right) = \sin\dfrac{1}{x} - \dfrac{1}{x}\cos\dfrac{1}{x}$

25. $y = \tan^2(x^3) \Longrightarrow y' = 2\tan(x^3)\sec^2(x^3)(3x^2) = 6x^2\tan(x^3)\sec^2(x^3)$

Section 3.5   The Chain Rule

27. $y = \sqrt{x + \sqrt{x}} \implies y' = \frac{1}{2}(x + \sqrt{x})^{-1/2}(1 + \frac{1}{2}x^{-1/2}) = \dfrac{1}{2\sqrt{x + \sqrt{x}}}\left(1 + \dfrac{1}{2\sqrt{x}}\right)$

29. $y = \sin\left(\tan\sqrt{\sin x}\right) \implies y' = \cos\left(\tan\sqrt{\sin x}\right)\left(\sec^2\sqrt{\sin x}\right)\left(\dfrac{1}{2\sqrt{\sin x}}\right)(\cos x)$

31. $y = f(x) = \dfrac{8}{\sqrt{4 + 3x}} = 8(4 + 3x)^{-1/2} \implies f'(x) = 8\left(-\frac{1}{2}\right)(4 + 3x)^{-3/2}(3) = -12(4 + 3x)^{-3/2}.$

The slope of the tangent at $(4, 2)$ is $f'(4) = -\frac{12}{64} = -\frac{3}{16}$ and its equation is $y - 2 = -\frac{3}{16}(x - 4)$ or

$y = -\frac{3}{16}x + \frac{11}{4}.$

33. (a) $y = \dfrac{2}{1 + e^{-x}} \implies y' = \dfrac{(1 + e^{-x})(0) - 2(-e^{-x})}{(1 + e^{-x})^2} = \dfrac{2e^{-x}}{(1 + e^{-x})^2}.$ At $(0, 1)$, $y' = \dfrac{2}{2^2} = \dfrac{1}{2}.$

So an equation of the tangent line is $y - 1 = \frac{1}{2}(x - 0)$ or $y = \frac{1}{2}x + 1.$

(b)

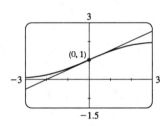

35. (a) $f(x) = \dfrac{\sqrt{1 - x^2}}{x} \implies$

$$f'(x) = \dfrac{x \cdot \frac{1}{2}(1 - x^2)^{-1/2}(-2x) - \sqrt{1 - x^2}}{x^2} \cdot \dfrac{\sqrt{1 - x^2}}{\sqrt{1 - x^2}}$$

$$= \dfrac{-x^2 - (1 - x^2)}{x^2\sqrt{1 - x^2}} = \dfrac{-1}{x^2\sqrt{1 - x^2}}$$

(b)

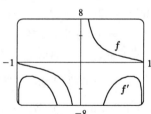

Notice that all tangents to the graph of f have negative slopes and $f'(x) < 0$ always.

37. $F(x) = f(g(x)) \implies$

$F'(x) = f'(g(x))g'(x)$, so $F'(3) = f'(g(3))g'(3) = f'(6)g'(3) = 7 \cdot 4 = 28.$

39. (a) $u(x) = f(g(x)) \implies u'(x) = f'(g(x))g'(x)$.

So $u'(1) = f'(g(1))g'(1) = f'(3)g'(1) = \left(-\frac{1}{4}\right)(-3) = \frac{3}{4}$.

(b) $v(x) = g(f(x)) \implies v'(x) = g'(f(x))f'(x)$. So $v'(1) = g'(f(1))f'(1) = g'(2)f'(1)$,
which does not exist since $g'(2)$ does not exist.

(c) $w(x) = g(g(x)) \implies w'(x) = g'(g(x))g'(x)$.

So $w'(1) = g'(g(1))g'(1) = g'(3)g'(1) = \left(\frac{2}{3}\right)(-3) = -2$.

41. $h(x) = f(g(x)) \implies h'(x) = f'(g(x))g'(x)$. So $h'(0.5) = f'(g(0.5))g'(0.5) = f'(0.1)g'(0.5)$.
We can estimate the derivatives by taking the average of 2 secant slopes.

For $f'(0.1)$: $m_1 = \dfrac{14.8 - 12.6}{0.1 - 0} = 22$, $m_2 = \dfrac{18.4 - 14.8}{0.2 - 0.1} = 36$. So $f'(0.1) \approx \dfrac{m_1 + m_2}{2} = 29$.

For $g'(0.5)$: $m_1 = \dfrac{0.10 - 0.17}{0.5 - 0.4} = -0.7$, $m_2 = \dfrac{0.05 - 0.10}{0.6 - 0.5} = -0.5$.

So $g'(0.5) \approx (m_1 + m_2)/2 = -0.6$. Hence, $h'(0.5) \approx (29)(-0.6) = -17.4$.

43. (a) Since h is differentiable on $[0, \infty)$ and $\sqrt{x}$ is differentiable on $(0, \infty)$, it follows that
$G(x) = h(\sqrt{x})$ is differentiable on $(0, \infty)$.

(b) By the Chain Rule, $G'(x) = h'(\sqrt{x})\dfrac{d}{dx}\sqrt{x} = \dfrac{h'(\sqrt{x})}{2\sqrt{x}}$.

45. (a) $F(x) = f(e^x) \implies F'(x) = f'(e^x)\dfrac{d}{dx}(e^x) = f'(e^x)e^x$

(b) $G(x) = e^{f(x)} \implies G'(x) = e^{f(x)}\dfrac{d}{dx}f(x) = e^{f(x)}f'(x)$

47. $f(x) = \sin 2x - 2\sin x \implies$
$f'(x) = 2\cos 2x - 2\cos x = 2(2\cos^2 x - 1) - 2\cos x = 4\cos^2 x - 2\cos x - 2$
$\quad = (\cos x - 1)(4\cos x + 2)$.
$f'(x) = 0 \iff \cos x = 1$ or $\cos x = -\frac{1}{2}$. So $x = 2n\pi$ or $(2n+1)\pi \pm \frac{\pi}{3}$, n any integer.

49. $y = Ae^{-x} + Bxe^{-x} \implies y' = -Ae^{-x} + Be^{-x} - Bxe^{-x} = (B - A)e^{-x} - Bxe^{-x} \implies$
$y'' = (A - B)e^{-x} - Be^{-x} + Bxe^{-x} = (A - 2B)e^{-x} + Bxe^{-x}$, so
$y'' + 2y' + y = (A - 2B)e^{-x} + Bxe^{-x} + 2[(B - A)e^{-x} - Bxe^{-x}] + Ae^{-x} + Bxe^{-x} = 0$.

51. In general, $Df(2x) = 2f'(2x)$, $D^2 f(2x) = 4f''(2x)$, $\cdots$, $D^n f(2x) = 2^n f^{(n)}(2x)$.

Since $f(x) = \cos x$ and $50 = 4(12) + 2$, we have $f^{(50)}(x) = f^{(2)}(x) = -\cos x$, so
$D^{50}\cos 2x = -2^{50}\cos 2x$.

53. $s(t) = 10 + \frac{1}{4}\sin(10\pi t) \implies$ the velocity after t seconds is

$v(t) = s'(t) = \frac{1}{4}\cos(10\pi t)(10\pi) = \frac{5\pi}{2}\cos(10\pi t)$ cm/s. The acceleration after t seconds is

$a(t) = v'(t) = \frac{5\pi}{2}[-\sin(10\pi t)](10\pi) = -25\pi^2\sin(10\pi t)$ cm/s^2.

55. (a) $B(t) = 4.0 + 0.35 \sin \dfrac{2\pi t}{5.4} \implies \dfrac{dB}{dt} = \left(0.35 \cos \dfrac{2\pi t}{5.4}\right)\left(\dfrac{2\pi}{5.4}\right) = \dfrac{7\pi}{54}\cos \dfrac{2\pi t}{5.4}$

(b) At $t = 1$, $\dfrac{dB}{dt} = \dfrac{7\pi}{54}\cos \dfrac{2\pi}{5.4} \approx 0.16$.

57. $a(t) = \dfrac{dv}{dt} = \dfrac{dv}{ds}\dfrac{ds}{dt} = v(t)\dfrac{dv}{ds}$. $\dfrac{dv}{dt}$ is the rate of change of the velocity with respect to time (in other words, the acceleration) whereas dv/ds is the rate of change of the velocity with respect to the displacement.

59. (a) Using a calculator or CAS, we obtain the model $Q = ab^t$ with $a = 100.0124369$ and $b = 0.000045145933$. We can change this model to one with base e and exponent $\ln b$:

$Q = ae^{t \ln b} = 100.0124369 e^{-10.00553063t}$.

(b) $Q'(t) = ab^t \ln b$. $Q'(0.04) \approx -670.63$ μA. The result of Example 2 in Section 2.1 was -670 μA.

61. $x = t \sin t$, $y = t \cos t$. To find the value of t that corresponds to the point $(0, -\pi)$, we can solve $0 = t \sin t$, which says that either $t = 0$ or $\sin t = 0$. If $\sin t = 0$, then $t = 0, \pm\pi, \ldots$. Since $y = -\pi = t \cos t$, we see that $t = \pi$ is the desired value. $\dfrac{dy}{dt} = \cos t - t \sin t$, $\dfrac{dx}{dt} = \sin t + t \cos t$, and

$\dfrac{dy}{dx} = \dfrac{dy/dt}{dx/dt} = \dfrac{\cos t - t \sin t}{\sin t + t \cos t}$. When $t = \pi$, $(x, y) = (0, -\pi)$ and $\dfrac{dy}{dx} = \dfrac{-1}{-\pi} = \dfrac{1}{\pi}$, so an equation of

the tangent is $y + \pi = \dfrac{1}{\pi}(x - 0)$ or $y = \dfrac{1}{\pi}x - \pi$.

63. (a) $x = t^2$, $y = t^3 - 3t \implies \dfrac{dy}{dx} = \dfrac{3t^2 - 3}{2t}$. $x = 3 \implies t^2 = 3 \implies t = \pm\sqrt{3} \implies \dfrac{dy}{dx} = \mp\sqrt{3}$. When $t = \sqrt{3}$, the equation is $y - 0 = \sqrt{3}(x - 3)$ or $y = \sqrt{3}x - 3\sqrt{3}$. When $t = -\sqrt{3}$, the equation is $y - 0 = -\sqrt{3}(x - 3)$ or $y = -\sqrt{3}x + 3\sqrt{3}$.

(b) Horizontal tangent: $dy/dx = 0 \iff 3t^2 - 3 = 0 \iff t = \pm 1$. $t = 1$ corresponds to the point $(1, -2)$ and $t = -1$ to $(1, 2)$.

Vertical tangent: dy/dx is undefined $\iff 2t = 0 \iff t = 0$. $t = 0$ corresponds to the origin.

(c)

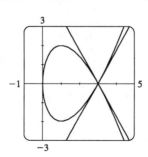

65. (a) Derive gives $g'(t) = \dfrac{45(t-2)^8}{(2t+1)^{10}}$ without simplifying. With both Maple and Mathematica, we first

get $g'(t) = 9\dfrac{(t-2)^8}{(2t+1)^9} - 18\dfrac{(t-2)^9}{(2t+1)^{10}}$, and the simplification command results in the above

expression.

(b) Derive gives $y' = 2\left(x^3 - x + 1\right)^3 (2x+1)^4 \left(17x^3 + 6x^2 - 9x + 3\right)$ without simplifying.

With both Maple and Mathematica, we first get

$$y' = 10(2x+1)^4 \left(x^3 - x + 1\right)^4 + 4(2x+1)^5 \left(x^3 - x + 1\right)^3 \left(3x^2 - 1\right).$$ If we use Mathematica's

Factor or Simplify, or Maple's factor, we get the above expression, but Maple's simplify

gives the polynomial expansion instead. For locating horizontal tangents, the factored form is the

most helpful.

67. (a) $\dfrac{d}{dx}\left(\sin^n x \cos nx\right) = n\sin^{n-1} x \cos x \cos nx + \sin^n x \left(-n\sin nx\right)$

$$= n\sin^{n-1} x \left(\cos nx \cos x - \sin nx \sin x\right) = n\sin^{n-1} x \cos(nx + x)$$

$$= n\sin^{n-1} x \cos\left[(n+1)x\right]$$

(b) $\dfrac{d}{dx}\left(\cos^n x \cos nx\right) = n\cos^{n-1} x \left(-\sin x\right) \cos nx + \cos^n x \left(-n\sin nx\right)$

$$= -n\cos^{n-1} x \left(\cos nx \sin x + \sin nx \cos x\right) = -n\cos^{n-1} x \sin(nx + x)$$

$$= -n\cos^{n-1} x \sin\left[(n+1)x\right]$$

69. Since $\theta^\circ = \left(\frac{\pi}{180}\right)\theta$ rad, we have $\dfrac{d}{d\theta}\left(\sin\theta^\circ\right) = \dfrac{d}{d\theta}\left(\sin\frac{\pi}{180}\theta\right) = \frac{\pi}{180}\cos\frac{\pi}{180}\theta = \frac{\pi}{180}\cos\theta^\circ$.

71. The Chain Rule says that $\dfrac{dy}{dx} = \dfrac{dy}{du}\dfrac{du}{dx}$, so

$$\dfrac{d^2y}{dx^2} = \dfrac{d}{dx}\left(\dfrac{dy}{dx}\right) = \dfrac{d}{dx}\left(\dfrac{dy}{du}\dfrac{du}{dx}\right) = \left[\dfrac{d}{dx}\left(\dfrac{dy}{du}\right)\right]\dfrac{du}{dx} + \dfrac{dy}{du}\dfrac{d}{dx}\left(\dfrac{du}{dx}\right) \quad \text{(Product Rule)}$$

$$= \left[\dfrac{d}{du}\left(\dfrac{dy}{du}\right)\dfrac{du}{dx}\right]\dfrac{du}{dx} + \dfrac{dy}{du}\dfrac{d^2u}{dx^2} = \dfrac{d^2y}{du^2}\left(\dfrac{du}{dx}\right)^2 + \dfrac{dy}{du}\dfrac{d^2u}{dx^2}.$$

Section 3.6 Implicit Differentiation

1. (a) $x^2 + 3x + xy = 5 \Longrightarrow \dfrac{d}{dx}\left(x^2 + 3x + xy\right) = \dfrac{d}{dx}(5) \;\Longrightarrow\; \dfrac{d}{dx}\left(x^2\right) + \dfrac{d}{dx}(3x) + \dfrac{d}{dx}(xy) = 0 \Longrightarrow$

$2x + 3 + y + xy' = 0 \Longrightarrow y' = -\dfrac{2x + y + 3}{x}$

(b) $x^2 + 3x + xy = 5 \Longrightarrow y = \dfrac{5 - x^2 - 3x}{x} = \dfrac{5}{x} - x - 3 \Longrightarrow y' = -\dfrac{5}{x^2} - 1$

(c) $y' = -\dfrac{2x + y + 3}{x} = \dfrac{-2x - 3 - (-3 - x + 5/x)}{x} = \dfrac{-x - 5/x}{x} = -1 - \dfrac{5}{x^2}$

3. $x^2 - xy + y^3 = 8 \Longrightarrow \dfrac{d}{dx}\left(x^2 - xy + y^3\right) = \dfrac{d}{dx}(8) \Longrightarrow 2x - y - xy' + 3y^2 y' = 0 \Longrightarrow$

$3y^2 y' - xy' = y - 2x \Longrightarrow \left(3y^2 - x\right) y' = y - 2x \Longrightarrow y' = \dfrac{y - 2x}{3y^2 - x}$

5. $x^4 + y^4 = 16 \Longrightarrow 4x^3 + 4y^3 y' = 0 \Longrightarrow y' = -x^3 / y^3$

7. $\cos(x - y) = xe^x \Longrightarrow -\sin(x - y)(1 - y') = e^x + xe^x \Longrightarrow$

$-\sin(x - y) + \sin(x - y)\, y' = e^x (1 + x) \Longrightarrow y' = 1 + \dfrac{e^x (1 + x)}{\sin(x - y)}$

9. $xy = \cot(xy) \Longrightarrow y + xy' = -\csc^2(xy)(y + xy') \Longrightarrow (y + xy')\left[1 + \csc^2(xy)\right] = 0 \Longrightarrow$

$y + xy' = 0 \text{ [since } 1 + \csc^2(xy) > 0] \Longrightarrow y' = -y/x$

11. $\dfrac{x^2}{16} - \dfrac{y^2}{9} = 1 \Longrightarrow \dfrac{x}{8} - \dfrac{2yy'}{9} = 0 \Longrightarrow y' = \dfrac{9x}{16y}$. When $x = -5$ and $y = \frac{9}{4}$ we have

$y' = \dfrac{9(-5)}{16(9/4)} = -\dfrac{5}{4}$ so an equation of the tangent is $y - \frac{9}{4} = -\frac{5}{4}(x + 5)$ or $y = \frac{5}{4}x - 4$.

13. $y^2 = x^3 (2 - x) = 2x^3 - x^4 \Longrightarrow 2yy' = 6x^2 - 4x^3 \Longrightarrow y' = \dfrac{3x^2 - 2x^3}{y}$. When $x = y = 1$,

$y' = \dfrac{3(1)^2 - 2(1)^3}{1} = 1$, so an equation of the tangent line is $y - 1 = 1(x - 1)$ or $y = x$.

15. $2\left(x^2 + y^2\right)^2 = 25\left(x^2 - y^2\right) \Longrightarrow 4\left(x^2 + y^2\right)(2x + 2yy') = 25(2x - 2yy') \Longrightarrow$

$4yy'\left(x^2 + y^2\right) + 25yy' = 25x - 4x\left(x^2 + y^2\right) \Longrightarrow y' = \dfrac{25x - 4x\left(x^2 + y^2\right)}{25y + 4y\left(x^2 + y^2\right)}$. When $x = 3$ and

$y = 1$, $y' = \dfrac{75 - 120}{25 + 40} = -\dfrac{9}{13}$ so an equation of the tangent is $y - 1 = -\dfrac{9}{13}(x - 3)$ or $y = -\frac{9}{13}x + \frac{40}{13}$.

17. (a) $y^2 = 5x^4 - x^2 \Longrightarrow 2yy' = 5\left(4x^3\right) - 2x \Longrightarrow$

$y' = \dfrac{10x^3 - x}{y}$. So at the point $(1, 2)$ we have

$y' = \dfrac{10(1)^3 - 1}{2} = \dfrac{9}{2}$, and an equation of the tangent line

is $y - 2 = \frac{9}{2}(x - 1)$ or $y = \frac{9}{2}x - \frac{5}{2}$.

(b)

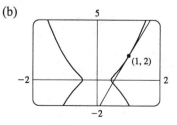

19. (a)

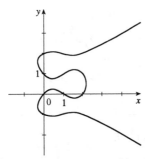

(b) $y' = \dfrac{3x^2 - 6x + 2}{2\left(2y^3 - 3y^2 - y + 1\right)} \implies y' = -1$

at $(0, 1)$ and $y' = \frac{1}{3}$ at $(0, 2)$.

Equations of the tangent lines are

$y = -x + 1$ and $y = \frac{1}{3}x + 2$.

(c) $y' = 0 \implies 3x^2 - 6x + 2 = 0 \implies$

$x = 1 \pm \frac{1}{3}\sqrt{3}$

There are 8 points with horizontal tangents:

4 at $x \approx 1.57735$ and 4 at $x \approx 0.42265$.

(d) By multiplying the right side of the equation by $x - 3$, we

obtain the graph at right.

By modifying the equation in other ways, we can generate the

graphs shown below.

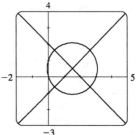

$y\left(y^2 - 1\right)(y - 2) =$
$x\left(x - 1\right)(x - 2)(x - 3)$

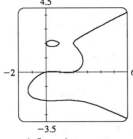

$y\left(y^2 - 4\right)(y - 2) =$
$x\left(x - 1\right)(x - 2)$

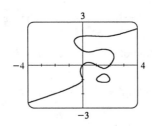

$y\left(y + 1\right)\left(y^2 - 1\right)(y - 2) =$
$x\left(x - 1\right)(x - 2)$

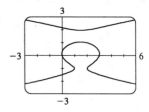

$\left(y + 1\right)\left(y^2 - 1\right)(y - 2) =$
$\left(x - 1\right)(x - 2)$

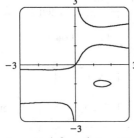

$x\left(y + 1\right)\left(y^2 - 1\right)(y - 2) =$
$y\left(x - 1\right)(x - 2)$

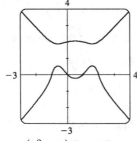

$y\left(y^2 + 1\right)(y - 2) =$
$x\left(x^2 - 1\right)(x - 2)$

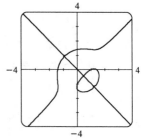

$y\left(y + 1\right)\left(y^2 - 2\right) =$
$x\left(x - 1\right)\left(x^2 - 2\right)$

21. From Exercise 15, a tangent to the lemniscate will be horizontal $\Longrightarrow y' = 0 \Longrightarrow$

$25x - 4x\left(x^2 + y^2\right) = 0 \Longrightarrow x^2 + y^2 = \frac{25}{4}$. (Note that $x = 0 \Longrightarrow y = 0$ and there is no horizontal

tangent at the origin.) Putting this in the equation of the lemniscate, we get $x^2 - y^2 = \frac{25}{8}$. Solving

these two equations we have $x^2 = \frac{75}{16}$ and $y^2 = \frac{25}{16}$, so the four points are $\left(\pm\frac{5\sqrt{3}}{4}, \pm\frac{5}{4}\right)$.

23. (a) $x^4 + y^4 = 16 \Longrightarrow 4x^3 + 4y^3 y' = 0 \Longrightarrow y' = -x^3/y^3$

(b) $y'' = -\dfrac{y^3\left(3x^2\right) - \left(x^3\right)\left(3y^2 y'\right)}{\left(y^3\right)^2} = -\dfrac{3x^2 y^3 - 3x^3 y^2\left(-x^3/y^3\right)}{y^6} = -\dfrac{3x^2 y^4 + 3x^6}{y^7}$

(c) $y'' = -\dfrac{3x^2\left(y^4 + x^4\right)}{y^7} = -\dfrac{3x^2\,(16)}{y^7} = -48\dfrac{x^2}{y^7}$

25. $y = \sin^{-1}\left(x^2\right) \Longrightarrow y' = \dfrac{1}{\sqrt{1 - \left(x^2\right)^2}}\dfrac{d}{dx}\left(x^2\right) = \dfrac{2x}{\sqrt{1 - x^4}}$

27. $y = \tan^{-1}\left(e^x\right) \Longrightarrow y' = \dfrac{1}{1 + \left(e^x\right)^2}\dfrac{d}{dx}\left(e^x\right) = \dfrac{e^x}{1 + e^{2x}}$

29. $H\left(x\right) = \left(1 + x^2\right)\arctan x \Longrightarrow H'\left(x\right) = (2x)\arctan x + \left(1 + x^2\right)\dfrac{1}{1 + x^2} = 1 + 2x\arctan x$

31. $f\left(x\right) = e^x - x^2\arctan x \Longrightarrow$

$f'\left(x\right) = e^x - \left[x^2\left(\dfrac{1}{1 + x^2}\right) + 2x\arctan x\right]$

$= e^x - \dfrac{x^2}{1 + x^2} - 2x\arctan x.$

This is reasonable because the graphs show that f is increasing
when $f'\left(x\right)$ is positive.

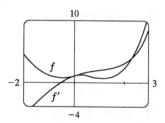

33. $2x^2 + y^2 = 3$ and $x = y^2$ intersect when $2x^2 + x - 3 = (2x + 3)(x - 1) = 0 \Longleftrightarrow x = -\frac{3}{2}$ or 1, but

$-\frac{3}{2}$ is extraneous. $2x^2 + y^2 = 3 \Longrightarrow 4x + 2yy' = 0 \Longrightarrow y' = -2x/y$ and $x = y^2 \Longrightarrow 1 = 2yy' \Longrightarrow$

$y' = 1/(2y)$. At $(1, 1)$ the slopes are $m_1 = -2$ and $m_2 = \frac{1}{2}$, so the curves are orthogonal there. By
symmetry they are also orthogonal at $(1, -1)$.

35.

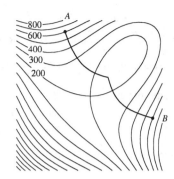

36. The orthogonal family represents the direction of the wind.

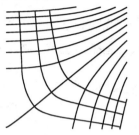

37. $x^2 + y^2 = r^2$ is a circle with center O and $ax + by = 0$ is a line through O.
$x^2 + y^2 = r^2 \Longrightarrow 2x + 2yy' = 0 \Longrightarrow y' = -x/y$, so the slope of the tangent
line at $P_0\,(x_0, y_0)$ is $-x_0/y_0$. The slope of the line OP is y_0/x_0, which is
the negative reciprocal of $-x_0/y_0$. Hence, the curves are orthogonal.

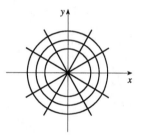

39. $y = cx^2 \Longrightarrow y' = 2cx$ and $x^2 + 2y^2 = k \Longrightarrow 2x + 4yy' = 0 \Longrightarrow$
$y' = -\dfrac{x}{2y} = -\dfrac{x}{2cx^2} = -\dfrac{1}{2cx}$, so the curves are orthogonal.

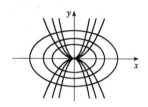

41. If the circle has radius r, its equation is $x^2 + y^2 = r^2 \Longrightarrow 2x + 2yy' = 0 \Longrightarrow y' = -x/y$, so the slope
of the tangent line at $P\,(x_0, y_0)$ is $-\dfrac{x_0}{y_0}$. The slope of line OP is $\dfrac{y_0}{x_0} = \dfrac{-1}{-x_0/y_0}$, so the tangent line at
P is perpendicular to OP.

43. $y = 0 \Longrightarrow x^2 - x\,(0) + 0^2 = 3 \Longleftrightarrow x = \pm\sqrt{3}$. So the graph of the ellipse crosses the x-axis at
the points $(\pm\sqrt{3}, 0)$. Using implicit differentiation to find y', we get $2x - xy' - y + 2yy' = 0 \Longrightarrow$

$y'\,(2y - x) = y - 2x \Longleftrightarrow y' = \dfrac{y - 2x}{2y - x}$. So $y'\left(\sqrt{3}, 0\right) = \dfrac{0 - 2\sqrt{3}}{2\,(0) - \sqrt{3}} = 2$, and

$y'\left(-\sqrt{3}, 0\right) = \dfrac{0 + 2\sqrt{3}}{2\,(0) + \sqrt{3}} = 2 = y'\left(\sqrt{3}, 0\right)$. So the tangent lines at these points are parallel.

45. $x^2y^2 + xy = 2 \Longrightarrow 2xy^2 + 2x^2yy' + y + xy' = 0 \Longleftrightarrow y'\left(2x^2y + x\right) = -2xy^2 - y \Longleftrightarrow$

$y' = -\dfrac{2xy^2 + y}{2x^2y + x}.$ So $-\dfrac{2xy^2 + y}{2x^2y + x} = -1 \Longleftrightarrow 2xy^2 + y = 2x^2y + x \Longleftrightarrow$

$y\left(2xy + 1\right) = x\left(2xy + 1\right) \Longleftrightarrow \left(2xy + 1\right)\left(y - x\right) = 0 \Longleftrightarrow y = x$ or $xy = -\frac{1}{2}$. But $xy = -\frac{1}{2} \Longrightarrow$

$x^2y^2 + xy = \frac{1}{4} - \frac{1}{2} \neq 2$ so we must have $x = y$. Then $x^2y^2 + xy = 2 \Longrightarrow x^4 + x^2 = 2 \Longleftrightarrow$

$x^4 + x^2 - 2 = 0 \Longleftrightarrow \left(x^2 + 2\right)\left(x^2 - 1\right) = 0$. So $x^2 = -2$, which is impossible, or $x^2 = 1 \Longleftrightarrow$

$x = \pm 1$. So the points on the curve where the tangent line has a slope of -1 are $(-1, -1)$ and $(1, 1)$.

47. (a) If $y = f^{-1}(x)$, then $f(y) = x$. Differentiating implicitly with respect to x and remembering that y

is a function of x, we get $f'(y)\dfrac{dy}{dx} = 1$, so $\dfrac{dy}{dx} = \dfrac{1}{f'(y)} = \dfrac{1}{f'(f^{-1}(x))}$.

(b) $f(4) = 5 \Longrightarrow f^{-1}(5) = 4$. $\left(f^{-1}\right)'(5) = 1/f'\left(f^{-1}(5)\right) = 1/f'(4) = 1/\left(\frac{2}{3}\right) = \frac{3}{2}$.

49. (a) $y = J(x)$ and $xy'' + y' + xy = 0 \Longrightarrow xJ''(x) + J'(x) + xJ(x) = 0$. If $x = 0$ we have

$0 + J'(0) + 0 = 0$, so $J'(0) = 0$.

(b) Differentiating $xy'' + y' + xy = 0$ implicitly, we get $xy''' + y'' + y'' + xy' + y = 0$, so

$xJ'''(x) + 2J''(x) + xJ'(x) + J(x) = 0 \Longrightarrow 0 + 2J''(0) + 0 + 1 = 0 \Longrightarrow 2J''(0) = -1 \Longrightarrow$

$J''(0) = -\frac{1}{2}$.

Section 3.7   Derivatives of Logarithmic Functions

1. The differentiation formula for logarithmic functions, $\dfrac{d}{dx}\left(\log_a x\right) = \dfrac{1}{x \ln a}$, is simplest when $a = e$ because $\ln e = 1$.

3. $f\left(\theta\right) = \ln\left(\cos\theta\right) \Longrightarrow f'\left(\theta\right) = \dfrac{1}{\cos\theta} \dfrac{d}{d\theta}\left(\cos\theta\right) = \dfrac{-\sin\theta}{\cos\theta} = -\tan\theta$

5. $f\left(x\right) = \log_3\left(x^2 - 4\right) \Longrightarrow f'\left(x\right) = \dfrac{1}{\left(x^2 - 4\right)\ln 3}\left(2x\right) = \dfrac{2x}{\left(x^2 - 4\right)\ln 3}$

7. $g\left(x\right) = \ln\dfrac{a - x}{a + x} = \ln\left(a - x\right) - \ln\left(a + x\right) \Longrightarrow$

$g'\left(x\right) = \dfrac{1}{a - x}\left(-1\right) - \dfrac{1}{a + x} = \dfrac{-\left(a + x\right) - \left(a - x\right)}{\left(a - x\right)\left(a + x\right)} = \dfrac{-2a}{a^2 - x^2}$

9. $F\left(x\right) = \ln\sqrt{x} = \ln x^{1/2} = \frac{1}{2}\ln x \Longrightarrow F'\left(x\right) = \dfrac{1}{2}\left(\dfrac{1}{x}\right) = \dfrac{1}{2x}$

11. $f\left(x\right) = \sqrt{x}\ln x \Longrightarrow f'\left(x\right) = \dfrac{1}{2\sqrt{x}}\ln x + \sqrt{x}\left(\dfrac{1}{x}\right) = \dfrac{\ln x}{2\sqrt{x}} + \dfrac{1}{\sqrt{x}} = \dfrac{\ln x + 2}{2\sqrt{x}}$

13. $y = \ln\left|x^3 - x^2\right| \Longrightarrow y' = \dfrac{1}{x^3 - x^2}\left(3x^2 - 2x\right) = \dfrac{x\left(3x - 2\right)}{x^2\left(x - 1\right)} = \dfrac{3x - 2}{x\left(x - 1\right)}$

15. $y = \ln\left[e^{-x}\left(1 + x\right)\right] = \ln\left(e^{-x}\right) + \ln\left(1 + x\right) = -x + \ln\left(1 + x\right) \Longrightarrow y' = -1 + \dfrac{1}{1 + x} = -\dfrac{x}{1 + x}$

17. $y = \log_{10} x \Longrightarrow y' = \dfrac{1}{x\ln 10} = \dfrac{1}{\ln 10}\left(\dfrac{1}{x}\right) \Longrightarrow y'' = \dfrac{1}{\ln 10}\left(-\dfrac{1}{x^2}\right) = -\dfrac{1}{x^2\ln 10}$

19. $f\left(x\right) = x^2\ln\left(1 - x^2\right) \Longrightarrow f'\left(x\right) = 2x\ln\left(1 - x^2\right) + \dfrac{x^2\left(-2x\right)}{1 - x^2} = 2x\ln\left(1 - x^2\right) - \dfrac{2x^3}{1 - x^2}$.

$\text{Dom}(f) = \left\{x \mid 1 - x^2 > 0\right\} = \left\{x \mid |x| < 1\right\} = \left(-1, 1\right)$.

21. $f\left(x\right) = \dfrac{x}{\ln x} \Longrightarrow f'\left(x\right) = \dfrac{\ln x - x\left(1/x\right)}{\left(\ln x\right)^2} = \dfrac{\ln x - 1}{\left(\ln x\right)^2} \Longrightarrow f'\left(e\right) = \dfrac{1 - 1}{1^2} = 0$

23. (a) The domain of f is $\left(0, \infty\right)$. $f\left(x\right) = x\ln x \Longrightarrow f'\left(x\right) = x\left(1/x\right) + \ln x = 1 + \ln x$. So $f'\left(x\right) < 0$ when $1 + \ln x < 0 \Longleftrightarrow \ln x < -1 \Longleftrightarrow x < e^{-1}$. Therefore, f is decreasing on $\left(0, 1/e\right)$.

(b) $y' = 1 + \ln x \Longrightarrow y'' = 1/x > 0$ for $x > 0$. So the curve is concave upward on $\left(0, \infty\right)$.

25. $y = \left(3x - 7\right)^4\left(8x^2 - 1\right)^3 \Longrightarrow \ln y = 4\ln\left(3x - 7\right) + 3\ln\left(8x^2 - 1\right) \Longrightarrow$

$\dfrac{y'}{y} = \dfrac{12}{3x - 7} + \dfrac{48x}{8x^2 - 1} \Longrightarrow y' = \left(3x - 7\right)^4\left(8x^2 - 1\right)^3\left(\dfrac{12}{3x - 7} + \dfrac{48x}{8x^2 - 1}\right)$

27. $y = \dfrac{\left(x + 1\right)^4\left(x - 5\right)^3}{\left(x - 3\right)^8} \Longrightarrow \ln y = 4\ln\left(x + 1\right) + 3\ln\left(x - 5\right) - 8\ln\left(x - 3\right) \Longrightarrow$

$\dfrac{y'}{y} = \dfrac{4}{x + 1} + \dfrac{3}{x - 5} - \dfrac{8}{x - 3} \Longrightarrow y' = \dfrac{\left(x + 1\right)^4\left(x - 5\right)^3}{\left(x - 3\right)^8}\left(\dfrac{4}{x + 1} + \dfrac{3}{x - 5} - \dfrac{8}{x - 3}\right)$

29. $y = x^x \implies \ln y = x \ln x \implies \dfrac{y'}{y} = \ln x + x \left(\dfrac{1}{x} \right) \implies y' = x^x (\ln x + 1)$

31. $y = x^{\sin x} \implies \ln y = \sin x \ln x \implies \dfrac{y'}{y} = \cos x \ln x + \dfrac{\sin x}{x} \implies y' = x^{\sin x} \left(\cos x \ln x + \dfrac{\sin x}{x} \right)$

33. $y = (\ln x)^x \implies \ln y = x \ln \ln x \implies \dfrac{y'}{y} = \ln \ln x + x \cdot \dfrac{1}{\ln x} \cdot \dfrac{1}{x} \implies y' = (\ln x)^x \left(\ln \ln x + \dfrac{1}{\ln x} \right)$

35. $y = \ln \left(x^2 + y^2 \right) \implies y' = \dfrac{2x + 2yy'}{x^2 + y^2} \implies x^2 y' + y^2 y' = 2x + 2yy' \implies y' = \dfrac{2x}{x^2 + y^2 - 2y}$

37. $f(x) = \ln(x - 1) \implies f'(x) = \dfrac{1}{x - 1} = (x - 1)^{-1} \implies f''(x) = -(x - 1)^{-2} \implies$

$f'''(x) = 2(x - 1)^{-3} \implies f^{(4)}(x) = -2 \cdot 3 (x - 1)^{-4} \implies \cdots \implies$

$f^{(n)}(x) = (-1)^{n-1} \cdot 2 \cdot 3 \cdot 4 \cdots \cdot (n - 1)(x - 1)^{-n} = (-1)^{n-1} \dfrac{(n - 1)!}{(x - 1)^n}$

39. If $f(x) = \ln(1 + x)$, then $f'(x) = \dfrac{1}{1 + x}$, so $f'(0) = 1$.

Thus, $\displaystyle\lim_{x \to 0} \dfrac{\ln(1 + x)}{x} = \lim_{x \to 0} \dfrac{f(x)}{x} = \lim_{x \to 0} \dfrac{f(x) - f(0)}{x - 0} = f'(0) = 1$.

Section 3.8 Linear Approximations and Differentials

1. $f(x) = x^3 \implies f'(x) = 3x^2$ so $f(1) = 1$ and $f'(1) = 3$. Thus, $L(x) = f(1) + f'(1)(x-1) = 1 + 3(x-1) = 3x - 2$.

3. $f(x) = e^{-2x} \implies f'(x) = -2e^{-2x}$ so $f(0) = 1$ and $f'(0) = -2$.
Thus, $L(x) = f(0) + f'(0)(x-0) = 1 + (-2)(x-0) = 1 - 2x$.

5. $f(x) = \sqrt{1-x} \implies f'(x) = \dfrac{-1}{2\sqrt{1-x}}$ so $f(0) = 1$ and

$f'(0) = -\frac{1}{2}$. Therefore,
$$\sqrt{1-x} = f(x) \approx f(0) + f'(0)(x-0)$$
$$= 1 + \left(-\tfrac{1}{2}\right)(x-0) = 1 - \tfrac{1}{2}x.$$
So $\sqrt{0.9} = \sqrt{1-0.1} \approx 1 - \frac{1}{2}(0.1) = 0.95$ and

$\sqrt{0.99} = \sqrt{1 - 0.01} \approx 1 - \frac{1}{2}(0.01) = 0.995$.

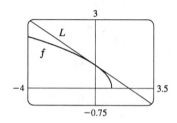

7. $f(x) = \sqrt{1+x} \implies f'(x) = \dfrac{1}{2\sqrt{1+x}}$ so $f(0) = 1$ and

$f'(0) = \frac{1}{2}$.

Thus, $f(x) \approx f(0) + f'(0)(x-0) = 1 + \frac{1}{2}(x-0) = 1 + \frac{1}{2}x$.

We need $\sqrt{1+x} - 0.1 < 1 + \frac{1}{2}x < \sqrt{1+x} + 0.1$. By zooming in
or using an intersect feature, we see that this is true when
$-0.69 < x < 1.09$.

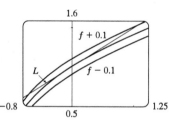

9. $f(x) = \dfrac{1}{(1+2x)^4} \implies f'(x) = \dfrac{-8}{(1+2x)^5}$ so $f(0) = 1$ and
$f'(0) = -8$.
Thus, $f(x) \approx f(0) + f'(0)(x-0) = 1 + (-8)(x-0) = 1 - 8x$.
We need $1/(1+2x)^4 - 0.1 < 1 - 8x < 1/(1+2x)^4 + 0.1$,
which is true when $-0.045 < x < 0.055$.

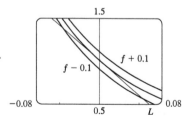

11. (a) $f(x) = \sin x \Longrightarrow f'(x) = \cos x$ so $f(0) = 0$ and $f'(0) = 1$.

Thus, $f(x) \approx f(0) + f'(0)(x - 0) = 0 + 1(x - 0) = x$.

(b)

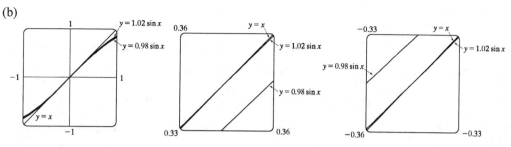

We want to know the values of x for which $y = x$ approximates $y = \sin x$ with less than a 2% difference; that is, the values of x for which

$$\left| \frac{x - \sin x}{\sin x} \right| < 0.02 \Longleftrightarrow \begin{cases} 0.98 \sin x < x < 1.02 \sin x & \text{where } \sin x > 0 \\ 1.02 \sin x < x < 0.98 \sin x & \text{where } \sin x < 0 \end{cases}$$

In the first figure, we see that the graphs are very close to each other near $x = 0$. Changing the viewing rectangle and using an intersect feature (see the second figure,) we find that $y = x$ intersects $y = 1.02 \sin x$ at $x \approx 0.344$. By symmetry, they also intersect at $x \approx -0.344$ (see the third figure.). Converting 0.344 radians to degrees, we get $0.344 \left(\frac{180°}{\pi} \right) \approx 19.7° \approx 20°$, which verifies the statement.

13. (a) $y = \cos x \Longrightarrow dy = -\sin x \, dx$

(b) When $x = \frac{\pi}{6}$ and $dx = 0.05$, $dy = -\frac{1}{2}(0.05) = -0.025$.

$\Delta y = f(x + \Delta x) - f(x) = \cos\left(\frac{\pi}{6} + 0.05\right) - \cos\frac{\pi}{6} \approx -0.02607$.

15. (a) If x is the edge length, then $V = x^3 \Longrightarrow dV = 3x^2 dx$. When $x = 30$ and $dx = 0.1$, $dV = 3(30)^2(0.1) = 270$, so the maximum error is about 270 cm^3.

(b) $S = 6x^2 \Longrightarrow dS = 12x \, dx$. When $x = 30$ and $dx = 0.1$, $dS = 12(30)(0.1) = 36$, so the maximum error is about 36 cm^2.

17. $V = \frac{2}{3}\pi r^3 \Longrightarrow dV = 2\pi r^2 dr$. When $r = 25$ and $dr = 0.05$ cm $= 0.0005$ m, $dV = 2\pi(25)^2(0.0005) = \frac{5\pi}{8}$, so the amount of paint is about $\frac{5\pi}{8} \approx 2$ m^3.

Chapter 3 Review

Concept Check

1. (a) The Power Rule: If n is any real number, then $\dfrac{d}{dx}(x^n) = nx^{n-1}$. The derivative of a variable base raised to a constant power is the power times the base raised to the power minus one.

(b) The Constant Multiple Rule: If c is a constant and f is a differentiable function, then $\dfrac{d}{dx}[cf(x)] = c\dfrac{d}{dx}f(x)$. The derivative of a constant times a function is the constant times the derivative of the function.

(c) The Sum Rule: If f and g are both differentiable, then $\dfrac{d}{dx}[f(x) + g(x)] = \dfrac{d}{dx}f(x) + \dfrac{d}{dx}g(x)$. The derivative of a sum of functions is the sum of the derivatives.

(d) The Difference Rule: If f and g are both differentiable, then $\dfrac{d}{dx}[f(x) - g(x)] = \dfrac{d}{dx}f(x) - \dfrac{d}{dx}g(x)$. The derivative of a difference of functions is the difference of the derivatives.

(e) The Product Rule: If f and g are both differentiable, then $\dfrac{d}{dx}[f(x)g(x)] = f(x)\dfrac{d}{dx}g(x) + g(x)\dfrac{d}{dx}f(x)$. The derivative of a product of two functions is the first function times the derivative of the second function plus the second function times the derivative of the first function.

(f) The Quotient Rule: If f and g are both differentiable, then $\dfrac{d}{dx}\left[\dfrac{f(x)}{g(x)}\right] = \dfrac{g(x)\dfrac{d}{dx}f(x) - f(x)\dfrac{d}{dx}g(x)}{[g(x)]^2}$. The derivative of a quotient of functions is the denominator times the derivative of the numerator minus the numerator times the derivative of the denominator, all divided by the square of the denominator.

(g) The Chain Rule: If f and g are both differentiable and $F = f \circ g$ is the composite function defined by $F(x) = f(g(x))$, then F is differentiable and F' is given by the product $F'(x) = f'(g(x))g'(x)$. The derivative of a composite function is the derivative of the outer function evaluated at the inner function times the derivative of the inner function.

2. (a) $y = x^n \Longrightarrow y' = nx^{n-1}$

(b) $y = e^x \Longrightarrow y' = e^x$

(c) $y = a^x \Longrightarrow y' = a^x \ln a$

(d) $y = \ln x \Longrightarrow y' = 1/x$

(e) $y = \log_a x \Longrightarrow y' = 1/(x \ln a)$

(f) $y = \sin x \Longrightarrow y' = \cos x$

(g) $y = \cos x \Longrightarrow y' = -\sin x$

(h) $y = \tan x \Longrightarrow y' = \sec^2 x$

(i) $y = \csc x \Longrightarrow y' = -\csc x \cot x$

(j) $y = \sec x \Longrightarrow y' = \sec x \tan x$

(k) $y = \cot x \Longrightarrow y' = -\csc^2 x$

(l) $y = \sin^{-1} x \Longrightarrow y' = 1/\sqrt{1 - x^2}$

(m) $y = \tan^{-1} x \Longrightarrow y' = 1/(1 + x^2)$

3. (a) e is the number such that $\lim\limits_{h \to 0} \dfrac{e^h - 1}{h} = 1$.

(b) $e = \lim\limits_{x \to 0} (1 + x)^{1/x}$

(c) The differentiation formula for $y = a^x$ ($y' = a^x \ln a$) is simplest when $a = e$ because $\ln e = 1$.

(d) The differentiation formula for $y = \log_a x$ [$y' = 1/(x \ln a)$] is simplest when $a = e$ because $\ln e = 1$.

4. (a) Implicit differentiation consists of differentiating both sides of an equation involving x and y with respect to x, and then solving the resulting equation for y'.

(b) Logarithmic differentiation consists of taking natural logarithms of both sides of an equation $y = f(x)$, simplifying, differentiating implicitly with respect to x, and then solving the resulting equation for y'.

5. The linearization L of f at $x = a$ is $L(x) = f(a) + f'(a)(x - a)$.

True-False Quiz

1. True. This is the Sum Rule.

3. True. This is the Chain Rule.

5. False. $\dfrac{d}{dx} f(\sqrt{x}) = \dfrac{f'(\sqrt{x})}{2\sqrt{x}}$ by the Chain Rule.

7. False. $\dfrac{d}{dx} 10^x = 10^x \ln 10$ by Equation 3.5.5.

9. True. $D(\tan^2 x) = 2 \tan x \sec^2 x$, and $D(\sec^2 x) = 2 \sec x (\sec x \tan x) = 2 \tan x \sec^2 x$. We can also show this by differentiating the identity $\tan^2 x + 1 = \sec^2 x$: we get

$$\frac{d}{dx}(\tan^2 x + 1) = \frac{d}{dx} \tan^2 x = \frac{d}{dx} \sec^2 x.$$

11. True. $g(x) = x^5 \Longrightarrow g'(x) = 5x^4 \Longrightarrow g'(2) = 5(2)^4 = 80$, and by the definition of the derivative, $\lim\limits_{x \to 2} \dfrac{g(x) - g(2)}{x - 2} = g'(2) = 80$.

Exercises

1. $y = (x+2)^8 (x+3)^6 \implies$

$\quad y' = (x+2)^8 \, 6 \, (x+3)^5 + (x+3)^6 \, 8 \, (x+2)^7$

$\quad\quad = 2 \, (x+2)^7 \, (x+3)^5 \, [3 \, (x+2) + 4 \, (x+3)] = 2 \, (x+2)^7 \, (x+3)^5 \, (7x+18)$

3. $y = \dfrac{x}{\sqrt{9-4x}} \implies$

$\quad y' = \dfrac{\sqrt{9-4x} - x \left[-4 / \left(2\sqrt{9-4x} \right) \right]}{9 - 4x} \cdot \dfrac{\sqrt{9-4x}}{\sqrt{9-4x}} = \dfrac{(9-4x) + 2x}{(9-4x)^{3/2}} = \dfrac{9 - 2x}{(9-4x)^{3/2}}$

5. $y = \sin(\cos x) \implies y' = \cos(\cos x)(-\sin x) = -\sin x \cos (\cos x)$

7. $y = xe^{-1/x} \implies y' = e^{-1/x} + xe^{-1/x} \left(1/x^2 \right) = e^{-1/x} \left(1 + 1/x \right)$

9. $y = \tan \sqrt{1-x} \implies y' = \left(\sec^2 \sqrt{1-x} \right) \left(\dfrac{1}{2\sqrt{1-x}} \right) (-1) = -\dfrac{\sec^2 \sqrt{1-x}}{2\sqrt{1-x}}$

11. $y = \dfrac{x}{8-3x} \implies y' = \dfrac{(8-3x)(1) - x(-3)}{(8-3x)^2} = \dfrac{8}{(8-3x)^2}$

13. $y = e^{cx} (c\sin x - \cos x) \implies$

$\quad y' = e^{cx} (c\cos x + \sin x) + ce^{cx} (c\sin x - \cos x)$

$\quad\quad = e^{cx} \left(c^2 \sin x - c\cos x + c\cos x + \sin x \right) = e^{cx} \sin x \left(c^2 + 1 \right)$

15. $y = e^{e^x} \implies y' = e^{e^x} e^x = e^{x + e^x}$

17. $x^2 y^3 + 3y^2 = x - 4y \implies x^2 \left(3y^2 y' \right) + y^3 (2x) + 6yy' = 1 - 4y' \implies$

$\quad 3x^2 y^2 y' + 6yy' + 4y' = 1 - 2xy^3 \implies y' = \dfrac{1 - 2xy^3}{3x^2 y^2 + 6y + 4}$

19. $y = \log_{10} \left(x^2 - x \right) \implies y' = \dfrac{1}{\left(x^2 - x \right) \ln 10} (2x - 1) = \dfrac{2x - 1}{(\ln 10) \left(x^2 - x \right)}$

21. $y = \ln \sin x - \frac{1}{2} \sin^2 x \implies y' = \dfrac{\cos x}{\sin x} - \sin x \cos x = \cot x - \sin x \cos x$

23. $y = \sin \left(\tan \sqrt{1+x^3} \right) \implies y' = \cos \left(\tan \sqrt{1+x^3} \right) \left(\sec^2 \sqrt{1+x^3} \right) \left[3x^2 / \left(2\sqrt{1+x^3} \right) \right]$

25. $y = \dfrac{\sqrt{x+1} \, (2-x)^5}{(x+3)^7} \implies \ln |y| = \frac{1}{2} \ln (x+1) + 5 \ln |2 - x| - 7 \ln (x+3) \implies$

$\quad \dfrac{y'}{y} = \dfrac{1}{2(x+1)} + \dfrac{-5}{2-x} - \dfrac{7}{x+3} \implies y' = \dfrac{\sqrt{x+1} \, (2-x)^5}{(x+3)^7} \left[\dfrac{1}{2(x+1)} - \dfrac{5}{2-x} - \dfrac{7}{x+3} \right]$ or

$\quad y' = \dfrac{(2-x)^4 \left(3x^2 - 55x - 52 \right)}{2\sqrt{x+1} \, (x+3)^8}.$

27. $f(x) = (2x-1)^{-5} \implies f'(x) = -5 (2x-1)^{-6} (2) = -10 (2x-1)^{-6} \implies$

$\quad f''(x) = 60 (2x-1)^{-7} (2) = 120 (2x-1)^{-7}. \; f''(0) = 120 (-1)^{-7} = -120.$

29. $f(x) = 2^x \Longrightarrow f'(x) = 2^x \ln 2 \Longrightarrow f''(x) = 2^x (\ln 2)^2 \Longrightarrow \cdots \Longrightarrow f^{(n)}(x) = 2^x (\ln 2)^n$

31. (a) $f(x) = x\sqrt{5-x} \Longrightarrow$

$$f'(x) = x\left[\tfrac{1}{2}(5-x)^{-1/2}(-1)\right] + \sqrt{5-x} = \frac{-x}{2\sqrt{5-x}} + \frac{2(5-x)}{2\sqrt{5-x}} = \frac{10-3x}{2\sqrt{5-x}}$$

(b) At $(1,2)$: $f'(1) = \tfrac{7}{4}$. So an equation of the tangent is $y - 2 = \tfrac{7}{4}(x-1)$ or $y = \tfrac{7}{4}x + \tfrac{1}{4}$.

At $(4,4)$: $f'(4) = -\tfrac{2}{2} = -1$. So an equation of the tangent is $y - 4 = -(x-4)$ or $y = -x + 8$.

(c)

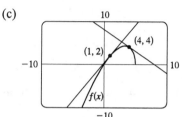

(d)

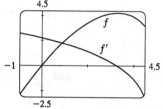

The graphs look reasonable, since f' is positive where f has tangents with positive slope, and f' is negative where f has tangents with negative slope.

33. $f(x) = xe^{\sin x} \Longrightarrow f'(x) = x\left[e^{\sin x}(\cos x)\right] + e^{\sin x}(1) = e^{\sin x}(x\cos x + 1)$. As a check on our work, we notice from the graphs that $f'(x) > 0$ when f is increasing. Also, we see in the larger viewing rectangle a certain similarity in the graphs of f and f': the sizes of the oscillations of f and f' are linked.

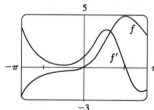

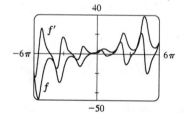

35. (a) $h(x) = f(x)g(x) \Longrightarrow$

$h'(x) = f'(x)g(x) + f(x)g'(x) \Longrightarrow$

$h'(2) = f'(2)g(2) + f(2)g'(2) = (-2)(5) + (3)(4) = 2$

(b) $F(x) = f(g(x)) \Longrightarrow F'(x) = f'(g(x))g'(x) \Longrightarrow$

$F'(2) = f'(g(2))g'(2) = f'(5)(4) = 11 \cdot 4 = 44$

37. $f(x) = x^2 g(x) \Longrightarrow f'(x) = 2xg(x) + x^2 g'(x) = x[2g(x) + xg'(x)]$

39. $f(x) = [g(x)]^2 \Longrightarrow f'(x) = 2g(x)g'(x)$

41. $f(x) = g(e^x) \Longrightarrow f'(x) = g'(e^x)e^x$

43. $f(x) = \ln|g(x)| \Longrightarrow f'(x) = \dfrac{1}{g(x)}g'(x) = \dfrac{g'(x)}{g(x)}$

45. $h\left(x\right)=\dfrac{f\left(x\right)g\left(x\right)}{f\left(x\right)+g\left(x\right)}\implies$

$h'\left(x\right)=\dfrac{\left[f\left(x\right)+g\left(x\right)\right]\left[f\left(x\right)g'\left(x\right)+g\left(x\right)f'\left(x\right)\right]-f\left(x\right)g\left(x\right)\left[f'\left(x\right)+g'\left(x\right)\right]}{\left[f\left(x\right)+g\left(x\right)\right]^{2}}$

$=\dfrac{\begin{array}{l}\left[f\left(x\right)\right]^{2}g'\left(x\right)+f\left(x\right)g\left(x\right)f'\left(x\right)+f\left(x\right)g\left(x\right)g'\left(x\right)\\\qquad+\left[g\left(x\right)\right]^{2}f'\left(x\right)-f\left(x\right)g\left(x\right)f'\left(x\right)-f\left(x\right)g\left(x\right)g'\left(x\right)\end{array}}{\left[f\left(x\right)+g\left(x\right)\right]^{2}}$

$=\dfrac{f'\left(x\right)\left[g\left(x\right)\right]^{2}+g'\left(x\right)\left[f\left(x\right)\right]^{2}}{\left[f\left(x\right)+g\left(x\right)\right]^{2}}$

47. $y=\left[\ln\left(x+4\right)\right]^{2}\implies y'=2\dfrac{\ln\left(x+4\right)}{x+4}=0\iff\ln\left(x+4\right)=0\iff x+4=1\iff x=-3,$ so the tangent is horizontal at $(-3,0)$.

49. $x^{2}+2y^{2}=1\implies 2x+4yy'=0\implies y'=-x/\left(2y\right)=1\iff x=-2y.$ Since the points lie on the ellipse, we have $\left(-2y\right)^{2}+2y^{2}=1\implies 6y^{2}=1\implies y=\pm\frac{1}{\sqrt{6}}.$ The points are $\left(-\frac{2}{\sqrt{6}},\frac{1}{\sqrt{6}}\right)$ and $\left(\frac{2}{\sqrt{6}},-\frac{1}{\sqrt{6}}\right).$

51. $s\left(t\right)=Ae^{-ct}\cos\left(\omega t+\delta\right)\implies$

$v\left(t\right)=s'\left(t\right)$
$=-cAe^{-ct}\cos\left(\omega t+\delta\right)+Ae^{-ct}\left[-\omega\sin\left(\omega t+\delta\right)\right]$
$=-Ae^{-ct}\left[c\cos\left(\omega t+\delta\right)+\omega\sin\left(\omega t+\delta\right)\right]\implies$

$a\left(t\right)=v'\left(t\right)$
$=cAe^{-ct}\left[c\cos\left(\omega t+\delta\right)+\omega\sin\left(\omega t+\delta\right)\right]+\left(-Ae^{-ct}\right)\left[-\omega c\sin\left(\omega t+\delta\right)+\omega^{2}\cos\left(\omega t+\delta\right)\right]$
$=Ae^{-ct}\left[\left(c^{2}-\omega^{2}\right)\cos\left(\omega t+\delta\right)+2c\omega\sin\left(\omega t+\delta\right)\right]$

53. The linear density ρ is the rate of change of mass m with respect to length x.

$m=x\left(1+\sqrt{x}\right)=x+x^{3/2}\implies\rho=dm/dx=1+\frac{3}{2}\sqrt{x},$ so the linear density when $x=4$ is $1+\frac{3}{2}\sqrt{4}=4\text{ kg/m}.$

55. (a) $C\left(x\right)=920+2x-0.02x^{2}+0.00007x^{3}\implies C'\left(x\right)=2-0.04x+0.00021x^{2}$

(b) $C'\left(100\right)=\$0.10$/unit. This value represents the rate at which costs are increasing as the hundredth unit is produced, and is the approximate cost of producing the 101st unit.

(c) $C\left(101\right)-C\left(100\right)=990.10107-990=\$0.10107,$ slightly larger than $C'\left(100\right).$

(d) $C''\left(x\right)=-0.04+0.00042x=0\implies x=\frac{2000}{21}\approx95.24$ and C'' changes from negative to positive at this value of x. This is the value of x at which the marginal cost is minimized.

57. (a) $f(x) = \sqrt[3]{1 + 3x} = (1 + 3x)^{1/3} \implies f'(x) = (1 + 3x)^{-2/3}$
so $L(x) = f(0) + f'(0)(x - 0) = 1^{1/3} + 1^{-2/3}x = 1 + x$.

Thus, $\sqrt[3]{1 + 3x} \approx 1 + x \implies \sqrt[3]{1.03} = \sqrt[3]{1 + 3(0.01)} \approx 1 + (0.01) = 1.01$.

(b) The linear approximation is $\sqrt[3]{1 + 3x} \approx 1 + x$, so for the required accuracy we want

$\sqrt[3]{1 + 3x} - 0.1 < 1 + x < \sqrt[3]{1 + 3x} + 0.1$. From the graph, it appears that this is true when
$-0.23 < x < 0.40$.

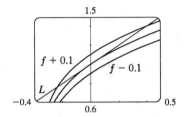

59. $\displaystyle\lim_{\theta \to \pi/3} \frac{\cos\theta - 0.5}{\theta - \pi/3} = \frac{d}{d\theta}\cos\theta \bigg|_{\theta = \pi/3} = -\sin\frac{\pi}{3} = -\frac{\sqrt{3}}{2}$

61. $\displaystyle\lim_{x \to 0} \frac{\sqrt{1 + \tan x} - \sqrt{1 + \sin x}}{x^3} = \lim_{x \to 0} \frac{\left(\sqrt{1 + \tan x} - \sqrt{1 + \sin x}\right)\left(\sqrt{1 + \tan x} + \sqrt{1 + \sin x}\right)}{x^3\left(\sqrt{1 + \tan x} + \sqrt{1 + \sin x}\right)}$

$\displaystyle= \lim_{x \to 0} \frac{(1 + \tan x) - (1 + \sin x)}{x^3\left(\sqrt{1 + \tan x} + \sqrt{1 + \sin x}\right)}$

$\displaystyle= \lim_{x \to 0} \frac{\sin x\,(1/\cos x - 1)\cos x}{x^3\left(\sqrt{1 + \tan x} + \sqrt{1 + \sin x}\right)\cos x}$

$\displaystyle= \lim_{x \to 0} \frac{\sin x\,(1 - \cos x)(1 + \cos x)}{x^3\left(\sqrt{1 + \tan x} + \sqrt{1 + \sin x}\right)\cos x\,(1 + \cos x)}$

$\displaystyle= \lim_{x \to 0} \frac{\sin x \cdot \sin^2 x}{x^3\left(\sqrt{1 + \tan x} + \sqrt{1 + \sin x}\right)\cos x\,(1 + \cos x)}$

$\displaystyle= \left(\lim_{x \to 0} \frac{\sin x}{x}\right)^3 \lim_{x \to 0} \frac{1}{\left(\sqrt{1 + \tan x} + \sqrt{1 + \sin x}\right)\cos x\,(1 + \cos x)}$

$\displaystyle= 1^3 \cdot \frac{1}{\left(\sqrt{1} + \sqrt{1}\right) \cdot 1 \cdot (1 + 1)} = \frac{1}{4}$

Focus on Problem Solving

1. We must find a value x_0 such that the normal lines to the parabola $y = x^2$ at $x = \pm x_0$ intersect at a

point one unit from the points $\left(\pm x_0, x_0^2\right)$. The normals to $y = x^2$ at $x = \pm x_0$ have slopes $-\dfrac{1}{\pm 2x_0}$ and

pass through $\left(\pm x_0, x_0^2\right)$ respectively, so the normals have the equations

$$y - x_0^2 = -\frac{1}{2x_0}\left(x - x_0\right) \qquad \text{and} \qquad y - x_0^2 = \frac{1}{2x_0}\left(x + x_0\right)$$

The common y-intercept is $x_0^2 + \frac{1}{2}$. We want to find the value of x_0 for which

the distance from $\left(0, x_0^2 + \frac{1}{2}\right)$ to $\left(x_0, x_0^2\right)$ equals 1. The square of the distance is

$(x_0 - 0)^2 + \left[x_0^2 - \left(x_0^2 + \frac{1}{2}\right)\right]^2 = x_0^2 + \frac{1}{4} = 1 \Longleftrightarrow x_0 = \pm\frac{\sqrt{3}}{2}$. For these values of x_0, the y-intercept

is $x_0^2 + \frac{1}{2} = \frac{5}{4}$, so the center of the circle is at $\left(0, \frac{5}{4}\right)$.

Another Solution: Let the center of the circle be $(0, a)$. Then the equation of the circle is

$x^2 + (y - a)^2 = 1$. Solving with the equation of the parabola, $y = x^2$, we get

$$x^2 + \left(x^2 - a\right)^2 = 1$$
$$x^2 + x^4 - 2ax^2 + a^2 = 1$$
$$x^4 + (1 - 2a)\,x^2 + a^2 - 1 = 0$$

The parabola and the circle will be tangent to each other when this quadratic equation in x^2 has equal

roots, that is, when the discriminant is 0. Thus

$$(1 - 2a)^2 - 4\left(a^2 - 1\right) = 0$$
$$1 - 4a + 4a^2 - 4a^2 + 4 = 0$$
$$4a = 5$$
$$a = \frac{5}{4}$$

The center of the circle is $\left(0, \frac{5}{4}\right)$.

3. (a)

$$D = \left\{ x \mid 3 - x \geq 0, 2 - \sqrt{3 - x} \geq 0, 1 - \sqrt{2 - \sqrt{3 - x}} \geq 0 \right\}$$

$$= \left\{ x \mid 3 \geq x, 2 \geq \sqrt{3 - x}, 1 \geq \sqrt{2 - \sqrt{3 - x}} \right\}$$

$$= \left\{ x \mid 3 \geq x, \ 4 \geq 3 - x, 1 \geq 2 - \sqrt{3 - x} \right\}$$

$$= \left\{ x \mid x \leq 3, x \geq -1, 1 \leq \sqrt{3 - x} \right\}$$

$$= \left\{ x \mid x \leq 3, x \geq -1, 1 \leq 3 - x \right\}$$

$$= \left\{ x \mid x \leq 3, x \geq -1, x \leq 2 \right\} = \left\{ x \mid -1 \leq x \leq 2 \right\}$$

$$= [-1, 2]$$

(b) $f(x) = \sqrt{1 - \sqrt{2 - \sqrt{3 - x}}} \implies$

$$f'(x) = \frac{1}{2\sqrt{1 - \sqrt{2 - \sqrt{3 - x}}}} \frac{d}{dx}\left(1 - \sqrt{2 - \sqrt{3 - x}} \right)$$

$$= \frac{1}{2\sqrt{1 - \sqrt{2 - \sqrt{3 - x}}}} \cdot \frac{-1}{2\sqrt{2 - \sqrt{3 - x}}} \frac{d}{dx}\left(2 - \sqrt{3 - x} \right)$$

$$= -\frac{1}{8\sqrt{1 - \sqrt{2 - \sqrt{3 - x}}}\sqrt{2 - \sqrt{3 - x}}\sqrt{3 - x}}$$

(c)

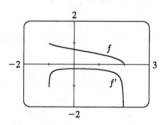

Note that f is always decreasing and f' is always negative.

5. We can assume without loss of generality that $\theta = 0$ at time $t = 0$, so that $\theta = 12\pi t$ rad. [The angular velocity of the wheel is 360 rpm $= 360 \cdot (2\pi \text{ rad}) / (60 \text{ s}) = 12\pi$ rad/s.] Then the position of A as a function of time is $A = (40 \cos \theta, 40 \sin \theta) = (40 \cos 12\pi t, 40 \sin 12\pi t)$, so

$$\sin \alpha = \frac{40 \sin \theta}{120} = \frac{\sin \theta}{3} = \tfrac{1}{3} \sin 12\pi t.$$

(a) Differentiating the expression for $\sin \alpha$, we get $\cos \alpha \cdot \dfrac{d\alpha}{dt} = \dfrac{1}{3} \cdot 12\pi \cdot \cos 12\pi t = 4\pi \cos \theta$.

When $\theta = \frac{\pi}{3}$, we have $\sin \alpha = \frac{1}{3} \sin \theta = \frac{\sqrt{3}}{6}$, so $\cos \alpha = \sqrt{1 - \left(\frac{\sqrt{3}}{6}\right)^2} = \sqrt{\frac{11}{12}}$ and

$$\frac{d\alpha}{dt} = \frac{4\pi \cos \frac{\pi}{3}}{\cos \alpha} = \frac{2\pi}{\sqrt{11/12}} = \frac{4\pi\sqrt{3}}{\sqrt{11}} \approx 6.56 \text{ rad/s}.$$

(b) By the Law of Cosines, $|AP|^2 = |OA|^2 + |OP|^2 - 2\,|OA|\,|OP| \cos \theta \implies$

$$120^2 = 40^2 + |OP|^2 - 2 \cdot 40\,|OP| \cos \theta \implies |OP|^2 - (80 \cos \theta)\,|OP| - 12{,}800 = 0 \implies$$

$$|OP| = \tfrac{1}{2}\left(80 \cos \theta \pm \sqrt{6400 \cos^2 \theta + 51{,}200}\right)$$

$$= 40 \cos \theta \pm 40\sqrt{\cos^2 \theta + 8}$$

$$= 40\left(\cos \theta + \sqrt{8 + \cos^2 \theta}\right) \text{ cm (since } |OP| > 0\text{).}$$

As a check, note that $|OP| = 160$ cm when $\theta = 0$ and $|OP| = 80\sqrt{2}$ cm when $\theta = \frac{\pi}{2}$.

(c) By part (b), the x-coordinate of P is given by $x = 40\left(\cos \theta + \sqrt{8 + \cos^2 \theta}\right)$, so

$$\frac{dx}{dt} = \frac{dx}{d\theta}\frac{d\theta}{dt} = 40\left(-\sin \theta - \frac{2 \cos \theta \sin \theta}{2\sqrt{8 + \cos^2 \theta}}\right) \cdot 12\pi$$

$$= -480\pi \sin \theta \left(1 + \frac{\sin \theta \cos \theta}{\sqrt{8 + \cos^2 \theta}}\right) \text{ cm/s}$$

In particular, $\dfrac{dx}{dt} = 0$ cm/s when $\theta = 0$ and $\dfrac{dx}{dt} = -480\pi$ cm/s when $\theta = \frac{\pi}{2}$.

7. Consider the statement that $\dfrac{d^n}{dx^n}\left(e^{ax}\sin bx\right)=r^n e^{ax}\sin\left(bx+n\theta\right)$.

For $n=1$, $\dfrac{d}{dx}\left(e^{ax}\sin bx\right)=ae^{ax}\sin bx+be^{ax}\cos bx$, and

$$re^{ax}\sin\left(bx+\theta\right)=re^{ax}\left[\sin bx\cos\theta+\cos bx\sin\theta\right]$$
$$=re^{ax}\left(\frac{a}{r}\sin bx+\frac{b}{r}\cos bx\right)$$
$$=ae^{ax}\sin bx+be^{ax}\cos bx$$

since $\tan\theta=\dfrac{b}{a}\implies\sin\theta=\dfrac{b}{r}$ and $\cos\theta=\dfrac{a}{r}$.

So the statement is true for $n=1$. Assume it is true for $n=k$. Then

$$\frac{d^{k+1}}{dx^{k+1}}\left(e^{ax}\sin bx\right)=\frac{d}{dx}\left[r^k e^{ax}\sin\left(bx+k\theta\right)\right]$$
$$=r^k ae^{ax}\sin\left(bx+k\theta\right)+r^k e^{ax}b\cos\left(bx+k\theta\right)$$
$$=r^k e^{ax}\left[a\sin\left(bx+k\theta\right)+b\cos\left(bx+k\theta\right)\right]$$

But

$$\sin\left[bx+\left(k+1\right)\theta\right]=\sin\left[\left(bx+k\theta\right)+\theta\right]$$
$$=\sin\left(bx+k\theta\right)\cos\theta+\sin\theta\cos\left(bx+k\theta\right)$$
$$=\frac{a}{r}\sin\left(bx+k\theta\right)+\frac{b}{r}\cos\left(bx+k\theta\right)$$

Hence, $a\sin\left(bx+k\theta\right)+b\cos\left(bx+k\theta\right)=r\sin\left[bx+\left(k+1\right)\theta\right]$. So

$$\frac{d^{k+1}}{dx^{k+1}}\left(e^{ax}\sin bx\right)=r^k e^{ax}\left[a\sin\left(bx+k\theta\right)+b\cos\left(bx+k\theta\right)\right]$$
$$=r^k e^{ax}\left[r\sin\left(bx+\left(k+1\right)\theta\right)\right]$$
$$=r^{k+1}e^{ax}\left[\sin\left(bx+\left(k+1\right)\theta\right)\right]$$

Therefore, the statement is true for all n by mathematical induction.

9. It seems from the figure that as P approaches the point $(0, 2)$ from the right, $x_T \to \infty$ and $y_T \to 2^+$. As P approaches the point $(3, 0)$ from the left, it appears that $x_T \to 3^+$ and $y_T \to \infty$. So we guess that $x_T \in (3, \infty)$ and $y_T \in (2, \infty)$. It is more difficult to estimate the range of values for x_N and y_N. We might perhaps guess that $x_N \in (0, 3)$, and $y_N \in (-\infty, 0)$ or $(-2, 0)$.

In order to actually solve the problem, we implicitly differentiate the equation of the ellipse to find the equation of the tangent line: $\dfrac{x^2}{9} + \dfrac{y^2}{4} = 1 \implies \dfrac{2x}{9} + \dfrac{2y}{4}y' = 0$, so $y' = -\dfrac{4\,x}{9\,y}$. So at the point (x_0, y_0) on the ellipse, an equation of the tangent line is $y - y_0 = -\dfrac{4\,x_0}{9\,y_0}(x - x_0)$ or $4x_0 x + 9y_0 y = 4x_0^2 + 9y_0^2$. This can be written as $\dfrac{x_0 x}{9} + \dfrac{y_0 y}{4} = \dfrac{x_0^2}{9} + \dfrac{y_0^2}{4} = 1$, because (x_0, y_0) lies on the ellipse. So an equation of the tangent line is $\dfrac{x_0 x}{9} + \dfrac{y_0 y}{4} = 1$.

Therefore, the x-intercept x_T for the tangent line is given by $\dfrac{x_0 x_T}{9} = 1 \iff x_T = \dfrac{9}{x_0}$, and the y-intercept y_T is given by $\dfrac{y_0 y_T}{4} = 1 \iff y_T = \dfrac{4}{y_0}$.

So as x_0 takes on all values in $(0, 3)$, x_T takes on all values in $(3, \infty)$, and as y_0 takes on all values in $(0, 2)$, y_T takes on all values in $(2, \infty)$.

At the point (x_0, y_0) on the ellipse, the slope of the normal line is $-\dfrac{1}{y'(x_0, y_0)} = \dfrac{9\,y_0}{4\,x_0}$, and its equation is $y - y_0 = \dfrac{9\,y_0}{4\,x_0}(x - x_0)$. So the x-intercept x_N for the normal line is given by

$$0 - y_0 = \dfrac{9\,y_0}{4\,x_0}(x_N - x_0)$$
$$x_N = -\dfrac{4x_0}{9} + x_0$$
$$= \dfrac{5x_0}{9}$$

and the y-intercept y_N is given by

$$y_N - y_0 = \dfrac{9\,y_0}{4\,x_0}(0 - x_0)$$
$$y_N = -\dfrac{9y_0}{4} + y_0$$
$$= -\dfrac{5y_0}{4}$$

So as x_0 takes on all values in $(0, 3)$, x_N takes on all values in $\left(0, \frac{5}{3}\right)$, and as y_0 takes on all values in $(0, 2)$, y_N takes on all values in $\left(-\frac{5}{2}, 0\right)$.

11. If we divide $1 - x$ into x^n by long division, we find that

$f(x) = \dfrac{x^n}{1-x} = -x^{n-1} - x^{n-2} - \cdots - x - 1 + \dfrac{1}{1-x}$. This can also be seen by multiplying the last

expression by $1-x$ and canceling terms on the right-hand side. So we let $g(x) = 1 + x + x^2 + \cdots + x^{n-1}$,

so that $f(x) = \dfrac{1}{1-x} - g(x) \implies f^{(n)}(x) = \left(\dfrac{1}{1-x}\right)^{(n)} - g^{(n)}(x)$. But g is a polynomial of

degree $(n-1)$, so its nth derivative will be 0, and therefore $f^{(n)}(x) = \left(\dfrac{1}{1-x}\right)^{(n)}$. Now

$\dfrac{d}{dx}(1-x)^{-1} = (-1)(1-x)^{-2}(-1) = (1-x)^{-2}$,

$\dfrac{d^2}{dx^2}(1-x)^{-1} = (-2)(1-x)^{-3}(-1) = 2(1-x)^{-3}$,

$\dfrac{d^3}{dx^3}(1-x)^{-1} = (-3)\cdot 2(1-x)^{-4}(-1) = 3\cdot 2(1-x)^{-4}$, $\dfrac{d^4}{dx^4}(1-x)^{-1} = 4\cdot 3\cdot 2(1-x)^{-5}$,

and so on. So after n differentiations, we will have $f^{(n)}(x) = \left(\dfrac{1}{1-x}\right)^{(n)} = \dfrac{n!}{(1-x)^{n+1}}$.

13. (a)

If the two lines L_1 and L_2 have slopes m_1 and m_2 and angles of inclination ϕ_1 and ϕ_2, then $m_1 = \tan\phi_1$ and $m_2 = \tan\phi_2$. The triangle in the figure shows that $\phi_1 + \alpha + (180° - \phi_2) = 180°$ and so $\alpha = \phi_2 - \phi_1$. Therefore, using the identity for $\tan(x-y)$, we have $\tan\alpha = \tan(\phi_2 - \phi_1) = \dfrac{\tan\phi_2 - \tan\phi_1}{1 + \tan\phi_2\tan\phi_1}$ and so

$\tan\alpha = \dfrac{m_2 - m_1}{1 + m_1 m_2}$.

(b) (i) The parabolas intersect when $x^2 = (x-2)^2 \implies x = 1$. If $y = x^2$, then $y' = 2x$, so the slope

of the tangent to $y = x^2$ at $(1,1)$ is $m_1 = 2(1) = 2$. If $y = (x-2)^2$, then $y' = 2(x-2)$,

so the slope of the tangent to $y = (x-2)^2$ at $(1,1)$ is $m_2 = 2(1-2) = -2$. Therefore,

$\tan\alpha = \dfrac{m_2 - m_1}{1 + m_1 m_2} = \dfrac{-2-2}{1 + 2(-2)} = \dfrac{4}{3}$ and so $\alpha = \tan^{-1}\frac{4}{3} \approx 53°$.

(ii) $x^2 - y^2 = 3$ and $x^2 - 4x + y^2 + 3 = 0$ intersect when $x^2 - 4x + (x^2 - 3) + 3 = 0 \iff$

$2x(x-2) = 0 \implies x = 0$ or 2, but 0 is extraneous. If $x = 2$, then $y = \pm 1$. If $x^2 - y^2 = 3$ then

$2x - 2yy' = 0 \implies y' = x/y$ and $x^2 - 4x + y^2 + 3 = 0 \implies 2x - 4 + 2yy' = 0 \implies y' = \dfrac{2-x}{y}$.

At $(2,1)$ the slopes are $m_1 = 2$ and $m_2 = 0$, so $\tan\alpha = \frac{0-2}{1+2\cdot 0} = -2 \implies \alpha \approx 117°$. At

$(2,-1)$ the slopes are $m_1 = -2$ and $m_2 = 0$, so $\tan\alpha = \frac{0-(-2)}{1+(-2)(0)} = 2 \implies \alpha \approx 63°$.

15.

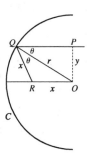

Since $\angle ROQ = \angle OQP = \theta$, the triangle QOR is isosceles, so $|QR| = |RO| = x$. By the Law of Cosines, $x^2 = x^2 + r^2 - 2rx\cos\theta$. Hence, $2rx\cos\theta = r^2$, so $x = \dfrac{r^2}{2r\cos\theta} = \dfrac{r}{2\cos\theta}$. Note that as $y \to 0^+$, $\theta \to 0^+$ (since $\sin\theta = y/r$), and hence $x \to \dfrac{r}{2\cos 0} = \dfrac{r}{2}$. Thus, as P is taken closer and closer to the x-axis, the point R approaches the midpoint of the radius AO.

17. $y = x^4 - 2x^2 - x \Longrightarrow y' = 4x^3 - 4x - 1$. The equation of the tangent line at $x = a$ is
$y - (a^4 - 2a^2 - a) = (4a^3 - 4a - 1)(x - a)$ or $y = (4a^3 - 4a - 1)x + (-3a^4 + 2a^2)$
and similarly for $x = b$. So if at $x = a$ and $x = b$ we have the same tangent line, then
$4a^3 - 4a - 1 = 4b^3 - 4b - 1$ and $-3a^4 + 2a^2 = -3b^4 + 2b^2$. The first equation gives
$a^3 - b^3 = a - b \Longrightarrow (a - b)(a^2 + ab + b^2) = (a - b)$. Assuming $a \neq b$, we have $1 = a^2 + ab + b^2$.
The second equation gives $3(a^4 - b^4) = 2(a^2 - b^2) \Longrightarrow 3(a^2 - b^2)(a^2 + b^2) = 2(a^2 - b^2)$ which
is true if $a = -b$. Substituting into $1 = a^2 + ab + b^2$ gives $1 = a^2 - a^2 + a^2 \Longrightarrow a = \pm 1$ so that $a = 1$
and $b = -1$ or vice versa. Thus, the points $(1, -2)$ and $(-1, 0)$ have a common tangent line.

As long as there are only two such points, we are done. So we show that these are in fact the only two such points. Suppose that $a^2 - b^2 \neq 0$. Then $3(a^2 - b^2)(a^2 + b^2) = 2(a^2 - b^2)$ gives
$3(a^2 + b^2) = 2$ or $a^2 + b^2 = \frac{2}{3}$. Thus, $ab = (a^2 + ab + b^2) - (a^2 + b^2) = 1 - \frac{2}{3} = \frac{1}{3}$, so $b = \dfrac{1}{3a}$.
Hence, $a^2 + \dfrac{1}{9a^2} = \dfrac{2}{3}$, so $9a^4 + 1 = 6a^2 \Longrightarrow 0 = 9a^4 - 6a^2 + 1 = (3a^2 - 1)^2$. So $3a^2 - 1 = 0 \Longrightarrow$
$a^2 = \frac{1}{3} \Longrightarrow b^2 = \dfrac{1}{9a^2} = \dfrac{1}{3} = a^2$, contradicting our assumption that $a^2 \neq b^2$.

19.

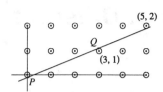

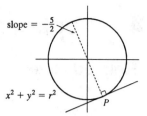

Because of the periodic nature of the lattice points, it suffices to consider the points in the 5×2 grid shown. We can see that the minimum value of r occurs when there is a line with slope $\frac{2}{5}$ which touches the circle centered at $(3, 1)$ and the circles centered at $(0, 0)$ and $(5, 2)$. To find P, the point at which the line is tangent to the circle at $(0, 0)$, we simultaneously solve $x^2 + y^2 = r^2$ and $y = -\frac{5}{2}x \implies$ $x^2 + \frac{25}{4}x^2 = r^2 \implies x^2 = \frac{4}{29}r^2 \implies x = \frac{2}{\sqrt{29}}r$, $y = -\frac{5}{\sqrt{29}}r$. To find Q, we either use symmetry or solve $(x - 3)^2 + (y - 1)^2 = r^2$ and $y - 1 = -\frac{5}{2}(x - 3)$. As above, we get $x = 3 - \frac{2}{\sqrt{29}}r$, $y = 1 + \frac{5}{\sqrt{29}}r$. Now the slope of the line PQ is $\frac{2}{5}$, so

$$m_{PQ} = \frac{1 + \frac{5}{\sqrt{29}}r - \left(-\frac{5}{\sqrt{29}}r\right)}{3 - \frac{2}{\sqrt{29}}r - \frac{2}{\sqrt{29}}r}$$

$$= \frac{1 + \frac{10}{\sqrt{29}}r}{3 - \frac{4}{\sqrt{29}}r}$$

$$= \frac{\sqrt{29} + 10r}{3\sqrt{29} - 4r} = \frac{2}{5}$$

and hence

$$5\sqrt{29} + 50r = 6\sqrt{29} - 8r$$

$$58r = \sqrt{29}$$

$$r = \frac{\sqrt{29}}{58}$$

So the minimum value of r for which any line with slope $\frac{2}{5}$ intersects circles with radius r centered at the lattice points on the plane is $r = \frac{\sqrt{29}}{58} \approx 0.093$.

Chapter 4   Applications of Differentiation

Section 4.1   Related Rates

1. $V = x^3 \implies \dfrac{dV}{dt} = 3x^2 \dfrac{dx}{dt}$

3. (a) Given: the rate of decrease of the surface area is 1 cm^2/min. If
we let t be time (in minutes) and S be the surface area (in
cm^2), then we are given that $dS/dt = -1$ cm^2/s.

(c)

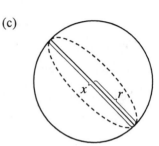

(b) Unknown: the rate of decrease of the diameter when the
diameter is 10 cm. If we let x be the diameter, then we want to
find dx/dt when $x = 10$ cm.

(d) If the radius is r and the diameter x, then $S = 4\pi r^2 = 4\pi \left(\frac{1}{2}x\right)^2 = \pi x^2 \implies dS/dx = 2\pi x \, (dx/dt)$.

(e) $-1 = \dfrac{dS}{dt} = 2\pi x \dfrac{dx}{dt} \implies \dfrac{dx}{dt} = -\dfrac{1}{2\pi x}$. When $x = 10$, $\dfrac{dx}{dt} = -\dfrac{1}{20\pi}$. So the rate of decrease is
$\frac{1}{20\pi}$ cm/min.

5. (a) Given: a plane flying horizontally at an altitude of 1 mi and a speed of 500 mi/h passes directly
over a radar station. If we let t be time (in hours) and x be the horizontal distance traveled by the
plane (in mi), then we are given that $dx/dt = 500$ mi/h.

(b) Unknown: the rate at which the distance from the plane to the
station is increasing when it is 2 mi from the station. If we let y be
the distance from the plane to the station, then we want to find
dy/dt when $y = 2$ mi.

(c)

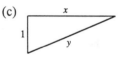

(d) By the Pythagorean Theorem, $y^2 = x^2 + 1 \implies 2y \, (dy/dt) = 2x \, (dx/dt)$.

(e) $\dfrac{dy}{dt} = \dfrac{x}{y}\dfrac{dx}{dt} = 500\dfrac{x}{y}$. When $y = 2$, $x = \sqrt{3}$, so $\dfrac{dy}{dt} = 500\left(\frac{\sqrt{3}}{2}\right) = 250\sqrt{3} \approx 433$ mi/h.

7.

We are given that $\dfrac{dx}{dt} = 60$ mi/h and $\dfrac{dy}{dt} = 25$ mi/h. $z^2 = x^2 + y^2 \implies$

$2z\dfrac{dz}{dt} = 2x\dfrac{dx}{dt} + 2y\dfrac{dy}{dt}$. After 2 hours, $x = 2\,(60) = 120$ and

$y = 2\,(25) = 50 \implies z = \sqrt{120^2 + 50^2} = 130$, so

$\dfrac{dz}{dt} = \dfrac{1}{z}\left(x\dfrac{dx}{dt} + y\dfrac{dy}{dt}\right) = \dfrac{120\,(60) + 50\,(25)}{130} = 65$ mi/h.

9.

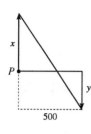

We are given that $\frac{dx}{dt} = 4$ ft/s and $\frac{dy}{dt} = 5$ ft/s. $z^2 = (x+y)^2 + 500^2 \implies$

$2z\frac{dz}{dt} = 2(x+y)\left(\frac{dx}{dt} + \frac{dy}{dt}\right)$. 15 minutes after the woman starts, we

have $x = (4 \text{ ft/s})(20 \text{ min})(60 \text{ s/min}) = 4800$ ft and

$y = 5 \cdot 15 \cdot 60 = 4500 \implies z = \sqrt{(4800 + 4500)^2 + 500^2}$, so

$\frac{dz}{dt} = \frac{x+y}{z}\left(\frac{dx}{dt} + \frac{dy}{dt}\right) = \frac{4800 + 4500}{\sqrt{86,740,000}}(5+4) = \frac{837}{\sqrt{8674}} \approx 8.99$ ft/s.

11. $A = \frac{1}{2}bh$, where b is the base and h is the altitude. We are given that $\frac{dh}{dt} = 1$ cm/min and

$\frac{dA}{dt} = 2$ cm^2/min. Using the Product Rule, we have $\frac{dA}{dt} = \frac{1}{2}\left(b\frac{dh}{dt} + h\frac{db}{dt}\right)$. When $h = 10$

and $A = 100$, we have $b = 20$, so $2 = \frac{1}{2}\left(20 \cdot 1 + 10\frac{db}{dt}\right) \implies 4 = 20 + 10\frac{db}{dt} \implies$

$\frac{db}{dt} = \frac{4-20}{10} = -1.6$ cm/min.

13.

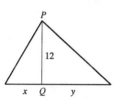

Using Q for the origin, we are given $\frac{dx}{dt} = -2$ ft/s and need to find

$\frac{dy}{dt}$ when $x = -5$. Using the Pythagorean Theorem twice, we have

$\sqrt{x^2 + 12^2} + \sqrt{y^2 + 12^2} = 39$, the total length of the rope.

Differentiating with respect to t, we get

$\frac{x}{\sqrt{x^2+12^2}}\frac{dx}{dt} + \frac{y}{\sqrt{y^2+12^2}}\frac{dy}{dt} = 0$, so $\frac{dy}{dt} = -\frac{x\sqrt{y^2+12^2}}{y\sqrt{x^2+12^2}}\frac{dx}{dt}$. Now when $x = -5$,

$39 = \sqrt{(-5)^2 + 12^2} + \sqrt{y^2+12^2} = 13 + \sqrt{y^2+12^2} \iff \sqrt{y^2+12^2} = 26$, and

$y = \sqrt{26^2 - 12^2} = \sqrt{532}$. So when $x = -5$, $\frac{dy}{dt} = -\frac{(-5)(26)}{\sqrt{532}(13)}(-2) \approx -0.87$ ft/s. So cart B is

moving towards Q at about 0.87 ft/s.

15.

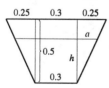

The figure is labeled in meters. The area A of a trapezoid is

$\frac{1}{2}(\text{base}_1 + \text{base}_2)(\text{height})$, so the volume V of the trough is $10A$.

Thus, $V = \frac{1}{2}[0.3 + (0.3 + 2a)]h(10)$, where $\frac{a}{h} = \frac{0.25}{0.5} = \frac{1}{2}$ so

$2a = h \implies V = 5(0.6 + h)h = 3h + 5h^2 \implies$

$0.2 = \frac{dV}{dt} = (3 + 10h)\frac{dh}{dt} \implies \frac{dh}{dt} = \frac{0.2}{3 + 10h}$. When $h = 0.3$,

$\frac{dh}{dt} = \frac{0.2}{3 + 10(0.3)} = \frac{0.2}{6}$ m/min $= \frac{10}{3}$ cm/min.

17.

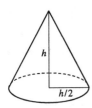

We are given that $\dfrac{dV}{dt} = 30$ ft^3/min.

$$V = \tfrac{1}{3}\pi r^2 h = \tfrac{1}{3}\pi \left(\dfrac{h}{2}\right)^2 h = \dfrac{\pi h^3}{12} \implies 30 = \dfrac{dV}{dt} = \dfrac{\pi h^2}{4}\dfrac{dh}{dt} \implies$$

$$\dfrac{dh}{dt} = \dfrac{120}{\pi h^2}.\ \text{When } h = 10 \text{ ft, } \dfrac{dh}{dt} = \dfrac{120}{10^2\pi} = \dfrac{6}{5\pi} \approx 0.38 \text{ ft/min.}$$

19.

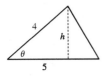

$A = \tfrac{1}{2}bh$, but $b = 5$ m and $h = 4\sin\theta$ so $A = 10\sin\theta$. We are

given $\dfrac{d\theta}{dt} = 0.06$ rad/s. $\dfrac{dA}{dt} = 10\cos\theta\dfrac{d\theta}{dt} = 0.6\cos\theta$. When

$\theta = \tfrac{\pi}{3}, \dfrac{dA}{dt} = 0.6\left(\cos\dfrac{\pi}{3}\right) = (0.6)\left(\dfrac{1}{2}\right) = 0.3$ m^2/s.

21. Differentiating both sides of $PV = C$ with respect to t and using the Product Rule gives us

$$P\dfrac{dV}{dt} + V\dfrac{dP}{dt} = 0 \implies \dfrac{dV}{dt} = -\dfrac{V}{P}\dfrac{dP}{dt}.\ \text{When } V = 600, P = 150 \text{ and } \dfrac{dP}{dt} = 20, \text{ we have}$$

$$\dfrac{dV}{dt} = -\dfrac{600}{150}(20) = -80, \text{ so the volume is decreasing at a rate of 80 cm}^3/\text{min.}$$

23. (a)

By the Pythagorean Theorem, $4000^2 + y^2 = \ell^2$. Differentiating with respect

to t, we obtain $2y\dfrac{dy}{dt} = 2\ell\dfrac{d\ell}{dt}$. We know that $\dfrac{dy}{dt} = 600$, so when $y = 3000$

and $\ell = 5000, \dfrac{d\ell}{dt} = \dfrac{y\,(dy/dt)}{\ell} = \dfrac{3000\,(600)}{5000} = \dfrac{1800}{5} = 360$ ft/s.

(b) Here $\tan\theta = y/4000$, so $\sec^2\theta\dfrac{d\theta}{dt} = \dfrac{1}{4000}\dfrac{dy}{dt} \implies \dfrac{d\theta}{dt} = \dfrac{\cos^2\theta}{4000}\dfrac{dy}{dt}$. When $y = 3000, \dfrac{dy}{dt} = 600$,

$\ell = 5000$ and $\cos\theta = \dfrac{4000}{\ell} = \dfrac{4000}{5000} = \dfrac{4}{5}$, so $\dfrac{d\theta}{dt} = \dfrac{(4/5)^2}{4000}(600) = 0.096$ rad/s.

25.

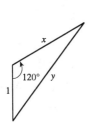

We are given that $\dfrac{dx}{dt} = 300$ km/h. By the Law of Cosines,

$$y^2 = x^2 + 1 - 2(1)\,x\cos 120° = x^2 + 1 - 2x\left(-\tfrac{1}{2}\right) = x^2 + x + 1, \text{ so}$$

$$2y\dfrac{dy}{dt} = 2x\dfrac{dx}{dt} + \dfrac{dx}{dt} \implies \dfrac{dy}{dt} = \dfrac{2x+1}{2y}\dfrac{dx}{dt}.\ \text{After 1 minute,}$$

$$x = \tfrac{300}{60} = 5 \implies y = \sqrt{5^2 + 5 + 1} = \sqrt{31} \implies$$

$$\dfrac{dy}{dt} = \dfrac{2\,(5) + 1}{2\sqrt{31}}(300) = \dfrac{1650}{\sqrt{31}} \approx 296 \text{ km/h.}$$

27.

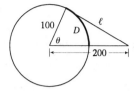

Let the distance between the runner and the friend be ℓ. Then by the Law of Cosines,

$$\ell^2 = 200^2 + 100^2 - 2 \cdot 200 \cdot 100 \cdot \cos\theta = 50{,}000 - 40{,}000\cos\theta \; (\bigstar).$$

Differentiating implicitly with respect to t, we obtain

$$2\ell\frac{d\ell}{dt} = -40{,}000\,(-\sin\theta)\,\frac{d\theta}{dt}.$$ Now if D is the distance run when

the angle is θ radians, then by the formula for the length of an arc on a circle, $s = r\theta$, we have

$D = 100\theta$, so $\theta = \dfrac{1}{100}D \Longrightarrow \dfrac{d\theta}{dt} = \dfrac{1}{100}\dfrac{dD}{dt} = \dfrac{7}{100}$. To substitute into the expression for $\dfrac{d\ell}{dt}$, we must

know $\sin\theta$ at the time when $\ell = 200$, which we find from $(\bigstar)$: $200^2 = 50{,}000 - 40{,}000\cos\theta \Longleftrightarrow$

$\cos\theta = \frac{1}{4} \Longrightarrow \sin\theta = \sqrt{1 - \left(\frac{1}{4}\right)^2} = \frac{\sqrt{15}}{4}$. Substituting, we get $2\,(200)\,\dfrac{d\ell}{dt} = 40{,}000\frac{\sqrt{15}}{4}\left(\frac{7}{100}\right) \Longrightarrow$

$d\ell/dt = \frac{7\sqrt{15}}{4} \approx 6.78$ m/s. Whether the distance between them is increasing or decreasing depends on the direction in which the runner is running.

Section 4.2 Maximum and Minimum Values

1. A function f has an *absolute minimum* at $x = c$ if $f(c)$ is the smallest function value on the entire domain of f, whereas f has a *local minimum* at c if $f(c)$ is the smallest function value when x is near c.

3. Absolute maximum at b; absolute minimum at d; local maxima at b, e; local minima at d, s; neither a maximum nor a minimum at a, c, r, and t.

5. Absolute maximum value is $f(4) = 4$; absolute minimum value is $f(7) = 0$; local maximum values are $f(4) = 4$ and $f(6) = 3$; local minimum values are $f(2) = 1$ and $f(5) = 2$.

7.

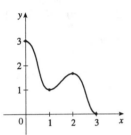

9.

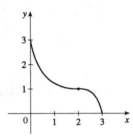

11. (a)

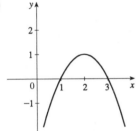

(b)

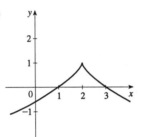

(c)

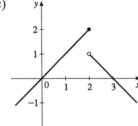

13. (a) *Note:* By the Extreme Value Theorem, f must *not* be continuous; because if it were, it would attain an absolute minimum.

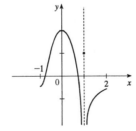

(b)

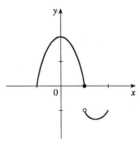

15. $f(x) = 1 + 2x$, $x \geq -1$. Absolute minimum $f(-1) = -1$; no local minimum. No local or absolute maximum.

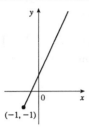

17. $f(x) = 1 - x^2$, $-2 \leq x \leq 1$. Absolute and local maximum $f(0) = 1$. Absolute minimum $f(-2) = -3$; no local minimum.

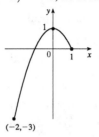

19. $f(\theta) = \sin\theta$, $-2\pi \leq \theta \leq 2\pi$. Absolute and local maxima $f\left(-\frac{3\pi}{2}\right) = f\left(\frac{\pi}{2}\right) = 1$. Absolute and local minima $f\left(-\frac{\pi}{2}\right) = f\left(\frac{3\pi}{2}\right) = -1$.

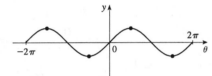

21. $f(x) = x^5$. No maximum or minimum.

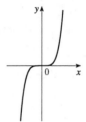

23. $f(x) = 1 - e^{-x}$, $x \geq 0$. Absolute minimum $f(0) = 0$; no local minimum. No absolute or local maximum.

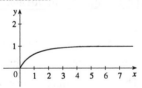

25. $f(x) = 4x^3 - 9x^2 - 12x + 3 \Longrightarrow$
$f'(x) = 12x^2 - 18x - 12 = 6\left(2x^2 - 3x - 2\right)$
$= 6\left(2x + 1\right)\left(x - 2\right)$.

$f'(x) = 0 \Longrightarrow x = -\frac{1}{2}, 2$; so the critical numbers are $x = -\frac{1}{2}, 2$.

27. $s(t) = t^4 + 4t^3 + 2t^2 \Longrightarrow s'(t) = 4t^3 + 12t^2 + 4t = 4t\left(t^2 + 3t + 1\right) = 0$ when $t = 0$ or $t^2 + 3t + 1 = 0$. By the quadratic formula, the critical numbers are $t = 0, \frac{-3 \pm \sqrt{5}}{2}$.

29. $f(r) = \dfrac{r}{r^2 + 1} \Longrightarrow f'(r) = \dfrac{\left(r^2 + 1\right)1 - r\left(2r\right)}{\left(r^2 + 1\right)^2} = \dfrac{-r^2 + 1}{\left(r^2 + 1\right)^2} = 0 \Longleftrightarrow r^2 = 1 \Longleftrightarrow r = \pm 1$, so these are the critical numbers. Note that $f'(x)$ always exists since $r^2 + 1 \neq 0$.

31. $F(x) = x^{4/5}\left(x - 4\right)^2 \Longrightarrow$
$F'(x) = \frac{4}{5}x^{-1/5}\left(x - 4\right)^2 + 2x^{4/5}\left(x - 4\right) = \frac{1}{5}x^{-1/5}\left(x - 4\right)\left[4\left(x - 4\right) + 10x\right]$
$= \dfrac{\left(x - 4\right)\left(14x - 16\right)}{5x^{1/5}} = \dfrac{2\left(x - 4\right)\left(7x - 8\right)}{5x^{1/5}} = 0$ when $x = 4, \frac{8}{7}$; and $F'(0)$ does not exist.

Critical numbers are $0, \frac{8}{7}, 4$.

33. $f(\theta) = \sin^2(2\theta) \implies f'(\theta) = 2\sin(2\theta)\cos(2\theta)(2) = 2(2\sin 2\theta \cos 2\theta) = 2[\sin(2 \cdot 2\theta)] = 2\sin 4\theta = 0 \iff \sin 4\theta = 0 \iff 4\theta = n\pi$, n an integer. So $\theta = n\pi/4$ are the critical numbers.

35. $f(x) = x\ln x \implies f'(x) = x(1/x) + \ln x = \ln x + 1 = 0 \iff \ln x = -1 \iff x = e^{-1} = 1/e$. Therefore, the only critical number is $x = 1/e$.

37. $f(x) = x^2 - 2x + 2$, $[0, 3]$. $f'(x) = 2x - 2 = 0 \iff x = 1$. $f(0) = 2$, $f(1) = 1$, $f(3) = 5$. So $f(3) = 5$ is the absolute maximum and $f(1) = 1$ is the absolute minimum.

39. $f(x) = 3x^5 - 5x^3 - 1$, $[-2, 2]$. $f'(x) = 15x^4 - 15x^2 = 15x^2(x+1)(x-1) = 0 \iff x = -1, 0, 1$. $f(-2) = -57$, $f(-1) = 1$, $f(0) = -1$, $f(1) = -3$, $f(2) = 55$. So $f(-2) = -57$ is the absolute minimum and $f(2) = 55$ is the absolute maximum.

41. $f(x) = x^2 + \dfrac{2}{x}$, $[\frac{1}{2}, 2]$. $f'(x) = 2x - \dfrac{2}{x^2} = 2\dfrac{x^3 - 1}{x^2} = 0 \iff x^3 - 1 = 0 \iff$ $(x-1)(x^2 + x + 1) = 0$, but $x^2 + x + 1 \neq 0$, so $x = 1$. The denominator is 0 at $x = 0$, but not in the desired interval. $f\left(\frac{1}{2}\right) = \frac{17}{4}$, $f(1) = 3$, $f(2) = 5$. So $f(1) = 3$ is the absolute minimum and $f(2) = 5$ is the absolute maximum.

43. $f(x) = \sin x + \cos x$, $\left[0, \frac{\pi}{3}\right]$. $f'(x) = \cos x - \sin x = 0 \iff \sin x = \cos x \implies \dfrac{\sin x}{\cos x} = 1 \implies \tan x = 1 \implies x = \frac{\pi}{4}$. $f(0) = 1$, $f\left(\frac{\pi}{4}\right) = \sqrt{2} \approx 1.41$, $f\left(\frac{\pi}{3}\right) = \frac{\sqrt{3}+1}{2} \approx 1.37$. So $f(0) = 1$ is the absolute minimum and $f\left(\frac{\pi}{4}\right) = \sqrt{2}$ is the absolute maximum.

45. $f(x) = xe^{-x}$, $[0, 2]$. $f'(x) = x(-e^{-x}) + e^{-x} = e^{-x}(1 - x) = 0 \iff x = 1$. Now $f(0) = 0$, $f(1) = e^{-1} = 1/e$, and $f(2) = 2/e^2 \approx 0.27$, so $f(0) = 0$ is the absolute minimum and $f(1) = 1/e$ is the absolute maximum.

47. (a)

From the graph, it appears that the absolute maximum value is about $f(-1.63) = 9.71$, and the absolute minimum value is about $f(1.63) = -7.71$. These values make sense because the graph is symmetric about the point $(0, 1)$. [$y = x^3 - 8x$ is symmetric about the origin.]

(b) $f(x) = x^3 - 8x + 1 \implies f'(x) = 3x^2 - 8$. So $f'(x) = 0 \implies x = \pm\sqrt{\frac{8}{3}}$.

$$f\left(\pm\sqrt{\tfrac{8}{3}}\right) = \left(\pm\sqrt{\tfrac{8}{3}}\right)^3 - 8\left(\pm\sqrt{\tfrac{8}{3}}\right) + 1 = \pm\tfrac{8}{3}\sqrt{\tfrac{8}{3}} \mp 8\sqrt{\tfrac{8}{3}} + 1$$

$$= -\tfrac{16}{3}\sqrt{\tfrac{8}{3}} + 1 = 1 - \tfrac{32\sqrt{6}}{9} \text{ (minimum) or } \tfrac{16}{3}\sqrt{\tfrac{8}{3}} + 1 = 1 + \tfrac{32\sqrt{6}}{9} \text{ (maximum).}$$

(From the graph, we see that the extreme values do not occur at the endpoints.)

49. (a)

From the graph, it appears that the absolute maximum value is about $f(0.75) = 0.32$, and the absolute minimum value is $f(0) = f(1) = 0$, that is, at both endpoints.

(b) $f(x) = x\sqrt{x - x^2} \implies$

$$f'(x) = x \cdot \frac{1 - 2x}{2\sqrt{x - x^2}} + \sqrt{x - x^2} = \frac{(x - 2x^2) + (2x - 2x^2)}{2\sqrt{x - x^2}} = \frac{3x - 4x^2}{2\sqrt{x - x^2}}. \text{ So } f'(x) = 0 \implies$$

$3x - 4x^2 = 0 \implies x = 0 \text{ or } \frac{3}{4}. \ f(0) = f(1) = 0 \text{ (minima), and } f\left(\frac{3}{4}\right) = \frac{3}{4}\sqrt{\frac{3}{4} - \left(\frac{3}{4}\right)^2} = \frac{3\sqrt{3}}{16}$

(maximum).

51. The density is defined as $\rho = \dfrac{\text{mass}}{\text{volume}} = \dfrac{1000}{V(T)}$ (in g/cm^3). But a critical point of ρ will

also be a critical point of V $\left[\text{since } \dfrac{d\rho}{dT} = -1000V^{-2}\dfrac{dV}{dT} \text{ and } V \text{ is never } 0\right]$, and V is easier

to differentiate than ρ. $V(T) = 999.87 - 0.06426T + 0.0085043T^2 - 0.0000679T^3 \implies$

$V'(T) = -0.06426 + 0.0170086T - 0.0002037T^2$. Setting this equal to 0 and using the quadratic

formula to find T, we get $T = \dfrac{-0.0170086 \pm \sqrt{0.0170086^2 - 4 \cdot 0.0002037 \cdot 0.06426}}{2(-0.0002037)} \approx 3.9665°$

or $79.5318°$. Since we are only interested in the region $0° \le T \le 30°$, we check the density ρ

at the endpoints and at $3.9665°$: $\rho(0) \approx \dfrac{1000}{999.87} \approx 1.00013$; $\rho(30) \approx \dfrac{1000}{1003.7641} \approx 0.99625$;

$\rho(3.9665) \approx \dfrac{1000}{999.7447} \approx 1.000255$. So water has its maximum density at about $3.9665°$C.

53. $v(t) = 0.001302t^3 - 0.09029t^2 + 23.61t - 3.083 \implies$

$a(t) = v'(t) = 0.003906t^2 - 0.18058t + 23.61 \implies a'(t) = 0.007812t - 0.18058. \ a'(t) = 0 \implies$

$t_1 = \dfrac{0.18058}{0.007812} \approx 23.12$. Evaluating $a(t)$ at the critical number and the endpoints gives us:

$a(0) = 23.61$, $a(t_1) \approx 21.52$, and $a(126) \approx 62.87$. The absolute maximum is about 62.87 ft/s^2 and

the absolute minimum is about 21.52 ft/s^2.

55. (a) $v(r) = k(r_0 - r)r^2 = kr_0r^2 - kr^3 \implies$

$v'(r) = 2kr_0r - 3kr^2. \ v'(r) = 0 \implies kr(2r_0 - 3r) = 0 \implies$

$r = 0 \text{ or } \frac{2}{3}r_0 \text{ (but 0 is not in the interval). Evaluating } v \text{ at } \frac{1}{2}r_0,$

$\frac{2}{3}r_0, \text{ and } r_0, \text{ we get } v\left(\frac{1}{2}r_0\right) = \frac{1}{8}kr_0^3, \ v\left(\frac{2}{3}r_0\right) = \frac{4}{27}kr_0^3, \text{ and}$

$v(r_0) = 0$. Since $\frac{4}{27} > \frac{1}{8}$, v attains its maximum value at

$r = \frac{2}{3}r_0$. This supports the statement in the text.

(c) From part (a), the maximum value of v is $\frac{4}{27}kr_0^3$.

(c)

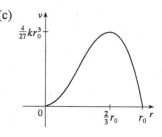

Section 4.3 Derivatives and the Shapes of Curves

Note: We will use the following abbreviations:

D	the domain of f	VA	vertical asymptote(s)
HA	horizontal asymptote(s)	IP	inflection point(s)
CU	concave upward	CD	concave downward

1. Geometrically, we are looking for the x-coordinates at which the slope of the tangent line equals the slope of the line segment connecting the endpoints. $\dfrac{f(8) - f(0)}{8 - 0} = \dfrac{6 - 4}{8} = \dfrac{1}{4}$. The values of c which satisfy $f'(c) = \frac{1}{4}$ seem to be about $c = 0.8, 3.2, 4.4,$ and 6.1.

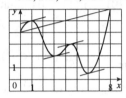

3. (a) Use the Increasing/Decreasing (I/D) Test on page 285.

(b) Use the Concavity Test on page 287.

(c) At any value of x where the concavity changes, we have an inflection point at $(x, f(x))$.

5. There is an inflection point at $x = 1$ because $f''(x)$ changes from negative to positive there, and one at $x = 7$ because $f''(x)$ changes from positive to negative there.

7. (a) $f(x) = x^6 + 192x + 17 \implies f'(x) = 6x^5 + 192 = 6(x^5 + 32)$. So $f'(x) > 0 \iff x^5 > -32 \iff x > -2$ and $f'(x) < 0 \iff x < -2$. So f is increasing on $(-2, \infty)$ and decreasing on $(-\infty, -2)$.

(b) f changes from decreasing to increasing at its only critical number, $x = -2$. Thus, $f(-2) = -303$ is a local minimum.

(c) $f''(x) = 30x^4 \geq 0$ for all x, so the concavity of f doesn't change and there are no inflection points. f is concave upward on $(-\infty, \infty)$.

9. (a) $y = f(x) = xe^x \implies f'(x) = xe^x + e^x = e^x(x + 1)$. So $f'(x) > 0 \iff x + 1 > 0 \iff x > -1$. Thus, f is increasing on $(-1, \infty)$ and decreasing on $(-\infty, -1)$.

(b) f changes from decreasing to increasing at its only critical number, $x = -1$. Thus, $f(-1) = -e^{-1}$ is a local minimum.

(c) $f'(x) = e^x(x + 1) \implies f''(x) = e^x(1) + (x + 1)e^x = e^x(x + 2)$. So $f''(x) > 0 \iff x + 2 > 0 \iff x > -2$. Thus, f is concave upward on $(-2, \infty)$ and concave downward on $(-\infty, -2)$. Since the concavity changes direction at $x = -2$, the point $(-2, -2e^{-2})$ is an inflection point.

11. (a) $y = f(x) = \dfrac{\ln x}{\sqrt{x}}$.   (Note that f is only defined for $x > 0$.)

$$f'(x) = \frac{\sqrt{x}\,(1/x) - \ln x \left(\frac{1}{2}x^{-1/2}\right)}{x} = \frac{\dfrac{1}{\sqrt{x}} - \dfrac{\ln x}{2\sqrt{x}}}{x} \cdot \frac{2\sqrt{x}}{2\sqrt{x}} = \frac{2 - \ln x}{2x^{3/2}} > 0 \iff \ln x < 2 \iff$$

$x < e^2$. Therefore f is increasing on $(0, e^2)$ and decreasing on (e^2, ∞).

(b) f changes from increasing to decreasing at $x = e^2$, so $f\left(e^2\right) = \dfrac{\ln e^2}{\sqrt{e^2}} = \dfrac{2}{e}$ is a local maximum.

(c) $f''(x) = \dfrac{2x^{3/2}\,(-1/x) - (2 - \ln x)\left(3x^{1/2}\right)}{\left(2x^{3/2}\right)^2} = \dfrac{-2x^{1/2} + 3x^{1/2}\,(\ln x - 2)}{4x^3}$

$= \dfrac{x^{1/2}\,(-2 + 3\ln x - 6)}{4x^3} = \dfrac{3\ln x - 8}{4x^{5/2}}$

$f''(x) = 0 \iff \ln x = \frac{8}{3} \iff x = e^{8/3}$. $f''(x) > 0 \iff x > e^{8/3}$, so f is concave upward on $\left(e^{8/3}, \infty\right)$ and concave downward on $\left(0, e^{8/3}\right)$. There is an inflection point at $\left(e^{8/3}, 8/\left(3e^{4/3}\right)\right) \approx (14.39, 0.70)$.

13. (a) $f(x) = 1 - 3x + 5x^2 - x^3 \implies f'(x) = -3 + 10x - 3x^2 = -\left(3x^2 - 10x + 3\right) = -(3x - 1)(x - 3)$. $f'(x) = 0 \iff x = \frac{1}{3}, 3$. $f'(x) > 0 \iff \frac{1}{3} < x < 3$ [the graph of f' is a parabola opening down]. Thus, f is increasing on $\left(\frac{1}{3}, 3\right)$ and decreasing on $\left(-\infty, \frac{1}{3}\right)$ and $(3, \infty)$.

(b) f changes from decreasing to increasing at $x = \frac{1}{3}$, so $f\left(\frac{1}{3}\right) = \frac{14}{27}$ is a local minimum. f changes from increasing to decreasing at $x = 3$, so $f(3) = 10$ is a local maximum.

(c) $f'(x) = -3x^2 + 10x - 3 \implies f''(x) = -6x + 10 = 0 \iff x = \frac{5}{3}$. $f''(x) > 0 \iff x < \frac{5}{3}$, so f is CU on $\left(-\infty, \frac{5}{3}\right)$ and CD on $\left(\frac{5}{3}, \infty\right)$. There is an IP at $\left(\frac{5}{3}, \frac{142}{27}\right) \approx (1.67, 5.26)$.

(d)

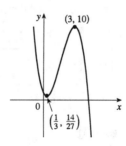

15. (a) $f(x) = (x^2 - 1)^3 \implies f'(x) = 6x(x^2 - 1)^2 \geq 0 \iff x > 0 \ (x \neq 1)$, so f is increasing on $(0, \infty)$ and decreasing on $(-\infty, 0)$.

(b) $f(0) = -1$ is a local minimum since f changes from decreasing to increasing at $x = 0$.

(c) $f''(x) = 6(x^2 - 1)^2 + 24x^2(x^2 - 1) = 6(x^2 - 1)(5x^2 - 1)$. The roots ± 1 and $\pm\frac{1}{\sqrt{5}}$ divide $\mathbb{R}$ into five intervals.

Interval	$x^2 - 1$	$5x^2 - 1$	$f''(x)$	Concavity
$x < -1$	$+$	$+$	$+$	upward
$-1 < x < -\frac{1}{\sqrt{5}}$	$-$	$+$	$-$	downward
$-\frac{1}{\sqrt{5}} < x < \frac{1}{\sqrt{5}}$	$-$	$-$	$+$	upward
$\frac{1}{\sqrt{5}} < x < 1$	$-$	$+$	$-$	downward
$x > 1$	$+$	$+$	$+$	upward

From the table, we see that f is CU on $(-\infty, -1)$, $\left(-\frac{1}{\sqrt{5}}, \frac{1}{\sqrt{5}}\right)$ and $(1, \infty)$, and CD on $\left(-1, -\frac{1}{\sqrt{5}}\right)$ and $\left(\frac{1}{\sqrt{5}}, 1\right)$. There are inflection points at $x = \pm 1, \pm\frac{1}{\sqrt{5}}$.

(d)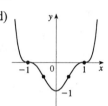

17. (a) $f(x) = x^{1/3}(x + 3)^{2/3} \implies$

$$f'(x) = \tfrac{1}{3}x^{-2/3}(x + 3)^{2/3} + x^{1/3}\left(\tfrac{2}{3}\right)(x + 3)^{-1/3}$$

$$= \frac{(x + 3)^{2/3}}{3x^{2/3}} + \frac{2x^{1/3}}{3(x + 3)^{1/3}} = \frac{(x + 3) + 2x}{3x^{2/3}(x + 3)^{1/3}} = \frac{x + 1}{x^{2/3}(x + 3)^{1/3}}.$$

The critical numbers are $-3, -1$, and 0. Note that $x^{2/3} \geq 0$ for all x. So $f'(x) > 0$ when $x < -3$ or $x > -1$ and $f'(x) < 0$ when $-3 < x < -1 \implies f$ is increasing on $(-\infty, -3)$ and $(-1, \infty)$ and decreasing on $(-3, -1)$.

(b) At $x = -3$, f changes from increasing to decreasing and at $x = -1$, vice versa, so $f(-3) = 0$ is a local maximum and $f(-1) = -4^{1/3} \approx -1.6$ is a local minimum.

(c) $f''(x) = \dfrac{x^{2/3}(x + 3)^{1/3} \cdot 1 - (x + 1)\left[x^{2/3}\frac{1}{3}(x + 3)^{-2/3} + (x + 3)^{1/3}\frac{2}{3}x^{-1/3}\right]}{\left[x^{2/3}(x + 3)^{1/3}\right]^2}$

$$= \frac{x^{2/3}(x + 3)^{1/3} - (x + 1)\left[\dfrac{x^{2/3}}{3(x + 3)^{2/3}} + \dfrac{2(x + 3)^{1/3}}{3x^{1/3}}\right]}{x^{4/3}(x + 3)^{2/3}}$$

$$= \frac{x^{2/3}(x + 3)^{1/3} - (x + 1)\left[\dfrac{x + 2(x + 3)}{3x^{1/3}(x + 3)^{2/3}}\right]}{x^{4/3}(x + 3)^{2/3}} = \frac{3x(x + 3) - 3(x + 1)(x + 2)}{3x^{5/3}(x + 3)^{4/3}}$$

$$= -\frac{2}{x^{5/3}(x + 3)^{4/3}}$$

Note that $(x+3)^{4/3} > 0$ for $x \neq -3$, so the sign of $f''(x)$ is the same as the sign of $-x^{5/3}$. Thus, $f''(x) > 0$ when $x < 0$, so f is CU on $(-\infty, -3)$ and $(-3, 0)$ and CD on $(0, \infty)$. There is an IP at $x = 0$.

(d)

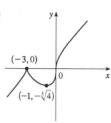

19. (a) $f(x) = 2\cos x + \sin^2 x \implies f'(x) = -2\sin x + 2\sin x \cos x = 2\sin x(\cos x - 1)$. Since $\cos x \leq 1$, $\cos x - 1 \leq 0$, so the sign of $f'(x)$ is the opposite of the sign of $\sin x$. Thus, $f'(x) > 0 \iff \sin x < 0 \iff (2n-1)\pi < x < 2n\pi$, so f is increasing on $((2n-1)\pi, 2n\pi)$ and decreasing on $(2n\pi, (2n+1)\pi)$.

(b) f changes from increasing to decreasing at $x = 2n\pi$, so $f(2n\pi) = 2$ are local maxima. f changes from decreasing to increasing when $x = 2n\pi + \pi$, so $f((2n+1)\pi) = -2$ are local minima.

(c) $f'(x) = -2\sin x + 2\sin x \cos x = -2\sin x + \sin 2x \implies$
$f''(x) = -2\cos x + 2\cos 2x = 2(2\cos^2 x - \cos x - 1)$
$= 2(2\cos x + 1)(\cos x - 1) > 0 \iff$
$\cos x < -\frac{1}{2} \iff x \in \left(2n\pi + \frac{2\pi}{3}, 2n\pi + \frac{4\pi}{3}\right)$, so f is CU on these intervals and CD on $\left(2n\pi - \frac{2\pi}{3}, 2n\pi + \frac{2\pi}{3}\right)$. There are IP at $\left(2n\pi \pm \frac{2\pi}{3}, -\frac{1}{4}\right)$.

(d)

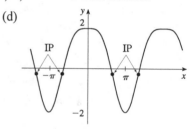

21. (a) $\displaystyle\lim_{x \to \pm\infty} \frac{1+x^2}{1-x^2} = \lim_{x \to \pm\infty} \frac{(1/x^2)+1}{(1/x^2)-1} = -1$, so $y = -1$ is a HA. $\displaystyle\lim_{x \to 1^-} \frac{1+x^2}{1-x^2} = \infty$,
$\displaystyle\lim_{x \to 1^+} \frac{1+x^2}{1-x^2} = -\infty$, $\displaystyle\lim_{x \to -1^-} \frac{1+x^2}{1-x^2} = -\infty$, $\displaystyle\lim_{x \to -1^+} \frac{1+x^2}{1-x^2} = \infty$. So $x = 1$ and $x = -1$ are VA.

(b) $f(x) = \dfrac{1+x^2}{1-x^2} = -1 + \dfrac{2}{1-x^2} \implies f'(x) = \dfrac{4x}{(1-x^2)^2} > 0 \iff x > 0 \; (x \neq 1)$, so f increases
on $(0, 1)$, $(1, \infty)$ and decreases on $(-\infty, -1)$, $(-1, 0)$.

(c) $f(0) = 1$ is a local minimum.

(d) $f''(x) = \dfrac{4(1-x^2)^2 - 4x \cdot 2(1-x^2)(-2x)}{(1-x^2)^4} = \dfrac{4(1+3x^2)}{(1-x^2)^3}$.

Since the numerator is always positive, the sign of $f''(x)$ is the same as the sign of $1 - x^2$. Thus, $f''(x) > 0 \iff 1 - x^2 > 0 \iff$ $x^2 < 1 \iff -1 < x < 1$, so f is CU on $(-1, 1)$ and CD on $(-\infty, -1)$ and $(1, \infty)$. There is no IP since $x = \pm 1$ are not in the domain of f.

(e)

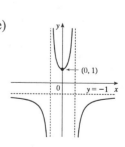

23. (a) $\lim\limits_{x \to -\infty} \left(\sqrt{x^2+1} - x\right) = \infty$ and

$$\lim_{x \to \infty} \left(\sqrt{x^2+1} - x\right) = \lim_{x \to \infty} \left(\sqrt{x^2+1} - x\right) \frac{\sqrt{x^2+1}+x}{\sqrt{x^2+1}+x} = \lim_{x \to \infty} \frac{1}{\sqrt{x^2+1}+x} = 0, \text{ so } y = 0 \text{ is}$$
a HA.

(b) $f(x) = \sqrt{x^2+1} - x \implies f'(x) = \dfrac{x}{\sqrt{x^2+1}} - 1.$ Since $\dfrac{x}{\sqrt{x^2+1}} < 1$ for all x, $f'(x) < 0$, so f

is decreasing on $\mathbb{R}$.

(c) No minimum or maximum

(d) $f''(x) = \dfrac{\left(x^2+1\right)^{1/2}(1) - x \cdot \frac{1}{2}\left(x^2+1\right)^{-1/2}(2x)}{\left(\sqrt{x^2+1}\right)^2}$

$= \dfrac{\left(x^2+1\right)^{1/2} - \dfrac{x^2}{\left(x^2+1\right)^{1/2}}}{x^2+1} = \dfrac{\left(x^2+1\right) - x^2}{\left(x^2+1\right)^{3/2}}$

$= \dfrac{1}{\left(x^2+1\right)^{3/2}} > 0,$ so f is CU on $\mathbb{R}$. No IP

(e)

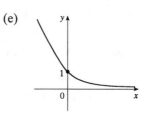

25. (a) $\lim\limits_{x \to \pm\infty} e^{-1/(x+1)} = 1$ since $-1/(x+1) \to 0,$

so $y = 1$ is a HA. $\lim\limits_{x \to -1^+} e^{-1/(x+1)} = 0$ since $-1/(x+1) \to -\infty,$ $\lim\limits_{x \to -1^-} e^{-1/(x+1)} = \infty$ since

$-1/(x+1) \to \infty,$ so $x = -1$ is a VA.

(b) $f(x) = e^{-1/(x+1)} \implies f'(x) = e^{-1/(x+1)}/(x+1)^2 \implies f'(x) > 0$ for all x except -1, so f is

increasing on $(-\infty, -1)$ and $(-1, \infty).$

(c) No local maximum or minimum

(d) $f''(x) = \dfrac{(x+1)^2 e^{-1/(x+1)}\left[1\big/(x+1)^2\right] - e^{-1/(x+1)}\left[2(x+1)\right]}{\left[(x+1)^2\right]^2} = \dfrac{e^{-1/(x+1)}\left[1 - (2x+2)\right]}{(x+1)^4}$

$= -\dfrac{e^{-1/(x+1)}(2x+1)}{(x+1)^4} \implies$

$f''(x) > 0 \iff 2x+1 < 0 \iff x < -\frac{1}{2},$ so f is CU on $(-\infty, -1)$ and $\left(-1, -\frac{1}{2}\right),$ and CD on

$\left(-\frac{1}{2}, \infty\right).$ f has an IP at $\left(-\frac{1}{2}, e^{-2}\right).$

(e)

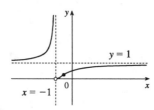

27. (a) From the graphs of $f(x) = 3x^5 - 40x^3 + 30x^2$, it seems that f is concave upward on $(-2, 0.25)$ and $(2, \infty)$, and concave downward on $(-\infty, -2)$ and $(0.25, 2)$, with inflection points at about $(-2, 350)$, $(0.25, 1)$, and $(2, -100)$.

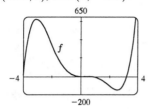
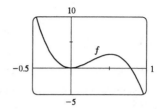

(b)

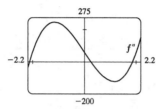

From the graph of $f''(x) = 60x^3 - 240x + 60$, it seems that f is CU on $(-2.1, 0.25)$ and $(1.9, \infty)$, and CD on $(-\infty, -2.1)$ and $(0.25, 2)$, with inflection points at about $(-2.1, 386)$, $(0.25, 1.3)$ and $(1.9, -87)$. (We have to check back on the graph of f to find the y-coordinates of the inflection points.)

29. (a)

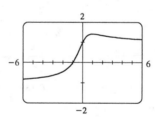

Tracing the graph gives us an estimate of $f(1) = 1.41$ as a local maximum, and no local minimum.

$f(x) = \dfrac{x+1}{\sqrt{x^2+1}} \implies f'(x) = \dfrac{1-x}{(x^2+1)^{3/2}}$. $f'(x) = 0 \iff x = 1$. $f(1) = \dfrac{2}{\sqrt{2}} = \sqrt{2}$ is the exact value.

(b) From the graph in part (a), f increases most rapidly somewhere between $x = -\frac{1}{2}$ and $x = -\frac{1}{4}$. To find the exact value, we need to find the maximum value of f', which we can do by finding the critical numbers of f'. $f''(x) = \dfrac{2x^2 - 3x - 1}{(x^2+1)^{5/2}} = 0 \iff x = \dfrac{3 \pm \sqrt{17}}{4}$. $x = \dfrac{3 + \sqrt{17}}{4}$ corresponds to the *minimum* value of f'. The maximum value of f' is at $\left(\dfrac{3 - \sqrt{17}}{4}, \sqrt{\dfrac{7}{6} - \dfrac{\sqrt{17}}{6}} \right) \approx (-0.28, 0.69)$.

31.

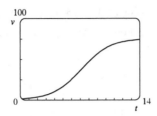

From the graph, we estimate that the most rapid increase in the number of VCRs occurs at about $t = 7$. To maximize the first derivative, we need to determine the values for which the second derivative is 0.

$$V(t) = \frac{75}{1 + 74e^{-0.6t}} \implies V'(t) = -\frac{75\left[74e^{-0.6t}(-0.6)\right]}{(1 + 74e^{-0.6t})^2} = \frac{3330e^{-0.6t}}{(1 + 74e^{-0.6t})^2} \implies$$

$$V''(t) = \frac{(1 + 74e^{-0.6t})^2\left[3330e^{-0.6t}(-0.6)\right] - (3330e^{-0.6t})\,2\,(1 + 74e^{-0.6t})\left[74e^{-0.6t}(-0.6)\right]}{\left[(1 + 74e^{-0.6t})^2\right]^2}$$

$$= \frac{(1 + 74e^{-0.6t})\left[3330e^{-0.6t}(-0.6)\right]\left[(1 + 74e^{-0.6t}) - 2\,(74e^{-0.6t})\right]}{(1 + 74e^{-0.6t})^4}$$

$$= \frac{-1998e^{-0.6t}\,(1 - 74e^{-0.6t})}{(1 + 74e^{-0.6t})^3}$$

$V''(t) = 0 \iff 1 = 74e^{-0.6t} \iff e^{0.6t} = 74 \iff 0.6t = \ln 74 \iff t = \frac{5}{3}\ln 74 \approx 7.173$ years, which corresponds to early September 1987.

33. In Maple, we define f and then use the command
`plot(diff(diff(f,x),x),x=-3..3);`. In Mathematica,
we define f and then use
`Plot[Dt[Dt[f,x],x],{x,-3,3}]`. We see that $f'' > 0$ for
$x > 0.1$ and $f'' < 0$ for $x < 0.1$. So f is concave up on
$(0.1, \infty)$ and concave down on $(-\infty, 0.1)$.

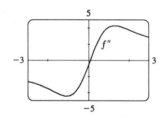

35. $f(x) = ax^3 + bx^2 + cx + d \implies f(1) = a + b + c + d = 0$,
$f(-2) = -8a + 4b - 2c + d = 3$, and
$f'(x) = 3ax^2 + 2bx + c$. Also $f'(1) = 3a + 2b + c = 0$ and
$f'(-2) = 12a - 4b + c = 0$ by Fermat's Theorem. Solving
these four equations, we get $a = \frac{2}{9}, b = \frac{1}{3}, c = -\frac{4}{3}, d = \frac{7}{9}$, so
the function is $f(x) = \frac{1}{9}\left(2x^3 + 3x^2 - 12x + 7\right)$.

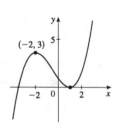

37. If f and g are CU on I, then $f'' > 0$ and $g'' > 0$ on I, so $(f + g)'' = f'' + g'' > 0$ on $I \implies f + g$ is CU on I.

39. Since f and g are positive, increasing, and CU on I, we have $f > 0$, $f' > 0$, $f'' > 0$, $g > 0$, $g' > 0$, $g'' > 0$ on I. Then $(fg)' = fg' + f'g \Longrightarrow$ $(fg)'' = (fg'' + g'f') + (gf'' + f'g') = fg'' + 2f'g' + f''g > 0 \Longrightarrow fg$ is CU on I.

41. $f(x) = \tan x - x \Longrightarrow f'(x) = \sec^2 x - 1 > 0$ for $0 < x < \frac{\pi}{2}$ since $\sec^2 x > 1$ for $0 < x < \frac{\pi}{2}$. So f is increasing on $\left(0, \frac{\pi}{2}\right)$. Thus, $f(x) > f(0) = 0$ for $0 < x < \frac{\pi}{2} \Longrightarrow \tan x - x > 0 \Longrightarrow \tan x > x$ for $0 < x < \frac{\pi}{2}$.

43. We are given that f is differentiable (and therefore continuous) everywhere. In particular, we can apply the Mean Value Theorem on the interval $[0, 4]$. There exists a number c in $(0, 4)$ such that $f(4) - f(0) = f'(c)(4 - 0)$, so $f(4) = f(0) + 4f'(c) = -3 + 4f'(c)$. We are given that $f'(x) \le 5$ for all x, so in particular we know that $f'(c) \le 5$. Multiplying both sides of this inequality by 4, we have $4f'(c) \le 20$, so $f(4) = -3 + 4f'(c) \le -3 + 20 = 17$. The largest possible value for $f(4)$ is 17.

45. Let $g(t)$ and $h(t)$ be the position functions of the two runners and let $f(t) = g(t) - h(t)$. By hypothesis, $f(0) = g(0) - h(0) = 0$ and $f(b) = g(b) - h(b) = 0$, where b is the finishing time. Then by the Mean Value Theorem, there is a time c, with $0 < c < b$, such that $f'(c) = \dfrac{f(b) - f(0)}{b - 0}$. But $f(b) = f(0) = 0$, so $f'(c) = 0$. Since $f'(c) = g'(c) - h'(c) = 0$, we have $g'(c) = h'(c)$. So at time c, both runners have the same velocity $g'(c) = h'(c)$.

47. Let the cubic function be $f(x) = ax^3 + bx^2 + cx + d \Longrightarrow f'(x) = 3ax^2 + 2bx + c \Longrightarrow$ $f''(x) = 6ax + 2b$. So f is CU when $6ax + 2b > 0 \Longleftrightarrow x > -\dfrac{b}{3a}$, and CD when $x < -\dfrac{b}{3a}$, and so the only point of inflection occurs when $x = -\dfrac{b}{3a}$. If the graph has three x-intercepts x_1, x_2 and x_3, then the equation of $f(x)$ must factor as
$$f(x) = a(x - x_1)(x - x_2)(x - x_3)$$
$$= a\left[x^3 - (x_1 + x_2 + x_3)x^2 + (x_1 x_2 + x_1 x_3 + x_2 x_3)x - x_1 x_2 x_3\right]$$
So $b = -a(x_1 + x_2 + x_3)$. Hence, the x-coordinate of the point of inflection is
$$-\frac{b}{3a} = -\frac{-a(x_1 + x_2 + x_3)}{3a}$$
$$= \frac{x_1 + x_2 + x_3}{3}$$

Section 4.4   Graphing with Calculus and Calculators

Abbreviations: HA   horizontal asymptote(s)   VA   vertical asymptote(s)
                            CU   concave upward   CD   concave downward
                            IP   inflection point(s)   FDT   First Derivative Test

1. $f(x) = 4x^4 - 7x^2 + 4x + 6 \implies f'(x) = 16x^3 - 14x + 4 \implies f''(x) = 48x^2 - 14$

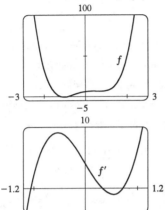

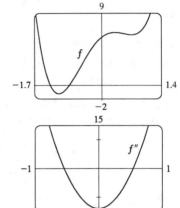

After finding suitable viewing rectangles (by ensuring that we have located all of the x-values where either $f' = 0$ or $f'' = 0$) we estimate from the graph of f' that f is increasing on $(-1.1, 0.3)$ and $(0.7, \infty)$ and decreasing on $(-\infty, -1.1)$ and $(0.3, 0.7)$, with a local maximum of $f(0.3) \approx 6.6$ and minima of $f(-1.1) \approx -1.0$ and $f(0.7) \approx 6.3$. We estimate from the graph of f'' that f is CU on $(-\infty, -0.5)$ and $(0.5, \infty)$ and CD on $(-0.5, 0.5)$, and that f has inflection points at about $(-0.5, 2.1)$ and $(0.5, 6.5)$.

3. $f(x) = \sqrt[3]{x^2 - 3x - 5} \implies f'(x) = \dfrac{1}{3}\dfrac{2x - 3}{(x^2 - 3x - 5)^{2/3}} \implies f''(x) = -\dfrac{2}{9}\dfrac{x^2 - 3x + 24}{(x^2 - 3x - 5)^{5/3}}$

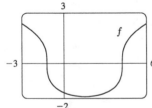

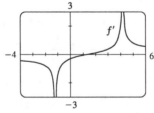

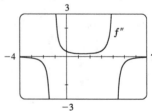

Note: With some CAS's, including Maple, it is necessary to define
$f(x) = \dfrac{x^2 - 3x - 5}{|x^2 - 3x - 5|}\,|x^2 - 3x - 5|^{1/3}$, since the CAS does not compute real cube roots of negative numbers. We estimate from the graph of f' that f is increasing on $(1.5, 4.2)$ and $(4.2, \infty)$, and decreasing on $(-\infty, -1.2)$ and $(-1.2, 1.5)$. f has no maximum. Minimum: $f(1.5) \approx -1.9$. From the graph of f'', we estimate that f is CU on $(-1.2, 4.2)$ and CD on $(-\infty, -1.2)$ and $(4.2, \infty)$. IP $(-1.2, 0)$ and $(4.2, 0)$.

5. $f(x) = x^2 \sin x \implies f'(x) = 2x \sin x + x^2 \cos x \implies f''(x) = 2 \sin x + 4x \cos x - x^2 \sin x$

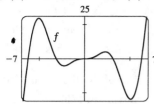 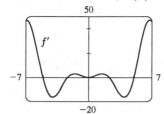 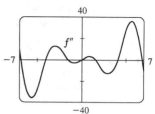

We estimate from the graph of f' that f is increasing on $(-7, -5.1)$, $(-2.3, 2.3)$, and $(5.1, 7)$ and decreasing on $(-5.1, -2.3)$, and $(2.3, 5.1)$. Local maxima: $f(-5.1) \approx 24.1$, $f(2.3) \approx 3.9$. Local minima: $f(-2.3) \approx -3.9$, $f(5.1) \approx -24.1$. From the graph of f'', we estimate that f is CU on $(-7, -6.8)$, $(-4.0, -1.5)$, $(0, 1.5)$, and $(4.0, 6.8)$, and CD on $(-6.8, -4.0)$, $(-1.5, 0)$, $(1.5, 4.0)$, and $(6.8, 7)$. f has IP at $(-6.8, -24.4)$, $(-4.0, 12.0)$, $(-1.5, -2.3)$, $(0, 0)$, $(1.5, 2.3)$, $(4.0, -12.0)$ and $(6.8, 24.4)$.

7.

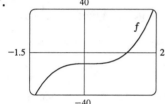 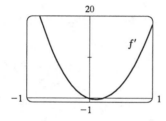 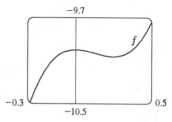

From the graphs, it appears that $f(x) = 8x^3 - 3x^2 - 10$ increases on $(-\infty, 0)$ and $(0.25, \infty)$ and decreases on $(0, 0.25)$; that f has a local maximum of $f(0) = -10.0$ and a local minimum of $f(0.25) \approx -10.1$; that f is CU on $(0.1, \infty)$ and CD on $(-\infty, 0.1)$; and that f has an IP at $(0.1, -10)$. $f(x) = 8x^3 - 3x^2 - 10 \implies f'(x) = 24x^2 - 6x = 6x(4x - 1)$, which is positive ($f$ is increasing) for $(-\infty, 0)$ and $(\frac{1}{4}, \infty)$, and negative (f is decreasing) on $(0, \frac{1}{4})$. By the FDT, f has a local maximum at $x = 0$: $f(0) = -10$; and f has a local minimum at $\frac{1}{4}$: $f(\frac{1}{4}) = \frac{1}{8} - \frac{3}{16} - 10 = -\frac{161}{16}$. $f'(x) = 24x^2 - 6x \implies f''(x) = 48x - 6 = 6(8x - 1)$, which is positive ($f$ is CU) on $(\frac{1}{8}, \infty)$, and negative (f is CD) on $(-\infty, \frac{1}{8})$. f has an IP at $(\frac{1}{8}, f(\frac{1}{8})) = (\frac{1}{8}, -\frac{321}{32})$.

9.

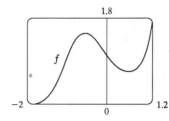

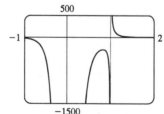

$f(x) = e^{x^3 - x} \to 0$ as $x \to -\infty$, and $f(x) \to \infty$ as $x \to \infty$. From the graph, it appears that f has a local minimum of about $f(0.58) = 0.68$, and a local maximum of about $f(-0.58) = 1.47$.

To find the exact values, we calculate $f'(x) = (3x^2 - 1)\, e^{x^3 - x}$, which is 0 when $3x^2 - 1 = 0 \iff x = \pm\frac{1}{\sqrt{3}}$. The negative root corresponds to the local maximum

$f\left(-\frac{1}{\sqrt{3}}\right) = e^{(-1/\sqrt{3})^3 - (-1/\sqrt{3})} = e^{2\sqrt{3}/9}$, and the positive root corresponds to the local

minimum $f\left(\frac{1}{\sqrt{3}}\right) = e^{(1/\sqrt{3})^3 - (1/\sqrt{3})} = e^{-2\sqrt{3}/9}$. To estimate the inflection points, we calculate

and graph $f''(x) = \dfrac{d}{dx}\left[(3x^2 - 1)\, e^{x^3 - x}\right] = (3x^2 - 1)\, e^{x^3 - x}\, (3x^2 - 1) + e^{x^3 - x}\, (6x) =$

$e^{x^3 - x}\left(9x^4 - 6x^2 + 6x + 1\right)$. From the graph, it appears that $f''(x)$ changes sign (and thus f has

inflection points) at $x \approx -0.15$ and $x \approx -1.09$. From the graph of f, we see that these x-values

correspond to inflection points at about $(-0.15, 1.15)$ and $(-1.09, 0.82)$.

11.

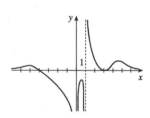

$f(x) = \dfrac{(x+4)(x-3)^2}{x^4(x-1)}$ has VA at $x = 0$ and at $x = 1$ since

$\lim\limits_{x \to 0} f(x) = -\infty$, $\lim\limits_{x \to 1^-} f(x) = -\infty$ and $\lim\limits_{x \to 1^+} f(x) = \infty$.

$f(x) = \dfrac{(1 + 4/x)(1 - 3/x)^2}{x(x-1)} \to 0^+$ as $x \to \pm\infty$, so f is

asymptotic to the x-axis. Since f is undefined at $x = 0$, it has no y-intercept.

$f(x) = 0 \implies (x+4)(x-3)^2 = 0 \implies x = -4$ or $x = 3$, so f has x-intercepts -4 and 3. Note, however, that the graph of f is only tangent to the x-axis and does not cross it at $x = 3$, since f is positive as $x \to 3^-$ and as $x \to 3^+$.

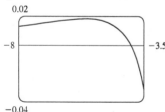

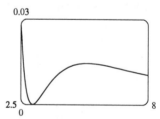

From these graphs, it appears that f has three maxima and one minimum. The maxima are approximately $f(-5.6) = 0.0182$, $f(0.82) = -281.5$ and $f(5.2) = 0.0145$ and we know (since the graph is tangent to the x-axis at $x = 3$) that the minimum is $f(3) = 0$.

13. $f(x) = \dfrac{x^2(x+1)^3}{(x-2)^2(x-4)^4} \implies f'(x) = -\dfrac{x(x+1)^2(x^3+18x^2-44x-16)}{(x-2)^3(x-4)^5}$ (from CAS).

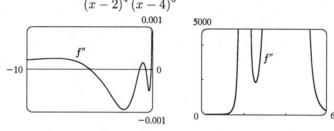

From the graphs of f', it seems that the critical points which indicate extrema occur at $x \approx -20$, -0.3, and 2.5, as estimated in Example 3. (There is another critical point at $x = -1$, but the sign of f' does not change there.) We differentiate again, obtaining

$$f''(x) = 2\frac{(x+1)(x^6+36x^5+6x^4-628x^3+684x^2+672x+64)}{(x-2)^4(x-4)^6}.$$

From the graphs of f'', it appears that f is CU on $(-\infty, -5.0)$, $(-1.0, -0.5)$, $(-0.1, 2.0)$, $(2.0, 4.0)$

and $(4.0, \infty)$ and CD on $(-5.0, -1.0)$ and $(-0.5, -0.1)$. We check back on the graphs of f to find the y-coordinates of the inflection points, and find that these points are approximately $(-5, -0.005)$, $(-1, 0)$, $(-0.5, 0.00001)$, and $(-0.1, 0.0000066)$.

15.

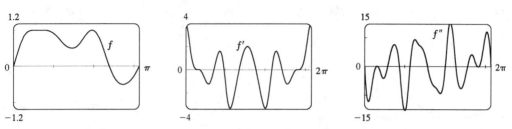

From the graph of $f(x) = \sin(x + \sin 3x)$ in the viewing rectangle $[0, \pi]$ by $[-1.2, 1.2]$, it looks like f has two maxima and two minima. If we calculate and graph $f'(x) = [\cos(x + \sin 3x)](1 + 3\cos 3x)$ on $[0, 2\pi]$, we see that the graph of f' appears to be almost tangent to the x-axis at about $x = 0.7$. The graph of $f'' = -[\sin(x + \sin 3x)](1 + 3\cos 3x)^2 + \cos(x + \sin 3x)(-9\sin 3x)$ is even more interesting near this x-value: it seems to just touch the x-axis.

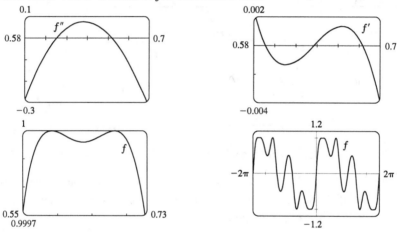

If we zoom in on this place on the graph of f'', we see that f'' actually does cross the axis twice near $x = 0.65$, indicating a change in concavity for a very short interval. If we look at the graph of f' on the same interval, we see that it changes sign three times near $x = 0.65$, indicating that what we had thought was a broad extremum at about $x = 0.7$ actually consists of three extrema (two maxima and a minimum). These maxima are roughly $f(0.59) = 1$ and $f(0.68) = 1$, and the minimum is roughly $f(0.64) = 0.99996$. There are also a maximum of about $f(1.96) = 1$ and minima of about $f(1.46) = 0.49$ and $f(2.73) = -0.51$. The points of inflection on $(0, \pi)$ are about $(0.61, 0.99998)$, $(0.66, 0.99998)$, $(1.17, 0.72)$, $(1.75, 0.77)$, and $(2.28, 0.34)$. On $(\pi, 2\pi)$, they are about $(4.01, -0.34)$, $(4.54, -0.77)$, $(5.11, -0.72)$, $(5.62, -0.99998)$, and $(5.67, -0.99998)$. There are also IP at $(0, 0)$ and $(\pi, 0)$. Note that the function is odd and periodic with period 2π, and it is also rotationally symmetric about all points of the form $((2n + 1)\pi, 0)$, n an integer.

17.

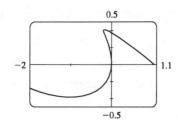

We graph the curve $x = t^4 - 2t^3 - 2t^2$, $y = t^3 - t$ in the viewing rectangle $[-2, 1.1]$ by

$[-0.5, 0.5]$. This rectangle corresponds approximately to $t \in [-1, 0.8]$. We estimate that the curve has horizontal tangents at about $(-1, -0.4)$ and $(-0.17, 0.39)$ and vertical tangents at about $(0, 0)$ and $(-0.19, 0.37)$. We calculate $\dfrac{dy}{dx} = \dfrac{dy/dt}{dx/dt} = \dfrac{3t^2 - 1}{4t^3 - 6t^2 - 4t}$. The horizontal tangents occur when $dy/dt = 3t^2 - 1 = 0 \iff t = \pm\frac{1}{\sqrt{3}}$, so both horizontal tangents are shown in our graph. $t = \frac{1}{\sqrt{3}}$ corresponds to the point $\left(\frac{-2\sqrt{3}-5}{9}, \frac{-2\sqrt{3}}{9}\right) \approx (-0.94, -0.38)$ and $t = -\frac{1}{\sqrt{3}}$ corresponds to $\left(\frac{2\sqrt{3}-5}{9}, \frac{2\sqrt{3}}{9}\right) \approx (-0.17, 0.38)$. The vertical tangents occur when $dx/dt = 2t\left(2t^2 - 3t - 2\right) = 0 \iff 2t(2t + 1)(t - 2) = 0 \iff t = 0, -\frac{1}{2}$ or 2. It seems that we have missed one vertical tangent, and indeed if we plot the curve on the t-interval $[-1.2, 2.2]$ we see that there is another vertical tangent at $(-8, 6)$. The t-values and points at which there are vertical tangents are $t = 0$, $(0, 0)$; $t = -\frac{1}{2}$, $\left(-\frac{3}{16}, \frac{3}{8}\right)$; and $t = 2$, $(-8, 6)$.

19. $x = t^3 - ct$, $y = t^2$. For $c = 0$, there is a cusp at $(0, 0)$. For $c < 0$, there is a local minimum at $(0, 0)$. For $c > 0$, there is a loop whose size increases as c increases ($c = \frac{1}{2}$ and $c = 1$ are shown in the figure). The curve intersects itself at the point $(0, c)$ when $t = \pm\sqrt{c}$ (solve $x = 0$.)

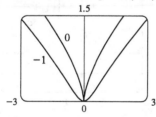
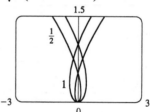

From the second figure, we see that the left- and rightmost points of the loop occur when there are vertical tangent lines. $dx/dt = 0 \implies 3t^2 - c = 0 \implies t = \pm\sqrt{c/3}$. The rightmost point occurs when $t = -\sqrt{c/3}$ and has coordinates $\left(\dfrac{2c\sqrt{3c}}{9}, \dfrac{c}{3}\right)$. The leftmost point occurs when $t = \sqrt{c/3}$ and has coordinates $\left(-\dfrac{2c\sqrt{3c}}{9}, \dfrac{c}{3}\right)$.

21. Note that $c = 0$ is a transitional value at which the graph consists of the x-axis. Also, we can see that if we substitute $-c$ for c, the function $f(x) = \dfrac{cx}{1 + c^2 x^2}$ will be reflected in the x-axis, so we investigate only positive values of c (except $c = -1$, as a demonstration of this reflective property). Also, f is an odd function. $\displaystyle\lim_{x \to \pm\infty} f(x) = 0$, so $y = 0$ is a horizontal asymptote for all c. We calculate

$$f'(x) = \frac{c\left(1 + c^2 x^2\right) - cx\left(2c^2 x\right)}{\left(1 + c^2 x^2\right)^2} = -\frac{c\left(c^2 x^2 - 1\right)}{\left(1 + c^2 x^2\right)^2}. \ f'(x) = 0 \Longleftrightarrow c^2 x^2 - 1 = 0 \Longleftrightarrow x = \pm 1/c.$$

So there is an absolute maximum of $f(1/c) = \frac{1}{2}$ and an absolute minimum of $f(-1/c) = -\frac{1}{2}$. These extrema have the same value regardless of c, but the maximum points move closer to the y-axis as c increases.

$$f''(x) = \frac{\left(-2c^3 x\right)\left(1 + c^2 x^2\right)^2 - \left(-c^3 x^2 + c\right)\left[2\left(1 + c^2 x^2\right)\left(2c^2 x\right)\right]}{\left(1 + c^2 x^2\right)^4}$$

$$= \frac{\left(-2c^3 x\right)\left(1 + c^2 x^2\right) + \left(c^3 x^2 - c\right)\left(4c^2 x\right)}{\left(1 + c^2 x^2\right)^3} = \frac{2c^3 x\left(c^2 x^2 - 3\right)}{\left(1 + c^2 x^2\right)^3}.$$

$f''(x) = 0 \Longleftrightarrow x = 0$ or $\pm\sqrt{3}/c$, so there are inflection points at $(0, 0)$ and at $\left(\pm\sqrt{3}/c, \pm\sqrt{3}/4\right)$. Again, the y-coordinate of the inflection points does not depend on c, but as c increases, both inflection points approach the y-axis.

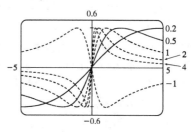

23.

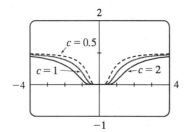

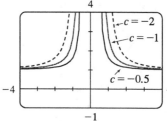

$c = 0$ is a transitional value — we get the graph of $y = 1$. For $c > 0$, we see that there is a HA at $y = 1$, and that the graph spreads out as c increases. At first glance there appears to be a minimum at $(0, 0)$, but $f(0)$ is undefined, so there is no minimum or maximum. For $c < 0$, we still have the HA at $y = 1$, but the range is $(1, \infty)$ rather than $(0, 1)$. We also have a VA at $x = 0$. $f(x) = e^{-c/x^2} \Longrightarrow$

$$f'(x) = e^{-c/x^2}\left(-2c/x^3\right) \Longrightarrow f''(x) = \frac{2c\left(2c - 3x^2\right)}{x^6 e^{c/x^2}}. \ f'(x) \neq 0 \text{ and } f'(x) \text{ exists for all } x \neq 0$$

(and 0 is not in the domain of f), so there are no maxima or minima. $f''(x) = 0 \Longrightarrow x = \pm\sqrt{2c/3}$, so if $c > 0$, the inflection points spread out as c increases, and if $c < 0$, there are no IP. For $c > 0$, there are IP at $\left(\pm\sqrt{2c/3}, e^{-3/2}\right)$. Note that the y-coordinate of the IP is constant.

25. $f(x) = x^4 + cx^2 = x^2 (x^2 + c)$. Note that f is an even function. For $c \geq 0$, the only x-intercept is the point $(0, 0)$. We calculate $f'(x) = 4x^3 + 2cx = 4x (x^2 + \frac{1}{2}c) \implies f''(x) = 12x^2 + 2c$. If $c \geq 0$, $x = 0$ is the only critical point and there are no inflection points. As we can see from the examples, there is no change in the basic shape of the graph for $c \geq 0$; it merely becomes steeper as c increases. For $c = 0$, the graph is the simple curve $y = x^4$. For $c < 0$, there are x-intercepts at 0 and at $\pm\sqrt{-c}$. Also, there is a maximum at $(0, 0)$, and there are minima at $\left(\pm\sqrt{-\frac{1}{2}c}, -\frac{1}{4}c^2\right)$. As $c \to -\infty$, the x-coordinates of these minima get larger in absolute value, and the minimum points move downward. There are inflection points at $\left(\pm\sqrt{-\frac{1}{6}c}, -\frac{5}{36}c^2\right)$, which also move away from the origin as $c \to -\infty$.

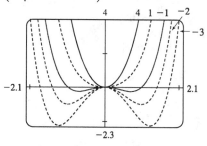

Section 4.5 Indeterminate Forms and l'Hospital's Rule

Note: The use of l'Hospital's Rule is indicated by an H above the equal sign: $\overset{H}{=}$

1. $\displaystyle\lim_{x\to2}\frac{x-2}{x^2-4}=\lim_{x\to2}\frac{x-2}{(x-2)(x+2)}=\lim_{x\to2}\frac{1}{x+2}=\frac{1}{4}$

3. $\displaystyle\lim_{x\to0}\frac{e^x-1}{\sin x}\overset{H}{=}\lim_{x\to0}\frac{e^x}{\cos x}=\frac{1}{1}=1$

5. $\displaystyle\lim_{x\to0}\frac{\tan x}{x+\sin x}\overset{H}{=}\lim_{x\to0}\frac{\sec^2 x}{1+\cos x}=\frac{1}{1+1}=\frac{1}{2}$

7. $\displaystyle\lim_{x\to\infty}\frac{e^x}{x^3}\overset{H}{=}\lim_{x\to\infty}\frac{e^x}{3x^2}\overset{H}{=}\lim_{x\to\infty}\frac{e^x}{6x}\overset{H}{=}\lim_{x\to\infty}\frac{e^x}{6}=\infty$

9. $\displaystyle\lim_{x\to0}\frac{e^x-1-x}{x^2}\overset{H}{=}\lim_{x\to0}\frac{e^x-1}{2x}\overset{H}{=}\lim_{x\to0}\frac{e^x}{2}=\frac{1}{2}$

11. $\displaystyle\lim_{x\to\infty}\frac{\ln\ln x}{\sqrt{x}}\overset{H}{=}\lim_{x\to\infty}\frac{1/(x\ln x)}{1/(2\sqrt{x})}=\lim_{x\to\infty}\frac{2}{\sqrt{x}\ln x}=0$

13. $\displaystyle\lim_{x\to0}\frac{\tan^{-1}(2x)}{3x}\overset{H}{=}\lim_{x\to0}\frac{2/(1+4x^2)}{3}=\frac{2}{3}$

15. $\displaystyle\lim_{x\to0^+}\sqrt{x}\ln x=\lim_{x\to0^+}\frac{\ln x}{x^{-1/2}}\overset{H}{=}\lim_{x\to0^+}\frac{1/x}{-\frac{1}{2}x^{-3/2}}=\lim_{x\to0^+}\left(-2\sqrt{x}\right)=0$

17. $\displaystyle\lim_{x\to\infty}e^{-x}\ln x=\lim_{x\to\infty}\frac{\ln x}{e^x}\overset{H}{=}\lim_{x\to\infty}\frac{1/x}{e^x}=\lim_{x\to\infty}\frac{1}{xe^x}=0$

19. $\displaystyle\lim_{x\to\infty}x^3e^{-x^2}=\lim_{x\to\infty}\frac{x^3}{e^{x^2}}\overset{H}{=}\lim_{x\to\infty}\frac{3x^2}{2xe^{x^2}}=\lim_{x\to\infty}\frac{3x}{2e^{x^2}}\overset{H}{=}\lim_{x\to\infty}\frac{3}{4xe^{x^2}}=0$

21. $\displaystyle\lim_{x\to0}\left(\frac{1}{x}-\csc x\right)=\lim_{x\to0}\left(\frac{1}{x}-\frac{1}{\sin x}\right)=\lim_{x\to0}\frac{\sin x-x}{x\sin x}$

$\displaystyle\overset{H}{=}\lim_{x\to0}\frac{\cos x-1}{\sin x+x\cos x}\overset{H}{=}\lim_{x\to0}\frac{-\sin x}{2\cos x-x\sin x}=\frac{0}{2}=0$

23. $\displaystyle\lim_{x\to\infty}\left(xe^{1/x}-x\right)=\lim_{x\to\infty}x\left(e^{1/x}-1\right)=\lim_{x\to\infty}\frac{e^{1/x}-1}{1/x}\overset{H}{=}\lim_{x\to\infty}\frac{e^{1/x}\left(-1/x^2\right)}{-1/x^2}=\lim_{x\to\infty}e^{1/x}$

$=e^0=1$

25. $y=x^{\sin x}\implies\ln y=\sin x\ln x$, so

$\displaystyle\lim_{x\to0^+}\ln y=\lim_{x\to0^+}\sin x\ln x=\lim_{x\to0^+}\frac{\ln x}{\csc x}\overset{H}{=}\lim_{x\to0^+}\frac{1/x}{-\csc x\cot x}=-\left(\lim_{x\to0^+}\frac{\sin x}{x}\right)\left(\lim_{x\to0^+}\tan x\right)$

$=-1\cdot0=0\implies$

$\displaystyle\lim_{x\to0^+}x^{\sin x}=\lim_{x\to0^+}e^{\ln y}=e^0=1$.

27. $y=(1-2x)^{1/x}\implies\ln y=\dfrac{1}{x}\ln(1-2x)$, so

$\displaystyle\lim_{x\to0}\ln y=\lim_{x\to0}\frac{\ln(1-2x)}{x}\overset{H}{=}\lim_{x\to0}\frac{-2/(1-2x)}{1}=-2\implies\lim_{x\to0}(1-2x)^{1/x}=\lim_{x\to0}e^{\ln y}=e^{-2}$.

29. $y = (-\ln x)^x \implies \ln y = x \ln(-\ln x)$, so

$$\lim_{x \to 0^+} \ln y = \lim_{x \to 0^+} x \ln(-\ln x) = \lim_{x \to 0^+} \frac{\ln(-\ln x)}{1/x} \overset{\text{H}}{=} \lim_{x \to 0^+} \frac{(1/-\ln x)(-1/x)}{-1/x^2} = \lim_{x \to 0^+} \frac{-x}{\ln x} = 0$$

$$\implies \lim_{x \to 0^+} (-\ln x)^x = e^0 = 1.$$

31.

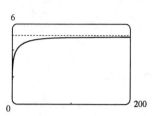

From the graph, it appears that $\lim\limits_{x \to \infty} x[\ln(x+5) - \ln x] = 5$. Now

$$\lim_{x \to \infty} x[\ln(x+5) - \ln x] = \lim_{x \to \infty} \frac{\ln(x+5) - \ln x}{1/x} \overset{\text{H}}{=} \lim_{x \to \infty} \frac{1/(x+5) - 1/x}{-1/x^2} = \lim_{x \to \infty} \frac{5x^2}{x(x+5)} = 5.$$

33.

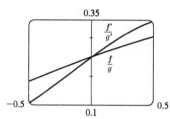

From the graph, it appears that $\lim\limits_{x \to 0} \dfrac{f(x)}{g(x)} = \lim\limits_{x \to 0} \dfrac{f'(x)}{g'(x)} = 0.25$. We calculate

$$\lim_{x \to 0} \frac{f(x)}{g(x)} = \lim_{x \to 0} \frac{e^x - 1}{x^3 + 4x} \overset{\text{H}}{=} \lim_{x \to 0} \frac{e^x}{3x^2 + 4} = \frac{1}{4}.$$

35. $\lim\limits_{x \to \infty} xe^{-x} = \lim\limits_{x \to \infty} (x/e^x) \overset{\text{H}}{=} \lim\limits_{x \to \infty} (1/e^x) = 0$, so $y = 0$ is a HA. $\lim\limits_{x \to -\infty} xe^{-x} = -\infty$.

$f(x) = xe^{-x} \implies f'(x) = e^{-x} - xe^{-x} = e^{-x}(1 - x) > 0 \iff x < 1$, so f is increasing

on $(-\infty, 1)$ and decreasing on $(1, \infty)$. By the FDT, $f(1) = 1/e$ is a local maximum.

$f''(x) = e^{-x}(x - 2) > 0 \iff x > 2$, so f is CU on $(2, \infty)$ and CD on $(-\infty, 2)$. IP is $\left(2, 2/e^2\right)$.

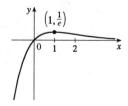

37. $\lim\limits_{x\to\infty}\dfrac{\ln x}{x}\overset{H}{=}\lim\limits_{x\to\infty}\dfrac{1/x}{1}=0$, so $y=0$ is a HA. Also $\lim\limits_{x\to0^+}\dfrac{\ln x}{x}=-\infty$ since $\ln x\to-\infty$ and

$x\to0^+$, so $x=0$ is a VA. $f(x)=\dfrac{\ln x}{x}\implies f'(x)=\dfrac{1-\ln x}{x^2}=0$ when $\ln x=1\iff x=e$.

$f'(x)>0\iff 1-\ln x>0\iff \ln x<1\iff 0<x<e$. $f'(x)<0\iff x>e$. So f
is increasing on $(0,e)$ and decreasing on (e,∞). By the FDT, $f(e)=1/e$ is a local maximum.

$f''(x)=\dfrac{(-1/x)x^2-(1-\ln x)(2x)}{x^4}=\dfrac{2\ln x-3}{x^3}$, so $f''(x)>0\iff 2\ln x-3>0\iff$

$\ln x>\frac{3}{2}\iff x>e^{3/2}$. $f''(x)<0\iff 0<x<e^{3/2}$. So f is CU on $\left(e^{3/2},\infty\right)$ and CD on
$\left(0,e^{3/2}\right)$. There is an inflection point at $\left(e^{3/2},\frac{3}{2}e^{-3/2}\right)$.

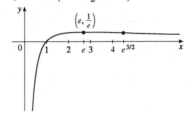

39. (a) $f(x)=x^2\ln x$

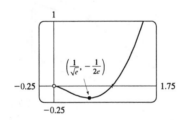

(b) $\lim\limits_{x\to\infty}x^2\ln x=\infty$, $\lim\limits_{x\to0^+}x^2\ln x=\lim\limits_{x\to0^+}\dfrac{\ln x}{1/x^2}\overset{H}{=}\lim\limits_{x\to0^+}\dfrac{1/x}{-2/x^3}=\lim\limits_{x\to0^+}\left(-\dfrac{x^2}{2}\right)=0$. There is a
hole at $(0,0)$.

(c) It appears that there is an IP at about $(0.2,-0.06)$ and a local minimum at $(0.6,-0.18)$.
$f(x)=x^2\ln x\implies f'(x)=2x\ln x+x=x(2\ln x+1)>0\iff \ln x>-\frac{1}{2}\iff x>e^{-1/2}$, so
f is increasing on $(1/\sqrt{e}\,,\infty)$, decreasing on $(0,1/\sqrt{e})$. By the FDT, $f(1/\sqrt{e})=-1/(2e)$ is a
local minimum. This point is approximately $(0.6065,-0.1839)$, which agrees with our estimate.
$f''(x)=x(2/x)+(2\ln x+1)=2\ln x+3>0\iff \ln x>-\frac{3}{2}\iff x>e^{-3/2}$, so f is CU on
$\left(e^{-3/2},\infty\right)$ and CD on $\left(0,e^{-3/2}\right)$. IP is $\left(e^{-3/2},-3/(2e^3)\right)\approx(0.2231,-0.0747)$.

41. (a) $f(x) = x^{1/x}$

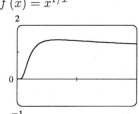

(b) Recall that $a^b = e^{b \ln a}$. $\lim\limits_{x \to 0^+} x^{1/x} = \lim\limits_{x \to 0^+} e^{(1/x) \ln x}$. As

$x \to 0^+$, $\dfrac{\ln x}{x} \to -\infty$, so $x^{1/x} \to 0$. This indicates that there

is a hole at $(0, 0)$. As $x \to \infty$, we have the indeterminate form

∞^0. $\lim\limits_{x \to \infty} x^{1/x} = \lim\limits_{x \to \infty} e^{(1/x) \ln x}$, but

$\lim\limits_{x \to \infty} \dfrac{\ln x}{x} \overset{\text{H}}{=} \lim\limits_{x \to \infty} \dfrac{1/x}{1} = 0$, so $\lim\limits_{x \to \infty} x^{1/x} = e^0 = 1$. This

indicates that $y = 1$ is a HA.

(c) Estimated maximum: $(2.72, 1.45)$. No estimated minimum. We use logarithmic differentiation to

find any critical numbers. $y = x^{1/x} \implies \ln y = \dfrac{1}{x} \ln x \implies \dfrac{y'}{y} = \dfrac{1}{x} \cdot \dfrac{1}{x} + (\ln x)\left(-\dfrac{1}{x^2}\right) \implies$

$y' = x^{1/x}\left(\dfrac{1 - \ln x}{x^2}\right) = 0 \implies \ln x = 1 \implies x = e$. For $0 < x < e$, $y' > 0$ and for $x > e$, $y' < 0$,

so $f(e) = e^{1/e}$ is a local maximum. This point is approximately $(2.7183, 1.4447)$, which agrees

with our estimate.

(d)

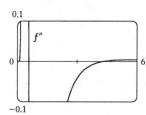

From the graph, we see that $f''(x) = 0$ at $x \approx 0.58$ and

$x \approx 4.37$. Since f'' changes sign at these values, they are

x-coordinates of inflection points.

43. If $c < 0$, then $\lim\limits_{x \to -\infty} f(x) = \lim\limits_{x \to -\infty} \dfrac{x}{e^{cx}} \overset{\text{H}}{=} \lim\limits_{x \to -\infty} \dfrac{1}{ce^{cx}} = 0$, and $\lim\limits_{x \to \infty} f(x) = \infty$.

If $c > 0$, then $\lim\limits_{x \to -\infty} f(x) = -\infty$, and $\lim\limits_{x \to \infty} f(x) \overset{\text{H}}{=} \lim\limits_{x \to \infty} \dfrac{1}{ce^{cx}} = 0$.

If $c = 0$, then $f(x) = x$, so $\lim\limits_{x \to \pm\infty} f(x) = \pm\infty$ respectively.

So we see that $c = 0$ is a transitional value. We now exclude the case $c = 0$, since we know

how the function behaves in that case. To find the maxima and minima of f, we differentiate:

$f(x) = xe^{-cx} \implies f'(x) = x(-ce^{-cx}) + e^{-cx} = (1 - cx)e^{-cx}$. This is 0 when $1 - cx = 0 \iff$

$x = 1/c$. If $c < 0$ then this represents a minimum of $f(1/c) = 1/(ce)$, since $f'(x)$ changes from

negative to positive at $x = 1/c$; and if $c > 0$, it represents a

maximum. As $|c|$ increases, the maximum or minimum gets closer

to the origin. To find the inflection points, we differentiate again:

$f'(x) = e^{-cx}(1 - cx) \implies$

$f''(x) = e^{-cx}(-c) + (1 - cx)(-ce^{-cx}) = (cx - 2)ce^{-cx}$. This

changes sign when $cx - 2 = 0 \iff x = 2/c$. So as $|c|$ increases,

the points of inflection get closer to the origin.

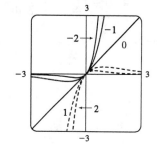

45. $\displaystyle\lim_{x\to\infty}\frac{e^x}{x^n}\overset{\text{H}}{=}\lim_{x\to\infty}\frac{e^x}{nx^{n-1}}\overset{\text{H}}{=}\lim_{x\to\infty}\frac{e^x}{n(n-1)x^{n-2}}\overset{\text{H}}{=}\cdots\overset{\text{H}}{=}\lim_{x\to\infty}\frac{e^x}{n!}=\infty$

47. First we will find $\displaystyle\lim_{n\to\infty}\left(1+\frac{i}{n}\right)^{nt}$, which is of the form 1^∞. $y=\left(1+\dfrac{i}{n}\right)^{nt}\implies$

$\ln y=nt\ln\left(1+\dfrac{i}{n}\right)$, so

$$\lim_{n\to\infty}\ln y=\lim_{n\to\infty}nt\ln\left(1+\frac{i}{n}\right)=t\lim_{n\to\infty}\frac{\ln(1+i/n)}{1/n}\overset{\text{H}}{=}t\lim_{n\to\infty}\frac{(-i/n^2)}{(1+i/n)(-1/n^2)}$$

$$=t\lim_{n\to\infty}\frac{i}{1+i/n}=ti\implies\lim_{n\to\infty}y=e^{it}.$$

Thus, as $n\to\infty$, $A=A_0\left(1+\dfrac{i}{n}\right)^{nt}\to A_0e^{it}$.

49. We see that both numerator and denominator approach 0, so we can use l'Hospital's Rule:

$$\lim_{x\to a}\frac{\sqrt{2a^3x-x^4}-a\sqrt[3]{aax}}{a-\sqrt[4]{ax^3}}\overset{\text{H}}{=}\lim_{x\to a}\frac{\frac{1}{2}(2a^3x-x^4)^{-1/2}(2a^3-4x^3)-a\left(\frac{1}{3}\right)(aax)^{-2/3}a^2}{-\frac{1}{4}(ax^3)^{-3/4}(3ax^2)}$$

$$=\frac{\frac{1}{2}(2a^3a-a^4)^{-1/2}(2a^3-4a^3)-\frac{1}{3}a^3(a^2a)^{-2/3}}{-\frac{1}{4}(aa^3)^{-3/4}(3aa^2)}$$

$$=\frac{(a^4)^{-1/2}(-a^3)-\frac{1}{3}a^3(a^3)^{-2/3}}{-\frac{3}{4}a^3(a^4)^{-3/4}}=\frac{-a-\frac{1}{3}a}{-3/4}=\frac{4}{3}\left(\frac{4}{3}a\right)=\frac{16}{9}a$$

51. Since $\displaystyle\lim_{h\to0}[f(x+h)-f(x-h)]=f(x)-f(x)=0$ (f is differentiable and hence continuous) and

$\displaystyle\lim_{h\to0}2h=0$, we use l'Hospital's Rule:

$$\lim_{h\to0}\frac{f(x+h)-f(x-h)}{2h}\overset{\text{H}}{=}\lim_{h\to0}\frac{f'(x+h)(1)-f'(x-h)(-1)}{2}=\frac{f'(x)+f'(x)}{2}=\frac{2f'(x)}{2}$$

$$=f'(x)$$

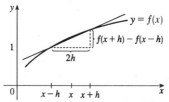

$\dfrac{f(x+h)-f(x-h)}{2h}$ is the slope of the secant line between $(x-h,f(x-h))$ and

$(x+h,f(x+h))$. As $h\to0$, this line gets closer to the tangent line and its slope approaches $f'(x)$.

Section 4.6 Optimization Problems

1. (a)

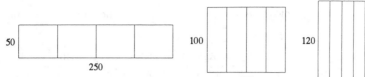

The areas of the three figures are 12,500, 12,500, and 9000 ft². There appears to be a maximum area of at least 12,500 ft².

(b)

Let x denote the length of each of two sides and three dividers. Let y denote the length of the other two sides.

(c) Area $A = \text{length} \times \text{width} = y \cdot x$

(d) Length of fencing $= 750 \implies 5x + 2y = 750$

(e) $5x + 2y = 750 \implies y = 375 - \frac{5}{2}x \implies A(x) = \left(375 - \frac{5}{2}x\right)x = 375x - \frac{5}{2}x^2$

(f) $A'(x) = 375 - 5x = 0 \implies x = 75$. Since $A''(x) = -5 < 0$ there is an absolute maximum when $x = 75$. Then $y = \frac{375}{2} = 187.5$. The largest area is $75\left(\frac{375}{2}\right) = 14{,}062.5$ ft². These values are between the values in the first and second figures in part (a). Our original estimate was low.

3. Let b be the base of the box and h the height. The surface area is $1200 = b^2 + 4hb \implies$
$h = \left(1200 - b^2\right)/(4b)$. The volume is $V = b^2h = b^2\left(1200 - b^2\right)/4b = 300b - b^3/4 \implies$
$V'(b) = 300 - \frac{3}{4}b^2$. $V'(b) = 0 \implies b = \sqrt{400} = 20$. Since $V'(b) > 0$ for $0 < b < 20$ and $V'(b) < 0$
for $b > 20$, there is an absolute maximum when $b = 20$ by the First Derivative Test for Absolute
Maximum or Minimum Values. If $b = 20$, then $h = \left(1200 - 20^2\right)/(4 \cdot 20) = 10$, so the largest
possible volume is $b^2h = (20)^2(10) = 4000$ cm³.

5. (a) Let the rectangle have sides x and y and area A, so $A = xy$ or $y = A/x$. The problem is to minimize
the perimeter $= 2x + 2y = 2x + 2A/x = P(x)$. Now $P'(x) = 2 - 2A/x^2 = 2\left(x^2 - A\right)/x^2$. So
the critical number is $x = \sqrt{A}$. Since $P'(x) < 0$ for $0 < x < \sqrt{A}$ and $P'(x) > 0$ for $x > \sqrt{A}$,
there is an absolute minimum at $x = \sqrt{A}$. The sides of the rectangle are $\sqrt{A}$ and $A/\sqrt{A} = \sqrt{A}$, so
the rectangle is a square.

(b) Let p be the perimeter and x and y the lengths of the sides, so $p = 2x + 2y \implies y = \frac{1}{2}p - x$.
The area is $A(x) = x\left(\frac{1}{2}p - x\right) = \frac{1}{2}px - x^2$. Now $0 = A'(x) = \frac{1}{2}p - 2x \implies x = \frac{1}{4}p$. Since
$A''(x) = -2 < 0$, there is an absolute maximum where $x = \frac{1}{4}p$ by the Second Derivative Test. The
sides of the rectangle are $\frac{1}{4}p$ and $\frac{1}{2}p - \frac{1}{4}p = \frac{1}{4}p$, so the rectangle is a square.

7. For (x, y) on the line $y = 2x - 3$, the distance to the origin is $\sqrt{(x - 0)^2 + (2x - 3)^2}$. We minimize the square of the distance, that is, $x^2 + (2x - 3)^2 = 5x^2 - 12x + 9 = D(x)$. $D'(x) = 10x - 12 = 0 \Longrightarrow$ $x = \frac{6}{5}$. Since there is a point closest to the origin, $x = \frac{6}{5}$ and hence $y = -\frac{3}{5}$. So the point is $\left(\frac{6}{5}, -\frac{3}{5}\right)$.

9.

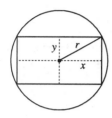

Area of rectangle is $(2x)(2y) = 4xy$. Also $r^2 = x^2 + y^2$ so $y = \sqrt{r^2 - x^2}$, so the area is $A(x) = 4x\sqrt{r^2 - x^2}$. Now

$$A'(x) = 4\left(\sqrt{r^2 - x^2} - \frac{x^2}{\sqrt{r^2 - x^2}}\right) = 4\frac{r^2 - 2x^2}{\sqrt{r^2 - x^2}}.$$ The critical

number is $x = \frac{1}{\sqrt{2}}r$. Clearly this gives a maximum.

$y = \sqrt{r^2 - \left(\frac{1}{\sqrt{2}}r\right)^2} = \sqrt{\frac{1}{2}r^2} = \frac{1}{\sqrt{2}}r = x$, which tells us that the

rectangle is a square. The dimensions are $2x = \sqrt{2}r$ and $2y = \sqrt{2}r$.

11.

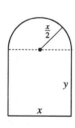

The rectangle has area xy. By similar triangles $\dfrac{3 - y}{x} = \dfrac{3}{4} \Longrightarrow$

$-4y + 12 = 3x$ or $y = -\frac{3}{4}x + 3$. So the area is

$A(x) = x\left[-\frac{3}{4}x + 3\right] = -\frac{3}{4}x^2 + 3x$ where $0 \le x \le 4$. Now

$0 = A'(x) = -\frac{3}{2}x + 3 \Longrightarrow x = 2$ and $y = \frac{3}{2}$. Since

$A(0) = A(4) = 0$, the maximum area is $A(2) = 2\left(\frac{3}{2}\right) = 3$ cm^2.

13.

We are given $2y + x + \pi\left(\dfrac{x}{2}\right) = 30$, so $y = \dfrac{1}{2}\left[30 - x - \dfrac{\pi x}{2}\right]$. The

area is $xy + \frac{1}{2}\pi\left(\dfrac{x}{2}\right)^2$, so

$$A(x) = x\left[15 - \frac{x}{2} - \frac{\pi x}{4}\right] + \frac{1}{8}\pi x^2 = 15x - \frac{1}{2}x^2 - \frac{\pi}{8}x^2.$$

$$A'(x) = 15 - \left(1 + \frac{\pi}{4}\right)x = 0 \Longrightarrow x = \frac{15}{1 + \pi/4} = \frac{60}{4 + \pi}.$$

Clearly this gives a maximum, so the dimensions are $x = \dfrac{60}{4 + \pi}$ ft and

$y = 15 - \dfrac{30}{4 + \pi} - \dfrac{15\pi}{4 + \pi} = \dfrac{30}{4 + \pi}$ ft, so the height of the rectangle is half the base.

15.

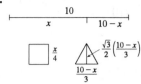

Let x be the length of the wire used for the square. The total area is

$$A\left(x\right) = \left(\frac{x}{4}\right)^2 + \frac{1}{2}\left(\frac{10-x}{3}\right)\frac{\sqrt{3}}{2}\left(\frac{10-x}{3}\right)$$

$$= \frac{1}{16}x^2 + \frac{\sqrt{3}}{36}\left(10-x\right)^2, 0 \le x \le 10.$$

$$A'\left(x\right) = \frac{1}{8}x - \frac{\sqrt{3}}{18}\left(10-x\right) = 0 \iff \frac{9}{72}x + \frac{4\sqrt{3}}{72}x - \frac{40\sqrt{3}}{72} = 0 \iff$$

$$x = \frac{40\sqrt{3}}{9+4\sqrt{3}}. \text{ Now } A\left(0\right) = \left(\frac{\sqrt{3}}{36}\right)100 \approx 4.81, A\left(10\right) = \frac{100}{16} = 6.25$$

and $A\left(\frac{40\sqrt{3}}{9+4\sqrt{3}}\right) \approx 2.72$, so

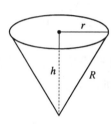

(a) The maximum area occurs when $x = 10$ m, and all the wire is used for the square.

(b) The minimum area occurs when $x = \frac{40\sqrt{3}}{9+4\sqrt{3}} \approx 4.35$ m.

17.

$$h^2 + r^2 = R^2 \implies$$

$$V = \frac{\pi}{3}r^2 h = \frac{\pi}{3}\left(R^2 - h^2\right)h = \frac{\pi}{3}\left(R^2 h - h^3\right).V'\left(h\right)$$

$$= \frac{\pi}{3}\left(R^2 - 3h^2\right) = 0 \text{ when } h = \frac{1}{\sqrt{3}}R.$$

This gives an absolute maximum since $V'\left(h\right) > 0$ for $0 < h < \frac{1}{\sqrt{3}}R$

and $V'\left(h\right) < 0$ for $h > \frac{1}{\sqrt{3}}R$. Maximum volume is

$$V\left(\frac{1}{\sqrt{3}}R\right) = \frac{\pi}{3}\left(\frac{1}{\sqrt{3}}R^3 - \frac{1}{3\sqrt{3}}R^3\right) = \frac{2}{9\sqrt{3}}\pi R^3.$$

19. $S = 6sh - \frac{3}{2}s^2 \cot\theta + 3s^2\frac{\sqrt{3}}{2}\csc\theta$

(a) $\frac{dS}{d\theta} = \frac{3}{2}s^2 \csc^2\theta - 3s^2\frac{\sqrt{3}}{2}\csc\theta\cot\theta$ or $\frac{3}{2}s^2 \csc\theta\left(\csc\theta - \sqrt{3}\cot\theta\right)$.

(b) $\frac{dS}{d\theta} = 0$ when $\csc\theta - \sqrt{3}\cot\theta = 0 \implies \frac{1}{\sin\theta} - \sqrt{3}\frac{\cos\theta}{\sin\theta} = 0 \implies \cos\theta = \frac{1}{\sqrt{3}}$. The First

Derivative Test shows that the minimum surface area occurs when $\theta = \cos^{-1}\frac{1}{\sqrt{3}} \approx 55°$.

(c)

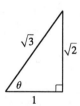

If $\cos\theta = \frac{1}{\sqrt{3}}$, then $\cot\theta = \frac{1}{\sqrt{2}}$ and $\csc\theta = \frac{\sqrt{3}}{\sqrt{2}}$, so the surface area

is $S = 6sh - \frac{3}{2}s^2\frac{1}{\sqrt{2}} + 3s^2\frac{\sqrt{3}}{2}\frac{\sqrt{3}}{\sqrt{2}} = 6sh - \frac{3}{2\sqrt{2}}s^2 + \frac{9}{2\sqrt{2}}s^2 = 6s\left(h + \frac{1}{2\sqrt{2}}s\right)$.

21.

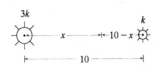

The total illumination is $I\left(x\right) = \dfrac{3k}{x^2} + \dfrac{k}{\left(10 - x\right)^2}, 0 < x < 10.$

Then $I'\left(x\right) = \dfrac{-6k}{x^3} + \dfrac{2k}{\left(10 - x\right)^3} = 0 \Longrightarrow$

$6k\left(10 - x\right)^3 = 2kx^3 \Longrightarrow \sqrt[3]{3}\left(10 - x\right) = x \Longrightarrow$

$x = \dfrac{10\sqrt[3]{3}}{1 + \sqrt[3]{3}} \approx 5.9\,\text{ft.}$ This gives a minimum since $I''\left(x\right) > 0$ for

$0 < x < 10.$

23.

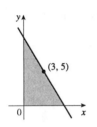

The line with slope m (where $m < 0$) through $\left(3, 5\right)$ has equation

$y - 5 = m\left(x - 3\right)$ or $y = mx + \left(5 - 3m\right).$ The y-intercept is

$5 - 3m$ and the x-intercept is $-5/m + 3.$ So the triangle has area

$A\left(m\right) = \frac{1}{2}\left(5 - 3m\right)\left(-5/m + 3\right) = 15 - 25/\left(2m\right) - \frac{9}{2}m.$ Now

$A'\left(m\right) = \dfrac{25}{2m^2} - \dfrac{9}{2} = 0 \Longleftrightarrow m = -\frac{5}{3}$ (since $m < 0$).

$A''\left(m\right) = -\dfrac{25}{m^3} > 0,$ so there is an absolute minimum when

$m = -\frac{5}{3}.$ Therefore, the equation of the line is

$y - 5 = -\frac{5}{3}\left(x - 3\right)$ or $y = -\frac{5}{3}x + 10.$

25.

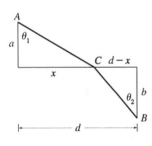

The total time is

$T\left(x\right) = \left(\text{time from } A \text{ to } C\right) + \left(\text{time from } C \text{ to } B\right)$

$\qquad = \dfrac{\sqrt{a^2 + x^2}}{v_1} + \dfrac{\sqrt{b^2 + \left(d - x\right)^2}}{v_2}, 0 < x < d.$

$T'\left(x\right) = \dfrac{x}{v_1\sqrt{a^2 + x^2}} - \dfrac{d - x}{v_2\sqrt{b^2 + \left(d - x\right)^2}}$

$\qquad = \dfrac{\sin\theta_1}{v_1} - \dfrac{\sin\theta_2}{v_2}.$

The minimum occurs when $T'\left(x\right) = 0 \Longrightarrow \dfrac{\sin\theta_1}{v_1} = \dfrac{\sin\theta_2}{v_2}.$

27.

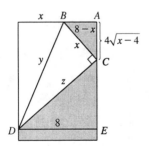

$y^2 = x^2 + z^2,$ but triangles CDE and BCA are similar, so

$z/8 = x / \left(4\sqrt{x - 4}\right) \Longrightarrow z = 2x / \sqrt{x - 4}.$ Thus, we minimize

$f\left(x\right) = y^2 = x^2 + 4x^2/\left(x - 4\right) = x^3/\left(x - 4\right), 4 < x \leq 8.$

$f'\left(x\right) = \dfrac{3x^2\left(x - 4\right) - x^3}{\left(x - 4\right)^2} = \dfrac{2x^2\left(x - 6\right)}{\left(x - 4\right)^2} = 0$ when $x = 6.$

$f'\left(x\right) < 0$ when $x < 6,$ $f'\left(x\right) > 0$ when $x > 6,$ so the minimum

occurs when $x = 6$ in.

29. $L(x) = |AP| + |BP| + |CP| = x + \sqrt{(5-x)^2 + 2^2} + \sqrt{(5-x)^2 + 3^2}$

$\qquad = x + \sqrt{x^2 - 10x + 29} + \sqrt{x^2 - 10x + 34} \Longrightarrow$

$L'(x) = 1 + \dfrac{x-5}{\sqrt{x^2 - 10x + 29}} + \dfrac{x-5}{\sqrt{x^2 - 10x + 34}}$

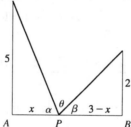

From the graphs of L and L', it seems that the minimum value of L is about $L(3.59) = 9.35$ m.

31.

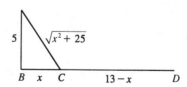

$\tan\alpha = \dfrac{5}{x}, \tan\beta = \dfrac{2}{3-x} \Longrightarrow \theta = \pi - \tan^{-1}\left(\dfrac{5}{x}\right) - \tan^{-1}\left(\dfrac{2}{3-x}\right) \Longrightarrow$

$\dfrac{d\theta}{dx} = -\dfrac{1}{1+\left(\dfrac{5}{x}\right)^2}\left(-\dfrac{5}{x^2}\right) - \dfrac{1}{1+\left(\dfrac{2}{3-x}\right)^2}\left[\dfrac{2}{(3-x)^2}\right]. \dfrac{d\theta}{dx} = 0 \Longrightarrow$

$\dfrac{5}{x^2+25} = \dfrac{2}{x^2-6x+13} \Longrightarrow 2x^2 + 50 = 5x^2 - 30x + 65 \Longrightarrow x^2 - 10x + 5 = 0 \Longrightarrow x = 5 \pm 2\sqrt{5}.$

We reject the root with the $+$ sign, since it is larger than 3. $d\theta/dx > 0$ for $x < 5 - 2\sqrt{5}$ and $d\theta/dx < 0$ for $x > 5 - 2\sqrt{5}$, so θ is maximized when $|AP| = x = 5 - 2\sqrt{5} \approx 0.53$.

33. (a)

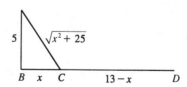

If k = energy/km over land, then energy/km over water = $1.4k$. So the total energy is

$E = 1.4k\sqrt{25 + x^2} + k(13 - x), 0 \le x \le 13$, and so $\dfrac{dE}{dx} = \dfrac{1.4kx}{(25 + x^2)^{1/2}} - k$. Set $\dfrac{dE}{dx} = 0$:

$1.4kx = k(25 + x^2)^{1/2} \Longrightarrow 1.96x^2 = x^2 + 25 \Longrightarrow 0.96x^2 = 25 \Longrightarrow x = \dfrac{5}{\sqrt{0.96}} \approx 5.1$. Testing

against the value of E at the endpoints: $E(0) = 1.4k(5) + 13k = 20k$, $E(5.1) \approx 17.9k$,

$E(13) \approx 19.5k$. Thus, to minimize energy, the bird should fly to a point about 5.1 km from B.

(b) If W/L is large, the bird would fly to a point C that is closer to B than to D to minimize the energy used flying over water. If W/L is small, the bird would fly to a point C that is closer to D than to B to minimize the distance of the flight. $E = W\sqrt{25 + x^2} + L(13 - x) \implies$

$\dfrac{dE}{dx} = \dfrac{Wx}{\sqrt{25 + x^2}} - L = 0$ when $\dfrac{W}{L} = \dfrac{\sqrt{25 + x^2}}{x}$. By the same sort of argument as in part (a), this ratio will give the minimal expenditure of energy if the bird heads for the point x km from B.

(c) For flight direct to D, $x = 13$, so from part (b), $W/L = \dfrac{\sqrt{25 + 13^2}}{13} \approx 1.07$. There is no value of W/L for which the bird should fly directly to B. But note that $\displaystyle\lim_{x \to 0^+}(W/L) = \infty$, so if the point at which E is a minimum is close to B, then W/L is large.

(d) Assuming that the birds instinctively choose the path that minimizes the energy expenditure, we can use the equation for $dE/dx = 0$ from part (a) with $1.4k = c$, $x = 4$, and $k = 1$:

$c(4) = 1 \cdot \left(25 + 4^2\right)^{1/2} \implies c = \sqrt{41}/4 \approx 1.6$.

35. (a) $T_1 = \dfrac{D}{c_1}$, $T_2 = \dfrac{2|PR|}{c_1} + \dfrac{|RS|}{c_2} = \dfrac{2h \sec\theta}{c_1} + \dfrac{D - 2h \tan\theta}{c_2}$, $T_3 = \dfrac{2\sqrt{h^2 + D^2/4}}{c_1} = \dfrac{\sqrt{4h^2 + D^2}}{c_1}$.

(b) $\dfrac{dT_2}{d\theta} = \dfrac{2h}{c_1} \cdot \sec\theta \tan\theta - \dfrac{2h}{c_2}\sec^2\theta = 0$ when $2h \sec\theta\left(\dfrac{1}{c_1}\tan\theta - \dfrac{1}{c_2}\sec\theta\right) = 0 \implies$

$\dfrac{1}{c_1}\dfrac{\sin\theta}{\cos\theta} - \dfrac{1}{c_2}\dfrac{1}{\cos\theta} = 0 \implies \sin\theta = \dfrac{c_1}{c_2}$. The First Derivative Test shows that this gives a minimum.

(c) Using part (a), we have $T_1 = \dfrac{D}{c_1} \implies c_1 = \dfrac{1}{0.26} \approx 3.85$ km/s. $T_3 = \dfrac{\sqrt{4h^2 + D^2}}{c_1} \implies$

$4h^2 + D^2 = T_3^2 c_1^2 \implies h = \frac{1}{2}\sqrt{T_3^2 c_1^2 - D^2} = \frac{1}{2}\sqrt{(0.34)^2 (1/0.26)^2 - 1^2} \approx 0.42$ km. To find c_2,

we use $\sin\theta = \dfrac{c_1}{c_2}$ from part (b) and $T_2 = \dfrac{2h \sec\theta}{c_1} + \dfrac{D - 2h \tan\theta}{c_2}$ from part (a).

$\sin\theta = \dfrac{c_1}{c_2} \implies \sec\theta = \dfrac{c_2}{\sqrt{c_2^2 - c_1^2}}$ and $\tan\theta = \dfrac{c_1}{\sqrt{c_2^2 - c_1^2}}$, so

$T_2 = \dfrac{2hc_2}{c_1\sqrt{c_2^2 - c_1^2}} + \dfrac{D\sqrt{c_2^2 - c_1^2} - 2hc_1}{c_2\sqrt{c_2^2 - c_1^2}}$.

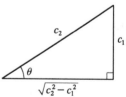

Using the values for T_2, h, c_1, and D, we can graph $Y_1 = T_2$ and

$Y_2 = \dfrac{2hc_2}{c_1\sqrt{c_2^2 - c_1^2}} + \dfrac{D\sqrt{c_2^2 - c_1^2} - 2hc_1}{c_2\sqrt{c_2^2 - c_1^2}}$ and find their intersection points. Doing so gives us

$c_2 \approx 4.10$ and 7.66, but if $c_2 = 4.10$, then $\theta = \arcsin(c_1/c_2) \approx 69.6°$, which implies that point S is to the left of point R in the diagram. So $c_2 = 7.66$ km/s.

Section 4.7 Applications to Economics

1. (a) $C(0)$ represents the fixed costs of production, such as rent, utilities, machinery etc., which are incurred even when nothing is produced.

(b) The inflection point is the point at which $C''(x)$ changes from negative to positive, that is, the marginal cost $C'(x)$ changes from decreasing to increasing. So the marginal cost is minimized.

(c) The marginal cost function is $C'(x)$. We graph it as in Example 1 in Section 2.8.

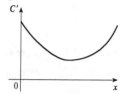

3. $c(x) = 21.4 - 0.002x$ and $c(x) = C(x)/x \Longrightarrow C(x) = 21.4x - 0.002x^2$. $C'(x) = 21.4 - 0.004x$ and $C'(1000) = 17.4$. This is an estimate of the cost of producing the 1001st unit.

5. (a) $C(x) = 1600 + 8x + 0.01x^2$, $C(1000) = \$19,600$. $c(x) = \dfrac{C(x)}{x} = \dfrac{1600}{x} + 8 + 0.01x$, $c(1000) = \$19.60$. $C'(x) = 8 + 0.02x$, $C'(1000) = \$28$.

(b) We must have $C'(x) = c(x) \Longleftrightarrow 8 + 0.02x = \dfrac{1600}{x} + 8 + 0.01x \Longleftrightarrow 0.01x = \dfrac{1600}{x} \Longleftrightarrow x^2 = \dfrac{1600}{0.01} = 160,000 \Longleftrightarrow x = 400$. This is a minimum since $c''(x) = \dfrac{3200}{x^3} > 0$ for $x > 0$.

(c) The minimum average cost is $c(400) = \$16$.

7. (a) $C(x) = 3700 + 5x - 0.04x^2 + 0.0003x^3 \Longrightarrow C'(x) = 5 - 0.08x + 0.0009x^2$ (marginal cost). $c(x) = \dfrac{C(x)}{x} = \dfrac{3700}{x} + 5 - 0.04x + 0.0003x^2$ (average cost).

(b)

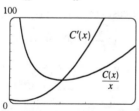

The graphs intersect at $(208.51, 27.45)$, so the production level that minimizes average cost is about 209 units.

(c) $c'(x) = -\dfrac{3700}{x^2} - 0.04 + 0.0006x = 0 \Longrightarrow x_1 = 208.51$. $c(x_1) \approx \$27.45/\text{unit}$.

(d) $C''(x) = -0.08 + 0.0018x = 0 \Longrightarrow x_1 = \dfrac{800}{18} = 44.4\overline{4}$. $C'(x_1) = \$3.22/\text{unit}$. $C''(x) = 0.0018 > 0$ for all x, so this is the minimum marginal cost.

9. $C(x) = 680 + 4x + 0.01x^2$, $p(x) = 12 - x/500$. Then $R(x) = xp(x) = 12x - x^2/500$. If the profit is maximum, then $R'(x) = C'(x) \Longleftrightarrow 12 - x/250 = 4 + 0.02x \Longleftrightarrow 8 = 0.024x \Longleftrightarrow x = 8/0.024 = \dfrac{1000}{3}$. The profit is maximized if $P''(x) < 0$, but since $P''(x) = R''(x) - C''(x)$, we can just check the condition $R''(x) < C''(x)$. Now $R''(x) = -\dfrac{1}{250} < 0.02 = C''(x)$, so $x = \dfrac{1000}{3}$ gives a maximum.

11. (a) $C(x) = 1200 + 12x - 0.1x^2 + 0.0005x^3$. $R(x) = xp(x) = 29x - 0.00021x^2$.

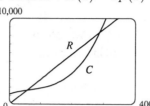

10,000

R

C

0   400

Since the profit is maximized when $R'(x) = C'(x)$, we examine the curves R and C in the figure, looking for x-values at which the slopes of the tangent lines are equal. It appears that $x = 200$ is a good estimate.

(b) $R'(x) = C'(x) \Longrightarrow 29 - 0.00042x = 12 - 0.2x + 0.0015x^2 \Longrightarrow$
$0.0015x^2 - 0.19958x - 17 = 0 \Longrightarrow x \approx 192.06$ (for $x > 0$). As in Exercise 9,
$R''(x) < C''(x) \Longrightarrow -0.00042 < -0.2 + 0.003x \Longleftrightarrow 0.003x > 0.19958 \Longleftrightarrow x > 66.5$. Our
value of 192 is in this range, so we have a maximum profit when we produce 192 yards of fabric.

13. (a) We are given that the demand function p is linear and $p(27,000) = 10$, $p(33,000) = 8$, so the slope
is $\dfrac{10 - 8}{27,000 - 33,000} = -\dfrac{1}{3000}$ and an equation of the line is $y - 10 = \left(-\dfrac{1}{3000}\right)(x - 27,000) \Longrightarrow$
$y = p(x) = -\dfrac{1}{3000}x + 19 = 19 - x/3000$.

(b) The revenue is $R(x) = xp(x) = 19x - x^2/3000 \Longrightarrow R'(x) = 19 - x/1500 = 0$ when
$x = 28,500$. Since $R''(x) = -1/1500 < 0$, the maximum revenue occurs when $x = 28,500 \Longrightarrow$
the price is $p(28,500) = \$9.50$.

15. (a) As in Example 3, we see that the demand function p is linear. We are given that $p(1000) = 450$ and
deduce that $p(1100) = 440$, since a $10 reduction in price increases sales by 100 per week. The slope
for p is $\dfrac{440 - 450}{1100 - 1000} = -\dfrac{1}{10}$, so an equation is $p - 450 = -\dfrac{1}{10}(x - 1000)$ or $p(x) = -\dfrac{1}{10}x + 550$.

(b) $R(x) = xp(x) = 500x - x^2/10$. $R'(x) = 550 - x/5 = 0$ when $x = 5(550) = 2750$.
$p(2750) = 275$, so the rebate should be $450 - 275 = \$175$.

(c) $C(x) = 68,000 + 150x \Longrightarrow P(x) = R(x) - C(x) = 550x - x^2/10 - 68,000 - 150x = $
$400x - x^2/10 - 68,000$, $P'(x) = 400 - x/5 = 0$ when $x = 2000$. $p(2000) = 350$. Therefore,
the rebate to maximize profits should be $450 - 350 = \$100$.

Section 4.8   Newton's Method

1.

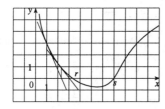

The tangent line at $x = 1$ intersects the x-axis at $x \approx 2.3$, so $x_2 \approx 2.3$. The tangent line at $x = 2.3$ intersects the x-axis at $x \approx 3$, so $x_3 \approx 3.0$.

3. Since $x_1 = 3$ and $y = 5x - 4$ is tangent to $y = f(x)$ at $x = 3$, we simply need to find where the tangent line intersects the x-axis. $y = 0 \Longrightarrow 5x_2 - 4 = 0 \Longrightarrow x_2 = \frac{4}{5}$.

5. $f(x) = x^3 + x + 1 \Longrightarrow f'(x) = 3x^2 + 1$, so $x_{n+1} = x_n - \dfrac{x_n^3 + x_n + 1}{3x_n^2 + 1}$. $x_1 = -1 \Longrightarrow$

$x_2 = -1 - \dfrac{-1 - 1 + 1}{3 \cdot 1 + 1} = -0.75 \Longrightarrow x_3 = -0.75 - \dfrac{(-0.75)^3 - 0.75 + 1}{3(-0.75)^2 + 1} \approx -0.6860$. Here is a

quick and easy method for finding the iterations on a programmable calculator. (The screens shown are from the TI-82, but the method is similar on other calculators.) Assign $x^3 + x + 1$ to Y_1 and $3x^2 + 1$ to X_2. Now store -1 in X and then enter $X - Y_1/Y_2 \to X$ to get -0.75. By successively pressing the ENTER key, you get the approximations $x_1, x_2, x_3, \ldots$.

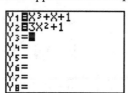

7. $x = \sqrt[4]{22} \Longrightarrow x^4 = 22 \Longrightarrow x^4 - 22 = 0$, so finding $\sqrt[4]{22}$ is equivalent to finding the positive root of

$x^4 - 22 = 0$. $f(x) = x^4 - 22 \Longrightarrow f'(x) = 4x^3$ and $x_{n+1} = x_n - \dfrac{x_n^4 - 22}{4x_n^3}$. Since $\sqrt[4]{16} = 2$ and

$\sqrt[4]{81} = 3$, it makes sense that $\sqrt[4]{22}$ is closer to 2, so we use 2 as our initial approximation. Taking $x_1 = 2$, we get $x_2 = 2.1875$, $x_3 \approx 2.16605940$, $x_4 \approx 2.16573684$ and $x_5 \approx 2.16573677 \approx x_6$. Thus, $\sqrt[4]{22} \approx 2.16573677$ to eight decimal places.

9.

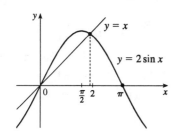

From the graph it appears that there is a root near 2, so we take $x_1 = 2$. Write the equation as $f(x) = 2 \sin x - x = 0$. Then $f'(x) = 2 \cos x - 1$, so

$$x_{n+1} = x_n - \dfrac{2 \sin x_n - x_n}{2 \cos x_n - 1} \Longrightarrow x_1 = 2,$$

$x_2 \approx 1.900996$, $x_3 \approx 1.895512$, $x_4 \approx 1.895494 \approx x_5$. So the root is 1.895494, to six decimal places.

11.

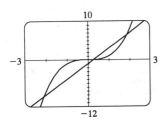

From the graph, we see that $y = x^3$ and $y = 4x - 1$ intersect three times. Good first approximations are $x = -2$, $x = 0$, and $x = 2$. $f(x) = x^3 - 4x + 1 \implies f'(x) = 3x^2 - 4$, so

$$x_{n+1} = x_n - \frac{x_n^3 - 4x_n + 1}{3x_n^2 - 4}.$$ We need to find approximations

until they agree to eight decimal places.

$x_1 = -2$

$x_2 = -2.125$

$x_3 \approx -2.11497545$

$x_4 \approx -2.11490754 \approx x_5$

$x_1 = 0$

$x_2 = 0.25$

$x_3 \approx 0.25409836$

$x_4 \approx 0.25410169 \approx x_5$

$x_1 = 2$

$x_2 = 1.875$

$x_3 \approx 1.86097852$

$x_4 \approx 1.86080588$

$x_5 \approx 1.86080585 \approx x_6$

To eight decimal places, the roots are -2.11490754, 0.25410169, and 1.86080585.

13.

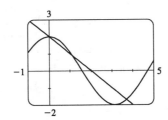

From the graph and by inspection, $x = 0$ is a root. Also, $y = 2\cos x$ and $y = 2 - x$ intersect at $x \approx 1$

and at $x \approx 3.5$. $f(x) = 2\cos x + x - 2 \implies f'(x) = -2\sin x + 1$, so $x_{n+1} = x_n - \dfrac{2\cos x_n + x_n - 2}{-2\sin x_n + 1}.$

$x_1 = 1$

$x_2 \approx 1.11802556$

$x_3 \approx 1.10918766$

$x_4 \approx 1.10914418 \approx x_5$

$x_1 = 3.5$

$x_2 \approx 3.71915887$

$x_3 \approx 3.69833117$

$x_4 \approx 3.69815369$

$x_5 \approx 3.69815367 \approx x_6$

To eight decimal places, the roots are 0, 1.10914418, and 3.69815367.

15.

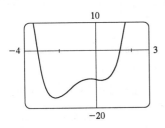

From the graph, there appear to be roots of

$f(x) = x^4 + 3x^3 - x - 10$ near -3.2 and 1.4.

$f'(x) = 4x^3 + 9x^2 - 1$, so

$$x_{n+1} = x_n - \frac{x_n^4 + 3x_n^3 - x_n - 10}{4x_n^3 + 9x_n^2 - 1}.$$ Taking $x_1 = -3.2$, we

get $x_2 \approx -3.20617358$, $x_3 \approx -3.20614267 \approx x_4$. Taking

$x_1 = 1.4$, we get $x_2 \approx 1.37560834$, $x_3 \approx 1.37506496$,

$x_4 \approx 1.37506470 \approx x_5$. To eight decimal places, the roots are

-3.20614267 and 1.37506470.

17.

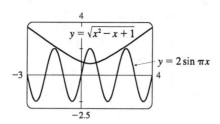

From the graph, we see that there are roots of this equation near 0.2 and 0.8.

$$f(x) = \sqrt{x^2 - x + 1} - 2\sin \pi x \implies f'(x) = \frac{2x - 1}{2\sqrt{x^2 - x + 1}} - 2\pi \cos \pi x, \text{ so}$$

$$x_{n+1} = x_n - \frac{\sqrt{x_n^2 - x_n + 1} - 2\sin \pi x_n}{\frac{2x_n - 1}{2\sqrt{x_n^2 - x_n + 1}} - 2\pi \cos \pi x_n}. \text{ Taking } x_1 = 0.2, \text{ we get } x_2 \approx 0.15212015,$$

$x_3 \approx 0.15438067$, $x_4 \approx 0.15438500 \approx x_5$. Taking $x_1 = 0.8$, we get $x_2 \approx 0.84787985$, $x_3 \approx 0.84561933$, $x_4 \approx 0.84561500 \approx x_5$. So, to eight decimal places, the roots of the equation are 0.15438500 and 0.84561500.

19. (a) $f(x) = x^2 - a \implies f'(x) = 2x$, so Newton's method gives

$$x_{n+1} = x_n - \frac{x_n^2 - a}{2x_n} = x_n - \frac{1}{2}x_n + \frac{a}{2x_n} = \frac{1}{2}x_n + \frac{a}{2x_n} = \frac{1}{2}\left(x_n + \frac{a}{x_n}\right).$$

(b) Using (a) with $a = 1000$ and $x_1 = \sqrt{900} = 30$, we get $x_2 \approx 31.666667$, $x_3 \approx 31.622807$, and $x_4 \approx 31.622777 \approx x_5$. So $\sqrt{1000} \approx 31.622777$.

21. $f(x) = x^3 - 3x + 6 \implies f'(x) = 3x^2 - 3$. If $x_1 = 1$, then $f'(x_1) = 0$ and the tangent line used for approximating x_2 is horizontal. Attempting to find x_2 results in trying to divide by zero.

23. For $f(x) = x^{1/3}$, $f'(x) = \frac{1}{3}x^{-2/3}$ and $x_{n+1} = x_n - \frac{f(x_n)}{f'(x_n)} = x_n - \frac{x_n^{1/3}}{\frac{1}{3}x_n^{-2/3}} = x_n - 3x_n = -2x_n$.

Therefore, each successive approximation becomes twice as large as the previous one in absolute value, so the sequence of approximations fails to converge to the root, which is 0.

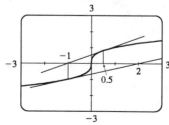

In the figure, we have $x_1 = 0.5$, $x_2 = -2(0.5) = -1$, and $x_3 = -2(-1) = 2$.

25.

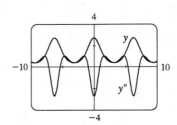

From the figure, we see that $y = f(x) = e^{\cos x}$ is periodic with period 2π. To find the x-coordinates of the IP, we only need to approximate the zeros of y''. There are two such values in $[-\pi, \pi]$.
$f'(x) = -e^{\cos x} \sin x \implies f''(x) = e^{\cos x} (\sin^2 x - \cos x)$. Since $e^{\cos x} \neq 0$, we will use Newton's method with $g(x) = \sin^2 x - \cos x$, $g'(x) = 2 \sin x \cos x + \sin x$, and $x_1 = 1$. $x_2 \approx 0.904173$, $x_3 \approx 0.904557 \approx x_4$. Since f is even, $(\pm 0.904557 + 2n\pi, 1.855277)$ are the IP.

27. The volume of the silo, in terms of its radius, is $V(r) = \pi r^2 (30) + \frac{1}{2} \left(\frac{4}{3} \pi r^3 \right) = 30 \pi r^2 + \frac{2}{3} \pi r^3$.

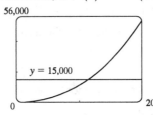

From a graph of V, we see that $V(r) = 15{,}000$ at $r \approx 11$ ft. Now we use Newton's method to solve the equation $V(r) - 15{,}000 = 0$. $\dfrac{dV}{dr} = 60 \pi r + 2 \pi r^2$, so $r_{n+1} = r_n - \dfrac{30 \pi r_n^2 + \frac{2}{3} \pi r_n^3 - 15{,}000}{60 \pi r_n + 2 \pi r_n^2}$. Taking $r_1 = 11$, we get $r_2 = 11.2853$, $r_3 = 11.2807 \approx r_4$. So in order for the silo to hold $15{,}000$ ft^3 of grain, its radius must be about 11.2807 ft.

29. In this case, $A = 18{,}000$, $R = 375$, and $n = 5(12) = 60$. So the formula becomes
$$18{,}000 = \frac{375}{x} \left[1 - (1+x)^{-60} \right] \iff 48x = 1 - (1+x)^{-60} \iff 48x(1+x)^{60} - (1+x)^{60} + 1 = 0.$$
Let the LHS be called $f(x)$, so that
$$f'(x) = 48x(60)(1+x)^{59} + 48(1+x)^{60} - 60(1+x)^{59}$$
$$= 12(1+x)^{59} [4x(60) + 4(1+x) - 5] = 12(1+x)^{59}(244x - 1),$$
$x_{n+1} = x_n - \dfrac{48x_n (1+x_n)^{60} - (1+x_n)^{60} + 1}{12(1+x_n)^{59}(244x_n - 1)}$. An interest rate of 1%/month seems like a reasonable estimate for $x = i$. So let $x_1 = 1\% = 0.01$, and we get $x_2 = 0.0082202$, $x_3 \approx 0.0076802$, $x_4 \approx 0.0076291$, $x_5 \approx 0.0076286 \approx x_6$. Thus, the dealer is charging a monthly interest rate of 0.76286% (or 9.55%/year, compounded monthly).

Section 4.9 Antiderivatives

1. $f(x) = 12x^2 + 6x - 5 \Longrightarrow F(x) = 12\left(\frac{1}{3}x^3\right) + 6\left(\frac{1}{2}x^2\right) - 5x + C = 4x^3 + 3x^2 - 5x + C$

3. $f(x) = 6/x^5 = 6x^{-5} \Longrightarrow F(x) = (6x^{-4})/(-4) + C = -3/(2x^4) + C$ if f is on an interval that

does not contain 0; otherwise, $F(x) = \begin{cases} -3/(2x^4) + C_1 & \text{if } x < 0 \\ -3/(2x^4) + C_2 & \text{if } x > 0 \end{cases}$

5. $g(t) = \dfrac{t^3 + 2t^2}{\sqrt{t}} = t^{5/2} + 2t^{3/2} \Longrightarrow G(t) = \dfrac{t^{7/2}}{7/2} + \dfrac{2t^{5/2}}{5/2} + C = \frac{2}{7}t^{7/2} + \frac{4}{5}t^{5/2} + C$

7. $f(t) = \sec^2 t + t^2 \Longrightarrow F(t) = \tan t + \frac{1}{3}t^3 + C_n$ on the interval $\left(n\pi - \frac{\pi}{2}, n\pi + \frac{\pi}{2}\right)$.

9. $f(x) = 2x + \dfrac{5}{\sqrt{1 - x^2}} \Longrightarrow F(x) = x^2 + 5\sin^{-1} x + C$

11. $f(x) = 5x^4 - 2x^5 \Longrightarrow F(x) = x^5 - \frac{1}{3}x^6 + C.$ $F(0) = 4 \Longrightarrow C = 4$, so $F(x) = x^5 - \frac{1}{3}x^6 + 4.$ The graph confirms our answer since $f(x) = 0$ when F has a local maximum, f is positive when F is increasing, and f is negative when F is decreasing.

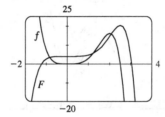

13. $f''(x) = x^2 + x^3 \Longrightarrow f'(x) = \frac{1}{3}x^3 + \frac{1}{4}x^4 + C \Longrightarrow f(x) = \frac{1}{12}x^4 + \frac{1}{20}x^5 + Cx + D$

15. $f'(x) = 2/x \Longrightarrow f(x) = 2\ln|x| + C = 2\ln(-x) + C$ (since $x < 0$). Now
$f(-1) = 2\ln 1 + C = 2(0) + C = 7 \Longrightarrow C = 7.$ Therefore, $f(x) = 2\ln(-x) + 7$, $x < 0.$

17. $f''(x) = 6x + 6 \Longrightarrow f'(x) = 3x^2 + 6x + C \Longrightarrow f(x) = x^3 + 3x^2 + Cx + D.$ $4 = f(0) = D$ and
$3 = f(1) = 1 + 3 + C + D = 4 + C + 4 \Longrightarrow C = -5$, so $f(x) = x^3 + 3x^2 - 5x + 4.$

19. $f''(x) = x^{-2}$, $x > 0 \Longrightarrow f'(x) = -1/x + C \Longrightarrow f(x) = -\ln|x| + Cx + D = -\ln x + Cx + D$
(since $x > 0$). $0 = f(1) = C + D$ and
$0 = f(2) = -\ln 2 + 2C + D = -\ln 2 + 2C - C$ (since $D = -C$) $= -\ln 2 + C \Longrightarrow C = \ln 2$ and
$D = -\ln 2.$ So $f(x) = -\ln x + (\ln 2)x - \ln 2.$

21. Given $f'(x) = 2x + 1$, we have $f(x) = x^2 + x + C.$ Since f passes through $(1, 6)$,
$6 = f(1) = 1^2 + 1 + C \Longrightarrow C = 4.$ Therefore, $f(x) = x^2 + x + 4$ and $f(2) = 2^2 + 2 + 4 = 10.$

23.

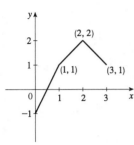

$$f'(x) = \begin{cases} 2 & \text{if } 0 \le x < 1 \\ 1 & \text{if } 1 < x < 2 \\ -1 & \text{if } 2 < x \le 3 \end{cases} \Longrightarrow f(x) = \begin{cases} 2x + C & \text{if } 0 \le x < 1 \\ x + D & \text{if } 1 < x < 2 \\ -x + E & \text{if } 2 < x \le 3 \end{cases}$$

$f(0) = -1 \Longrightarrow 2(0) + C = -1 \Longrightarrow C = -1$. Starting at the point $(0, -1)$ and moving to the right on a line with slope 2 gets us to the point $(1, 1)$. The slope for $1 < x < 2$ is 1, so we get to the point $(2, 2)$. The line connecting $(1, 1)$ to $(2, 2)$ is $y = x$, so $D = 0$. The slope for $2 < x \le 3$ is -1, so we get to $(3, 1)$. $f(3) = 1 \Longrightarrow -3 + E = 1 \Longrightarrow E = 4$.

Thus, $f(x) = \begin{cases} 2x - 1 & \text{if } 0 \le x < 1 \\ x & \text{if } 1 < x < 2 \\ -x + 4 & \text{if } 2 < x \le 3 \end{cases}$

Note that f is continuous, but $f'(x)$ does not exist at $x = 1$ or at $x = 2$.

25.

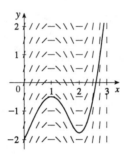

27.

x	$f(x)$
0	1
0.5	0.959
1.0	0.841
1.5	0.665
2.0	0.455
2.5	0.239
3.0	0.047

x	$f(x)$
3.5	−0.100
4.0	−0.189
4.5	−0.217
5.0	−0.192
5.5	−0.128
6.0	−0.047

We compute slopes (values of f) as in the table and draw a direction field as in Example 5. Then we use the direction field to graph F starting at $(0, 0)$.

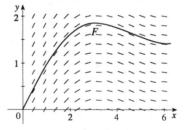

29.

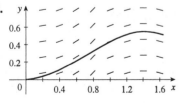

Remember that the values of f are the slopes of F at any x. For example, at $x = 1.4$, the slope of F is $f(1.4) = 0$.

31. $v(t) = s'(t) = 3\sqrt{t} = 3t^{1/2} \Longrightarrow s(t) = 3\left(\frac{2}{3}t^{3/2}\right) + C = 2t^{3/2} + C.$ $s(1) = 5 \Longrightarrow 2 + C = 5 \Longrightarrow$ $C = 3$, so $s(t) = 2t^{3/2} + 3$.

33. (a) We first observe that since the stone is dropped 450 m above the ground, $v(0) = 0$ and $s(0) = 450$.

$v'(t) = a(t) = -9.8 \Longrightarrow v(t) = -9.8t + C$, but $C = v(0) = 0$, so $v(t) = -9.8t \Longrightarrow$

$s(t) = -4.9t^2 + D \Longrightarrow D = s(0) = 450 \Longrightarrow s(t) = 450 - 4.9t^2$.

(b) It reaches the ground when $0 = s(t) = 450 - 4.9t^2 \Longrightarrow t^2 = 450/4.9 \Longrightarrow t_1 = \sqrt{450/4.9} \approx 9.58$ s.

(c) $v(t_1) = -9.8\sqrt{450/4.9} \approx -93.9$ m/s

(d) This is just reworking parts (a) and (b) with $v(0) = -5$. $v(t) = -9.8t + C \Longrightarrow -5 = 0 + C \Longrightarrow$

$v(t) = -9.8t - 5.$ $s(t) = -4.9t^2 - 5t + D \Longrightarrow 450 = s(0) = D \Longrightarrow s(t) = -4.9t^2 - 5t + 450.$

$s(t) = 0 \Longrightarrow t = \left(5 \pm \sqrt{8845}\right)/(-9.8) \Longrightarrow t_1 \approx 9.09$ s.

35. By Exercise 34, $s(t) = -4.9t^2 + v_0 t + s_0$ and $v(t) = s'(t) = -9.8t + v_0$. So

$[v(t)]^2 = (-9.8t + v_0)^2 = (9.8)^2 t^2 - 19.6 v_0 t + v_0^2 = v_0^2 - 19.6\left(v_0 t - 4.9t^2\right)$. But $-4.9t^2 + v_0 t$ is

just $s(t)$ without the s_0 term, that is, $s(t) - s_0$. Thus, $[v(t)]^2 = v_0^2 - 19.6\left[s(t) - s_0\right]$.

37. Using Exercise 34 with $a = -32$, $v_0 = 0$, and $s_0 = h$ (the height of the cliff), we know that the

height at time t is $s(t) = -16t^2 + h$. $v(t) = s'(t) = -32t \Longrightarrow -32t = -120 \Longrightarrow t = 3.75$, so

$0 = s(3.75) = -16(3.75)^2 + h \Longrightarrow h = 16(3.75)^2 = 225$ ft.

39. $a(t) = k$, the initial velocity is 30 mi/h$= 30 \cdot \frac{5280}{3600} = 44$ ft/s, and the final velocity is

50 mi/h$= 50 \cdot \frac{5280}{3600} = \frac{220}{3}$ ft/s. So $v(t) = kt + C$ and $v(0) = 44 \Longrightarrow C = 44$. Thus,

$v(t) = kt + 44 \Longrightarrow \frac{220}{3} = v(5) = 5k + 44 \Longrightarrow k = \frac{88}{15} \approx 5.87$ ft/s^2.

41. Marginal cost $= 1.92 - 0.002x = C'(x) \Longrightarrow C(x) = 1.92x - 0.001x^2 + K$. But

$C(1) = 1.92 - 0.001 + K = 562 \Longrightarrow K = 560.081$. Therefore, $C(x) = 1.92x - 0.001x^2 + 560.081 \Longrightarrow$

$C(100) = 742.081$, so the cost of producing 100 items is \$742.08.

43. (a) The Mean Value Theorem says that there exists a number c in the interval (x_1, x_2) such

that $H'(c) = \dfrac{H(x_2) - H(x_1)}{x_2 - x_1}$. Since $H = G - F$ and G and F are antiderivatives of

f, $H'(c) = G'(c) - F'(c) = f(c) - f(c) = 0$. So now $\dfrac{H(x_2) - H(x_1)}{x_2 - x_1} = 0 \Longrightarrow$

$H(x_2) - H(x_1) = 0 \ (x_2 \neq x_1) \Longrightarrow H(x_2) = H(x_1)$. Since this is true for any $x_1 < x_2$ in I, H

must be a constant function.

(b) We have $H = G - F$ and $H(x) = C$, so $C = G - F \Longrightarrow G(x) = F(x) + C$. Thus, any

antiderivative G can be expressed as $F(x) + C$.

45. (a) First note that $90 \, \text{mi/h} = 90 \times \frac{5280}{3600} \, \text{ft/s} = 132 \, \text{ft/s}$. Then $a(t) = 4 \, \text{ft/s}^2 \Longrightarrow v(t) = 4t + C$, but $v(0) = 0 \Longrightarrow C = 0$. Now $4t = 132$ when $t = \frac{132}{4} = 33 \, \text{s}$, so it takes $33 \, \text{s}$ to reach $132 \, \text{ft/s}$. Therefore, taking $s(0) = 0$, we have $s(t) = 2t^2, 0 \le t \le 33$. So $s(33) = 2178 \, \text{ft}$. $15 \, \text{minutes} = 15(60) = 900 \, \text{s}$, so for $33 \le t \le 933$ we have $v(t) = 132 \, \text{ft/s} \Longrightarrow$ $s(933) = 132(900) + 2178 = 120{,}978 \, \text{ft} = 22.9125 \, \text{mi}$.

(b) As in part (a), the train accelerates for $33 \, \text{s}$ and travels $2178 \, \text{ft}$ while doing so. Similarly, it decelerates for $33 \, \text{s}$ and travels $2178 \, \text{ft}$ at the end of its trip. During the remaining $900 - 66 = 834 \, \text{s}$ it travels at $132 \, \text{ft/s}$, so the distance traveled is $132 \cdot 834 = 110{,}088 \, \text{ft}$. Thus, the total distance is $2178 + 110{,}088 + 2178 = 114{,}444 \, \text{ft} = 21.675 \, \text{mi}$.

(c) $45 \, \text{mi} = 45(5280) = 237{,}600 \, \text{ft}$. Subtract $2(2178)$ to take care of the speeding up and slowing down, and we have $233{,}244 \, \text{ft}$ at $132 \, \text{ft/s}$ for a trip of $233{,}244/132 = 1767 \, \text{s}$ at $90 \, \text{mi/h}$. The total time is $1767 + 2(33) = 1833 \, \text{s}$ or $30.55 \, \text{min}$.

(d) $37.5(60) = 2250 \, \text{s}$. $2250 - 2(33) = 2184 \, \text{s}$ at maximum speed. $2184(132) + 2(2178) = 292{,}644$ total feet or $292{,}644/5280 = 55.425 \, \text{mi}$.

Chapter 4 Review

Concept Check

1. A function f has an *absolute maximum* at $x = c$ if $f(c)$ is the largest function value on the entire domain of f, whereas f has a *local maximum* at c if $f(c)$ is the largest function value when x is near c.

2. (a) See Theorem 4.2.3.

(b) See the Closed Interval Method before Example 6 in Section 4.2.

3. (a) See Theorem 4.2.4.

(b) See Definition 4.2.5.

4. See the Mean Value Theorem at the beginning of Section 4.3. Geometrical interpretation — there is some point P on the graph of a function f [on the interval (a, b)] where the tangent line is parallel to the secant line that connects $(a, f(a))$ and $(b, f(b))$.

5. (a) See the I/D Test after Example 1 in Section 4.3.

(b) See the Concavity Test after Example 3 in Section 4.3.

6. (a) See the First Derivative Test after Example 2 in Section 4.3.

(b) See the Second Derivative Test before Example 4 in Section 4.3.

(c) See the note before Example 5 in Section 4.3.

7. (a) See page 303.

(b) Write fg as $\dfrac{f}{1/g}$ or $\dfrac{g}{1/f}$.

(c) Convert the difference into a quotient by using a common denominator, rationalizing, factoring, etc.

(d) Convert the power to a product by taking the natural logarithm of both sides of $y = f^g$ or by writing f^g as $e^{g \ln f}$.

8. (a) See Figure 3 in Section 4.8.

(b) $x_2 = x_1 - \dfrac{f(x_1)}{f'(x_1)}$

(c) $x_{n+1} = x_n - \dfrac{f(x_n)}{f'(x_n)}$

(d) Newton's method is likely to fail or to work very slowly when $f'(x_1)$ is close to 0.

True-False Quiz

1. False. For example, take $f(x) = x^3$, then $f'(x) = 3x^2$ and $f'(0) = 0$, but $f(0) = 0$ is not a maximum or minimum; $(0,0)$ is an inflection point.

3. False. For example, $f(x) = x$ is continuous on $(0,1)$ but attains neither a maximum nor a minimum value on $(0,1)$. Don't confuse this with f being continuous on the *closed* interval $[a,b]$, which would make the statement true.

5. True by the ID Test.

7. False. $f'(x) = g'(x) \Longrightarrow f(x) = g(x) + C$. For example, $f(x) = x+2$, $g(x) = x+1 \Longrightarrow$ $f'(x) = g'(x) = 1$, but $f(x) \neq g(x)$.

9. True. The graph of one such function is sketched. [An example is $f(x) = e^{-x}$.]

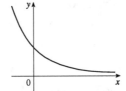

Exercises

1. $f(x) = x^3 - 12x + 5$, $-5 \leq x \leq 3$. $f'(x) = 3x^2 - 12 = 0 \Longrightarrow x^2 = 4 \Longrightarrow x = \pm 2$. $f''(x) = 6x \Longrightarrow$ $f''(-2) = -12 < 0$, so $f(-2) = 21$ is a local maximum, and $f''(2) = 12 > 0$, so $f(2) = -11$ is a local minimum. Also $f(-5) = -60$ and $f(3) = -4$, so $f(-2) = 21$ is the absolute maximum and $f(-5) = -60$ is the absolute minimum.

3. $f(x) = x - \sqrt{2}\sin x$, $0 \leq x \leq \pi$. $f'(x) = 1 - \sqrt{2}\cos x = 0 \Longrightarrow \cos x = \frac{1}{\sqrt{2}} \Longrightarrow x = \frac{\pi}{4}$. $f''(x) = \sqrt{2}\sin x \Longrightarrow f''\left(\frac{\pi}{4}\right) = \sqrt{2}\sin\frac{\pi}{4} = 1 > 0$, so $f\left(\frac{\pi}{4}\right) = \frac{\pi}{4} - 1 \approx -0.21$ is a local minimum. Also $f(0) = 0$ and $f(\pi) = \pi$, so the absolute minimum is $f\left(\frac{\pi}{4}\right) = \frac{\pi}{4} - 1$, and the absolute maximum is $f(\pi) = \pi$.

5. (a) f is a polynomial function and hence, it has no asymptotes.

(b) $f(x) = x^4 - 3x^3 + 3x^2 - x \Longrightarrow f'(x) = 4x^3 - 9x^2 + 6x - 1$. Since the sum of the coefficients is 0, 1 is a root of f', so $f'(x) = (x-1)(4x^2 - 5x + 1) = (x-1)^2(4x-1)$. $f'(x) < 0 \Longrightarrow x < \frac{1}{4}$, so f is decreasing on $\left(-\infty, \frac{1}{4}\right)$ and f is increasing on $\left(\frac{1}{4}, \infty\right)$.

(c) $f'(x)$ does not change sign at $x = 1$, so there is not a local maximum or minimum there. $f\left(\frac{1}{4}\right) = -\frac{27}{256}$ is a local minimum.

(d) $f''(x) = 12x^2 - 18x + 6 = 6(2x-1)(x-1)$. $f''(x) = 0 \Longrightarrow x = \frac{1}{2}$ or 1. $f''(x) < 0 \Longrightarrow \frac{1}{2} < x < 1 \Longrightarrow f$ is CD on $\left(\frac{1}{2}, 1\right)$ and CU on $\left(-\infty, \frac{1}{2}\right)$ and $(1, \infty)$. There are inflection points at $\left(\frac{1}{2}, -\frac{1}{16}\right)$ and $(1, 0)$.

(e)

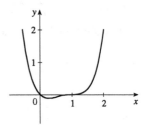

7. (a) No asymptote

(b) $y = f(x) = x + \sqrt{1-x} \implies f'(x) = 1 - 1/\left(2\sqrt{1-x}\right) = 0 \iff 2\sqrt{1-x} = 1 \iff$
$1 - x = \frac{1}{4} \iff x = \frac{3}{4}$ and $f'(x) > 0 \iff x < \frac{3}{4}$, so f is increasing on $\left(-\infty, \frac{3}{4}\right)$ and decreasing
on $\left(\frac{3}{4}, 1\right)$.

(c) $f\left(\frac{3}{4}\right) = \frac{5}{4}$ is a local maximum.

(e)

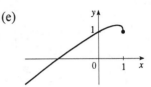

(d) $f''(x) = -\dfrac{1}{4(1-x)^{3/2}} < 0$ on the domain of f, so f is
CD on $(-\infty, 1)$. No IP.

9. (a) $\displaystyle\lim_{x \to \pm\infty} \sin^{-1}(1/x) = \sin^{-1}(0) = 0$, so $y = 0$ is a HA.

(b) $y = f(x) = \sin^{-1}(1/x) \implies f'(x) = \dfrac{1}{\sqrt{1 - (1/x)^2}}\left(-\dfrac{1}{x^2}\right) = \dfrac{-1}{\sqrt{x^4 - x^2}} < 0$, so f is

decreasing on $(-\infty, -1)$ and $(1, \infty)$. Note that the domain of f is $-1 \le x \le 1$.

(c) No local maximum or minimum, but $f(1) = \frac{\pi}{2}$ is the absolute maximum and $f(-1) = -\frac{\pi}{2}$ is the
absolute minimum.

(d) $f''(x) = \dfrac{4x^3 - 2x}{2(x^4 - x^2)^{3/2}} = \dfrac{x(2x^2 - 1)}{(x^4 - x^2)^{3/2}} > 0$ for $x > 1$
and $f''(x) < 0$ for $x < -1$, so f is CU on $(1, \infty)$ and CD
on $(-\infty, -1)$. No IP.

(e)

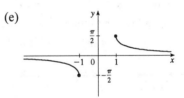

11. (a) $\displaystyle\lim_{x \to \pm\infty}\left(e^x + e^{-3x}\right) = \infty$, no asymptote.

(b) $y = f(x) = e^x + e^{-3x} \implies$
$f'(x) = e^x - 3e^{-3x} = e^{-3x}\left(e^{4x} - 3\right) > 0 \iff$
$e^{4x} > 3 \iff 4x > \ln 3 \iff x > \frac{1}{4}\ln 3$, so f is
increasing on $\left(\frac{1}{4}\ln 3, \infty\right)$ and decreasing on
$\left(-\infty, \frac{1}{4}\ln 3\right)$.

(c) Absolute minimum $f\left(\frac{1}{4}\ln 3\right) = 3^{1/4} + 3^{-3/4} \approx 1.75$.

(d) $f''(x) = e^x + 9e^{-3x} > 0$, so f is CU on $(-\infty, \infty)$.
No IP.

(e)

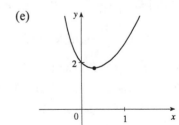

13. $f(x) = \dfrac{x^2-1}{x^3} \implies f'(x) = \dfrac{x^3(2x) - (x^2-1)3x^2}{x^6} = \dfrac{3-x^2}{x^4} \implies$

$f''(x) = \dfrac{x^4(-2x) - (3-x^2)4x^3}{x^8} = \dfrac{2x^2-12}{x^5}$

Estimates: From the graphs of f' and f'', it appears that f is increasing on $(-1.73, 0)$ and $(0, 1.73)$ and decreasing on $(-\infty, -1.73)$ and $(1.73, \infty)$; f has a local maximum of about $f(1.73) = 0.38$ and a local minimum of about $f(-1.7) = -0.38$; f is CU on $(-2.45, 0)$ and $(2.45, \infty)$, and CD on $(-\infty, -2.45)$ and $(0, 2.45)$; and f has inflection points at about $(-2.45, -0.34)$ and $(2.45, 0.34)$.

Exact: Now $f'(x) = \dfrac{3-x^2}{x^4}$ is positive for $0 < x^2 < 3$, that is, f is increasing on $(-\sqrt{3}, 0)$ and $(0, \sqrt{3})$; and $f'(x)$ is negative (and so f is decreasing) on $(-\infty, -\sqrt{3})$ and $(\sqrt{3}, \infty)$. $f'(x) = 0$ when $x = \pm\sqrt{3}$. f' goes from positive to negative at $x = \sqrt{3}$, so f has a local maximum of $f(\sqrt{3}) = \dfrac{(\sqrt{3})^2-1}{(\sqrt{3})^3} = \dfrac{2\sqrt{3}}{9}$; and since f is odd, we know that maxima on the interval $(0, \infty)$ correspond to minima on $(-\infty, 0)$, so f has a local minimum of $f(-\sqrt{3}) = -\dfrac{2\sqrt{3}}{9}$. Also, $f''(x) = \dfrac{2x^2-12}{x^5}$ is positive (so f is CU) on $(-\sqrt{6}, 0)$ and $(\sqrt{6}, \infty)$, and negative (so f is CD) on $(-\infty, -\sqrt{6})$ and $(0, \sqrt{6})$. There are IP at $\left(\sqrt{6}, \dfrac{5\sqrt{6}}{36}\right)$ and $\left(-\sqrt{6}, -\dfrac{5\sqrt{6}}{36}\right)$.

15. $f(x) = 3x^6 - 5x^5 + x^4 - 5x^3 - 2x^2 + 2$, $f'(x) = 18x^5 - 25x^4 + 4x^3 - 15x^2 - 4x$,
$f''(x) = 90x^4 - 100x^3 + 12x^2 - 30x - 4$

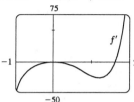

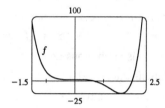

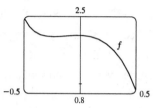

From the graphs of f' and f'', it appears that f is increasing on $(-0.23, 0)$ and $(1.62, \infty)$ and decreasing on $(-\infty, -0.23)$ and $(0, 1.62)$; f has a local maximum of about $f(0) = 2$ and local

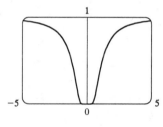

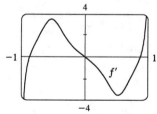

minima of about $f(-0.23) = 1.96$ and $f(1.62) = -19.2$; f is CU on $(-\infty, -0.12)$ and $(1.24, \infty)$ and CD on $(-0.12, 1.24)$; and f has inflection points at about $(-0.12, 1.98)$ and $(1.2, -12.1)$.

17.

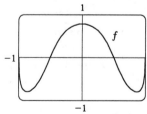

From the graph, we estimate the points of inflection to be about $(\pm 0.8, 0.2)$. $f(x) = e^{-1/x^2} \Longrightarrow f'(x) = 2x^{-3}e^{-1/x^2} \Longrightarrow$
$f''(x) = 2\left[x^{-3}\left(2x^{-3}\right)e^{-1/x^2} + e^{-1/x^2}\left(-3x^{-4}\right)\right] =$
$2x^{-6}e^{-1/x^2}\left(2 - 3x^2\right)$. This is 0 when $2 - 3x^2 = 0 \Longleftrightarrow$
$x = \pm\sqrt{\frac{2}{3}}$, so the inflection points are $\left(\pm\sqrt{\frac{2}{3}}, e^{-3/2}\right)$.

19. $f(x) = \arctan(\cos(3\arcsin x))$. We use a CAS to compute f' and f'', and to graph f, f', and f'':

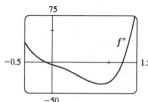

From the graph of f', it appears that the only maximum occurs at $x = 0$ and there are minima at $x = \pm 0.87$. From the graph of f'', it appears that there are inflection points at $x = \pm 0.52$.

21.

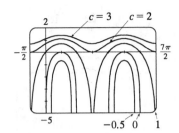

The family of functions $f(x) = \ln(\sin x + C)$ all have the same period and all have maximum values at $x = \frac{\pi}{2} + 2\pi n$. Since the domain of $\ln$ is $(0, \infty)$, f has a graph only if $\sin x + C > 0$ somewhere. Since $-1 \le \sin x \le 1$, this happens if $C > -1$, that is, f has no graph if $C \le -1$. Similarly, if $C > 1$, then $\sin x + C > 0$ and f is continuous on $(-\infty, \infty)$. As C increases, the graph of f is shifted vertically upward and flattens out. If $-1 < C \le 1$, f is defined where $\sin x + C > 0 \Longleftrightarrow$ $\sin x > -C \Longleftrightarrow \sin^{-1}(-C) < x < \pi - \sin^{-1}(-C)$. Since the period is 2π, the domain of f is $\left(2n\pi + \sin^{-1}(-C), (2n+1)\pi - \sin^{-1}(-C)\right)$, n an integer.

23. For $(1, 6)$ to be on the curve $y = x^3 + ax^2 + bx + 1$, we have that $6 = a + b + 2 \Longrightarrow b = 4 - a$. Now $y' = 3x^2 + 2ax + b$ and $y'' = 6x + 2a$. Also, for $(1, 6)$ to be an inflection point it must be true that $y''(1) = 6(1) + 2a = 0 \Longrightarrow a = -3 \Longrightarrow b = 4 - (-3) = 7$.

25. $\displaystyle\lim_{x \to \pi} \frac{\sin x}{x^2 - \pi^2} \overset{\text{H}}{=} \lim_{x \to \pi} \frac{\cos x}{2x} = -\frac{1}{2\pi}$

27. $\displaystyle\lim_{x \to \infty} \frac{\ln(\ln x)}{\ln x} \overset{\text{H}}{=} \lim_{x \to \infty} \frac{1/(x \ln x)}{1/x} = \lim_{x \to \infty} \frac{1}{\ln x} = 0$

29. $\displaystyle\lim_{x \to 0} \frac{\ln(1 - x) + x + \frac{1}{2}x^2}{x^3} \overset{\text{H}}{=} \lim_{x \to 0} \frac{-\dfrac{1}{1 - x} + 1 + x}{3x^2} \overset{\text{H}}{=} \lim_{x \to 0} \frac{-\dfrac{1}{(1 - x)^2} + 1}{6x} \overset{\text{H}}{=} \lim_{x \to 0} \frac{-\dfrac{2}{(1 - x)^3}}{6}$

$\displaystyle = -\frac{2}{6} = -\frac{1}{3}$

31. $\displaystyle\lim_{x \to 0} \left(\csc^2 x - x^{-2}\right) = \lim_{x \to 0}\left[\frac{1}{\sin^2 x} - \frac{1}{x^2}\right] = \lim_{x \to 0} \frac{x^2 - \sin^2 x}{x^2 \sin^2 x} \overset{\text{H}}{=} \lim_{x \to 0} \frac{2x - \sin 2x}{2x \sin^2 x + x^2 \sin 2x}$

$\displaystyle \overset{\text{H}}{=} \lim_{x \to 0} \frac{2 - 2\cos 2x}{2\sin^2 x + 4x \sin 2x + 2x^2 \cos 2x} \overset{\text{H}}{=} \lim_{x \to 0} \frac{4 \sin 2x}{6 \sin 2x + 12x \cos 2x - 4x^2 \sin 2x}$

$\displaystyle \overset{\text{H}}{=} \lim_{x \to 0} \frac{8 \cos 2x}{24 \cos 2x - 32x \sin 2x - 8x^2 \cos 2x} = \frac{8}{24} = \frac{1}{3}$

33. We are given $d\theta/dt = -0.25$ rad/h. $x = 400 \cot \theta \Longrightarrow$

$\dfrac{dx}{dt} = -400 \csc^2 \theta \dfrac{d\theta}{dt}$. When $\theta = \frac{\pi}{6}$,

$\dfrac{dx}{dt} = -400 (2)^2 (-0.25) = 400$ ft/h.

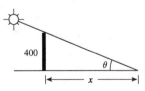

35. Given $dh/dt = 5$ and $dx/dt = 15$, find dz/dt. $z^2 = x^2 + h^2 \Longrightarrow 2z\dfrac{dz}{dt} = 2x\dfrac{dx}{dt} + 2h\dfrac{dh}{dt} \Longrightarrow$

$\dfrac{dz}{dt} = \dfrac{1}{z}\left(15x + 5h\right)$. When $t = 3$, $h = 45 + 3\,(5) = 60$ and $x = 15\,(3) = 45 \Longrightarrow z = 75$, so

$\dfrac{dz}{dt} = \dfrac{1}{75}\left[15\,(45) + 5\,(60)\right] = 13$ ft/s.

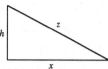

37. If $B = 0$, the line is vertical and the distance from $x = -\dfrac{C}{A}$ to (x_1, y_1) is

$\left|x_1 + \dfrac{C}{A}\right| = \dfrac{|Ax_1 + By_1 + C|}{\sqrt{A^2 + B^2}}$, so assume $B \neq 0$. The square of the distance from (x_1, y_1)

to the line is $f\,(x) = (x - x_1)^2 + (y - y_1)^2$ where $Ax + By + C = 0$, so we minimize

$f\,(x) = (x - x_1)^2 + \left(-\dfrac{A}{B}x - \dfrac{C}{B} - y_1\right)^2 \Longrightarrow f'\,(x) = 2\,(x - x_1) + 2\left(-\dfrac{A}{B}x - \dfrac{C}{B} - y_1\right)\left(-\dfrac{A}{B}\right).$

$f'\,(x) = 0 \Longrightarrow x = \dfrac{B^2x_1 - ABy_1 - AC}{A^2 + B^2}$ and this gives a minimum since $f''\,(x) = 2\left(1 + \dfrac{A^2}{B^2}\right) > 0$.

Substituting this value of x into $f\,(x)$ and simplifying gives $f\,(x) = \dfrac{(Ax_1 + By_1 + C)^2}{A^2 + B^2}$, so the

minimum distance is $\sqrt{f\,(x)} = \dfrac{|Ax_1 + By_1 + C|}{\sqrt{A^2 + B^2}}$.

39.

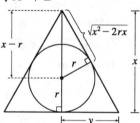

By similar triangles, $\dfrac{y}{x} = \dfrac{r}{\sqrt{x^2 - 2rx}}$, so the area of the triangle is

$A\,(x) = \tfrac{1}{2}\,(2y)\,x = xy = \dfrac{rx^2}{\sqrt{x^2 - 2rx}} \Longrightarrow$

$A'\,(x) = \dfrac{2rx\sqrt{x^2 - 2rx} - rx^2\,(x - r)\big/\sqrt{x^2 - 2rx}}{x^2 - 2rx} = \dfrac{rx^2\,(x - 3r)}{(x^2 - 2rx)^{3/2}} = 0$ when $x = 3r$.

$A'\,(x) < 0$ when $2r < x < 3r$, $A'\,(x) > 0$ when $x > 3r$. So $x = 3r$ gives a minimum and

$A\,(3r) = r\,(9r^2)\big/(\sqrt{3}r) = 3\sqrt{3}r^2$.

41.

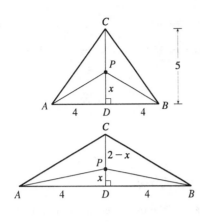

We minimize
$$L(x) = |PA| + |PB| + |PC| = 2\sqrt{x^2 + 16} + (5 - x),$$
$0 \le x \le 5$. $L'(x) = 2x / \sqrt{x^2 + 16} - 1 = 0 \iff$
$2x = \sqrt{x^2 + 16} \iff 4x^2 = x^2 + 16 \iff x = \frac{4}{\sqrt{3}}$.

$L(0) = 13$, $L\left(\frac{4}{\sqrt{3}}\right) \approx 11.9$, $L(5) \approx 12.8$, so the

minimum occurs when $x = \frac{4}{\sqrt{3}} \approx 2.3$.

If $|CD| = 2$, $L(x)$ changes from $(5 - x)$ to $(2 - x)$ with

$0 \le x \le 2$. But we still get $L'(x) = 0 \iff x = \frac{4}{\sqrt{3}}$, which

isn't in the interval $[0, 2]$. Now $L(0) = 10$ and

$L(2) = 2\sqrt{20} = 4\sqrt{5} \approx 8.9$. The minimum occurs when

$P = C$.

43. $v = K\sqrt{\dfrac{L}{C} + \dfrac{C}{L}} \implies \dfrac{dv}{dL} = \dfrac{K}{2\sqrt{(L/C) + (C/L)}} \left(\dfrac{1}{C} - \dfrac{C}{L^2}\right) = 0 \iff \dfrac{1}{C} = \dfrac{C}{L^2} \iff L^2 = C^2 \iff$

$L = C$. This gives the minimum velocity since $v' < 0$ for $0 < L < C$ and $v' > 0$ for $L > C$.

45. Let $x =$ selling price of ticket. Then $12 - x$ is the amount the ticket price has been lowered,

so the number of tickets sold is $11{,}000 + 1000(12 - x) = 23{,}000 - 1000x$. The revenue is

$R(x) = x(23{,}000 - 1000x) = 23{,}000x - 1000x^2$, so $R'(x) = 23{,}000 - 2000x = 0$ when $x = 11.5$.

Since $R''(x) = -2000 < 0$, the maximum revenue occurs when the ticket prices are $\$11.50$.

47. $f(x) = x^6 + 2x^2 - 8x + 3 \implies f'(x) = 6x^5 + 4x - 8$. We want to find the minimum of f, so we

examine the graph of f' looking for values at which f' changes from negative to positive.

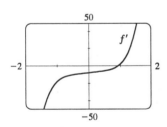

From the graph, we see that this occurs at $x \approx 1$. So we will

use Newton's method with $g(x) = f'(x) = 6x^5 + 4x - 8$,

$g'(x) = 30x^4 + 4$, and $x_1 = 1$. $x_{n+1} = x_n - \dfrac{6x_n^5 + 4x_n - 8}{30x_n^4 + 4}$

gives us $x_2 \approx 0.941176$, $x_3 \approx 0.934068$,

$x_4 \approx 0.933975 \approx x_5$. Thus, $f(x_5) \approx -2.063421$ is the

absolute minimum value of f.

49. $f(x) = e^x - 1/x \implies F(x) = \begin{cases} e^x - \ln|x| + C_1 & \text{if } x < 0 \\ e^x - \ln|x| + C_2 & \text{if } x > 0 \end{cases}$

51. $f'(x) = 2/(1 + x^2) \implies f(x) = 2\arctan x + C$. $f(0) = 2\arctan 0 + C = -1 \implies C = -1$.

Therefore, $f(x) = 2\arctan x - 1$.

53. $f''(x) = x^3 + x \implies f'(x) = \frac{1}{4}x^4 + \frac{1}{2}x^2 + C \implies 1 = f'(0) = C \implies f'(x) = \frac{1}{4}x^4 + \frac{1}{2}x^2 + 1 \implies$

$f(x) = \frac{1}{20}x^5 + \frac{1}{6}x^3 + x + D \implies -1 = f(0) = D \implies f(x) = \frac{1}{20}x^5 + \frac{1}{6}x^3 + x - 1$.

55. (a)

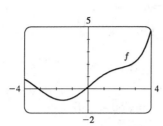 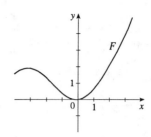

Since f is 0 just to the left of the y-axis, we must have a minimum of F at the same place since we are increasing through $(0,0)$ on F. There must be a local maximum to the left of $x = -3$, since f changes from positive to negative there.

(b) $f(x) = 0.1e^x + \sin x \implies F(x) = 0.1e^x - \cos x + C$. $F(0) = 0 \implies 0.1 - 1 + C = 0 \implies$ $C = 0.9$, so $F(x) = 0.1e^x - \cos x + 0.9$

(c)

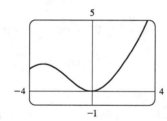

57. Choosing the positive direction to be upward, we have $a(t) = -9.8 \implies v(t) = -9.8t + v_0$, but $v(0) = 0 = v_0 \implies v(t) = -9.8t = s'(t) \implies s(t) = -4.9t^2 + s_0$, but $s(0) = s_0 = 500 \implies$ $s(t) = -4.9t^2 + 500$. When $s = 0$, $-4.9t^2 + 500 = 0 \implies t_1 = \sqrt{\frac{500}{4.9}} \approx 10.1 \implies$ $v(t_1) = -9.8\sqrt{\frac{500}{4.9}} \approx -98.995$ m/s. Therefore, the canister will not burst.

59. $f(x) = x^{101} + x^{51} + x - 1 = 0$. Since f is continuous and $f(0) = -1$ and $f(1) = 2$, the equation has at least one root in $(0,1)$, by the Intermediate Value Theorem. Suppose the equation has two roots, a and b, with $a < b$. Then $f(a) = 0 = f(b)$, so by the Mean Value Theorem, $f'(x) = \dfrac{f(b) - f(a)}{b - a} = \dfrac{0}{b - a} = 0$, so $f'(x)$ has a root in (a, b). But this is impossible since $f'(x) = 101x^{100} + 51x^{50} + 1 \geq 1$ for all x.

61. (a)

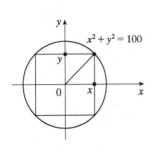

The cross-sectional area is
$A = 2x \cdot 2y = 4xy = 4x\sqrt{100 - x^2}, 0 \le x \le 10$, so
$$\frac{dA}{dx} = 4x\left(\tfrac{1}{2}\right)(100 - x^2)^{-1/2}(-2x) + (100 - x^2)^{1/2} \cdot 4$$
$$= \frac{-4x^2}{(100 - x^2)^{1/2}} + 4(100 - x^2)^{1/2}$$
$$= 0 \text{ when } -x^2 + (100 - x^2) = 0 \Longrightarrow$$
$$x^2 = 50 \Longrightarrow x = \sqrt{50} \approx 7.07 \Longrightarrow$$
$$y = \sqrt{100 - (\sqrt{50})^2} = \sqrt{50}. \text{ Since } A(0) = A(10) = 0, \text{ the}$$

rectangle of maximum area is a square.

(b)

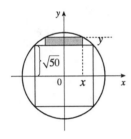

The cross-sectional area of each rectangular plank (shaded in
the figure) is $A = 2x\left(y - \sqrt{50}\right) = 2x\left[\sqrt{100 - x^2} - \sqrt{50}\right]$,
$0 \le x \le \sqrt{50}$, so
$$\frac{dA}{dx} = 2\left(\sqrt{100 - x^2} - \sqrt{50}\right) + 2x\left(\tfrac{1}{2}\right)(100 - x^2)^{-1/2}(-2x)$$
$$= 2(100 - x^2)^{1/2} - 2\sqrt{50} - \frac{2x^2}{(100 - x^2)^{1/2}}.$$

Set $\dfrac{dA}{dx} = 0$: $(100 - x^2) - \sqrt{50}(100 - x^2)^{1/2} - x^2 = 0 \Longrightarrow 100 - 2x^2 = \sqrt{50}(100 - x^2)^{1/2} \Longrightarrow$

$10,000 - 400x^2 + 4x^4 = 50(100 - x^2) \Longrightarrow 2500 - 175x^2 + 2x^4 = 0 \Longrightarrow$

$x^2 = \dfrac{175 \pm \sqrt{10,625}}{4} \approx 69.52 \text{ or } 17.98 \Longrightarrow x \approx 8.34 \text{ or } 4.24.$

But $8.34 > \sqrt{50}$, so $x_1 \approx 4.24 \Longrightarrow y - \sqrt{50} = \sqrt{100 - x_1^2} - \sqrt{50} \approx 1.99$. Each plank should
have dimensions about $8\frac{1}{2}$ inches by 2 inches.

(c) From the figure in part (a), the width is $2x$ and the depth is $2y$, so the strength is
$S = k(2x)(2y)^2 = 8kxy^2 = 8kx(100 - x^2) = 800kx - 8kx^3, 0 \le x \le 10.$
$dS/dx = 800k - 24kx^2 = 0$ when $24kx^2 = 800k \Longrightarrow x^2 = \frac{100}{3} \Longrightarrow x = \frac{10}{\sqrt{3}} \Longrightarrow$
$y = \sqrt{\frac{200}{3}} = \frac{10\sqrt{2}}{\sqrt{3}} = \sqrt{2}x$ and $S(0) = S(10) = 0$, so the maximum strength occurs when
$x = \frac{10}{\sqrt{3}}$. The dimensions should be $\frac{20}{\sqrt{3}} \approx 11.55$ inches by $\frac{20\sqrt{2}}{\sqrt{3}} \approx 16.33$ inches.

63. (a) $I = \dfrac{k\cos\theta}{d^2} = \dfrac{k\,(h/d)}{d^2} = k\dfrac{h}{d^3} = k\dfrac{h}{\left(\sqrt{40^2 + h^2}\right)^3} = k\dfrac{h}{(1600 + h^2)^{3/2}} \implies$

$$\frac{dI}{dh} = k\frac{(1600 + h^2)^{3/2} - h\frac{3}{2}(1600 + h^2)^{1/2}\cdot 2h}{(1600 + h^2)^3}$$

$$= \frac{k\,(1600 + h^2)^{1/2}\,(1600 + h^2 - 3h^2)}{(1600 + h^2)^3}$$

$$= \frac{k\,(1600 - 2h^2)}{(1600 + h^2)^{5/2}}$$

Set $dI/dh = 0$: $1600 - 2h^2 = 0 \implies h^2 = 800 \implies h = \sqrt{800} = 20\sqrt{2}$. By the First Derivative Test, I has a relative maximum at $h = 20\sqrt{2} \approx 28$ ft.

(b)

$\dfrac{dx}{dt} = 4$ ft/s

$$I = \frac{k\cos\theta}{d^2} = \frac{k\,[(h-4)/d]}{d^2} = \frac{k\,(h-4)}{d^3}$$

$$= \frac{k\,(h-4)}{\left[(h-4)^2 + x^2\right]^{3/2}}$$

$$= k\,(h-4)\left[(h-4)^2 + x^2\right]^{-3/2}$$

$$\frac{dI}{dt} = \frac{dI}{dx}\cdot\frac{dx}{dt} = k\,(h-4)\left(-\frac{3}{2}\right)\left[(h-4)^2 + x^2\right]^{-5/2}\cdot 2x \cdot \frac{dx}{dt}$$

$$= k\,(h-4)\,(-3x)\left[(h-4)^2 + x^2\right]^{-5/2}\cdot 4$$

$$= \frac{-12xk\,(h-4)}{\left[(h-4)^2 + x^2\right]^{5/2}}$$

$$\left.\frac{dI}{dt}\right|_{x=40} = -\frac{480k\,(h-4)}{\left[(h-4)^2 + 1600\right]^{5/2}}$$

Focus on Problem Solving

1. Let $y = f(x) = e^{-x^2}$. The area of the rectangle under the curve from $-x$ to x is $A(x) = 2xe^{-x^2}$ where $x \geq 0$. We maximize $A(x)$: $A'(x) = 2e^{-x^2} - 4x^2 e^{-x^2} = 2e^{-x^2}\left(1 - 2x^2\right) = 0 \Longrightarrow x = \frac{1}{\sqrt{2}}$. This gives a maximum since $A'(x) > 0$ for $0 \leq x < \frac{1}{\sqrt{2}}$ and $A'(x) < 0$ for $x > \frac{1}{\sqrt{2}}$. We next determine the points of inflection of $f(x)$. Notice that $f'(x) = -2xe^{-x^2} = -A(x)$. So $f''(x) = -A'(x)$ and hence, $f''(x) < 0$ for $-\frac{1}{\sqrt{2}} < x < \frac{1}{\sqrt{2}}$ and $f''(x) > 0$ for $x < -\frac{1}{\sqrt{2}}$ and $x > \frac{1}{\sqrt{2}}$. So $f(x)$ changes concavity at $x = \pm\frac{1}{\sqrt{2}}$, and the two vertices of the rectangle of largest area are at the inflection points.

3. First, we recognize some symmetry in the inequality: $\dfrac{e^{x+y}}{xy} \geq e^2 \Longleftrightarrow \dfrac{e^x}{x} \cdot \dfrac{e^y}{y} \geq e \cdot e$. This suggests that we need to show that $e^x/x \geq e$ for $x > 0$. If we can do this, then the inequality $e^y/y \geq e$ is true, and the given inequality follows.

$f(x) = e^x/x \Longrightarrow f'(x) = \dfrac{xe^x - e^x}{x^2} = \dfrac{e^x(x-1)}{x^2} = 0 \Longrightarrow x = 1$. By the First Derivative Test, we have a minimum of $f(1) = e$, so $e^x/x \geq e$ for all x.

5. Differentiating $x^2 + xy + y^2 = 12$ implicitly with respect to x gives $2x + y + x\dfrac{dy}{dx} + 2y\dfrac{dy}{dx} = 0$, so $\dfrac{dy}{dx} = -\dfrac{2x+y}{x+2y}$. At a highest or lowest point, $\dfrac{dy}{dx} = 0 \Longleftrightarrow y = -2x$. Substituting this into the original equation gives $x^2 + x(-2x) + (-2x)^2 = 12$, so $3x^2 = 12$ and $x = \pm 2$. If $x = 2$, then $y = -2x = -4$, and if $x = -2$ then $y = 4$. Thus, the highest and lowest points are $(-2, 4)$ and $(2, -4)$.

7. $f(x) = \dfrac{1}{1 + |x|} + \dfrac{1}{1 + |x - 2|}$

$= \begin{cases} \dfrac{1}{1-x} + \dfrac{1}{1-(x-2)} & \text{if } x < 0 \\[2mm] \dfrac{1}{1+x} + \dfrac{1}{1-(x-2)} & \text{if } 0 \leq x < 2 \\[2mm] \dfrac{1}{1+x} + \dfrac{1}{1+(x-2)} & \text{if } x \geq 2 \end{cases} \Longrightarrow f'(x) = \begin{cases} \dfrac{1}{(1-x)^2} + \dfrac{1}{(3-x)^2} & \text{if } x < 0 \\[2mm] \dfrac{-1}{(1+x)^2} + \dfrac{1}{(3-x)^2} & \text{if } 0 < x < 2 \\[2mm] \dfrac{-1}{(1+x)^2} - \dfrac{1}{(x-1)^2} & \text{if } x > 2 \end{cases}$

We see that $f'(x) > 0$ for $x < 0$ and $f'(x) < 0$ for $x > 2$. For $0 < x < 2$, we have

$f'(x) = \dfrac{1}{(3-x)^2} - \dfrac{1}{(x+1)^2} = \dfrac{(x^2 + 2x + 1) - (x^2 - 6x + 9)}{(3-x)^2(x+1)^2} = \dfrac{8(x-1)}{(3-x)^2(x+1)^2}$, so

$f'(x) < 0$ for $0 < x < 1$, $f'(1) = 0$ and $f'(x) > 0$ for $1 < x < 2$. We have shown that $f'(x) > 0$ for $x < 0$; $f'(x) < 0$ for $0 < x < 1$; $f'(x) > 0$ for $1 < x < 2$; and $f'(x) < 0$ for $x > 2$. Therefore, by the First Derivative Test, the local maxima of f are at $x = 0$ and $x = 2$, where f takes the value $\frac{4}{3}$. Therefore, $\frac{4}{3}$ is the absolute maximum value of f.

9. We first show that $\dfrac{x}{1+x^2} < \tan^{-1} x$ for $x > 0$. Let $f(x) = \tan^{-1} x - \dfrac{x}{1+x^2}$. Then

$$f'(x) = \frac{1}{1+x^2} - \frac{1\left(1+x^2\right) - x\left(2x\right)}{\left(1+x^2\right)^2} = \frac{\left(1+x^2\right) - \left(1-x^2\right)}{\left(1+x^2\right)^2} = \frac{2x^2}{\left(1+x^2\right)^2} > 0 \text{ for } x > 0. \text{ So}$$

$f(x)$ is increasing on $(0, \infty)$. Hence, $0 < x \Longrightarrow 0 = f(0) < f(x) = \tan^{-1} x - \dfrac{x}{1+x^2}$. So

$\dfrac{x}{1+x^2} < \tan^{-1} x$ for $0 < x$. We next show that $\tan^{-1} x < x$ for $x > 0$. Let $h(x) = x - \tan^{-1} x$.

Then $h'(x) = 1 - \dfrac{1}{1+x^2} = \dfrac{x^2}{1+x^2} > 0$. Hence, $h(x)$ is increasing on $(0, \infty)$. So for $0 < x$,

$0 = h(0) < h(x) = x - \tan^{-1} x$. Hence, $\tan^{-1} x < x$ for $x > 0$, and we conclude that

$\dfrac{x}{1+x^2} < \tan^{-1} x < x$ for $x > 0$.

11. $A = \left(x_1, x_1^2\right)$ and $B = \left(x_2, x_2^2\right)$, where x_1 and x_2 are the solutions of the quadratic equation $x^2 = mx + b$. Let $P = \left(x, x^2\right)$ and set $A_1 = (x_1, 0)$, $B_1 = (x_2, 0)$, and $P_1 = (x, 0)$. Let $f(x)$ denote the area of triangle PAB. Then $f(x)$ can be expressed in terms of the areas of three trapezoids as follows:

$$f(x) = \text{area}(A_1 A B B_1) - \text{area}(A_1 A P P_1) - \text{area}(B_1 B P P_1)$$
$$= \tfrac{1}{2}\left(x_1^2 + x_2^2\right)(x_2 - x_1) - \tfrac{1}{2}\left(x_1^2 + x^2\right)(x - x_1) - \tfrac{1}{2}\left(x^2 + x_2^2\right)(x_2 - x).$$

After expanding and canceling terms, we get

$$f(x) = \tfrac{1}{2}\left(x_2 x_1^2 - x_1 x_2^2 - xx_1^2 + x_1 x^2 - x_2 x^2 + xx_2^2\right)$$
$$= \tfrac{1}{2}\left[x_1^2 (x_2 - x) + x_2^2 (x - x_1) + x^2 (x_1 - x_2)\right].$$

$f'(x) = \tfrac{1}{2}\left[-x_1^2 + x_2^2 + 2x(x_1 - x_2)\right]$. $f''(x) = \tfrac{1}{2}\left[2(x_1 - x_2)\right] = x_1 - x_2 < 0$ since $x_2 > x_1$.

$f'(x) = 0 \Longrightarrow 2x(x_1 - x_2) = x_1^2 - x_2^2 \Longrightarrow x_P = \tfrac{1}{2}(x_1 + x_2)$.

$$f(x_P) = \tfrac{1}{2}\left(x_1^2 \left[\tfrac{1}{2}(x_2 - x_1)\right] + x_2^2 \left[\tfrac{1}{2}(x_2 - x_1)\right] + \tfrac{1}{4}(x_1 + x_2)^2 (x_1 - x_2)\right)$$

$$= \tfrac{1}{2}\left[\tfrac{1}{2}(x_2 - x_1)\left(x_1^2 + x_2^2\right) - \tfrac{1}{4}(x_2 - x_1)(x_1 + x_2)^2\right]$$

$$= \tfrac{1}{8}(x_2 - x_1)\left[2\left(x_1^2 + x_2^2\right) - \left(x_1^2 + 2x_1 x_2 + x_2^2\right)\right] = \tfrac{1}{8}(x_2 - x_1)\left(x_1^2 - 2x_1 x_2 + x_2^2\right)$$

$$= \tfrac{1}{8}(x_2 - x_1)(x_1 - x_2)^2 = \tfrac{1}{8}(x_2 - x_1)(x_2 - x_1)^2 = \tfrac{1}{8}(x_2 - x_1)^3$$

To put this in terms of m and b, we solve the system $y = x_1^2$ and $y = mx_1 + b$, giving us

$x_1^2 - mx_1 - b = 0 \Longrightarrow x_1 = \tfrac{1}{2}\left(m - \sqrt{m^2 + 4b}\right)$. Similarly, $x_2 = \tfrac{1}{2}\left(m + \sqrt{m^2 + 4b}\right)$. The area is

then $\tfrac{1}{8}(x_2 - x_1)^3 = \tfrac{1}{8}\left(\sqrt{m^2 + 4b}\right)^3$, and is attained at the point $P\left(x_P, x_P^2\right) = P\left(\tfrac{1}{2}m, \tfrac{1}{4}m^2\right)$.

Note: Another way to get an expression for $f(x)$ is to use the formula for an area of a triangle in terms of the coordinates of the vertices: $f(x) = \tfrac{1}{2}\left[\left(x_2 x_1^2 - x_1 x_2^2\right) + \left(x_1 x^2 - xx_1^2\right) + \left(xx_2^2 - x_2 x^2\right)\right].$

13. (a)

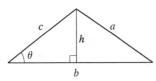

$A = \frac{1}{2}bh$ with $\sin\theta = h/c$, so $A = \frac{1}{2}bc\sin\theta$. But A is a constant, so differentiating this equation with respect to t, we

get $\dfrac{dA}{dt} = 0 = \dfrac{1}{2}\left[bc\cos\theta\dfrac{d\theta}{dt} + b\dfrac{dc}{dt}\sin\theta + \dfrac{db}{dt}c\sin\theta\right] \implies$

$bc\cos\theta\dfrac{d\theta}{dt} = -\sin\theta\left[b\dfrac{dc}{dt} + c\dfrac{db}{dt}\right] \implies$

$\dfrac{d\theta}{dt} = -\tan\theta\left[\dfrac{1}{c}\dfrac{dc}{dt} + \dfrac{1}{b}\dfrac{db}{dt}\right].$

(b) We use the Law of Cosines to get the length of side a in terms of those of b and c, and then we differentiate implicitly with respect to t: $\quad a^2 = b^2 + c^2 - 2bc\cos\theta \implies$

$2a\dfrac{da}{dt} = 2b\dfrac{db}{dt} + 2c\dfrac{dc}{dt} - 2\left[bc(-\sin\theta)\dfrac{d\theta}{dt} + b\dfrac{dc}{dt}\cos\theta + \dfrac{db}{dt}c\cos\theta\right] \implies$

$\dfrac{da}{dt} = \dfrac{1}{a}\left(b\dfrac{db}{dt} + c\dfrac{dc}{dt} + bc\sin\theta\dfrac{d\theta}{dt} - b\dfrac{dc}{dt}\cos\theta - c\dfrac{db}{dt}\cos\theta\right).$ Now we substitute our value

of a from the Law of Cosines and the value of $\dfrac{d\theta}{dt}$ from part (a), and simplify (primes signify differentiation by t):

$$\dfrac{da}{dt} = \dfrac{bb' + cc' + bc\sin\theta\left[-\tan\theta\left(c'/c + b'/b\right)\right] - (bc' + cb')(\cos\theta)}{\sqrt{b^2 + c^2 - 2bc\cos\theta}}$$

$$= \dfrac{bb' + cc' - \left[\sin^2\theta\left(bc' + cb'\right) + \cos^2\theta\left(bc' + cb'\right)\right]/\cos\theta}{\sqrt{b^2 + c^2 - 2bc\cos\theta}}$$

$$= \dfrac{bb' + cc' - (bc' + cb')\sec\theta}{\sqrt{b^2 + c^2 - 2bc\cos\theta}}$$

15. (a)

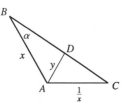

Let $y = |AD|$, $x = |AB|$, and $1/x = |AC|$, so that $|AB| \cdot |AC| = 1$. We compute the area $\mathcal{A}$ of $\triangle ABC$ in two

ways. First, $\mathcal{A} = \frac{1}{2}|AB||AC|\sin\frac{2\pi}{3} = \frac{1}{2}\cdot 1\cdot\frac{\sqrt{3}}{2} = \frac{\sqrt{3}}{4}$.

Second, $\mathcal{A} = (\text{area of } \triangle ABD) + (\text{area of } \triangle ACD)$

$= \frac{1}{2}|AB||AD|\sin\frac{\pi}{3} + \frac{1}{2}|AD||AC|\sin\frac{\pi}{3}$

$= \frac{1}{2}xy\frac{\sqrt{3}}{2} + \frac{1}{2}y(1/x)\frac{\sqrt{3}}{2} = \frac{\sqrt{3}}{4}y(x + 1/x).$

Equating the two expressions for the area, we get $\frac{\sqrt{3}}{4}y\left(x + \dfrac{1}{x}\right) = \dfrac{\sqrt{3}}{4} \iff$

$y = \dfrac{1}{x + 1/x} = \dfrac{x}{x^2 + 1}, x > 0.$

Another Method: Use the Law of Sines on the triangles ABD and ABC. In $\triangle ABD$, we have $\angle A + \angle B + \angle D = 180° \iff 60° + \alpha + \angle D = 180° \iff \angle D = 120° - \alpha.$

Thus, $\dfrac{x}{y} = \dfrac{\sin(120° - \alpha)}{\sin \alpha} = \dfrac{\sin 120° \cos \alpha - \cos 120° \sin \alpha}{\sin \alpha} = \dfrac{\frac{\sqrt{3}}{2} \cos \alpha + \frac{1}{2} \sin \alpha}{\sin \alpha} \Longrightarrow$

$\dfrac{x}{y} = \dfrac{\sqrt{3}}{2} \cot \alpha + \dfrac{1}{2}$, and by a similar argument with $\triangle ABC$, $\dfrac{\sqrt{3}}{2} \cot \alpha = x^2 + \dfrac{1}{2}$. Eliminating $\cot \alpha$

gives $\dfrac{x}{y} = \left(x^2 + \dfrac{1}{2}\right) + \dfrac{1}{2} \Longrightarrow y = \dfrac{x}{x^2 + 1}$, $x > 0$.

(b) We differentiate our expression for y with respect to x to find the maximum:

$\dfrac{dy}{dx} = \dfrac{(x^2 + 1) - x(2x)}{(x^2 + 1)^2} = \dfrac{1 - x^2}{(x^2 + 1)^2} = 0$ when $x = 1$. This indicates a maximum by the First

Derivative Test, since $y'(x) > 0$ for $0 < x < 1$ and $y'(x) < 0$ for $x > 1$, so the maximum value of

y is $y(1) = \dfrac{1}{2}$.

17. Let $s_A(t)$ and $s_B(t)$ be the position functions for cars A and B and let $f(t) = s_A(t) - s_B(t)$. Since A passed B twice (B passed A once), there must be three values of t such that $f(t) = 0$. Let these times be denoted t_1, t_2, and t_3. By the Mean Value Theorem, we know that for some number c_1 in (t_1, t_2),

$f'(c_1) = \dfrac{f(t_2) - f(t_1)}{t_2 - t_1}$, but $f(t_2) - f(t_1) = 0$, so $f'(c_1) = 0 \Longleftrightarrow s'_A(c_1) - s'_B(c_1) = 0 \Longleftrightarrow$

$v_A(c_1) - v_B(c_1) = 0 \Longrightarrow v_A(c_1) = v_B(c_1)$. By a similar argument, there exists some number c_2 in (t_2, t_3) such that $v_A(c_2) = v_B(c_2)$. Now let $g(t) = v_A(t) - v_B(t)$ and apply the Mean

Value Theorem on $[c_1, c_2]$ with $c_1 < c < c_2$. $g'(c) = \dfrac{g(c_2) - g(c_1)}{c_2 - c_1}$, but $g(c_2) = g(c_1) = 0$, so

$g'(c) = 0 \Longrightarrow v'_A(c) - v'_B(c) = 0 \Longrightarrow a_A(c) - a_B(c) = 0 \Longrightarrow a_A(c) = a_B(c)$, that is, A and B had equal accelerations at $t = c$.

19.

By similar triangles, $\dfrac{r}{5} = \dfrac{h}{16} \Longrightarrow r = \dfrac{5h}{16}$. The volume of the cone is

$V = \frac{1}{3}\pi r^2 h = \frac{1}{3}\pi \left(\dfrac{5h}{16}\right)^2 h = \dfrac{25\pi}{768}h^3$, so $\dfrac{dV}{dt} = \dfrac{25\pi}{256}h^2 \dfrac{dh}{dt}$. Now the

rate of change of the volume is also equal to the difference of what is being added ($2 \text{ cm}^3/\text{min}$) and what is oozing out ($k\pi r l$, where $\pi r l$ is the area of

the cone and k is a proportionality constant). Thus, $\dfrac{dV}{dt} = 2 - k\pi r l$.

Equating the two expressions for $\dfrac{dV}{dt}$ and substituting $h = 10$, $\dfrac{dh}{dt} = -0.3$, $r = \dfrac{5(10)}{16} = \dfrac{25}{8}$,

and $\dfrac{l}{\sqrt{281}} = \dfrac{10}{16} \Longleftrightarrow l = \frac{5}{8}\sqrt{281}$, we get $\frac{25\pi}{256}(10)^2(-0.3) = 2 - k\pi \frac{25}{8} \cdot \frac{5}{8}\sqrt{281} \Longleftrightarrow$

$\dfrac{125k\pi\sqrt{281}}{64} = 2 + \dfrac{750\pi}{256}$. Solving for k gives us $k = \dfrac{256 + 375\pi}{250\pi\sqrt{281}}$. To maintain a certain height,

the rate of oozing, $k\pi r l$, must equal the rate of the liquid being poured in, that is, $dV/dt = 0$.

$k\pi r l = \dfrac{256 + 375\pi}{250\pi\sqrt{281}} \cdot \pi \cdot \dfrac{25}{8} \cdot \dfrac{5\sqrt{281}}{8} = \dfrac{256 + 375\pi}{128} \approx 11.204 \text{ cm}^3/\text{min}.$

Chapter 5   Integrals

Section 5.1   Areas and Distances

1. (a) Since f is *increasing*, we can obtain a *lower* estimate by using *left* endpoints.

$$L_5 = \sum_{i=1}^{5} f(x_{i-1}) \, \Delta x \quad [\Delta x = \tfrac{10-0}{5} = 2]$$

$$= f(x_0) \cdot 2 + f(x_1) \cdot 2 + f(x_2) \cdot 2 + f(x_3) \cdot 2 + f(x_4) \cdot 2$$

$$= 2\,[f(0) + f(2) + f(4) + f(6) + f(8)]$$

$$\approx 2\,(1 + 3 + 4.3 + 5.4 + 6.3) = 2\,(20) = 40$$

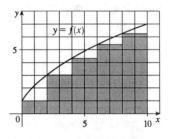

Since f is *increasing*, we can obtain an *upper* estimate by using *right* endpoints.

$$R_5 = \sum_{i=1}^{5} f(x_i) \, \Delta x$$

$$= 2\,[f(x_1) + f(x_2) + f(x_3) + f(x_4) + f(x_5)]$$

$$= 2\,[f(2) + f(4) + f(6) + f(8) + f(10)]$$

$$\approx 2\,(3 + 4.3 + 5.4 + 6.3 + 7) = 2\,(26) = 52$$

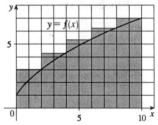

(b) $L_{10} = \displaystyle\sum_{i=1}^{10} f(x_{i-1}) \, \Delta x \quad [\Delta x = \tfrac{10-0}{10} = 1]$

$$= 1\,[f(x_0) + f(x_1) + \cdots + f(x_9)]$$

$$= f(0) + f(1) + \cdots + f(9)$$

$$\approx 1 + 2.1 + 3 + 3.7 + 4.3 + 4.9$$

$$+ 5.4 + 5.8 + 6.3 + 6.7 = 43.2$$

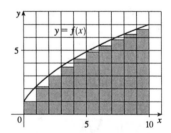

$$R_{10} = \sum_{i=1}^{10} f(x_i) \, \Delta x = f(1) + f(2) + \cdots + f(10)$$

$$= L_{10} + 1 \cdot f(10) - 1 \cdot f(0) \quad \begin{bmatrix} \text{add rightmost rectangle} \\ \text{subtract leftmost} \end{bmatrix}$$

$$= 43.2 + 7 - 1 = 49.2$$

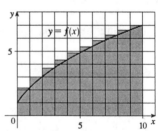

3. (a) $f(x) = x^3 + 2$ and $\Delta x = \frac{2-(-1)}{3} = 1 \Longrightarrow$

$\quad R_3 = 1 \cdot f(0) + 1 \cdot f(1) + 1 \cdot f(2) = 1 \cdot 2 + 1 \cdot 3 + 1 \cdot 10 = 15.$

$\quad \Delta x = \frac{2-(-1)}{6} = 0.5 \Longrightarrow$

$\quad R_6 = 0.5 \left[f(-0.5) + f(0) + f(0.5) + f(1) + f(1.5) + f(2) \right]$

$\qquad = 0.5 \left(1.875 + 2 + 2.125 + 3 + 5.375 + 10 \right)$

$\qquad = 0.5 \left(24.375 \right) = 12.1875$

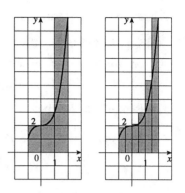

(b) $L_3 = 1 \cdot f(-1) + 1 \cdot f(0) + 1 \cdot f(1) = 1 \cdot 1 + 1 \cdot 2 + 1 \cdot 3 = 6.$

$\quad L_6 = 0.5 \left[f(-1) + f(-0.5) + f(0) + f(0.5) + f(1) + f(1.5) \right]$

$\qquad = 0.5 \left(1 + 1.875 + 2 + 2.125 + 3 + 5.375 \right)$

$\qquad = 0.5 \left(15.375 \right) = 7.6875$

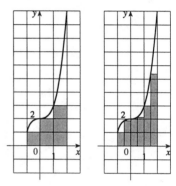

(c) $M_3 = 1 \cdot f(-0.5) + 1 \cdot f(0.5) + 1 \cdot f(1.5)$

$\qquad = 1 \cdot 1.875 + 1 \cdot 2.125 + 1 \cdot 5.375 = 9.375.$

$\quad M_6 = 0.5 \left[f(-0.75) + f(-0.25) + f(0.25) \right.$

$\qquad \qquad \left. + f(0.75) + f(1.25) + f(1.75) \right]$

$\qquad = 0.5 \left(1.578125 + 1.984375 + 2.015625 \right.$

$\qquad \qquad \left. + 2.421875 + 3.953125 + 7.359375 \right)$

$\qquad = 0.5 \left(19.3125 \right) = 9.65625$

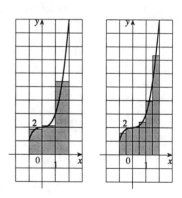

(d) M_6 appears to be the best estimate.

5. Here is one possible algorithm (ordered sequence of operations) for calculating the sums:

1   Let SUM $= 0$, X_MIN $= 0$, X_MAX $= \pi$, N $= 10$ (or 30 or 50, depending on which sum we are calculating), DELTA_X $= ($X_MAX $-$ X_MIN$)/$N, and RIGHT_ENDPOINT $=$ X_MIN $+$ DELTA_X.

2   Repeat steps 2a, 2b in sequence until RIGHT_ENDPOINT $>$ X_MAX.

   2a   Add $\sin$ (RIGHT_ENDPOINT) to SUM.

   2b   Add DELTA_X to RIGHT_ENDPOINT.

At the end of this procedure, (DELTA_X) $\cdot$ (SUM) is equal to the answer we are looking for. We find that

$$R_{10} = \frac{\pi}{10} \sum_{i=1}^{10} \sin\left(\frac{i\pi}{10}\right)$$
$$\approx 1.9835,$$

$$R_{30} = \frac{\pi}{30} \sum_{i=1}^{30} \sin\left(\frac{i\pi}{30}\right)$$
$$\approx 1.9982,$$

and

$$R_{50} = \frac{\pi}{50} \sum_{i=1}^{50} \sin\left(\frac{i\pi}{50}\right)$$
$$\approx 1.9993.$$

It appears that the exact area is 2.

7. In Maple, we have to perform a number of steps before getting a numerical answer. After loading the student package [command: `with(student);`] we use the command `leftsum(x^(1/2),x=1..4,10 [or 30, or 50]);` which gives us the expression in summation notation. To get a numerical approximation to the sum, we use `evalf(");`. Mathematica does not have a special command for these sums, so we must type them in manually. For example, the first left sum is given by `(3/10)*Sum[Sqrt[1+3(i-1)/10],{i,1,10}]`, and we use the `N` command on the resulting output to get a numerical approximation.

In Derive, we use the `LEFT_RIEMANN` command to get the left sums, but must define the right sums ourselves. (We can define a new function using `LEFT_RIEMANN` with k ranging from 1 to n instead of from 0 to $n - 1$.)

(a) With $f(x) = \sqrt{x}$, $1 \le x \le 4$, the left sums are of the form $L_n = \dfrac{3}{n} \displaystyle\sum_{i=1}^{n} \sqrt{1 + \dfrac{3(i-1)}{10}}$.

Specifically, $L_{10} \approx 4.5148$, $L_{30} \approx 4.6165$, and $L_{50} \approx 4.6366$. The right sums are of the form $R_n = \dfrac{3}{n} \displaystyle\sum_{i=1}^{n} \sqrt{1 + \dfrac{3i}{n}}$. Specifically, $R_{10} \approx 4.8148$, $R_{30} \approx 4.7165$, and $R_{50} \approx 4.6966$.

(b) In Maple, we use the `leftbox` and `rightbox` commands (with the same arguments as `leftsum` and `rightsum` above) to generate the graphs.

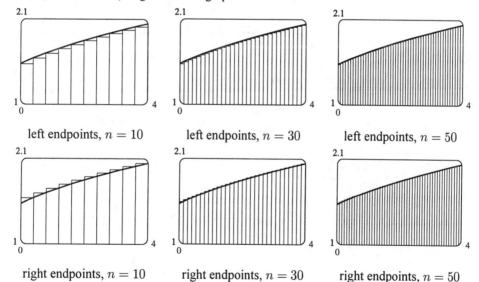

left endpoints, $n = 10$ left endpoints, $n = 30$ left endpoints, $n = 50$

right endpoints, $n = 10$ right endpoints, $n = 30$ right endpoints, $n = 50$

(c) We know that since $\sqrt{x}$ is an increasing function on $[1, 4]$, all of the left sums are smaller than the actual area, and all of the right sums are larger than the actual area. Since the left sum with $n = 50$ is about $4.637 > 4.6$ and the right sum with $n = 50$ is about $4.697 < 4.7$, we conclude that $4.6 < L_{50} <$ actual area $< R_{50} < 4.7$, so the actual area is between 4.6 and 4.7.

9. Since v is an increasing function, L_6 will give us a lower estimate and R_6 will give us an upper estimate.
$$L_6 = (0 \text{ ft/s})(0.5 \text{ s}) + (6.2)(0.5) + (10.8)(0.5) + (14.9)(0.5) + (18.1)(0.5) + (19.4)(0.5)$$
$$= 0.5(69.4) = 34.7 \text{ ft}$$
$$R_6 = 0.5(6.2 + 10.8 + 14.9 + 18.1 + 19.4 + 20.2) = 0.5(89.6) = 44.8 \text{ ft}$$

11. For a decreasing function, using left endpoints gives us an overestimate and using right endpoints results in an underestimate. We will use M_6 to get an estimate. $\Delta t = 1$, so
$$M_6 = 1 \cdot v(0.5) + 1 \cdot v(1.5) + 1 \cdot v(2.5) + 1 \cdot v(3.5) + 1 \cdot v(4.5) + 1 \cdot v(5.5)$$
$$\approx 55 + 40 + 28 + 18 + 10 + 4 = 155 \text{ ft}$$
For a very rough check on the above calculation, we can draw a line from $(0, 70)$ to $(6, 0)$ and calculate the area of the triangle: $\frac{1}{2}(70)(6) = 210$. This is clearly an overestimate, so our midpoint estimate of 155 is reasonable.

13. Let S_n be the given formula.

1. S_1 is true because $1^3 = \left[\dfrac{1(1+1)}{2}\right]^2$, that is, $1 = \dfrac{4}{4}$.

2. Assume that S_k is true; that is, $1^3 + 2^3 + 3^3 + \cdots + k^3 = \left[\dfrac{k(k+1)}{2}\right]^2$. Then

$$1^3 + 2^3 + 3^3 + \cdots + k^3 + (k+1)^3 = \left[\dfrac{k(k+1)}{2}\right]^2 + (k+1)^3$$

$$= \dfrac{(k+1)^2}{2^2}\left[k^2 + 4(k+1)\right] = \dfrac{(k+1)^2}{2^2}(k+2)^2 = \left[\dfrac{(k+1)\left[(k+1)+1\right]}{2}\right]^2$$

So S_{k+1} is true.

By the Principle of Mathematical Induction, S_n is true for all n.

15. (a) $\Delta x = (2-0)/n = 2/n$ and $x_i = 0 + i\Delta x = 2i/n$.
$$A = \lim_{n\to\infty} R_n = \lim_{n\to\infty} \sum_{i=1}^{n} f(x_i)\,\Delta x = \lim_{n\to\infty} \sum_{i=1}^{n} \left(\dfrac{2i}{n}\right)^5 \cdot \dfrac{2}{n} = \lim_{n\to\infty} \dfrac{64}{n^6}\sum_{i=1}^{n} i^5.$$

(b) $\displaystyle\sum_{i=1}^{n} i^5 = \dfrac{n^2(n+1)^2(2n^2+2n-1)}{12}$

(c) $\displaystyle\lim_{n\to\infty}\dfrac{64}{n^6}\cdot\dfrac{n^2(n+1)^2(2n^2+2n-1)}{12} = \dfrac{64}{12}\lim_{n\to\infty}\dfrac{(n^2+2n+1)(2n^2+2n-1)}{n^2\cdot n^2}$

$$= \dfrac{16}{3}\lim_{n\to\infty}\left(1+\dfrac{2}{n}+\dfrac{1}{n^2}\right)\left(2+\dfrac{2}{n}-\dfrac{1}{n^2}\right) = \tfrac{16}{3}\cdot 1\cdot 2 = \tfrac{32}{3}$$

17. $\Delta x = (b-0)/n = b/n$ and $x_i = 0 + i\Delta x = bi/n$.

$$A = \lim_{n\to\infty} R_n = \lim_{n\to\infty} \sum_{i=1}^{n} f(x_i)\,\Delta x = \lim_{n\to\infty} \sum_{i=1}^{n}\cos\left(\dfrac{bi}{n}\right)\cdot\dfrac{b}{n} = \lim_{n\to\infty}\left[\dfrac{b\sin\left(b\left(\dfrac{1}{2n}+1\right)\right)}{2n\sin\left(\dfrac{b}{2n}\right)} - \dfrac{b}{2n}\right]$$

$= \sin b$. If $b = \frac{\pi}{2}$, then $A = \sin\frac{\pi}{2} = 1$.

Section 5.2 The Definite Integral

1. $R_4 = \sum\limits_{i=1}^{4} f(x_i)\,\Delta x$ [$x_i^* = x_i$ is a right endpoint and $\Delta x = 0.5$]

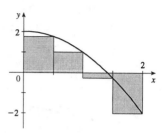

$= 0.5\,[f(0.5) + f(1) + f(1.5) + f(2)]$ [$f(x) = 2 - x^2$]

$= 0.5\,[1.75 + 1 + (-0.25) + (-2)]$

$= 0.5\,(0.5) = 0.25$

The Riemann sum represents the area of the two triangles above the x-axis minus the area of the two rectangles below the x-axis.

3. (a) Using the right endpoints to approximate $\int_0^8 f(x)\,dx$, we have

$$\sum_{i=1}^{4} f(x_i)\,\Delta x = 2\,[f(2) + f(4) + f(6) + f(8)]$$

$$= 2\,(1 + 2 - 2 + 1) = 4$$

(b) Using the left endpoints to approximate $\int_0^8 f(x)\,dx$, we have

$$\sum_{i=1}^{4} f(x_{i-1})\,\Delta x = 2\,[f(0) + f(2) + f(4) + f(6)]$$

$$= 2\,(2 + 1 + 2 - 2) = 6$$

(c) Using the midpoint of each interval to approximate $\int_0^8 f(x)\,dx$, we have

$$\sum_{i=1}^{4} f(\overline{x}_i)\,\Delta x = 2\,[f(1) + f(3) + f(5) + f(7)]$$

$$= 2\,(3 + 2 + 1 - 1) = 10$$

5. The width of the subintervals is $\Delta x = (5 - 0)/5 = 1$, so the endpoints of the five subintervals are $0, 1, 2, 3, 4, 5$ and the midpoints are $0.5, 1.5, 2.5, 3.5, 4.5$. The Midpoint Rule gives

$\int_0^5 x^3\,dx \approx \sum\limits_{i=1}^{5} f(\overline{x}_i)\,\Delta x = 1\left[(0.5)^3 + (1.5)^3 + (2.5)^3 + (3.5)^3 + (4.5)^3\right] = 153.1250$.

7. $\Delta x = (2 - 1)/10 = 0.1$ so the endpoints are $1.0, 1.1, \ldots, 2.0$ and the midpoints are $1.05, 1.15, \ldots, 1.95$. The Midpoint Rule gives

$$\int_1^2 \sqrt{1 + x^2}\,dx \approx \sum_{i=1}^{10} f(\overline{x}_i)\,\Delta x$$

$$= 0.1\left[\sqrt{1 + (1.05)^2} + \sqrt{1 + (1.15)^2} + \cdots + \sqrt{1 + (1.95)^2}\right] \approx 1.8100.$$

9. In Maple, we use the command `with(student);` to load the sum and box commands, then
`m:=middlesum(sqrt(1+x^2),x=1..2,10);` which gives us the sum in summation notation, then
`M:=evalf(m);` which gives $M_{10} \approx 1.81001414$, confirming the result of Exercise 7. The command
`middlebox(sqrt(1+x^2),x=1..2,10);` generates the graph. Repeating for $n = 20$ and $n = 30$
gives $M_{20} \approx 1.81007263$ and $M_{30} \approx 1.81008347$.

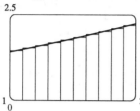
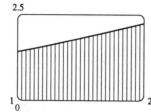
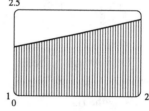

11. On $[0, \pi]$, $\displaystyle\lim_{n\to\infty} \sum_{i=1}^{n} \cos x_i \Delta x = \int_0^\pi \cos x \, dx$, by Definition 2 (or Equation 3) with $f(x) = \cos x$.

13. On $[0, 1]$, $\displaystyle\lim_{n\to\infty} \sum_{i=1}^{n} \left[2(x_i^*)^2 - 5x_i^* \right] \Delta x = \int_0^1 (2x^2 - 5x) \, dx$.

15. $\displaystyle\int_0^2 (2 - x^2) \, dx = \lim_{n\to\infty} \sum_{i=1}^{n} f(x_i) \Delta x \quad [\Delta x = 2/n \text{ and } x_i = 2i/n] \quad = \lim_{n\to\infty} \sum_{i=1}^{n} \left(2 - \frac{4i^2}{n^2} \right) \left(\frac{2}{n} \right)$

$$= \lim_{n\to\infty} \frac{2}{n} \left(2n - \frac{8}{n^3} \sum_{i=1}^{n} i^2 \right) = \lim_{n\to\infty} \left[4 - \frac{8}{n^3} \cdot \frac{n(n+1)(2n+1)}{6} \right]$$

$$= \lim_{n\to\infty} \left(4 - \frac{4}{3} \cdot \frac{n+1}{n} \cdot \frac{2n+1}{n} \right) = \lim_{n\to\infty} \left[4 - \frac{4}{3} \left(1 + \frac{1}{n} \right) \left(2 + \frac{1}{n} \right) \right]$$

$$= 4 - \tfrac{4}{3} \cdot 2 = \tfrac{4}{3}$$

17. Note that integrating from 1 to 2 makes a big change in our expression for x_i. Now
$x_i = 1 + i\Delta x = 1 + i(1/n) = 1 + i/n$.

$$\int_1^2 x^3 \, dx = \lim_{n\to\infty} \sum_{i=1}^{n} f(x_i) \Delta x = \lim_{n\to\infty} \sum_{i=1}^{n} \left(1 + \frac{i}{n} \right)^3 \left(\frac{1}{n} \right) = \lim_{n\to\infty} \frac{1}{n} \sum_{i=1}^{n} \left(\frac{n+i}{n} \right)^3$$

$$= \lim_{n\to\infty} \frac{1}{n^4} \sum_{i=1}^{n} (n^3 + 3n^2 i + 3ni^2 + i^3) = \lim_{n\to\infty} \frac{1}{n^4} \left[n \cdot n^3 + 3n^2 \sum_{i=1}^{n} i + 3n \sum_{i=1}^{n} i^2 + \sum_{i=1}^{n} i^3 \right]$$

$$= \lim_{n\to\infty} \left[1 + \frac{3}{n^2} \cdot \frac{n(n+1)}{2} + \frac{3}{n^3} \cdot \frac{n(n+1)(2n+1)}{6} + \frac{1}{n^4} \cdot \frac{n^2(n+1)^2}{4} \right]$$

$$= \lim_{n\to\infty} \left[1 + \frac{3}{2} \cdot \frac{n+1}{n} + \frac{1}{2} \cdot \frac{n+1}{n} \cdot \frac{2n+1}{n} + \frac{1}{4} \cdot \frac{(n+1)^2}{n^2} \right]$$

$$= \lim_{n\to\infty} \left[1 + \frac{3}{2} \left(1 + \frac{1}{n} \right) + \frac{1}{2} \left(1 + \frac{1}{n} \right) \left(2 + \frac{1}{n} \right) + \frac{1}{4} \left(1 + \frac{1}{n} \right)^2 \right]$$

$$= 1 + \tfrac{3}{2} + \tfrac{1}{2} \cdot 2 + \tfrac{1}{4} = 3.75$$

19. $\Delta x = (\pi - 0)/n = \pi/n$ and $x_i^* = x_i = \pi i/n$.

$$\int_0^\pi \sin 5x \, dx = \lim_{n \to \infty} \sum_{i=1}^n (\sin 5x_i) \left(\frac{\pi}{n}\right) = \pi \lim_{n \to \infty} \frac{1}{n} \sum_{i=1}^n \sin \left(\frac{5\pi i}{n}\right) = \pi \lim_{n \to \infty} \frac{1}{n} \cot \left(\frac{5\pi}{2n}\right)$$

$$= \pi \left(\frac{2}{5\pi}\right) = \frac{2}{5}$$

21. (a) Think of $\int_0^2 f(x) \, dx$ as the area of a trapezoid with bases 1 and 3 and height 2.

$\int_0^2 f(x) \, dx = \frac{1}{2}(1 + 3) \, 2 = 4$.

(b) $\int_0^5 f(x) \, dx = \underset{\text{trapezoid}}{\int_0^2 f(x) \, dx} + \underset{\text{rectangle}}{\int_2^3 f(x) \, dx} + \underset{\text{triangle}}{\int_3^5 f(x) \, dx}$

$= \frac{1}{2}(1 + 3) \, 2 + \quad 3 \cdot 1 \quad + \quad \frac{1}{2} \cdot 2 \cdot 3 \quad = 10$

(c) $\int_5^7 f(x) \, dx$ is the negative of the area of the triangle with base 2 and height 3.

$\int_5^7 f(x) \, dx = -\frac{1}{2} \cdot 2 \cdot 3 = -3$.

(d) As in parts (a) and (c), $\int_7^9 f(x) \, dx = -\frac{1}{2}(3 + 2) \, 2 = -5$. Now

$\int_0^9 f(x) \, dx = \int_0^5 f(x) \, dx + \int_5^7 f(x) \, dx + \int_7^9 f(x) \, dx = 10 - 3 - 5 = 2$.

23. $\int_1^3 (1 + 2x) \, dx$ can be interpreted as the area under the graph of
$f(x) = 1 + 2x$ between $x = 1$ and $x = 3$. This is equal to the area of
the rectangle plus the area of the triangle, so
$\int_1^3 (1 + 2x) \, dx = A = 2 \cdot 3 + \frac{1}{2} \cdot 2 \cdot 4 = 10$.
Or: Use the formula for the area of a trapezoid:
$a = \frac{1}{2}(2)(3 + 7) = 10$.

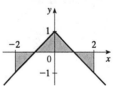

25. $\int_{-3}^0 \left(1 + \sqrt{9 - x^2}\right) dx$ can be interpreted as the area under the graph of
$f(x) = 1 + \sqrt{9 - x^2}$ between $x = -3$ and $x = 0$. This is equal to
one-quarter the area of the circle with radius 3, plus the area of the
rectangle, so $\int_{-3}^0 \left(1 + \sqrt{9 - x^2}\right) dx = \frac{1}{4}\pi \cdot 3^2 + 1 \cdot 3 = 3 + \frac{9}{4}\pi$.

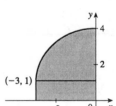

27. $\int_{-2}^2 (1 - |x|) \, dx$ can be interpreted as the area of the middle triangle
minus the areas of the outside ones, so
$\int_{-2}^2 (1 - |x|) \, dx = \frac{1}{2} \cdot 2 \cdot 1 - 2 \cdot \frac{1}{2} \cdot 1 \cdot 1 = 0$.

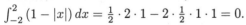

29. $\int_1^3 f(x) \, dx + \int_3^6 f(x) \, dx + \int_6^{12} f(x) \, dx = \int_1^6 f(x) \, dx + \int_6^{12} f(x) \, dx = \int_1^{12} f(x) \, dx$

31. $\int_2^5 f(x) \, dx + \int_5^8 f(x) \, dx = \int_2^8 f(x) \, dx \Longrightarrow I + 2.5 = 1.7 \Longrightarrow I = -0.8$

33. $\int_0^1 (5 - 6x^2) \, dx = \int_0^1 5 \, dx - 6 \int_0^1 x^2 \, dx = 5(1 - 0) - 6 \left(\frac{1}{3}\right) = 5 - 2 = 3$

35. $\int_1^3 e^{x+2} \, dx = e^2 \int_1^3 e^x \, dx = e^2 \left(e^3 - e\right) = e^5 - e^3$

Section 5.3 Evaluating Definite Integrals

1. If $w'(t)$ is the rate of change of weight in pounds per year, then $w(t)$ represents the weight in pounds of the child at age t. We know from the Total Change Theorem that $\int_5^{10} w'(t)\,dt = w(10) - w(5)$, so the integral represents the increase in the child's weight between the ages of 5 and 10.

3. Since $r(t)$ is the rate at which oil leaks, we can write $r(t) = -V'(t)$, where $V(t)$ is the volume of oil at time t. [Note that the minus sign is needed because V is decreasing, so $V'(t)$ is negative, but $r(t)$ is positive.] Thus, by the Total Change Theorem, $\int_0^{120} r(t)\,dt = -\int_0^{120} V'(t)\,dt = -[V(120) - V(0)] = V(0) - V(120)$, which is the number of gallons of oil that leaked from the tank in the first two hours.

5. By the Total Change Theorem, $\int_{1000}^{5000} R'(x)\,dx = R(5000) - R(1000)$, so it represents the increase in revenue when production is increased from 1000 units to 5000 units.

7. $\int_{-2}^4 (3x - 5)\,dx = \left[3 \cdot \frac{1}{2}x^2 - 5x\right]_{-2}^4 = (3 \cdot 8 - 5 \cdot 4) - [3 \cdot 2 - (-10)] = -12$

9. $\int_{-1}^0 (2x - e^x)\,dx = \left[x^2 - e^x\right]_{-1}^0 = (0 - 1) - (1 - e^{-1}) = -2 + 1/e$

11. $\int_0^4 \sqrt{x}\,dx = \int_0^4 x^{1/2}\,dx = \left[\frac{x^{3/2}}{3/2}\right]_0^4 = \left[\frac{2x^{3/2}}{3}\right]_0^4 = \frac{2(4)^{3/2}}{3} - 0 = \frac{16}{3}$

13. $\displaystyle\int_1^2 \frac{x^2 + 1}{\sqrt{x}}\,dx = \int_1^2 (x^{3/2} + x^{-1/2})\,dx = \left[\frac{x^{5/2}}{5/2} + \frac{x^{1/2}}{1/2}\right]_1^2 = \left[\frac{2}{5}x^{5/2} + 2x^{1/2}\right]_1^2$
$$= \left(\frac{2}{5}4\sqrt{2} + 2\sqrt{2}\right) - \left(\frac{2}{5} + 2\right) = \frac{18\sqrt{2} - 12}{5} = \frac{6}{5}\left(3\sqrt{2} - 2\right)$$

15. $\int_0^1 u(\sqrt{u} + \sqrt[3]{u})\,du = \int_0^1 (u^{3/2} + u^{4/3})\,du = \left[\frac{u^{5/2}}{5/2} + \frac{u^{7/3}}{7/3}\right]_0^1 = \left[\frac{2}{5}u^{5/2} + \frac{3}{7}u^{7/3}\right]_0^1 = \frac{2}{5} + \frac{3}{7} = \frac{29}{35}$

17. $\int_{\pi/4}^{\pi/3} \sin t\,dt = \left[-\cos t\right]_{\pi/4}^{\pi/3} = -\cos\frac{\pi}{3} + \cos\frac{\pi}{4} = -\frac{1}{2} + \frac{\sqrt{2}}{2} = \frac{\sqrt{2} - 1}{2}$

19. $\int_{\pi/6}^{\pi/3} \csc^2\theta\,d\theta = \left[-\cot\theta\right]_{\pi/6}^{\pi/3} = -\cot\frac{\pi}{3} + \cot\frac{\pi}{6} = -\frac{1}{3}\sqrt{3} + \sqrt{3} = \frac{2}{3}\sqrt{3}$

21. $\int_4^8 (1/x)\,dx = \left[\ln x\right]_4^8 = \ln 8 - \ln 4 = \ln\frac{8}{4} = \ln 2$

23. $\displaystyle\int_8^9 2^t\,dt = \left[\frac{1}{\ln 2}2^t\right]_8^9 = \frac{1}{\ln 2}\left(2^9 - 2^8\right) = \frac{2^8}{\ln 2}$

25. $\displaystyle\int_1^{\sqrt{3}} \frac{6}{1 + x^2}\,dx = 6\left[\tan^{-1}x\right]_1^{\sqrt{3}} = 6\left(\tan^{-1}\sqrt{3} - \tan^{-1}1\right) = 6\left(\frac{\pi}{3} - \frac{\pi}{4}\right) = \frac{\pi}{2}$

27. $\displaystyle\int_1^e \frac{x^2 + x + 1}{x}\,dx = \int_1^e \left[x + 1 + \frac{1}{x}\right]dx = \left[\frac{1}{2}x^2 + x + \ln x\right]_1^e$
$$= \left[\frac{1}{2}e^2 + e + \ln e\right] - \left[\frac{1}{2} + 1 + \ln 1\right] = \frac{1}{2}e^2 + e - \frac{1}{2}$$

29. $\int_{-2}^3 |x^2 - 1|\,dx = \int_{-2}^{-1} (x^2 - 1)\,dx + \int_{-1}^1 (1 - x^2)\,dx + \int_1^3 (x^2 - 1)\,dx$
$$= \left[\frac{x^3}{3} - x\right]_{-2}^{-1} + \left[x - \frac{x^3}{3}\right]_{-1}^1 + \left[\frac{x^3}{3} - x\right]_1^3$$
$$= \left(-\frac{1}{3} + 1\right) - \left(-\frac{8}{3} + 2\right) + \left(1 - \frac{1}{3}\right) - \left(-1 + \frac{1}{3}\right) + (9 - 3) - \left(\frac{1}{3} - 1\right) = \frac{28}{3}$$

31. The area of the viewing rectangle is $3 \cdot 27 = 81$, so it appears that the shaded area is about 60. The actual area is
$\int_0^{27} x^{1/3}\,dx = \left[\frac{3}{4}x^{4/3}\right]_0^{27} = \frac{3}{4}\cdot 81 - 0 = \frac{243}{4} = 60.75.$

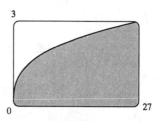

33. It appears that the area under the graph is about two-thirds of the area of the viewing rectangle, or about $\frac{2}{3}\pi \approx 2.1$. The actual area is
$\int_0^{\pi} \sin x\,dx = [-\cos x]_0^{\pi} = -\cos\pi + \cos 0 = -(-1) + 1 = 2.$

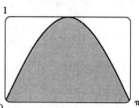

35. The graph shows that $y = x + x^2 - x^4$ has x-intercepts at $x = 0$ and at $x \approx 1.32$. So the area of the region below the curve and above the x-axis is about
$\int_0^{1.32}\left(x + x^2 - x^4\right)dx = \left[\frac{1}{2}x^2 + \frac{1}{3}x^3 - \frac{1}{5}x^5\right]_0^{1.32}$
$= \left[\frac{1}{2}(1.32)^2 + \frac{1}{3}(1.32)^3 - \frac{1}{5}(1.32)^5\right] - 0$
$\approx 0.84.$

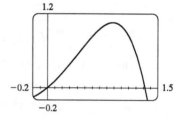

37. $\int_{-1}^{2} x^3\,dx = \left[\frac{1}{4}x^4\right]_{-1}^{2} = 4 - \frac{1}{4}$
$= \frac{15}{4} = 3.75$

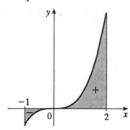

41. $\int x\sqrt{x}\,dx = \int x^{3/2}\,dx$
$= \frac{2}{5}x^{5/2} + C$

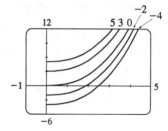

39. $\dfrac{d}{dx}\left(\dfrac{x}{2} - \dfrac{\sin 2x}{4} + C\right) = \frac{1}{2} - \frac{1}{4}(\cos 2x)(2) + 0 = \frac{1}{2} - \frac{1}{2}\cos 2x = \frac{1}{2} - \frac{1}{2}\left(1 - 2\sin^2 x\right) = \sin^2 x$

41. See above, to the right of Solution 39.

43. $\int \left(2 - \sqrt{x}\right)^2 dx = \int \left(4 - 4\sqrt{x} + x\right) dx = 4x - 4\dfrac{x^{3/2}}{3/2} + \dfrac{x^2}{2} + C = 4x - \frac{8}{3}x^{3/2} + \frac{1}{2}x^2 + C$

45. $\int \left(2x + \sec x \tan x\right) dx = x^2 + \sec x + C$

47. $A = \int_0^2 \left(2y - y^2\right) dy = \left[y^2 - \frac{1}{3}y^3\right]_0^2 = \left(4 - \frac{8}{3}\right) - 0 = \frac{4}{3}$

49. (a) displacement $= \int_0^3 \left(3t - 5\right) dt = \left[\frac{3}{2}t^2 - 5t\right]_0^3 = \frac{27}{2} - 15 = -\frac{3}{2}$ m

 (b) distance traveled $= \int_0^3 |3t - 5| dt = \int_0^{5/3} \left(5 - 3t\right) dt + \int_{5/3}^3 \left(3t - 5\right) dt$

$= \left[5t - \frac{3}{2}t^2\right]_0^{5/3} + \left[\frac{3}{2}t^2 - 5t\right]_{5/3}^3$

$= \frac{25}{3} - \frac{3}{2} \cdot \frac{25}{9} + \frac{27}{2} - 15 - \left(\frac{3}{2} \cdot \frac{25}{9} - \frac{25}{3}\right) = \frac{41}{6}$ m

51. (a) $v'(t) = a(t) = t + 4 \Longrightarrow v(t) = \frac{1}{2}t^2 + 4t + C \Longrightarrow v(0) = C = 5 \Longrightarrow v(t) = \frac{1}{2}t^2 + 4t + 5$ m/s

 (b) distance traveled $= \int_0^{10} |v(t)| dt = \int_0^{10} \left|\frac{1}{2}t^2 + 4t + 5\right| dt = \int_0^{10} \left(\frac{1}{2}t^2 + 4t + 5\right) dt$

$= \left[\frac{1}{6}t^3 + 2t^2 + 5t\right]_0^{10} = \frac{500}{3} + 200 + 50 = 416\frac{2}{3}$ m

53. Since $m'(x) = \rho(x)$,

$m = \int_0^4 \rho(x) dx = \int_0^4 \left(9 + 2\sqrt{x}\right) dx = \left[9x + \frac{4}{3}x^{3/2}\right]_0^4 = 36 + \frac{32}{3} - 0 = \frac{140}{3} = 46\frac{2}{3}$ kg.

55. Let s be the position of the car. We know from Equation 2 that $s(100) - s(0) = \int_0^{100} v(t) dt$. We use the Midpoint Rule for $0 \le t \le 100$ with $n = 5$. Note that the length of each of the five time intervals is 20 seconds $= \frac{1}{180}$ hour. So the distance traveled is

$\int\limits_0^{100} v(t) dt \approx \frac{1}{180} \left[v(10) + v(30) + v(50) + v(70) + v(90)\right] = \frac{1}{180} \left(38 + 58 + 51 + 53 + 47\right)$

$= \frac{247}{180} \approx 1.4$ miles.

57. $\int_{2000}^{4000} C'(x) dx = \int_{2000}^{4000} \left(3 - 0.01x + 0.000006x^2\right) dx = \left[3x - 0.005x^2 + 0.000002x^3\right]_{2000}^{4000}$

$= 60{,}000 - 2{,}000 = \$58{,}000$

59. (a) We can find the area between the Lorenz curve and the line $y = x$ by subtracting the area under $y = L(x)$ from the area under $y = x$. Thus, the

coefficient of inequality $= \dfrac{\text{area between Lorenz curve and straight line}}{\text{area under straight line}}$

$= \dfrac{\int_0^1 \left[x - L(x)\right] dx}{\int_0^1 x\, dx} = \dfrac{\int_0^1 \left[x - L(x)\right] dx}{\left[x^2/2\right]_0^1} = \dfrac{\int_0^1 \left[x - L(x)\right] dx}{1/2}$

$= 2 \int_0^1 \left[x - L(x)\right] dx$

 (b) $L(x) = \frac{5}{12}x^2 + \frac{7}{12}x \Longrightarrow L\left(\frac{1}{2}\right) = \frac{5}{48} + \frac{7}{24} = \frac{19}{48} = 0.3958\overline{3}$, so the bottom 50% of the households receive about 40% of the income.

coefficient of inequality $= 2 \int_0^1 \left[x - L(x)\right] dx = 2 \int_0^1 \left(x - \frac{5}{12}x^2 - \frac{7}{12}x\right) dx = 2 \int_0^1 \frac{5}{12} \left(x - x^2\right) dx$

$= \frac{5}{6} \left[\frac{1}{2}x^2 - \frac{1}{3}x^3\right]_0^1 = \frac{5}{36}$

61. By FTC2, $\int_1^2 \left(h'\right)'(u) du = h'(2) - h'(1) = 5 - 2 = 3$. The other information is unnecessary.

Section 5.4 The Fundamental Theorem of Calculus

1. The precise version of this statement is given by the Fundamental Theorem of Calculus. See the statement of this theorem and the paragraph that follows it on page 388.

3. On $[1, 3]$, $\ln 1 \leq \ln x \leq \ln 3 \Longrightarrow 0(3-1) \leq \int_1^3 \ln x \, dx \leq (\ln 3)(3-1) \Longrightarrow 0 \leq \int_1^3 \ln x \, dx \leq 2 \ln 3$.

5. $-1 \leq x \leq 1 \Longrightarrow 0 \leq x^2 \leq 1 \Longrightarrow e^0 \leq e^{x^2} \leq e^1 \Longrightarrow 1 \leq e^{x^2} \leq e$.

Thus, $1[1-(-1)] \leq \int_{-1}^{1} e^{x^2} \, dx \leq e[1-(-1)]$; that is, $2 \leq \int_{-1}^{1} e^{x^2} \, dx \leq 2e$.

7. (a) $g(0) = \int_0^0 f(t) \, dt = 0$, $g(1) = \int_0^1 f(t) \, dt = 1 \cdot 2 = 2$,

$g(2) = \int_0^2 f(t) \, dt = \int_0^1 f(t) \, dt + \int_1^2 f(t) \, dt = g(1) + \int_1^2 f(t) \, dt = 2 + 1 \cdot 2 + \frac{1}{2} \cdot 1 \cdot 2 = 5$;

$g(3) = \int_0^3 f(t) \, dt = g(2) + \int_2^3 f(t) \, dt = 5 + \frac{1}{2} \cdot 1 \cdot 4 = 7$,

$g(6) = g(3) + \int_3^6 f(t) \, dt = 7 + \left[-\left(\frac{1}{2} \cdot 2 \cdot 2 + 1 \cdot 2\right)\right] = 7 - 4 = 3$

(c) g is increasing on $[0, 3]$ because as x increases from 0 to 3, we keep adding more area.

(d) g has a maximum value when we start subtracting area, that is, at $x = 3$.

(d)

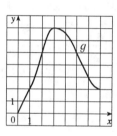

9.

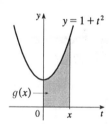

(a) By FTC1, $g(x) = \int_0^x (1+t^2) \, dt \Longrightarrow g'(x) = f(x) = 1 + x^2$.

(b) By FTC2, $g(x) = \int_0^x (1+t^2) \, dt = \left[t + \frac{1}{3}t^3\right]_0^x = x + \frac{1}{3}x^3 - 0 \Longrightarrow g'(x) = 1 + x^2$.

11. $g(x) = \int_1^x (t^2 - 1)^{20} \, dt \Longrightarrow g'(x) = f(x) = (x^2 - 1)^{20}$.

13. $g(u) = \int_\pi^u \dfrac{1}{1+t^4} \, dt \Longrightarrow g'(u) = f(u) = \dfrac{1}{1+u^4}$

15. Let $u = 1/x$. Then $du/dx = -1/x^2$, so

$$h'(x) = \frac{d}{dx} \int_2^{1/x} \sin^4 t \, dt = \frac{d}{du} \int_2^u \sin^4 t \, dt \cdot \frac{du}{dx} = \sin^4 u \, \frac{du}{dx} = \frac{-\sin^4(1/x)}{x^2}.$$

17. Let $u = \tan x$. Then $\dfrac{du}{dx} = \sec^2 x$, so

$$\frac{d}{dx} \int_{\tan x}^{17} \sin\left(t^4\right) dt = -\frac{d}{dx} \int_{17}^{\tan x} \sin\left(t^4\right) dt = -\frac{d}{du} \int_{17}^{u} \sin\left(t^4\right) dt \cdot \frac{du}{dx}$$

$$= -\sin\left(u^4\right) \frac{du}{dx} = -\sin\left(\tan^4 x\right) \sec^2 x$$

19. $g(x) = \displaystyle\int_{2x}^{3x} \frac{u-1}{u+1} du = \int_{2x}^{0} \frac{u-1}{u+1} du + \int_{0}^{3x} \frac{u-1}{u+1} du = -\int_{0}^{2x} \frac{u-1}{u+1} du + \int_{0}^{3x} \frac{u-1}{u+1} du \Longrightarrow$

$g'(x) = -\dfrac{2x-1}{2x+1} \cdot \dfrac{d}{dx}(2x) + \dfrac{3x-1}{3x+1} \cdot \dfrac{d}{dx}(3x) = -2 \cdot \dfrac{2x-1}{2x+1} + 3 \cdot \dfrac{3x-1}{3x+1}$

21. $F(x) = \int_{1}^{x} f(t)\, dt \Longrightarrow F'(x) = f(x) = \int_{1}^{x^2} \dfrac{\sqrt{1+u^4}}{u}\, du \Longrightarrow$

$F''(x) = f'(x) = \dfrac{\sqrt{1+(x^2)^4}}{x^2} \cdot \dfrac{d}{dx}(x^2) = \dfrac{2\sqrt{1+x^8}}{x}$. So $F''(2) = \sqrt{1+2^8} = \sqrt{257}$.

23. (a) By FTC1, $g'(x) = f(x)$. So $g'(x) = f(x) = 0$ at $x = 1, 3, 5, 7$, and 9. g has local maxima at $x = 1$ and 5 (since $f = g'$ changes from positive to negative there) and local minima at $x = 3$ and 7. There is no local maximum or minimum at $x = 9$, since f is not defined for $x > 9$.

(b) We can see from the graph that $\left|\int_0^1 f\, dt\right| < \left|\int_1^3 f\, dt\right| < \left|\int_3^5 f\, dt\right| < \left|\int_5^7 f\, dt\right| < \left|\int_7^9 f\, dt\right|$.

So $g(1) = \left|\int_0^1 f\, dt\right|$, $g(5) = \int_0^5 f\, dt = g(1) - \left|\int_1^3 f\, dt\right| + \left|\int_3^5 f\, dt\right|$, and

$g(9) = \int_0^9 f\, dt = g(5) - \left|\int_5^7 f\, dt\right| + \left|\int_7^9 f\, dt\right|$. Thus, $g(1) < g(5) < g(9)$, and so the absolute maximum of $g(x)$ occurs at $x = 9$.

(c) g is concave downward on those intervals where $g'' < 0$. But $g'(x) = f(x)$, so $g''(x) = f'(x)$, which is negative on (approximately) $\left(\frac{1}{2}, 2\right)$, $(4, 6)$ and $(8, 9)$. So g is concave downward on these intervals.

(d)

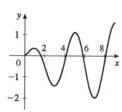

25. (a) The Fresnel Function $S(x) = \int_0^x \sin\left(\frac{\pi}{2} t^2\right) dt$ has local maximum values where $0 = S'(x) = \sin\left(\frac{\pi}{2} x^2\right)$ and S' changes from positive to negative. For $x > 0$, this happens when $\frac{\pi}{2} x^2 = (2n-1)\pi$ [odd multiples of π] $\Longleftrightarrow x = \sqrt{2(2n-1)}$, n any positive integer. For $x < 0$, S' changes from positive to negative where $\frac{\pi}{2} x^2 = 2n\pi$ [even multiples of π] $\Longleftrightarrow x = -2\sqrt{n}$, since if $x < 0$, then as x increases, x^2 decreases. S' does not change sign at $x = 0$.

(b) S is concave upward on those intervals where $S''(x) > 0$. Differentiating our expression for $S'(x)$, we get $S''(x) = \cos(\frac{\pi}{2}x^2)(2\frac{\pi}{2}x) = \pi x \cos(\frac{\pi}{2}x^2)$. For $x > 0$, $S''(x) > 0$ where $\cos(\frac{\pi}{2}x^2) > 0 \iff 0 < \frac{\pi}{2}x^2 < \frac{\pi}{2}$ or $(2n - \frac{1}{2})\pi < \frac{\pi}{2}x^2 < (2n + \frac{1}{2})\pi$, n any integer $\iff 0 < x < 1$ or $\sqrt{4n - 1} < x < \sqrt{4n + 1}$, n any positive integer. For $x < 0$, as x increases, x^2 decreases, so the intervals of upward concavity for $x < 0$ are $(-\sqrt{4n - 1}, -\sqrt{4n - 3})$, n any positive integer. To summarize: S is concave upward on the intervals $(0, 1)$, $(-\sqrt{3}, -1)$, $(\sqrt{3}, \sqrt{5})$, $(-\sqrt{7}, -\sqrt{5})$, $(\sqrt{7}, 3)$, ...

(c) In Maple, we use `plot({int(sin(Pi*t^2/2),t=0..x),0.2},x=0..2);`. Note that Maple recognizes the Fresnel function, calling it `FresnelS(x)`. In Mathematica, we use `Plot[{Integrate[Sin[Pi*t^2/2],{t,0,x}],0.2},{x,0,2}]`. In Derive, we load the utility file FRESNEL and plot FRESNEL_SIN(x). From the graphs, we see that $\int_0^x \sin(\frac{\pi}{2}t^2)dt = 0.2$ at $x \approx 0.74$.

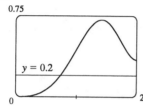

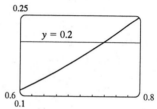

27. By FTC2, $\int_1^x f'(t)\,dt = f(x) - f(1) \implies f(x) = f(1) + \int_1^x f'(t)\,dt = f(1) + \int_1^x (2^t/t)\,dt$. This integral cannot be expressed in a simpler form. Since we want $f(1) = 0$, we have $f(x) = \int_1^x (2^t/t)\,dt$.

29. Using FTC1, we differentiate both sides of $6 + \displaystyle\int_a^x \frac{f(t)}{t^2}\,dt = 2\sqrt{x}$ to get $\dfrac{f(x)}{x^2} = 2\dfrac{1}{2\sqrt{x}} \implies f(x) = x^{3/2}$. To find a, we substitute $x = a$ in the original equation to obtain $6 + \displaystyle\int_a^a \frac{f(t)}{t^2}\,dt = 2\sqrt{a} \implies 6 + 0 = 2\sqrt{a} \implies a = 9$.

31. (a) Let $F(t) = \int_0^t f(s)\,ds$. Then, by FTC1, $F'(t) = f(t) = $ rate of depreciation, so $F(t)$ represents the loss in value over the interval $[0, t]$.

(b) $C(t) = [A + F(t)]/t$ represents the average expenditure over the interval $[0, t]$. The company wants to minimize average expenditure.

(c) $C(t) = \dfrac{1}{t}\left[A + \displaystyle\int_0^t f(s)\,ds\right]$.   Using FTC1, we have

$$C'(t) = -\frac{1}{t^2}\left[A + \int_0^t f(s)\,ds\right] + \frac{1}{t}f(t) \cdot C'(t) = 0 \implies tf(t) = A + \int_0^t f(s)\,ds \implies$$

$$f(t) = \frac{1}{t}\left[A + \int_0^t f(s)\,ds\right] = C(t).$$

Section 5.5   The Substitution Rule

1. $u = x^2 - 1 \Longrightarrow du = 2x\,dx$, so
$$\int x \left(x^2 - 1\right)^{99} dx = \int u^{99} \left(\tfrac{1}{2}\,du\right) = \tfrac{1}{2}\left(\tfrac{1}{100}u^{100}\right) + C = \tfrac{1}{200}\left(x^2 - 1\right)^{100} + C.$$

3. $u = 4x \Longrightarrow du = 4\,dx$, so $\int e^{4x}\,dx = \int e^u \left(\tfrac{1}{4}\,du\right) = \tfrac{1}{4}\left(e^u\right) + C = \tfrac{1}{4}e^{4x} + C.$

5. $u = x^2 + 6x \Longrightarrow du = 2\left(x + 3\right)dx$, so
$$\int \frac{x + 3}{\left(x^2 + 6x\right)^2}\,dx = \frac{1}{2}\int \frac{du}{u^2} = \tfrac{1}{2}\int u^{-2}\,du = -\tfrac{1}{2}u^{-1} + C = -\frac{1}{2\left(x^2 + 6x\right)} + C.$$

7. $u = \ln x \Longrightarrow du = \dfrac{dx}{x}$, so $\displaystyle\int \frac{\left(\ln x\right)^2}{x}\,dx = \int u^2\,du = \tfrac{1}{3}u^3 + C = \tfrac{1}{3}\left(\ln x\right)^3 + C.$

9. $u = x - 1 \Longrightarrow du = dx$, so $\int \sqrt{x - 1}\,dx = \int u^{1/2}\,du = \tfrac{2}{3}u^{3/2} + C = \tfrac{2}{3}\left(x - 1\right)^{3/2} + C.$

11. $u = 2 + x^4 \Longrightarrow du = 4x^3\,dx$, so
$$\int x^3 \sqrt{2 + x^4}\,dx = \int u^{1/2} \left(\tfrac{1}{4}\,du\right) = \frac{1}{4}\frac{u^{3/2}}{3/2} + C = \frac{1}{6}\left(2 + x^4\right)^{3/2} + C.$$

13. $u = t + 1 \Longrightarrow du = dt$, so $\displaystyle\int \frac{2}{\left(t + 1\right)^6}\,dt = 2\int u^{-6}\,du = -\tfrac{2}{5}u^{-5} + C = -\frac{2}{5\left(t + 1\right)^5} + C.$

15. $u = 1 + e^x \Longrightarrow du = e^x\,dx$, so $\int e^x \left(1 + e^x\right)^{10}\,dx = \int u^{10}\,du = \tfrac{1}{11}u^{11} + C = \tfrac{1}{11}\left(1 + e^x\right)^{11} + C.$

17. $u = 3\theta \Longrightarrow du = 3\,d\theta$, so $\int \sec^2 3\theta\,d\theta = \int \sec^2 u \left(\tfrac{1}{3}\,du\right) = \tfrac{1}{3}\tan u + C = \tfrac{1}{3}\tan 3\theta + C.$

19. $u = \cos x \Longrightarrow du = -\sin x\,dx$, so $\int \cos^4 x \sin x\,dx = \int u^4 \left(-du\right) = -\tfrac{1}{5}u^5 + C = -\tfrac{1}{5}\cos^5 x + C.$

21. $u = \cos x \Longrightarrow du = -\sin x\,dx$, so
$$\int \sin^5 x \cos^2 x\,dx = \int \left(\sin^2 x\right)^2 \cos^2 x\,dx = -\int \left(1 - u^2\right)^2 u^2\,du = -\int \left(u^2 - 2u^4 + u^6\right)du$$
$$= -\tfrac{1}{3}u^3 + \tfrac{2}{5}u^5 - \tfrac{1}{7}u^7 + C = -\tfrac{1}{3}\cos^3 x + \tfrac{2}{5}\cos^5 x - \tfrac{1}{7}\cos^7 x + C.$$

23. $u = 2x - 1 \Longrightarrow du = 2\,dx$, so $\displaystyle\int \frac{dx}{2x - 1} = \int \frac{\tfrac{1}{2}\,du}{u} = \tfrac{1}{2}\ln|u| + C = \tfrac{1}{2}\ln|2x - 1| + C.$

25. $u = \ln x \Longrightarrow du = \dfrac{dx}{x}$, so $\displaystyle\int \frac{dx}{x \ln x} = \int \frac{du}{u} = \ln|u| + C = \ln|\ln x| + C.$

27. $\displaystyle\int \frac{e^x + 1}{e^x}\,dx = \int \left(1 + e^{-x}\right)dx = x - e^{-x} + C$ [Substitute $u = -x$.]

29. $u = x^2 + 2x \Longrightarrow du = 2\left(x + 1\right)dx$, so
$$\int \frac{x + 1}{x^2 + 2x}\,dx = \int \frac{\tfrac{1}{2}\,du}{u} = \frac{1}{2}\ln|u| + C = \frac{1}{2}\ln\left|x^2 + 2x\right| + C.$$

31. $\displaystyle\int \frac{1 + x}{1 + x^2}\,dx = \int \frac{1}{1 + x^2}\,dx + \int \frac{x}{1 + x^2}\,dx = \tan^{-1} x + \frac{1}{2}\int \frac{2x\,dx}{1 + x^2}$
$$= \tan^{-1} x + \tfrac{1}{2}\ln\left(1 + x^2\right) + C$$

(In the last step, we evaluate $\int du/u$ where $u = 1 + x^2$.)

Note: In Exercises 33 and 35, let $f(x)$ denote the integrand and $F(x)$ its antiderivative (with $C = 0$).

33. $f(x) = \dfrac{3x - 1}{(3x^2 - 2x + 1)^4}$. $u = 3x^2 - 2x + 1 \Longrightarrow$

$du = 2(3x - 1)\, dx$, so

$$\int \frac{3x - 1}{(3x^2 - 2x + 1)^4}\, dx = \int \frac{1}{u^4}\left(\frac{1}{2}\, du\right) = \tfrac{1}{2}\int u^{-4}\, du$$

$$= -\tfrac{1}{6}u^{-3} + C = -\frac{1}{6(3x^2 - 2x + 1)^3} + C$$

Notice that at $x = \tfrac{1}{3}$, f changes from negative to positive, and F has a local minimum.

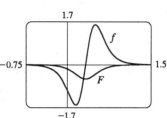

35. $f(x) = \sin^3 x \cos x$. $u = \sin x \Longrightarrow du = \cos x\, dx$, so
$\int \sin^3 x \cos x\, dx = \int u^3\, du = \tfrac{1}{4}u^4 + C$
$= \tfrac{1}{4}\sin^4 x + C.$

Note that at $x = \tfrac{\pi}{2}$, f changes from positive to negative and F has a local maximum. Also, both f and F are periodic with period π, so at $x = 0$ and at $x = \pi$, f changes from negative to positive and F has local minima.

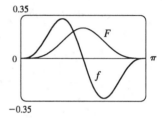

37. $u = 2x - 1 \Longrightarrow du = 2\, dx$ (when $x = 0$, $u = -1$ and when $x = 1$, $u = 1$), so
$\int_0^1 (2x - 1)^{100}\, dx = \int_{-1}^1 u^{100}\left(\tfrac{1}{2}\, du\right) = 2 \cdot \tfrac{1}{2}\int_0^1 u^{100}\, du$ [since the integrand is an even function]
$$= \left[\tfrac{1}{101}u^{101}\right]_0^1 = \tfrac{1}{101}.$$

39. $u = 2x \Longrightarrow du = 2\, dx$, so $\displaystyle\int_0^{1/2} \frac{1}{1 + 4x^2}\, dx = \frac{1}{2}\int_0^1 \frac{1}{1 + u^2}\, du = \tfrac{1}{2}\left[\tan^{-1} u\right]_0^1 = \tfrac{1}{2}\left(\tfrac{\pi}{4} - 0\right) = \tfrac{\pi}{8}$

41. $u = x - 1 \Longrightarrow du = dx$, so $\int_1^2 x\sqrt{x - 1}\, dx = \int_0^1 (u + 1)\sqrt{u}\, du = \int_0^1 \left(u^{3/2} + u^{1/2}\right) du =$
$\left[\tfrac{2}{5}u^{5/2} + \tfrac{2}{3}u^{3/2}\right]_0^1 = \tfrac{2}{5} + \tfrac{2}{3} = \tfrac{16}{15}.$

43. $u = \pi t \Longrightarrow du = \pi\, dt$, so $\int_0^1 \cos \pi t\, dt = \int_0^\pi \cos u\left(\tfrac{1}{\pi}\, du\right) = \tfrac{1}{\pi}\left[\sin u\right]_0^\pi = \tfrac{1}{\pi}(0 - 0) = 0.$

45. $u = 1 + \dfrac{1}{x} \Longrightarrow du = -\dfrac{dx}{x^2}$, so

$$\int_1^4 \frac{1}{x^2}\sqrt{1 + \frac{1}{x}}\, dx = \int_2^{5/4} u^{1/2}\,(-du) = \int_{5/4}^2 u^{1/2}\, du$$

$$= \left[\tfrac{2}{3}u^{3/2}\right]_{5/4}^2 = \tfrac{2}{3}\left(2\sqrt{2} - \tfrac{5\sqrt{5}}{8}\right) = \tfrac{4\sqrt{2}}{3} - \tfrac{5\sqrt{5}}{12}.$$

47. $\int_{-a}^a x\sqrt{x^2 + a^2}\, dx = 0$ by Theorem 6(b), since $f(x) = x\sqrt{x^2 + a^2}$ is an odd function.

49. $u = 2x + 3 \Longrightarrow du = 2\, dx$, so
$$\int_0^3 \frac{dx}{2x + 3} = \int_3^9 \frac{\tfrac{1}{2}\, du}{u} = \left[\tfrac{1}{2}\ln u\right]_3^9 = \tfrac{1}{2}(\ln 9 - \ln 3) = \tfrac{1}{2}\ln\tfrac{9}{3} = \tfrac{1}{2}\ln 3 \quad (\text{or } \ln\sqrt{3}).$$

51. $u = \ln x \Longrightarrow du = \dfrac{dx}{x}$, so $\displaystyle\int_e^{e^4} \frac{dx}{x\sqrt{\ln x}} = \int_1^4 u^{-1/2}\, du = 2\left[u^{1/2}\right]_1^4 = 2(2 - 1) = 2.$

53. From the graph, it appears that the area under the curve is about

$1 + \left(\text{a little more than } \frac{1}{2} \cdot 1 \cdot 0.7\right)$, or about 1.4. The exact area is given

by $A = \int_0^1 \sqrt{2x + 1}\, dx$. Let $u = 2x + 1$, so $du = 2\, dx$, the limits

change to $2 \cdot 0 + 1 = 1$ and $2 \cdot 1 + 1 = 3$, and

$A = \int_1^3 \sqrt{u}\left(\frac{1}{2}\, du\right) = \left[\frac{1}{3} u^{3/2}\right]_1^3 = \frac{1}{3}\left(3\sqrt{3} - 1\right) = \sqrt{3} - \frac{1}{3} \approx 1.399.$

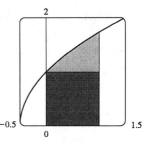

55. We split the integral: $\int_{-2}^{2} (x + 3)\sqrt{4 - x^2}\, dx = \int_{-2}^{2} x\sqrt{4 - x^2}\, dx + \int_{-2}^{2} 3\sqrt{4 - x^2}\, dx$. The first

integral is 0 by Theorem 6(b), since $f(x) = x\sqrt{4 - x^2}$ is an odd function and we are integrating from

$x = -2$ to $x = 2$. The second integral we interpret as three times the area of a semicircle with radius 2,

so the original integral is equal to $0 + 3 \cdot \frac{1}{2}\left(\pi \cdot 2^2\right) = 6\pi$.

57. $\frac{1}{2}\left(\frac{1}{x - 1} - \frac{1}{x + 1}\right) = \frac{1}{2}\left[\frac{x + 1 - (x - 1)}{(x - 1)(x + 1)}\right] = \frac{1}{2}\left(\frac{2}{x^2 - 1}\right) = \frac{1}{x^2 - 1}$. Thus,

$\int \frac{1}{x^2 - 1}\, dx = \frac{1}{2}\int\left(\frac{1}{x - 1} - \frac{1}{x + 1}\right) dx = \frac{1}{2}\left(\ln|x - 1| - \ln|x + 1|\right) + C = \frac{1}{2}\ln\left|\frac{x - 1}{x + 1}\right| + C.$

59. $\int \frac{x^2 + 2x - 1}{2x^3 + 3x^2 - 2x}\, dx = \int\left[\frac{1}{5(2x - 1)} - \frac{1}{10(x + 2)} + \frac{1}{2x}\right] dx$

$= \frac{1}{5} \cdot \frac{1}{2}\ln|2x - 1| - \frac{1}{10}\ln|x + 2| + \frac{1}{2}\ln|x| + C$

$= \frac{1}{10}\ln|2x - 1| - \frac{1}{10}\ln|x + 2| + \frac{1}{2}\ln|x| + C$

61. Solving $\frac{x^2}{a^2} + \frac{y^2}{b^2} = 1$ for y, we get $\frac{y^2}{b^2} = 1 - \frac{x^2}{a^2} = \frac{a^2 - x^2}{a^2}$ or

$y = \pm\frac{b}{a}\sqrt{a^2 - x^2}$. Because the ellipse is symmetric with respect to

both axes, the total area A is four times the area in the first quadrant.

The part of the ellipse in the first quadrant is given by the function

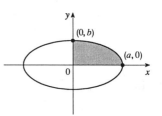

$y = \frac{b}{a}\sqrt{a^2 - x^2}$, $0 \le x \le a$, and so $\frac{1}{4}A = \int_0^a \frac{b}{a}\sqrt{a^2 - x^2}\, dx$. To evaluate this integral

we substitute $x = a\sin\theta$. Then $dx = a\cos\theta\, d\theta$. To change the limits of integration we

note that when $x = 0$, $\sin\theta = 0$, so $\theta = 0$. When $x = a$, $\sin\theta = 1$, so $\theta = \frac{\pi}{2}$. Also,

$\sqrt{a^2 - x^2} = \sqrt{a^2 - a^2\sin^2\theta} = \sqrt{a^2\cos^2\theta} = a|\cos\theta| = a\cos\theta$ since $0 \le \theta \le \frac{\pi}{2}$. Therefore,

$$A = 4\frac{b}{a}\int_0^a \sqrt{a^2 - x^2}\, dx = 4\frac{b}{a}\int_0^{\pi/2} a\cos\theta \cdot a\cos\theta\, d\theta$$

$$= 4ab\int_0^{\pi/2}\cos^2\theta\, d\theta = 4ab\int_0^{\pi/2}\frac{1}{2}(1 + \cos 2\theta)\, d\theta$$

$$= 2ab\left[\theta + \frac{1}{2}\sin 2\theta\right]_0^{\pi/2} = 2ab\left[\frac{\pi}{2} + 0 - 0\right] = \pi ab.$$

We have shown that the area of an ellipse with semiaxes a and b is πab.

63.

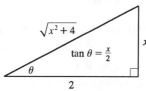

$x = 2\tan\theta \implies dx = 2\sec^2\theta\, d\theta$, so

$$\int \frac{1}{x^2\sqrt{x^2+4}}\, dx = \int \frac{2\sec^2\theta}{4\tan^2\theta\sqrt{4\tan^2\theta+4}}\, d\theta = \frac{1}{4}\int \frac{1}{\sin^2\theta\sqrt{\tan^2\theta+1}}\, d\theta$$

$$= \frac{1}{4}\int \frac{1}{\sin^2\theta\sqrt{\sec^2\theta}}\, d\theta$$

$$= \frac{1}{4}\int \frac{1}{\sin^2\theta\sec\theta}\, d\theta \quad (\text{since } -\tfrac{\pi}{2} < \theta < \tfrac{\pi}{2}, \sec\theta > 0)$$

$$= \frac{1}{4}\int \left(\frac{\cos\theta}{\sin\theta}\cdot\frac{1}{\sin\theta}\right) d\theta = \tfrac{1}{4}\int \cot\theta\csc\theta\, d\theta = -\tfrac{1}{4}\csc\theta + C$$

$$= -\frac{1}{4}\cdot\frac{\sqrt{x^2+4}}{x} + C = -\frac{\sqrt{x^2+4}}{4x} + C.$$

65. The volume of inhaled air in the lungs at time t is

$$V(t) = \int_0^t f(u)\, du = \int_0^t \tfrac{1}{2}\sin\left(\tfrac{2}{5}\pi u\right) du = \int_0^{2\pi t/5} \tfrac{1}{2}\sin v\left(\tfrac{5}{2\pi}\, dv\right) \text{ [substitute } v = \tfrac{2\pi}{5}u,\ dv = \tfrac{2\pi}{5}\, du]$$

$$= \tfrac{5}{4\pi}\left[-\cos v\right]_0^{2\pi t/5} = \tfrac{5}{4\pi}\left[-\cos\left(\tfrac{2}{5}\pi t\right) + 1\right] = \tfrac{5}{4\pi}\left[1 - \cos\left(\tfrac{2}{5}\pi t\right)\right] \text{ liters.}$$

67. $u = 2x \implies du = 2\, dx$, so $\int_0^2 f(2x)\, dx = \int_0^4 f(u)\left(\tfrac{1}{2}\, du\right) = \tfrac{1}{2}\int_0^4 f(u)\, du = \tfrac{1}{2}(10) = 5.$

69. $u = -x \implies du = -dx$, so $\int_a^b f(-x)\, dx = \int_{-a}^{-b} f(u)\,(-du) = \int_{-b}^{-a} f(u)\, du = \int_{-b}^{-a} f(x)\, dx.$ From the diagram, we see that the equality follows from the fact that we are reflecting the graph of f, and the limits of integration, about the y-axis.

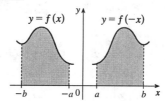

71. $u = 1 - x \implies du = -dx$, so

$$\int_0^1 x^a (1-x)^b\, dx = -\int_1^0 (1-u)^a u^b\, du = \int_0^1 u^b (1-u)^a\, du = \int_0^1 x^b (1-x)^a\, dx.$$

Section 5.6 Integration by Parts

Note: A mnemonic device which is helpful for selecting u when using integration by parts is the LIATE principle of precedence for u:

$$\underline{\text{L}}\text{ogarithmic}$$
$$\underline{\text{I}}\text{nverse trigonometric}$$
$$\underline{\text{A}}\text{lgebraic}$$
$$\underline{\text{T}}\text{rigonometric}$$
$$\underline{\text{E}}\text{xponential}$$

If the integrand has several factors, then we try to choose among them a u which appears as high as possible on the list. For example, in Exercise 1 the integrand is xe^{2x}, which is the product of an algebraic function (x) and an exponential function (e^{2x}). Since $\underline{\text{A}}$lgebraic appears before $\underline{\text{E}}$xponential, we choose $u = x$. Sometimes the integration turns out to be similar regardless of the selection of u and dv, but it is good to refer to LIATE when in doubt.

1. Let $u = x$, $dv = e^{2x}\,dx \Longrightarrow du = dx$, $v = \frac{1}{2}e^{2x}$. Then by Equation 2,
$\int xe^{2x}\,dx = \frac{1}{2}xe^{2x} - \int \frac{1}{2}e^{2x}\,dx = \frac{1}{2}xe^{2x} - \frac{1}{4}e^{2x} + C.$

3. Let $u = x$, $dv = \sin 4x\,dx \Longrightarrow du = dx$, $v = -\frac{1}{4}\cos 4x$.
Then $\int x \sin 4x\,dx = -\frac{1}{4}x \cos 4x - \int \left(-\frac{1}{4}\cos 4x\right)dx = -\frac{1}{4}x \cos 4x + \frac{1}{16}\sin 4x + C.$

5. Let $u = x^2$, $dv = \cos 3x\,dx \Longrightarrow du = 2x\,dx$, $v = \frac{1}{3}\sin 3x$.
Then $I = \int x^2 \cos 3x\,dx = \frac{1}{3}x^2 \sin 3x - \frac{2}{3}\int x \sin 3x\,dx$ by Equation 2.
Next let $U = x$, $dV = \sin 3x\,dx \Longrightarrow dU = dx$, $V = -\frac{1}{3}\cos 3x$ to get
$\int x \sin 3x\,dx = -\frac{1}{3}x \cos 3x + \frac{1}{3}\int \cos 3x\,dx = -\frac{1}{3}x \cos 3x + \frac{1}{9}\sin 3x + C_1.$
Substituting for $\int x \sin 3x\,dx$, we get $I = \frac{1}{3}x^2 \sin 3x - \frac{2}{3}\left(-\frac{1}{3}x \cos 3x + \frac{1}{9}\sin 3x + C_1\right) =$
$\frac{1}{3}x^2 \sin 3x + \frac{2}{9}x \cos 3x - \frac{2}{27}\sin 3x + C$, where $C = -\frac{2}{3}C_1$.

7. Let $u = (\ln x)^2$, $dv = dx \Longrightarrow du = 2\ln x \cdot \frac{1}{x}\,dx$, $v = x$.
Then $I = \int (\ln x)^2\,dx = x(\ln x)^2 - 2\int \ln x\,dx$. Next let $U = \ln x$, $dV = dx \Longrightarrow dU = 1/x\,dx$,
$V = x$ to get $\int \ln x\,dx = x \ln x - \int x \cdot (1/x)\,dx = x \ln x - x + C_1.$
Thus, $I = x(\ln x)^2 - 2x \ln x + 2x + C$, where $C = -2C_1$.

9. $I = \int \theta \sin\theta \cos\theta\,d\theta = \frac{1}{4}\int 2\theta \sin 2\theta\,d\theta = \frac{1}{8}\int t \sin t\,dt$ (with $t = 2\theta$ and $dt = 2\,d\theta$).
Let $u = t$, $dv = \sin t\,dt \Longrightarrow du = dt$, $v = -\cos t$.
Then $I = \frac{1}{8}\left(-t \cos t + \int \cos t\,dt\right) = \frac{1}{8}\left(-t \cos t + \sin t\right) + C = \frac{1}{8}\left(\sin 2\theta - 2\theta \cos 2\theta\right) + C.$

11. Let $u = \ln t$, $dv = t^2\,dt \Longrightarrow du = dt/t$, $v = \frac{1}{3}t^3$.
Then $\int t^2 \ln t\,dt = \frac{1}{3}t^3 \ln t - \int \frac{1}{3}t^3 (1/t)\,dt = \frac{1}{3}t^3 \ln t - \frac{1}{9}t^3 + C = \frac{1}{9}t^3 (3 \ln t - 1) + C.$

13. First let $u = \sin 3\theta$, $dv = e^{2\theta}\, d\theta \Longrightarrow du = 3\cos 3\theta\, d\theta$, $v = \frac{1}{2}e^{2\theta}$. Then
$I = \int e^{2\theta}\sin 3\theta\, d\theta = \frac{1}{2}e^{2\theta}\sin 3\theta - \frac{3}{2}\int e^{2\theta}\cos 3\theta\, d\theta$.
Next let $U = \cos 3\theta$, $dV = e^{2\theta}\, d\theta \Longrightarrow dU = -3\sin 3\theta\, d\theta$, $V = \frac{1}{2}e^{2\theta}$ to get
$\int e^{2\theta}\cos 3\theta\, d\theta = \frac{1}{2}e^{2\theta}\cos 3\theta + \frac{3}{2}\int e^{2\theta}\sin 3\theta\, d\theta$. Substituting in the previous formula gives
$I = \frac{1}{2}e^{2\theta}\sin 3\theta - \frac{3}{4}e^{2\theta}\cos 3\theta - \frac{9}{4}\int e^{2\theta}\sin 3\theta\, d\theta = \frac{1}{2}e^{2\theta}\sin 3\theta - \frac{3}{4}e^{2\theta}\cos 3\theta - \frac{9}{4}I \Longrightarrow$
$\frac{13}{4}I = \frac{1}{2}e^{2\theta}\sin 3\theta - \frac{3}{4}e^{2\theta}\cos 3\theta + C_1$. Hence, $I = \frac{1}{13}e^{2\theta}\left(2\sin 3\theta - 3\cos 3\theta\right) + C$, where $C = \frac{4}{13}C_1$.

15. Let $u = t$, $dv = e^{-t}\, dt \Longrightarrow du = dt$, $v = -e^{-t}$. By Formula 6,
$\int_0^1 te^{-t}\, dt = \left[-te^{-t}\right]_0^1 + \int_0^1 e^{-t}\, dt = -1/e + \left[-e^{-t}\right]_0^1 = -1/e - 1/e + 1 = 1 - 2/e$.

17. Let $u = x$, $dv = \cos 2x\, dx \Longrightarrow du = dx$, $v = \frac{1}{2}\sin 2x\, dx$.
Then $\int_0^{\pi/2} x\cos 2x\, dx = \left[\frac{1}{2}x\sin 2x\right]_0^{\pi/2} - \frac{1}{2}\int_0^{\pi/2}\sin 2x\, dx = 0 + \left[\frac{1}{4}\cos 2x\right]_0^{\pi/2} = \frac{1}{4}\left(-1-1\right) = -\frac{1}{2}$.

19. Let $u = \sin^{-1} x$, $dv = dx \Longrightarrow du = \dfrac{dx}{\sqrt{1-x^2}}$, $v = x$.
Then $I = \int_0^{1/2}\sin^{-1} x\, dx = \left[x\sin^{-1} x\right]_0^{1/2} - \int_0^{1/2}\dfrac{x\, dx}{\sqrt{1-x^2}} = \frac{1}{2}\cdot\frac{\pi}{6} - \int_1^{3/4}t^{-1/2}\left(-\frac{1}{2}\, dt\right)$, where
$t = 1 - x^2 \Longrightarrow dt = -2x\, dx$.
Thus, $I = \frac{\pi}{12} - \frac{1}{2}\int_{3/4}^1 t^{-1/2}\, dt = \frac{\pi}{12} - \left[\sqrt{t}\right]_{3/4}^1 = \frac{\pi}{12} - 1 + \frac{\sqrt{3}}{2} = \frac{1}{12}\left(\pi - 12 + 6\sqrt{3}\right)$.

21. $I = \int_1^4 \ln\sqrt{x}\, dx = \frac{1}{2}\int_1^4 \ln x\, dx = \frac{1}{2}\left[x\ln x - x\right]_1^4$ as in Example 2.
So $I = \frac{1}{2}\left[(4\ln 4 - 4) - (0 - 1)\right] = 2\ln 4 - \frac{3}{2}$.

23. Let $u = x^2 - 1$, $dv = e^x\, dx \Longrightarrow du = 2x\, dx$, $v = e^x$.
Then $I = \int_0^1 \left(x^2 - 1\right)e^x\, dx = \left[(x^2 - 1)e^x\right]_0^1 - 2\int_0^1 xe^x\, dx$. Now let $U = x$, $dV = e^x\, dx \Longrightarrow$
$dU = dx$, $V = e^x$, so $\int_0^1 xe^x\, dx = \left[xe^x\right]_0^1 - \int_0^1 e^x\, dx = e - \left[e^x\right]_0^1 = e - (e - 1) = 1$.
So $I = \left[0 - (-1)\right] - 2\,(1) = -1$.

25. Let $w = \sqrt{x}$, so that $x = w^2$ and $dx = 2w\, dw$. Thus, $\int \sin\sqrt{x}\, dx = \int 2w\sin w\, dw$.
Now use parts with $u = 2w$, $dv = \sin w\, dw$, $du = 2\, dw$, $v = -\cos w$ to get
$\int 2w\sin w\, dw = -2w\cos w + \int 2\cos w\, dw = -2w\cos w + 2\sin w + C$
$\qquad = -2\sqrt{x}\cos\sqrt{x} + 2\sin\sqrt{x} + C$.

27. $\int x^5 e^{x^2}\, dx = \int \left(x^2\right)^2 e^{x^2} x\, dx = \int t^2 e^t \frac{1}{2}\, dt$ (where $t = x^2 \Longrightarrow \frac{1}{2}\, dt = x\, dx$)
$\qquad = \frac{1}{2}\left(t^2 - 2t + 2\right)e^t + C$ (by Example 3) $= \frac{1}{2}\left(x^4 - 2x^2 + 2\right)e^{x^2} + C$

Note: In Exercises 29 and 31, let $f(x)$ denote the integrand and $F(x)$ its antiderivative (with $C = 0$).

29. Let $u = x$, $dv = \cos \pi x \, dx \Longrightarrow du = dx$,
$v = (\sin \pi x)/x$. Then

$$\int x \cos \pi x \, dx = x \cdot \frac{\sin \pi x}{\pi} - \int \frac{\sin \pi x}{\pi} \, dx$$

$$= \frac{x \sin \pi x}{\pi} + \frac{\cos \pi x}{\pi^2} + C.$$

We see from the graph that this is reasonable, since F has extreme values where f is 0.

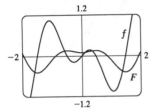

31. Let $u = 2x + 3$, $dv = e^x \, dx \Longrightarrow du = 2 \, dx$,
$v = e^x$. Then

$$\int (2x + 3) e^x \, dx = (2x + 3) e^x - 2 \int e^x \, dx$$
$$= (2x + 3) e^x - 2e^x + C$$
$$= (2x + 1) e^x + C.$$

We see from the graph that this is reasonable, since F has a minimum where f changes from negative to positive.

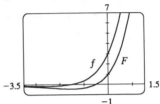

33. Let $u = \cos^{n-1} x$, $dv = \cos x \, dx \Longrightarrow du = -(n-1) \cos^{n-2} x \sin x \, dx$, $v = \sin x$. Then

$$I = \int \cos^n x \, dx = \cos^{n-1} x \sin x + (n-1) \int \cos^{n-2} x \sin^2 x \, dx$$
$$= \cos^{n-1} x \sin x + (n-1) \int \cos^{n-2} x \left(1 - \cos^2 x\right) dx$$
$$= \cos^{n-1} x \sin x + (n-1) \int \cos^{n-2} x \, dx - (n-1) \int \cos^n x \, dx$$
$$= \cos^{n-1} x \sin x + (n-1) \int \cos^{n-2} x \, dx - (n-1) I.$$

Rearranging terms gives $nI = \cos^{n-1} x \sin x + (n-1) \int \cos^{n-2} x \, dx$ or

$$I = \frac{1}{n} \cos^{n-1} x \sin x + \frac{n-1}{n} \int \cos^{n-2} x \, dx.$$

35. (a) $\displaystyle\int_0^{\pi/2} \sin^n x \, dx = \left[-\frac{\cos x \sin^{n-1} x}{n} \right]_0^{\pi/2} + \frac{n-1}{n} \int_0^{\pi/2} \sin^{n-2} x \, dx = 0$
$$+ \frac{n-1}{n} \int_0^{\pi/2} \sin^{n-2} x \, dx = \frac{n-1}{n} \int_0^{\pi/2} \sin^{n-2} x \, dx$$

(b) $\int_0^{\pi/2} \sin^3 x \, dx = \frac{2}{3} \int_0^{\pi/2} \sin x \, dx = \left[-\frac{2}{3} \cos x \right]_0^{\pi/2} = \frac{2}{3}$;
$\int_0^{\pi/2} \sin^5 x \, dx = \frac{4}{5} \int_0^{\pi/2} \sin^3 x \, dx = \frac{4}{5} \cdot \frac{2}{3} = \frac{8}{15}$

(c) The formula holds for $n = 1$ (that is, $2n + 1 = 3$) by part (b). Assume it holds for some $k \geq 1$.

Then $\displaystyle\int_0^{\pi/2} \sin^{2k+1} x \, dx = \frac{2 \cdot 4 \cdot 6 \cdots \cdots (2k)}{3 \cdot 5 \cdot 7 \cdots \cdots (2k+1)}$. By part (a),

$$\int_0^{\pi/2} \sin^{2k+3} x \, dx = \frac{2k+2}{2k+3} \int_0^{\pi/2} \sin^{2k+1} x \, dx = \frac{2 \cdot 4 \cdot 6 \cdots \cdots (2k)(2k+2)}{3 \cdot 5 \cdot 7 \cdots \cdots (2k+1)(2k+3)}$$

$$= \frac{2 \cdot 4 \cdot 6 \cdots \cdots [2(k+1)]}{3 \cdot 5 \cdot 7 \cdots \cdots [2(k+1)+1]}, \text{ as desired.}$$

By induction, the formula holds for all $n \geq 1$.

37. Let $u = (\ln x)^n$, $dv = dx \Longrightarrow du = n (\ln x)^{n-1} (dx/x)$, $v = x$.
By Equation 2, $\int (\ln x)^n \ dx = x (\ln x)^n - n \int (\ln x)^{n-1} \ dx$.

39. Take $n = 3$ in Exercise 37 to get
$\int (\ln x)^3 \ dx = x (\ln x)^3 - 3 \int (\ln x)^2 \ dx$
$\qquad = x (\ln x)^3 - 3x (\ln x)^2 + 6x \ln x - 6x + C$ (by Exercise 7).
Or: Instead of using Exercise 7, apply Exercise 37 again with $n = 2$.

41. Since $v (t) > 0$ for all t, the desired distance $s (t) = \int_0^t v (w) \, dw = \int_0^t w^2 e^{-w} \, dw$. Let $u = w^2$,
$dv = e^{-w} \, dw \Longrightarrow du = 2w \, dw$, $v = -e^{-w}$. Then $s (t) = \left[-w^2 e^{-w} \right]_0^t + 2 \int_0^t w e^{-w} \, dw$.
Now let $U = w$, $dV = e^{-w} \, dw \Longrightarrow dU = dw$, $V = -e^{-w}$. Then
$s (t) = -t^2 e^{-t} + 2 \left(\left[-w e^{-w} \right]_0^t + \int_0^t e^{-w} \, dw \right) = -t^2 e^{-t} - 2t e^{-t} - 2 e^{-t} + 2$
$\qquad = 2 - e^{-t} \left(t^2 + 2t + 2 \right)$ meters.

43. Take $g (x) = x$ and $g' (x) = 1$ in Equation 1.

45. Suppose $f (0) = g (0) = 0$ and let $u = f (x)$, $dv = g'' (x) \, dx \Longrightarrow du = f' (x) \, dx$, $v = g' (x)$.
Then $\int_0^a f (x) g'' (x) \, dx = [f (x) g' (x)]_0^a - \int_0^a f' (x) g' (x) \, dx = f (a) g' (a) - \int_0^a f' (x) g' (x) \, dx$.
Now let $U = f' (x)$, $dV = g' (x) \, dx \Longrightarrow dU = f'' (x) \, dx$ and $V = g (x)$, so
$\int_0^a f' (x) g' (x) \, dx = [f' (x) g (x)]_0^a - \int_0^a f'' (x) g (x) \, dx = f' (a) g (a) - \int_0^a f'' (x) g (x) \, dx$.
Combining the two results, we get $\int_0^a f (x) g'' (x) \, dx = f (a) g' (a) - f' (a) g (a) + \int_0^a f'' (x) g (x) \, dx$.

Section 5.7 Integration Using Tables and Computer Algebra Systems

1. Using long division, $\dfrac{x^3 - x^2 + x - 1}{x^2 + 9} = x - 1 - 8\dfrac{x - 1}{x^2 + 9}$.

$I = \displaystyle\int \left(x - 1 - 8\frac{x-1}{x^2+9}\right) dx = \int (x-1)\, dx - 8\left[\int \frac{x}{x^2+9}\, dx - \int \frac{1}{x^2+9}\, dx\right].$

Using Formula 17 with $a = 3$, we have

$I = \frac{1}{2}x^2 - x - 8 \cdot \frac{1}{2}\ln\left(x^2+9\right) + 8 \cdot \frac{1}{3}\tan^{-1}(x/3) + C = \frac{1}{2}x^2 - x - 4\ln\left(x^2+9\right) + \frac{8}{3}\tan^{-1}(x/3) + C.$

3. By Formula 99 with $a = -3$ and $b = 4$,

$\displaystyle\int e^{-3x}\cos 4x\, dx = \frac{e^{-3x}}{(-3)^2 + 4^2}(-3\cos 4x + 4\sin 4x) + C = \frac{e^{-3x}}{25}(-3\cos 4x + 4\sin 4x) + C.$

5. Let $u = 3x$. Then $du = 3\, dx$, so

$$\int \frac{\sqrt{9x^2 - 1}}{x^2}\, dx = \int \frac{\sqrt{u^2 - 1}}{u^2/9}\frac{du}{3} = 3\int \frac{\sqrt{u^2 - 1}}{u^2}\, du$$

$$= -\frac{3\sqrt{u^2 - 1}}{u} + 3\ln\left|u + \sqrt{u^2 - 1}\right| + C \quad \text{(by Formula 42)}$$

$$= -\frac{\sqrt{9x^2 - 1}}{x} + 3\ln\left|3x + \sqrt{9x^2 - 1}\right| + C.$$

7. Let $u = x^2$. Then $du = 2x\, dx$, so

$\displaystyle\int x\sin^{-1}\left(x^2\right) dx = \frac{1}{2}\int \sin^{-1} u\, du = \frac{1}{2}\left(u\sin^{-1} u + \sqrt{1 - u^2}\right) + C \quad \text{(Formula 87)}$

$= \frac{1}{2}\left(x^2\sin^{-1}\left(x^2\right) + \sqrt{1 - x^4}\right) + C.$

9. $5 - 4x - x^2 = 5 - \left(x^2 + 4x + 4\right) + 4 = 9 - (x + 2)^2$, so

$\displaystyle\int \sqrt{5 - 4x - x^2}\, dx = \int \sqrt{9 - (x + 2)^2}\, dx$

$= \dfrac{x + 2}{2}\sqrt{9 - (x + 2)^2} + \frac{9}{2}\sin^{-1}\dfrac{x + 2}{3} + C \quad \text{(Formula 30)}$

$= \dfrac{x + 2}{2}\sqrt{5 - 4x - x^2} + \frac{9}{2}\sin^{-1}\dfrac{x + 2}{3} + C$

11. $\displaystyle\int \sec^5 x\, dx = \frac{1}{4}\tan x\sec^3 x + \frac{3}{4}\int \sec^3 x\, dx \quad \text{(Formula 77)}$

$= \frac{1}{4}\tan x\sec^3 x + \frac{3}{4}\left(\frac{1}{2}\tan x\sec x + \frac{1}{2}\int \sec x\, dx\right) \quad \text{(Formula 77 again)}$

$= \frac{1}{4}\tan x\sec^3 x + \frac{3}{8}\tan x\sec x + \frac{3}{8}\ln|\sec x + \tan x| + C \quad \text{(Formula 14)}$

13. Let $u = \sin x$. Then $du = \cos x\, dx$, so

$\displaystyle\int \sin^2 x\cos x\ln(\sin x)\, dx = \int u^2\ln u\, du = \frac{1}{9}u^3(3\ln u - 1) + C \quad \text{(Formula 101)}$

$= \frac{1}{9}\sin^3 x\left[3\ln(\sin x) - 1\right] + C.$

15. $\int_0^{\pi/2}\cos^5 x\, dx = \frac{1}{5}\left[\cos^4 x\sin x\right]_0^{\pi/2} + \frac{4}{5}\int_0^{\pi/2}\cos^3 x\, dx \quad \text{(Formula 74)}$

$= 0 + \frac{4}{5}\left[\frac{1}{3}\left(2 + \cos^2 x\right)\sin x\right]_0^{\pi/2} \quad \text{(Formula 68)} \quad = \frac{4}{15}(2 - 0) = \frac{8}{15}$

17. Let $u = x^5$, $du = 5x^4 \, dx$.

$$\int \frac{x^4 \, dx}{\sqrt{x^{10} - 2}} = \frac{1}{5} \int \frac{du}{\sqrt{u^2 - 2}} = \frac{1}{5} \ln \left| u + \sqrt{u^2 - 2} \right| + C \quad \text{(Formula 43)}$$
$$= \frac{1}{5} \ln \left| x^5 + \sqrt{x^{10} - 2} \right| + C.$$

19. Let $u = 1 + e^x$, so $du = e^x \, dx$. Then

$$\int e^x \ln(1 + e^x) \, dx = \int \ln u \, du = u \ln u - u + C \quad \text{(Formula 100)}$$
$$= (1 + e^x) \ln(1 + e^x) - e^x - 1 + C = (1 + e^x) \ln(1 + e^x) - e^x + C_1.$$

21. Let $u = e^x \Longrightarrow \ln u = x \Longrightarrow dx = \dfrac{du}{u}$. Then

$$\int \sqrt{e^{2x} - 1} \, dx = \int \frac{\sqrt{u^2 - 1}}{u} \, du = \sqrt{u^2 - 1} - \cos^{-1}(1/u) + C \quad \text{(Formula 41)}$$
$$= \sqrt{e^{2x} - 1} - \cos^{-1}(e^{-x}) + C$$

23. (a) $\dfrac{d}{du} \left[\dfrac{1}{b^3} \left(a + bu - \dfrac{a^2}{a + bu} - 2a \ln |a + bu| \right) + C \right] = \dfrac{1}{b^3} \left[b + \dfrac{ba^2}{(a + bu)^2} - \dfrac{2ab}{(a + bu)} \right]$

$$= \frac{1}{b^3} \left[\frac{b(a + bu)^2 + ba^2 - (a + bu) 2ab}{(a + bu)^2} \right] = \frac{1}{b^3} \left[\frac{b^3 u^2}{(a + bu)^2} \right] = \frac{u^2}{(a + bu)^2}$$

(b) Let $t = a + bu \Longrightarrow dt = b \, du$.

$$\int \frac{u^2 \, du}{(a + bu)^2} = \frac{1}{b^3} \int \frac{(t - a)^2}{t^2} \, dt = \frac{1}{b^3} \int \left(1 - \frac{2a}{t} + \frac{a^2}{t^2} \right) dt = \frac{1}{b^3} \left(t - 2a \ln |t| - \frac{a^2}{t} \right) + C$$
$$= \frac{1}{b^3} \left(a + bu - \frac{a^2}{a + bu} - 2a \ln |a + bu| \right) + C$$

25. Maple, Mathematica and Derive all give

$\int x^2 \sqrt{5 - x^2} \, dx = -\frac{1}{4} x \left(5 - x^2 \right)^{3/2} + \frac{5}{8} x \sqrt{5 - x^2} + \frac{25}{8} \sin^{-1} \left(\frac{1}{\sqrt{5}} x \right)$. Using Formula 31,

we get $\int x^2 \sqrt{5 - x^2} \, dx = \frac{1}{8} x \left(2x^2 - 5 \right) \sqrt{5 - x^2} + \frac{1}{8} \left(5^2 \right) \sin^{-1} \left(\frac{1}{\sqrt{5}} x \right) + C$. But

$-\frac{1}{4} x \left(5 - x^2 \right)^{3/2} + \frac{5}{8} x \sqrt{5 - x^2} = \frac{1}{8} x \sqrt{5 - x^2} \left[5 - 2 \left(5 - x^2 \right) \right] = \frac{1}{8} x \left(2x^2 - 5 \right) \sqrt{5 - x^2}$, and the

$\sin^{-1}$ terms are the same in each expression, so the answers are equivalent.

27. Maple and Derive both give $\int \sin^3 x \cos^2 x \, dx = -\frac{1}{5} \sin^2 x \cos^3 x - \frac{2}{15} \cos^3 x$ (although Derive factors the expression), and Mathematica gives $\int \sin^3 x \cos^2 x \, dx = -\frac{1}{8} \cos x - \frac{1}{48} \cos 3x + \frac{1}{80} \cos 5x$. We can use a CAS to show that both of these expressions are equal to $-\frac{1}{3} \cos^3 x + \frac{1}{5} \cos^5 x$. Using Formula 86, we write

$$\int \sin^3 x \cos^2 x \, dx = -\frac{1}{5} \sin^2 x \cos^3 x + \frac{2}{5} \int \sin x \cos^2 x \, dx = -\frac{1}{5} \sin^2 x \cos^3 x + \frac{2}{5} \left(-\frac{1}{3} \cos^3 x \right) + C$$
$$= -\frac{1}{5} \sin^2 x \cos^3 x - \frac{2}{15} \cos^3 x + C.$$

29. Maple gives $\int x\sqrt{1+2x}\,dx = \frac{1}{10}(1+2x)^{5/2} - \frac{1}{6}(1+2x)^{3/2}$, Mathematica gives
$\sqrt{1+2x}\left(\frac{2}{5}x^2 + \frac{1}{15}x - \frac{1}{15}\right)$, and Derive gives $\frac{1}{15}(1+2x)^{3/2}(3x-1)$.   The first two expressions can
be simplified to Derive's result.   If we use Formula 54, we get

$$\int x\sqrt{1+2x}\,dx = \frac{2}{15(2)^2}(3\cdot 2x - 2\cdot 1)(1+2x)^{3/2} + C = \frac{1}{30}(6x-2)(1+2x)^{3/2} + C$$
$$= \frac{1}{15}(3x-1)(1+2x)^{3/2}.$$

31. Maple gives $\int \tan^3 x\,dx = \frac{1}{2}\tan^2 x - \frac{1}{2}\ln\left(1+\tan^2 x\right)$, while Mathematica and
Derive both give $\ln\cos x + \frac{1}{2}\tan^2 x$.   These expressions are equivalent, since

$$-\frac{1}{2}\ln\left(1+\tan^2 x\right) = \ln\left[\left(\sec^2 x\right)^{-1/2}\right] = \ln\cos x.$$ Using Formula 69, we get
$\int \tan^3 x\,dx = \frac{1}{2}\tan^2 x + \ln|\cos x| + C.$

33. Derive gives $I = \int 2^x\sqrt{4^x - 1}\,dx = \dfrac{2^{x-1}\sqrt{2^{2x}-1}}{\ln 2} - \dfrac{\ln\left(\sqrt{2^{2x}-1}+2^x\right)}{2\ln 2}$ immediately.   Neither
Maple nor Mathematica is able to evaluate I in its given form.   However, if we instead write I as
$\int 2^x\sqrt{\left(2^x\right)^2 - 1}\,dx$, both systems give the same answer as Derive (after minor simplification).   Our
trick works because the CAS now recognizes 2^x as a promising substitution.

35. Maple gives the antiderivative $F(x) = \displaystyle\int \frac{x^2 - 1}{x^4 + x^2 + 1}\,dx = -\frac{1}{2}\ln\left(x^2 + x + 1\right) + \frac{1}{2}\ln\left(x^2 - x + 1\right).$
We can see that at 0, this antiderivative is 0.   From the graphs, it appears that F has a maximum at
$x = -1$ and a minimum at $x = 1$ [since $F'(x) = f(x)$ changes sign at these x-values], and that F has
inflection points at $x \approx -1.7$, $x = 0$ and $x \approx 1.7$ [since $f(x)$ has extrema at these x-values].

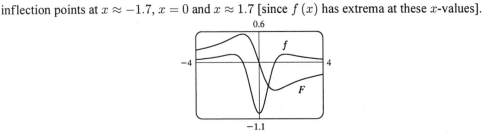

37. Since f is everywhere positive, we know that its antiderivative F is increasing.   The antiderivative
given by Maple is

$\int \sin^4 x\cos^6 x\,dx = -\frac{1}{10}\sin^3 x\cos^7 x - \frac{3}{80}\sin x\cos^7 x + \frac{1}{160}\cos^5 x\sin x$
$$+\frac{1}{128}\cos^3 x\sin x + \frac{3}{256}\cos x\sin x + \frac{3}{256}x,$$

and this is 0 at $x = 0$.

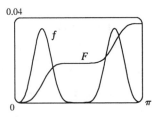

Section 5.8 Approximate Integration

1. (a) $L_2 = \sum\limits_{i=1}^{2} f\left(x_{i-1}\right) \Delta x = f\left(x_0\right) \cdot 2 + f\left(x_1\right) \cdot 2 = 2\left[f\left(0\right) + f\left(2\right)\right] = 2\left(0.5 + 2.5\right) = 6$

$R_2 = \sum\limits_{i=1}^{2} f\left(x_i\right) \Delta x = f\left(x_1\right) \cdot 2 + f\left(x_2\right) \cdot 2 = 2\left[f\left(2\right) + f\left(4\right)\right] = 2\left(2.5 + 3.5\right) = 12$

$M_2 = \sum\limits_{i=1}^{2} f\left(\overline{x}_i\right) \Delta x = f\left(\overline{x}_1\right) \cdot 2 + f\left(\overline{x}_2\right) \cdot 2 = 2\left[f\left(1\right) + f\left(3\right)\right] \approx 2\left(1.7 + 3.2\right) = 9.8$

(b)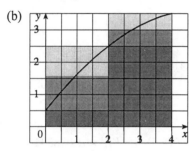

L_2 is an underestimate, since the area under the small rectangles is less than the area under the curve, and R_2 is an overestimate, since the area under the large rectangles is greater than the area under the curve. It appears that M_2 is an overestimate, though it is fairly close to I. See the solution to Exercise 29 for a proof of the fact that if f is concave down on (a, b), then the Midpoint Rule is an overestimate of $\int_a^b f\left(x\right) dx$.

(c) $T_2 = \left(\frac{1}{2}\Delta x\right)\left[f\left(x_0\right) + 2f\left(x_1\right) + f\left(x_2\right)\right] = \frac{2}{2}\left[f\left(0\right) + 2f\left(2\right) + f\left(4\right)\right] = 0.5 + 2\left(2.5\right) + 3.5 = 9.$
This approximation is an underestimate, since the graph is concave down. See the solution to Exercise 29 for a general proof of this conclusion.

(d) For any n, we will have $L_n < T_n < I < M_n < R_n$.

3. $f\left(x\right) = \cos\left(x^2\right)$, $\Delta x = \frac{1-0}{4} = \frac{1}{4}$

(a) $T_4 = \frac{1}{4 \cdot 2}\left[f\left(0\right) + 2f\left(\frac{1}{4}\right) + 2f\left(\frac{2}{4}\right) + 2f\left(\frac{3}{4}\right) + f\left(1\right)\right] \approx 0.895759$

(b) $M_4 = \frac{1}{4}\left[f\left(\frac{1}{8}\right) + f\left(\frac{3}{8}\right) + f\left(\frac{5}{8}\right) + f\left(\frac{7}{8}\right)\right] \approx 0.908907$

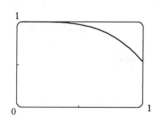

The graph shows that f is concave down on $[0, 1]$. So T_4 is an underestimate and M_4 is an overestimate. We can conclude that $0.895759 < \int_0^1 \cos\left(x^2\right) dx < 0.908907$.

5. $f\left(x\right) = e^{-x^2}$, $\Delta x = \frac{1-0}{10} = \frac{1}{10}$

(a) $T_{10} = \frac{1}{10 \cdot 2}\left[f\left(0\right) + 2f\left(0.1\right) + 2f\left(0.2\right) + \cdots + 2f\left(0.8\right) + 2f\left(0.9\right) + f\left(1\right)\right] \approx 0.746211$

(b) $M_{10} = \frac{1}{10}\left[f\left(0.05\right) + f\left(0.15\right) + f\left(0.25\right) + \cdots + f\left(0.75\right) + f\left(0.85\right) + f\left(0.95\right)\right] \approx 0.747131$

(c) $S_{10} = \frac{1}{10 \cdot 3}\left[f\left(0\right) + 4f\left(0.1\right) + 2f\left(0.2\right) + 4f\left(0.3\right) + 2f\left(0.4\right) + 4f\left(0.5\right)\right.$
$\left. + 2f\left(0.6\right) + 4f\left(0.7\right) + 2f\left(0.8\right) + 4f\left(0.9\right) + f\left(1\right)\right] \approx 0.746825$

7. $f(x) = \cos(e^x)$, $\Delta x = \frac{1/2 - 0}{8} = \frac{1}{16}$

(a) $T_8 = \frac{1}{16 \cdot 2} \left[f(0) + 2f\left(\frac{1}{16}\right) + 2f\left(\frac{1}{8}\right) + \cdots + 2f\left(\frac{7}{16}\right) + f\left(\frac{1}{2}\right) \right] \approx 0.132465$

(b) $M_8 = \frac{1}{16} \left[f\left(\frac{1}{32}\right) + f\left(\frac{3}{32}\right) + f\left(\frac{5}{32}\right) + \cdots + f\left(\frac{13}{32}\right) + f\left(\frac{15}{32}\right) \right] \approx 0.132857$

(c) $S_8 = \frac{1}{16 \cdot 3} \left[f(0) + 4f\left(\frac{1}{16}\right) + 2f\left(\frac{1}{8}\right) + 4f\left(\frac{3}{16}\right) + 2f\left(\frac{1}{4}\right) \right.$
$$\left. + 4f\left(\frac{5}{16}\right) + 2f\left(\frac{3}{8}\right) + 4f\left(\frac{7}{16}\right) + f\left(\frac{1}{2}\right) \right] \approx 0.132727$$

9. $f(x) = x^5 e^x$, $\Delta x = \frac{1-0}{10} = \frac{1}{10}$

(a) $T_{10} = \frac{1}{10 \cdot 2} \left[f(0) + 2f(0.1) + 2f(0.2) + \cdots + 2f(0.9) + f(1) \right] \approx 0.409140$

(b) $M_{10} = \frac{1}{10} \left[f(0.05) + f(0.15) + f(0.25) + \cdots + f(0.95) \right] \approx 0.388849$

(c) $S_{10} = \frac{1}{10 \cdot 3} \left[f(0) + 4f(0.1) + 2f(0.2) + 4f(0.3) + 2f(0.4) + 4f(0.5) \right.$
$$\left. + 2f(0.6) + 4f(0.7) + 2f(0.8) + 4f(0.9) + f(1) \right] \approx 0.395802$$

11. $f(x) = \dfrac{1}{1+x^4}$, $\Delta x = \dfrac{3-0}{6} = \dfrac{1}{2}$

(a) $T_6 = \frac{1}{2 \cdot 2} \left[f(0) + 2f(0.5) + 2f(1) + 2f(1.5) + 2f(2) + 2f(2.5) + f(3) \right] \approx 1.098004$

(b) $M_6 = \frac{1}{2} \left[f(0.25) + f(0.75) + f(1.25) + f(1.75) + f(2.25) + f(2.75) \right] \approx 1.098709$

(c) $S_6 = \frac{1}{2 \cdot 3} \left[f(0) + 4f(0.5) + 2f(1) + 4f(1.5) + 2f(2) + 4f(2.5) + f(3) \right] = 1.109031$

13. $f(x) = e^{-x^2}$, $\Delta x = \frac{2-0}{10} = \frac{1}{5}$

(a) $T_{10} = \frac{1}{5 \cdot 2} \left\{ f(0) + 2\left[f(0.2) + f(0.4) + \cdots + f(1.8) \right] + f(2) \right\} \approx 0.881839$
$M_{10} = \frac{1}{5} \left[f(0.1) + f(0.3) + f(0.5) + \cdots + f(1.7) + f(1.9) \right] \approx 0.882202$

(b) $f(x) = e^{-x^2}$, $f'(x) = -2xe^{-x^2}$, $f''(x) = \left(4x^2 - 2\right)e^{-x^2}$, $f'''(x) = 4x\left(3 - 2x^2\right)e^{-x^2}$.
$f'''(x) = 0 \Longleftrightarrow x = 0$ or $x = \pm\sqrt{\frac{3}{2}}$. So to find the maximum value of $|f''(x)|$ on $[0, 2]$, we need only consider its values at $x = 0$, $x = 2$, and $x = \sqrt{\frac{3}{2}}$. $|f''(0)| = 2$, $|f''(2)| \approx 0.2564$ and $\left| f''\left(\sqrt{\frac{3}{2}}\right) \right| \approx 0.8925$. Thus, taking $K = 2$, $a = 0$, $b = 2$, and $n = 10$ in Theorem 3, we get
$|E_T| \leq 2 \cdot 2^3 \Big/ \left[12 \, (10)^2 \right] = \frac{1}{75} = 0.01\overline{3}$, and $|E_M| \leq |E_T|/2 \leq 0.00\overline{6}$.

(c) Take $K = 2$ [as in part (b)] in Theorem 3. $|E_T| \leq \dfrac{K(b-a)^3}{12n^2} \leq 10^{-5} \Longleftrightarrow$
$\dfrac{2(2-0)^3}{12n^2} \leq 10^{-5} \Longleftrightarrow \frac{3}{4}n^2 \geq 10^5 \Longleftrightarrow n \geq 365.1\ldots \Longleftrightarrow n \geq 366$. Take $n = 366$ for T_n. For E_M, again take $K = 2$ in Theorem 3 to get $|E_M| \leq 10^{-5} \Longleftrightarrow \frac{3}{2}n^2 \geq 10^5 \Longleftrightarrow n \geq 258.2 \Longrightarrow n \geq 259$. Take $n = 259$ for M_n.

15. (a) $T_{10} = \frac{1}{10 \cdot 2} \{f(0) + 2[f(0.1) + f(0.2) + \cdots + f(0.9)] + f(1)\} \approx 1.71971349$

$S_{10} = \frac{1}{10 \cdot 3}[f(0) + 4f(0.1) + 2f(0.2) + 4f(0.3) + \cdots + 4f(0.9) + f(1)] \approx 1.71828278$

Since $I = \int_0^1 e^x dx = [e^x]_0^1 = e - 1 \approx 1.71828183$, $E_T = I - T_{10} \approx -0.00143166$ and

$E_S = I - S_{10} \approx -0.00000095$.

(b) $f(x) = e^x \implies f''(x) = e^x \le e$ for $0 \le x \le 1$. Taking $K = e$, $a = 0$, $b = 1$, and

$n = 10$ in Theorem 3, we get $|E_T| \le \dfrac{e(1)^3}{12(10)^2} \approx 0.002265 > 0.00143166$ [actual

$|E_T|$ from (a)]. $f^{(4)}(x) = e^x < e$ for $0 \le x \le 1$. Using Theorem 4, we have

$|E_S| \le e(1)^5 \big/ [180(10)^4] \approx 0.0000015 > 0.00000095$ [actual $|E_S|$ from (a)]. We see that the

actual errors are about two-thirds the size of the error estimates.

(c) From part (b), we take $K = e$ to get

$$|E_T| \le \frac{K(b-a)^3}{12n^2} \le 0.00001 \implies$$

$$n^2 \ge \frac{e(1^3)}{12(0.00001)} \implies$$

$$n \ge 150.5$$

Take $n = 151$ for T_n. Now

$$|E_M| \le \frac{K(b-a)^3}{24n^2} \le 0.00001 \implies$$

$$n \ge 106.4$$

Take $n = 107$ for M_n. Finally,

$$|E_S| \le \frac{K(b-a)^5}{180n^4} \le 0.00001 \implies$$

$$n^4 \ge \frac{e(1^5)}{180(0.00001)} \implies$$

$$n \ge 6.23$$

Take $n = 8$ for S_n (since n has to be even for Simpson's Rule).

17. (a) Using the CAS, we differentiate $f(x) = e^{\cos x}$ twice, and find that $f''(x) = e^{\cos x}\left(\sin^2 x - \cos x\right)$. From the graph, we see that the maximum value of $|f''(x)|$ occurs at the endpoints of the interval $[0, 2\pi]$. Since $f''(0) = -e$, we can use $K = e$ or $K = 2.8$.

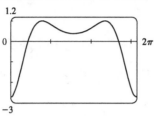

(b) A CAS gives $M_{10} \approx 7.954926518$. (In Maple, use student[middlesum].)

(c) Using Theorem 3 for the Midpoint Rule, with $K = e$, we get $|E_M| \le \dfrac{e(2\pi - 0)^3}{24 \cdot 10^2} \approx 0.280945995$.

With $K = 2.8$, we get $|E_M| \le \dfrac{2.8(2\pi - 0)^3}{24 \cdot 10^2} = 0.289391916$.

(d) A CAS gives $I \approx 7.954926521$.

(e) The actual error is only about 3×10^{-9}, much less than the estimate in part (c).

(f) We use the CAS to differentiate twice more, and then graph
$f^{(4)}(x) = e^{\cos x}\left(\sin^4 x - 6\sin^2 x \cos x + 3 - 7\sin^2 x + \cos x\right)$. From the graph, we see that the maximum value of $\left|f^{(4)}(x)\right|$ occurs at the endpoints of the interval $[0, 2\pi]$. Since $f^{(4)}(0) = 4e$, we can use $K = 4e$ or $K = 10.9$.

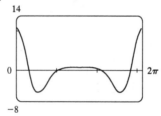

(g) A CAS gives $S_{10} \approx 7.953789422$. (In Maple, use student[simpson].)

(h) Using Theorem 4 with $K = 4e$, we get $|E_S| \le \dfrac{4e(2\pi - 0)^5}{180 \cdot 10^4} \approx 0.059153618$. With $K = 10.9$, we get $|E_S| \le \dfrac{10.9(2\pi - 0)^5}{180 \cdot 10^4} = 0.059299814$.

(i) The actual error is about $7.954926521 - 7.953789422 \approx 0.00114$. This is quite a bit smaller than the estimate in part (h), though the difference is not nearly as great as it was in the case of the Midpoint Rule.

(j) To ensure that $|E_S| \le 0.0001$, we use Theorem 4: $|E_S| \le \dfrac{4e(2\pi)^5}{180 \cdot n^4} \le 0.0001 \implies$

$\dfrac{4e(2\pi)^5}{180 \cdot 0.0001} \le n^4 \implies n^4 \ge 5{,}915{,}362 \iff n \ge 49.3$. So we must take $n \ge 50$ to ensure that $|I - S_n| \le 0.0001$. ($K = 10.9$ leads to the same value of n.)

19. $I = \int_0^1 x^3 dx = \left[\frac{1}{4}x^4\right]_0^1 = 0.25.$ $f(x) = x^3.$

$n = 4$: $\quad L_4 = \frac{1}{4}\left[0^3 + \left(\frac{1}{4}\right)^3 + \left(\frac{1}{2}\right)^3 + \left(\frac{3}{4}\right)^3\right] = 0.140625$

$\qquad R_4 = \frac{1}{4}\left[\left(\frac{1}{4}\right)^3 + \left(\frac{1}{2}\right)^3 + \left(\frac{3}{4}\right)^3 + 1^3\right] = 0.390625$

$\qquad T_4 = \frac{1}{4 \cdot 2}\left[0^3 + 2\left(\frac{1}{4}\right)^3 + 2\left(\frac{1}{2}\right)^3 + 2\left(\frac{3}{4}\right)^3 + 1^3\right] = 0.265625,$

$\qquad M_4 = \frac{1}{4}\left[\left(\frac{1}{8}\right)^3 + \left(\frac{3}{8}\right)^3 + \left(\frac{5}{8}\right)^3 + \left(\frac{7}{8}\right)^3\right] = 0.2421875,$

$\qquad E_L = I - L_4 = \frac{1}{4} - 0.140625 = 0.109375,$ $E_R = \frac{1}{4} - 0.390625 = -0.140625,$

$\qquad E_T = \frac{1}{4} - 0.265625 = -0.015625,$ $E_M = \frac{1}{4} - 0.2421875 = 0.0078125$

$n = 8$: $\quad L_8 = \frac{1}{8}\left[f(0) + f\left(\frac{1}{8}\right) + f\left(\frac{2}{8}\right) + \cdots + f\left(\frac{7}{8}\right)\right] \approx 0.191406$

$\qquad R_8 = \frac{1}{8}\left[f\left(\frac{1}{8}\right) + f\left(\frac{2}{8}\right) + \cdots + f\left(\frac{7}{8}\right) + f(1)\right] \approx 0.316406$

$\qquad T_8 = \frac{1}{8 \cdot 2}\left\{\left[f(0) + 2\left[f\left(\frac{1}{8}\right) + f\left(\frac{2}{8}\right) + \cdots + f\left(\frac{7}{8}\right)\right] + f(1)\right]\right\} \approx 0.253906$

$\qquad M_8 = \frac{1}{8}\left[f\left(\frac{1}{16}\right) + f\left(\frac{3}{16}\right) + \cdots + f\left(\frac{13}{16}\right) + f\left(\frac{15}{16}\right)\right] = 0.248047$

$\qquad E_L \approx \frac{1}{4} - 0.191406 \approx 0.058594,$ $E_R \approx \frac{1}{4} - 0.316406 \approx -0.066406,$

$\qquad E_T \approx \frac{1}{4} - 0.253906 \approx -0.003906,$ $E_M \approx \frac{1}{4} - 0.248047 \approx 0.001953.$

$n = 16$: $\quad L_{16} = \frac{1}{16}\left[f(0) + f\left(\frac{1}{16}\right) + f\left(\frac{2}{16}\right) + \cdots + f\left(\frac{15}{16}\right)\right] \approx 0.219727$

$\qquad R_{16} = \frac{1}{16}\left[f\left(\frac{1}{16}\right) + f\left(\frac{2}{16}\right) + \cdots + f\left(\frac{15}{16}\right) + f(1)\right] \approx 0.282227$

$\qquad T_{16} = \frac{1}{16 \cdot 2}\left\{f(0) + 2\left[f\left(\frac{1}{16}\right) + f\left(\frac{2}{16}\right) + \cdots + f\left(\frac{15}{16}\right)\right] + f(1)\right\} \approx 0.250977$

$\qquad M_{16} = \frac{1}{16}\left[f\left(\frac{1}{32}\right) + f\left(\frac{3}{32}\right) + \cdots + f\left(\frac{31}{32}\right)\right] \approx 0.249512$

$\qquad E_L \approx \frac{1}{4} - 0.219727 \approx 0.030273,$ $E_R \approx \frac{1}{4} - 0.282227 \approx -0.032227,$

$\qquad E_T \approx \frac{1}{4} - 0.250977 \approx -0.000977,$ $E_M \approx \frac{1}{4} - 0.249512 \approx 0.000488.$

n	L_n	R_n	T_n	M_n
4	0.140625	0.390625	0.265625	0.242188
8	0.191406	0.316406	0.253906	0.248047
16	0.219727	0.282227	0.250977	0.249512

n	E_L	E_R	E_T	E_M
4	0.109375	−0.140625	−0.015625	0.007813
8	0.058594	−0.066406	−0.003906	0.001953
16	0.030273	−0.032227	−0.000977	0.000488

Observations:

1. E_L and E_R are always opposite in sign, as are E_T and E_M.

2. As n is doubled, E_L and E_R are decreased by about a factor of 2, and E_T and E_M are decreased by a factor of about 4.

3. The Midpoint approximation is about twice as accurate as the Trapezoidal approximation.

4. All the approximations become more accurate as the value of n increases.

5. The Midpoint and Trapezoidal approximations are much more accurate than the endpoint approximations.

21. $\Delta x = (4 - 0)/4 = 1$

(a) $T_4 = \frac{1}{2}[f(0) + 2f(1) + 2f(2) + 2f(3) + f(4)] \approx \frac{1}{2}[0 + 2(3) + 2(5) + 2(3) + 1] = 11.5$

(b) $M_4 = 1 \cdot [f(0.5) + f(1.5) + f(2.5) + f(3.5)] \approx 1 + 4.5 + 4.5 + 2 = 12$

(c) $S_4 = \frac{1}{3}[f(0) + 4f(1) + 2f(2) + 4f(3) + f(4)] \approx \frac{1}{3}[0 + 4(3) + 2(5) + 4(3) + 1] = 11.\overline{6}$

23. $\Delta t = \left(\frac{10}{60} - 0\right)/10 = \frac{1}{60}$.

Distance traveled $= \int_0^{10} v\, dt \approx S_{10} = \frac{1}{60 \cdot 3}[40 + 4(42) + 2(45) + 4(49) + 2(52)$
$$+4(54) + 2(56) + 4(57) + 2(57) + 4(55) + 56]$$
$$= \frac{1}{180}(1544) = 8.5\overline{7} \text{ mi.}$$

25. By the Total Change Theorem, the increase in velocity is equal to $\int_0^6 a(t)\, dt$. We use Simpson's Rule
with $n = 6$ and $\Delta t = 1$ to estimate this integral:
$$\int_0^6 a(t)\, dt \approx S_6 = \frac{1}{1 \cdot 3}[a(0) + 4a(1) + 2a(2) + 4a(3) + 2a(4) + 4a(5) + a(6)]$$
$$\approx \frac{1}{3}[0 + 4(0.5) + 2(4.1) + 4(9.8) + 2(12.9) + 4(9.5) + 0]$$
$$= \frac{1}{3}(113.2) = 37.7\overline{3} \text{ ft/s.}$$

27. By the Total Change Theorem, the total percentage increase is equal to $\int_{1986}^{1994} r(t)\, dt$. We use Simpson's
Rule with $n = 8$ and $\Delta t = 1$ to estimate this integral:
$$\int_{1986}^{1994} r(t)\, dt \approx S_8 = \frac{1}{1 \cdot 3}[r(1986) + 4r(1987) + 2r(1988) + \cdots + 4r(1993) + r(1994)]$$
$$\approx \frac{1}{3}[3.2 + 4(4.1) + 2(4.1) + 4(5.8) + 2(5.8) + 4(2.9) + 2(1.2) + 4(2.2) + 2.4]$$
$$= \frac{1}{3}(87.8) = 29.2\overline{6}\%$$

29. Since the Trapezoidal and Midpoint approximations on the interval $[a, b]$ are the sums of the Trapezoidal
and Midpoint approximations on the subintervals $[x_{i-1}, x_i]$, $i = 1, 2, \ldots, n$, we can focus our attention
on one such interval. The condition $f''(x) < 0$ for $a \le x \le b$ means that the graph of f is concave
down as in Figure 5. In that figure, T_n is the area of the trapezoid $AQRD$, $\int_a^b f(x)\, dx$ is the area of
the region $AQPRD$, and M_n is the area of the trapezoid $ABCD$, so $T_n < \int_a^b f(x)\, dx < M_n$. In
general, the condition $f'' < 0$ implies that the graph of f on $[a, b]$ lies above the chord joining the points
$(a, f(a))$ and $(b, f(b))$. Thus, $\int_a^b f(x)\, dx > T_n$. Since M_n is the area under a tangent to the graph,
and since $f'' < 0$ implies that the tangent lies above the graph, we also have $M_n > \int_a^b f(x)\, dx$. Thus,
$T_n < \int_a^b f(x)\, dx < M_n$.

31. $T_n = \frac{1}{2}\Delta x[f(x_0) + 2f(x_1) + \cdots + 2f(x_{n-1}) + f(x_n)]$ and
$M_n = \Delta x[f(\overline{x}_1) + f(\overline{x}_2) + \cdots + f(\overline{x}_{n-1}) + f(\overline{x}_n)]$, where $\overline{x}_i = \frac{1}{2}(x_{i-1} + x_i)$. Now
$T_{2n} = \frac{1}{2}\left(\frac{1}{2}\Delta x\right)[f(x_0) + 2f(\overline{x}_1) + 2f(x_1) + 2f(\overline{x}_2) + 2f(x_2) + \cdots$
$$+2f(\overline{x}_{n-1}) + 2f(x_{n-1}) + 2f(\overline{x}_n) + f(x_n)], \text{ so}$$

$\frac{1}{2}(T_n + M_n) = \frac{1}{2}T_n + \frac{1}{2}M_n$
$$= \frac{1}{4}\Delta x[f(x_0) + 2f(x_1) + \cdots + 2f(x_{n-1}) + f(x_n)]$$
$$+\frac{1}{4}\Delta x[2f(\overline{x}_1) + 2f(\overline{x}_2) + \cdots + 2f(\overline{x}_{n-1}) + 2f(\overline{x}_n)]$$
$$= T_{2n}.$$

Section 5.9 Improper Integrals

1. (a) Since $\int_1^\infty x^4 e^{-x^4}\,dx$ has an infinite interval of integration, this is an improper integral of Type I.

(b) Since $y = \sec x$ has an infinite discontinuity at $x = \frac{\pi}{2}$, $\int_0^{\pi/2} \sec x\,dx$ is a Type II improper integral.

(c) Since $y = \dfrac{x}{(x-2)(x-3)}$ has an infinite discontinuity at $x = 2$, $\displaystyle\int_0^2 \dfrac{x}{x^2 - 5x + 6}\,dx$ is a Type II improper integral.

(d) Since $\displaystyle\int_{-\infty}^0 \dfrac{1}{x^2 + 25}\,dx$ has an infinite interval of integration, it is an improper integral of Type I.

3. (a) The area under the graph of $y = 1/x^3 = x^{-3}$ between $x = 1$ and $x = t$ is
$A(t) = \int_1^t x^{-3}\,dx = \left[-\frac{1}{2}x^{-2}\right]_1^t = \frac{1}{2} - 1/(2t^2)$. So the area for $1 \le x \le 10$ is
$A(10) = 0.5 - 0.005 = 0.495$, the area for $1 \le x \le 100$ is $A(100) = 0.5 - 0.00005 = 0.49995$,
and the area for $1 \le x \le 1000$ is $A(1000) = 0.5 - 0.0000005 = 0.4999995$. The total area under
the curve for $x \ge 1$ is $\displaystyle\lim_{t\to\infty} A(t) = \lim_{t\to\infty}\left[\frac{1}{2} - 1/(2t^2)\right] = \frac{1}{2}$.

5. $\int_0^\infty e^{-x}\,dx = \displaystyle\lim_{t\to\infty}\int_0^t e^{-x}\,dx = \lim_{t\to\infty}\left[-e^{-x}\right]_0^t = \lim_{t\to\infty}(-e^{-t} + 1) = 1$

7. $\displaystyle\int_{-\infty}^1 \frac{dx}{(2x-3)^2} = \lim_{t\to-\infty}\frac{1}{2}\int_t^1 \frac{2\,dx}{(2x-3)^2} = \lim_{t\to-\infty}\frac{1}{2}\left[-\frac{1}{2x-3}\right]_t^1 = \lim_{t\to-\infty}\left[\frac{1}{2} + \frac{1}{2(2t-3)}\right] = \frac{1}{2}$

9. $\int_{-\infty}^\infty x^3\,dx = \int_{-\infty}^0 x^3\,dx + \int_0^\infty x^3\,dx$. $\int_{-\infty}^0 x^3\,dx = \displaystyle\lim_{t\to-\infty}\left[\frac{1}{4}x^4\right]_t^0 = \lim_{t\to-\infty}\left(-\frac{1}{4}t^4\right) = -\infty$.
Divergent

11. $\int_{-\infty}^\infty xe^{-x^2}\,dx = \int_{-\infty}^0 xe^{-x^2}\,dx + \int_0^\infty xe^{-x^2}\,dx$. $\int_{-\infty}^0 xe^{-x^2}\,dx = \displaystyle\lim_{t\to-\infty}\left(-\frac{1}{2}\right)\left[e^{-x^2}\right]_t^0$
$= \displaystyle\lim_{t\to-\infty}\left(-\frac{1}{2}\right)\left(1 - e^{-t^2}\right) = -\frac{1}{2}$, and
$\int_0^\infty xe^{-x^2}\,dx = \displaystyle\lim_{t\to\infty}\left(-\frac{1}{2}\right)\left[e^{-x^2}\right]_0^t = \lim_{t\to\infty}\left(-\frac{1}{2}\right)\left(e^{-t^2} - 1\right) = \frac{1}{2}$. Therefore,
$\int_{-\infty}^\infty xe^{-x^2}\,dx = -\frac{1}{2} + \frac{1}{2} = 0$.

13. $\int_0^\infty \cos x\,dx = \displaystyle\lim_{t\to\infty}\left[\sin x\right]_0^t = \lim_{t\to\infty}\sin t$, which does not exist. Divergent

15. $\displaystyle\int_0^\infty \frac{5\,dx}{2x+3} = \frac{5}{2}\lim_{t\to\infty}\int_0^t \frac{2\,dx}{2x+3} = \frac{5}{2}\lim_{t\to\infty}\left[\ln(2x+3)\right]_0^t = \frac{5}{2}\lim_{t\to\infty}\left[\ln(2t+3) - \ln 3\right] = \infty$.
Divergent

17. $\int_{-\infty}^1 xe^{2x}\,dx = \displaystyle\lim_{t\to-\infty}\int_t^1 xe^{2x}\,dx = \lim_{t\to-\infty}\left[\frac{1}{2}xe^{2x} - \frac{1}{4}e^{2x}\right]_t^1$ (by parts)
$= \displaystyle\lim_{t\to-\infty}\left[\frac{1}{2}e^2 - \frac{1}{4}e^2 - \frac{1}{2}te^{2t} + \frac{1}{4}e^{2t}\right] = \frac{1}{4}e^2 - 0 + 0 = \frac{1}{4}e^2$,

since $\displaystyle\lim_{t\to-\infty} te^{2t} = \lim_{t\to-\infty}\frac{t}{e^{-2t}} \overset{H}{=} \lim_{t\to-\infty}\frac{1}{-2e^{-2t}} = \lim_{t\to-\infty} -\frac{1}{2}e^{2t} = 0$.

207

19. $\displaystyle\int_1^\infty \frac{\ln x}{x}\, dx = \lim_{t\to\infty} \left[\frac{(\ln x)^2}{2}\right]_1^t = \lim_{t\to\infty} \frac{(\ln t)^2}{2} = \infty.$ Divergent

21. $\displaystyle\int_{-\infty}^\infty \frac{x\, dx}{1+x^2} = \int_{-\infty}^0 \frac{x\, dx}{1+x^2} + \int_0^\infty \frac{x\, dx}{1+x^2}.$

$\displaystyle\int_{-\infty}^0 \frac{x}{1+x^2} = \lim_{t\to-\infty} \left[\tfrac{1}{2}\ln\left(1+x^2\right)\right]_t^0 = \lim_{t\to-\infty} \left[0 - \frac{1}{2}\ln\left(1+t^2\right)\right] = -\infty.$ Divergent

23. $\displaystyle\int_0^3 \frac{dx}{\sqrt{x}} = \lim_{t\to 0^+} \int_t^3 \frac{dx}{\sqrt{x}} = \lim_{t\to 0^+} \left[2\sqrt{x}\right]_t^3 = \lim_{t\to 0^+}\left(2\sqrt{3} - 2\sqrt{t}\right) = 2\sqrt{3}$

25. $\displaystyle\int_{-1}^0 \frac{dx}{x^2} = \lim_{t\to 0^-} \int_{-1}^t \frac{dx}{x^2} = \lim_{t\to 0^-} \left[\frac{-1}{x}\right]_{-1}^t = \lim_{t\to 0^-}\left[-\frac{1}{t} + \frac{1}{-1}\right] = \infty.$ Divergent

27. $\displaystyle\int_{-2}^3 \frac{dx}{x^4} = \int_{-2}^0 \frac{dx}{x^4} + \int_0^3 \frac{dx}{x^4}.$ $\displaystyle\int_{-2}^0 \frac{dx}{x^4} = \lim_{t\to 0^-} \left[\frac{x^{-3}}{3}\right]_{-2}^t = \lim_{t\to 0^-}\left[-\frac{1}{3t^3} - \frac{1}{24}\right] = \infty.$ Divergent

29. $\displaystyle\int_4^5 \frac{dx}{(5-x)^{2/5}} = \lim_{t\to 5^-} \left[-\tfrac{5}{3}(5-x)^{3/5}\right]_4^t = \lim_{t\to 5^-}\left[-\frac{5}{3}(5-t)^{3/5} + \frac{5}{3}\right] = 0 + \frac{5}{3} = \frac{5}{3}$

31. Integrate by parts with $u = \ln x$, $dv = x\, dx$, $du = dx/x$, $v = \tfrac{1}{2}x^2$:

$\displaystyle\int_0^1 x\ln x\, dx = \lim_{t\to 0^+} \int_t^1 x\ln x\, dx = \lim_{t\to 0^+} \left[\tfrac{1}{2}x^2 \ln x - \tfrac{1}{4}x^2\right]_t^1 = -\tfrac{1}{4} - \lim_{t\to 0^+} \tfrac{1}{2}t^2 \ln t$

$\displaystyle = -\tfrac{1}{4} - \tfrac{1}{2}\lim_{t\to 0^+} \frac{\ln t}{1/t^2} \overset{\text{H}}{=} -\tfrac{1}{4} - \tfrac{1}{2}\lim_{t\to 0^+} \frac{1/t}{-2/t^3} = -\tfrac{1}{4} + \tfrac{1}{4}\lim_{t\to 0^+} t^2 = -\tfrac{1}{4}$

33.

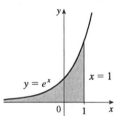

Area $= \displaystyle\int_{-\infty}^1 e^x\, dx = \lim_{t\to-\infty} [e^x]_t^1$

$= e - \lim_{t\to-\infty} e^t = e$

35.

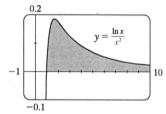

We integrate by parts with $u = \ln x$,

$dv = \left(1/x^2\right) dx$, $du = dx/x$, $v = -1/x$:

$\displaystyle\int_1^\infty \frac{\ln x}{x^2}\, dx = \lim_{t\to\infty}\left(\left[-\frac{1}{x}\ln x\right]_1^t + \int_1^t \frac{1}{x^2}\, dx\right)$

$\displaystyle = \lim_{t\to\infty}\left(-\frac{\ln t}{t} - \frac{1}{t} + 1\right) = 1$

37.

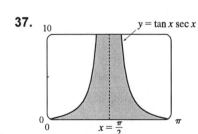

$\displaystyle\int_0^\pi \tan x \sec x\, dx = \int_0^{\pi/2} \tan x \sec x\, dx + \int_{\pi/2}^\pi \tan x \sec x\, dx.$

$\displaystyle\int_0^{\pi/2} \tan x \sec x\, dx = \lim_{t\to(\pi/2)^-} \int_0^t \tan x \sec x\, dx$

$\displaystyle = \lim_{t\to(\pi/2)^-} [\sec x]_0^t = \lim_{t\to(\pi/2)^-} (\sec t - 1)$

$= \infty.$ Divergent.

39. (a)

t	$\int_1^t g(x)\,dx$
2	0.447453
5	0.577101
10	0.621306
100	0.668479
1000	0.672957
10,000	0.673407

$g(x) = \dfrac{\sin^2 x}{x^2}$. It appears that the integral is

convergent.

(c)

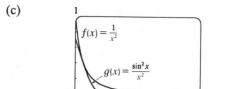

Since $\int_1^\infty f(x)\,dx$ is finite and the area under $g(x)$ is less than the area under $f(x)$ on any interval $[1, t]$, $\int_1^\infty g(x)\,dx$ must be finite; that is, the integral is convergent.

(b) $-1 \le \sin x \le 1 \Longrightarrow 0 \le \sin^2 x \le 1 \Longrightarrow 0 \le \dfrac{\sin^2 x}{x^2} \le \dfrac{1}{x^2}$. Since $\displaystyle\int_1^\infty \dfrac{1}{x^2}\,dx$ is convergent

(Equation 2 with $p = 2 > 1$), $\displaystyle\int_1^\infty \dfrac{\sin^2 x}{x^2}\,dx$ is convergent by the Comparison Theorem.

41. For $x \ge 1$, $\dfrac{\cos^2 x}{1 + x^2} \le \dfrac{1}{1 + x^2} < \dfrac{1}{x^2}$. $\displaystyle\int_1^\infty \dfrac{1}{x^2}\,dx$ is convergent by Equation 2 with $p = 2 > 1$, so

$\displaystyle\int_1^\infty \dfrac{\cos^2 x}{1 + x^2}\,dx$ is convergent by the Comparison Theorem.

43. For $x \ge 1$, $x + e^{2x} > e^{2x} > 0 \Longrightarrow \dfrac{1}{x + e^{2x}} \le \dfrac{1}{e^{2x}} = e^{-2x}$ on $[1, \infty)$.

$\displaystyle\int_1^\infty e^{-2x}\,dx = \lim_{t\to\infty} \left[-\tfrac{1}{2}e^{-2x}\right]_1^t = \lim_{t\to\infty}\left[-\tfrac{1}{2}e^{-2t} + \tfrac{1}{2}e^{-2}\right] = \tfrac{1}{2}e^{-2}$. Therefore, $\int_1^\infty e^{-2x}\,dx$ is

convergent, and by the Comparison Theorem, $\displaystyle\int_1^\infty \dfrac{dx}{x + e^{2x}}$ is also convergent.

45. $\dfrac{1}{x\sin x} \ge \dfrac{1}{x}$ on $\left(0, \tfrac{\pi}{2}\right]$ since $0 \le \sin x \le 1$. $\displaystyle\int_0^{\pi/2} \dfrac{dx}{x} = \lim_{t\to 0^+}\int_t^{\pi/2} \dfrac{dx}{x} = \lim_{t\to 0^+}\left[\ln x\right]_t^{\pi/2}$.

But $\ln t \to -\infty$ as $t \to 0^+$, so $\displaystyle\int_0^{\pi/2} \dfrac{dx}{x}$ is divergent, and by the Comparison Theorem, $\displaystyle\int_0^{\pi/2} \dfrac{dx}{x\sin x}$ is also divergent.

47. $\displaystyle\int_0^\infty \dfrac{dx}{\sqrt{x}\,(1 + x)} = \int_0^1 \dfrac{dx}{\sqrt{x}\,(1 + x)} + \int_1^\infty \dfrac{dx}{\sqrt{x}\,(1 + x)} = \lim_{t\to 0^+}\int_t^1 \dfrac{dx}{\sqrt{x}\,(1 + x)} + \lim_{t\to\infty}\int_1^t \dfrac{dx}{\sqrt{x}\,(1 + x)}$

$= \displaystyle\lim_{t\to 0^+}\int_{\sqrt{t}}^1 \dfrac{2\,du}{1 + u^2} + \lim_{t\to\infty}\int_1^{\sqrt{t}} \dfrac{2\,du}{1 + u^2}$ (let $u = \sqrt{x}$, $x = u^2$)

$= \displaystyle\lim_{t\to 0^+}\left[2\tan^{-1} u\right]_{\sqrt{t}}^1 + \lim_{t\to\infty}\left[2\tan^{-1} u\right]_1^{\sqrt{t}}$

$= \displaystyle\lim_{t\to 0^+}\left[2\left(\tfrac{\pi}{4}\right) - 2\tan^{-1}\sqrt{t}\right] + \lim_{t\to\infty}\left[2\tan^{-1}\sqrt{t} - 2\left(\tfrac{\pi}{4}\right)\right]$

$= \tfrac{\pi}{2} - 0 + 2\left(\tfrac{\pi}{2}\right) - \tfrac{\pi}{2} = \pi$

49. If $p = 1$, then $\int_0^1 \frac{dx}{x^p} = \lim_{t \to 0^+} [\ln x]_t^1 = \infty$. Divergent. If $p \neq 1$, then

$$\int_0^1 \frac{dx}{x^p} = \lim_{t \to 0^+} \int_t^1 \frac{dx}{x^p} \quad \text{(note that the integral is not improper if } p < 0)$$

$$= \lim_{t \to 0^+} \left[\frac{x^{-p+1}}{-p+1} \right]_t^1 = \lim_{t \to 0^+} \frac{1}{1-p} \left[1 - \frac{1}{t^{p-1}} \right].$$

If $p > 1$, then $p - 1 > 0$, so $\frac{1}{t^{p-1}} \to \infty$ as $t \to 0^+$, and the integral diverges. Finally, if $p < 1$, then

$$\int_0^1 \frac{dx}{x^p} = \frac{1}{1-p} \left[\lim_{t \to 0^+} (1 - t^{1-p}) \right] = \frac{1}{1-p}. \text{ Thus, the integral converges if and only if } p < 1, \text{ and in}$$

that case its value is $\frac{1}{1-p}$.

51. (a) $\int_{-\infty}^{\infty} x \, dx = \int_{-\infty}^0 x \, dx + \int_0^{\infty} x \, dx$, and $\int_0^{\infty} x \, dx = \lim_{t \to \infty} \int_0^t x \, dx = \lim_{t \to \infty} \left[\frac{1}{2} t^2 - \frac{1}{2} (0^2) \right] = \infty$, so
the integral is divergent.

(b) $\int_{-t}^t x \, dx = \left[\frac{1}{2} x^2 \right]_{-t}^t = \frac{1}{2} t^2 - \frac{1}{2} t^2 = 0$, so $\lim_{t \to \infty} \int_{-t}^t x \, dx = 0$. Therefore,
$\int_{-\infty}^{\infty} x \, dx \neq \lim_{t \to \infty} \int_{-t}^t x \, dx$.

53. (a)

(b) $r(t) = F'(t)$ is the rate at which the fraction $F(t)$ of burnt-out bulbs increases as t increases. This could be interpreted as a fractional burnout rate.

(c) $\int_0^{\infty} r(t) \, dt = \lim_{x \to \infty} F(x) = 1$, since all of the bulbs will eventually burn out.

55. $I = \int_0^{\infty} t e^{kt} \, dt = \lim_{s \to \infty} \left[\frac{1}{k^2} (kt - 1) e^{kt} \right]_0^s \quad \text{(Formula 96, or integrate by parts)}$

$$= \lim_{s \to \infty} \left[\left(\frac{1}{k} s e^{ks} - \frac{1}{k^2} e^{ks} \right) - \left(-\frac{1}{k^2} \right) \right]$$

Since $k < 0$ the first two terms approach 0 (you can verify that the first term
does so with l'Hospital's Rule) so the whole expression is equal to $1/k^2$. Thus,
$M = -kI = -k \left(1/k^2 \right) = -1/k = -1/(-0.000121) \approx 8264.5$ years.

57. We use integration by parts: let $u = x$, $dv = xe^{-x^2} \, dx$, $du = dx$, $v = -\frac{1}{2} e^{-x^2}$. So

$$\int_0^{\infty} x^2 e^{-x^2} \, dx = \lim_{t \to \infty} \left[-\frac{x}{2} e^{-x^2} \right]_0^t + \frac{1}{2} \int_0^{\infty} e^{-x^2} \, dx = \lim_{t \to \infty} -\frac{t}{2 e^{t^2}} + \frac{1}{2} \int_0^{\infty} e^{-x^2} \, dx = \frac{1}{2} \int_0^{\infty} e^{-x^2} \, dx.$$

(The limit is 0 by l'Hospital's Rule.)

59. $I = \int_a^{\infty} \frac{1}{x^2 + 1} \, dx = \lim_{t \to \infty} \int_a^t \frac{1}{x^2 + 1} \, dx = \lim_{t \to \infty} \left[\tan^{-1} x \right]_a^t = \lim_{t \to \infty} (\tan^{-1} t - \tan^{-1} a) =$
$\frac{\pi}{2} - \tan^{-1} a$. $I < 0.001 \implies \frac{\pi}{2} - \tan^{-1} a < 0.01 \implies \tan^{-1} a > \frac{\pi}{2} - 0.001 \implies$
$a > \tan \left(\frac{\pi}{2} - 0.001 \right) \approx 1000$.

Chapter 5 Review

Concept Check

1. (a) $\sum_{i=1}^{n} f(x_i^*) \Delta x$ is an expression for a Riemann sum of a function f. x_i^* is a point in the ith subinterval $[x_{i-1}, x_i]$ and Δx is the length of the subintervals.

   (b) See Figure 1 in Section 5.2.

   (c) In Section 5.2, see Figure 3 and the paragraph that precedes it.

2. (a) See Definition 5.2.2.

   (b) See Figure 2 in Section 5.2.

   (c) In Section 5.2, see Figure 4 and the paragraph that precedes it.

3. (a) See the Evaluation Theorem at the beginning of Section 5.3.

   (b) See the Total Change Theorem after Example 6 in Section 5.3.

4. $\int_{t_1}^{t_2} r(t)\, dt$ represents the change in the amount of water in the reservoir between time t_1 and time t_2.

5. (a) $\int_{60}^{120} v(t)\, dt$ represents the change in position of the particle from $t = 60$ to $t = 120$ seconds.

   (b) $\int_{60}^{120} |v(t)|\, dt$ represents the total distance traveled by the particle from $t = 60$ to 120 seconds.

   (c) $\int_{60}^{120} a(t)\, dt$ represents the change in the velocity of the particle from $t = 60$ to $t = 120$ seconds.

6. (a) $\int f(x)\, dx$ is a *function* F that has the property $F' = f$.

   (b) The connection is given by the Evaluation Theorem: $\int_a^b f(x)\, dx = \left[\int f(x)\, dx \right]_a^b$ if f is continuous..

7. See page 388.

8. (a) See the Substitution Rule (5.5.4). This says that it is permissible to operate with the dx after an integral sign as if it were a differential.

   (b) See Formula 5.6.1 or 5.6.2. We try to choose $u = f(x)$ to be a function that becomes simpler when differentiated (or at least not more complicated) as long as $dv = g'(x)\, dx$ can be readily integrated to give v.

9. (a) See the Midpoint Rule in Section 5.8.

   (b) See the Trapezoidal Rule in Section 5.8.

   (c) See Simpson's Rule in Section 5.8.

10. See Definitions 1(a), (b), and (c) in Section 5.9.

11. See Definitions 3(b), (a), and (c) in Section 5.9.

12. See the Comparison Theorem that follows Example 8 in Section 5.9.

True-False Quiz

1. True by Property 2 of the Integral in Section 5.2.

3. False. For example, let $f(x) = x^2$. Then $\int_0^1 \sqrt{x^2}\, dx = \int_0^1 x\, dx = \frac{1}{2}$, but

$$\sqrt{\int_0^1 x^2\, dx} = \sqrt{\frac{1}{3}} = \frac{1}{\sqrt{3}}.$$

5. True by Comparison Property 2 of the Integral in Section 5.4.

7. True. The integrand is an odd function that is continuous on $[-1, 1]$, so the result follows from Equation 5.5.6(b).

9. True by Theorem 5.9.2 with $p = \sqrt{2} > 1$.

11. False. For example, the function $y = |x|$ is continuous on $\mathbb{R}$, but has no derivative at $x = 0$.

Exercises

1. (a)

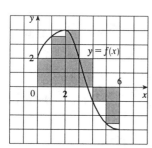

$$L_6 = \sum_{i=1}^{6} f(x_{i-1})\, \Delta x \quad [\Delta x = \tfrac{6-0}{6} = 1]$$
$$= f(x_0) \cdot 1 + f(x_1) \cdot 1 + f(x_2) \cdot 1$$
$$+ f(x_3) \cdot 1 + f(x_4) \cdot 1 + f(x_5) \cdot 1$$
$$\approx 2 + 3.5 + 4 + 2 + (-1) + (-2.5)$$
$$= 8$$

The Riemann sum represents the sum of the areas of the first four rectangles and the negatives of the areas of the last two rectangles.

(b)

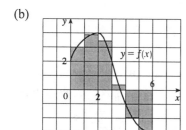

$$M_6 = \sum_{i=1}^{6} f(\overline{x}_i)\, \Delta x \quad [\Delta x = \tfrac{6-0}{6} = 1]$$
$$= f(\overline{x}_1) \cdot 1 + f(\overline{x}_2) \cdot 1 + f(\overline{x}_3) \cdot 1$$
$$+ f(\overline{x}_4) \cdot 1 + f(\overline{x}_5) \cdot 1 + f(\overline{x}_6) \cdot 1$$
$$= f(0.5) + f(1.5) + f(2.5) + f(3.5) + f(4.5) + f(5.5)$$
$$= 3 + 3.9 + 3.3 + 0.2 + (-2) + (-2.8) = 5.6$$

The Riemann sum represents the sum of the areas of the first four rectangles and the negatives of the areas of the last two rectangles.

3. $\int_0^1 \left(x + \sqrt{1 - x^2}\right) dx = \int_0^1 x\, dx + \int_0^1 \sqrt{1 - x^2}\, dx$
$$= I_1 + I_2.$$

I_1 can be interpreted as the area of the triangle shown in the figure and I_2 can be interpreted as the area of the quarter-circle.

Area $= \frac{1}{2}(1)(1) + \frac{1}{4}(\pi)(1)^2 = \frac{1}{2} + \frac{\pi}{4}.$

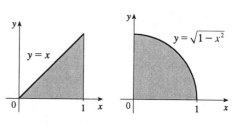

5. $\int_0^6 f(x)\,dx = \int_0^4 f(x)\,dx + \int_4^6 f(x)\,dx \Longrightarrow 10 = 7 + \int_4^6 f(x)\,dx \Longrightarrow \int_4^6 f(x)\,dx = 10 - 7 = 3$

7. First note that either a or b must be the graph of $\int_0^x f(t)\,dt$, since $\int_0^0 f(t)\,dt = 0$, and $c(0) \neq 0$. Now notice that $b > 0$ when c is increasing, and that $c > 0$ when a is increasing. It follows that c is the graph of $f(x)$, b is the graph of $f'(x)$, and a is the graph of $\int_0^x f(t)\,dt$.

9. $\int_0^1 (1 - x^9)\,dx = \left[x - \frac{1}{10}x^{10}\right]_0^1 = 1 - \frac{1}{10} = \frac{9}{10}$

11. $u = x^2 + 1$, $du = 2x\,dx$, so $\displaystyle\int_0^1 \frac{x}{x^2+1}\,dx = \int_1^2 \frac{1}{u}\left(\frac{1}{2}\,du\right) = \frac{1}{2}\left[\ln u\right]_1^2 = \frac{1}{2}\ln 2.$

13. $\int_1^8 \sqrt[3]{x}\,(x - 1)\,dx = \int_1^8 \left(x^{4/3} - x^{1/3}\right)dx = \left[\frac{3}{7}x^{7/3} - \frac{3}{4}x^{4/3}\right]_1^8 = \left(\frac{3}{7}\cdot 128 - \frac{3}{4}\cdot 16\right) - \left(\frac{3}{7} - \frac{3}{4}\right)$
$= \frac{1209}{28}$

15. $u = 2x + 3 \Longrightarrow du = 2\,dx$, so $\displaystyle\int_3^{11} \frac{dx}{\sqrt{2x+3}} = \int_9^{25} u^{-1/2}\left(\frac{1}{2}\,du\right) = \left[u^{1/2}\right]_9^{25} = 5 - 3 = 2.$

17. $\int_0^1 e^{\pi t}\,dt = \left[\frac{1}{\pi}e^{\pi t}\right]_0^1 = \frac{1}{\pi}(e^\pi - 1)$

19. Integrate by parts with $u = x$, $dv = \sec x \tan x\,dx \Longrightarrow du = dx$, $v = \sec x$:
$\int x \sec x \tan x\,dx = x \sec x - \int \sec x\,dx = x \sec x - \ln|\sec x + \tan x| + C.$

21. $u = e^x + 1 \Longrightarrow du = e^x\,dx$, so $\displaystyle\int \frac{e^x}{e^x+1}\,dx = \int \frac{du}{u} = \ln|u| + C = \ln(e^x + 1) + C.$

23. Let $u = \cos x$, $dv = e^x\,dx \Longrightarrow du = -\sin x\,dx$, $v = e^x$: $(*)\ I = \int e^x \cos x\,dx =$
$e^x \cos x + \int e^x \sin x\,dx$. To integrate $\int e^x \sin x\,dx$, let $U = \sin x$, $dV = e^x\,dx \Longrightarrow dU = \cos x\,dx$,
$V = e^x$. $\int e^x \sin x\,dx = e^x \sin x - \int e^x \cos x\,dx = e^x \sin x - I$. By substitution in $(*)$,
$I = e^x \cos x + e^x \sin x - I \Longrightarrow 2I = e^x(\cos x + \sin x) \Longrightarrow I = \frac{1}{2}e^x(\cos x + \sin x) + C.$

25. $u = \sqrt{x} \Longrightarrow du = \dfrac{dx}{2\sqrt{x}}$, so $\displaystyle\int \frac{e^{\sqrt{x}}}{\sqrt{x}}\,dx = 2\int e^u\,du = 2e^u + C = 2e^{\sqrt{x}} + C.$

27. Let $f(x)$ denote the integrand and $F(x)$ its antiderivative (with $C = 0$).
$u = 1 + \sin x \Longrightarrow du = \cos x\,dx$, so
$$\int \frac{\cos x\,dx}{\sqrt{1 + \sin x}} = \int u^{-1/2}\,du = 2u^{1/2} + C$$
$$= 2\sqrt{1 + \sin x} + C.$$

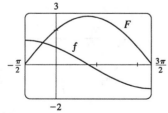

29. From the graph, it appears that the area under the curve $y = x\sqrt{x}$ between $x = 0$ and $x = 4$ is somewhat less than half the area of an 8×4 rectangle, so perhaps about 13 or 14. To find the exact value, we evaluate
$\int_0^4 x\sqrt{x}\,dx = \int_0^4 x^{3/2}\,dx = \left[\frac{2}{5}x^{5/2}\right]_0^4$
$= \frac{2}{5}(4)^{5/2} = \frac{64}{5} = 12.8.$

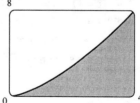

31. By FTC1, $F(x) = \int_1^x \sqrt{1+t^4}\,dt \implies F'(x) = \sqrt{1+x^4}$.

33. $g(x) = \int_0^{x^3} \dfrac{t\,dt}{\sqrt{1+t^3}}$. Let $y = g(u)$ and $u = x^3$.

Then $g'(x) = \dfrac{dy}{dx} = \dfrac{dy}{du}\dfrac{du}{dx} = \dfrac{u}{\sqrt{1+u^3}}3x^2 = \dfrac{x^3}{\sqrt{1+x^9}}3x^2 = \dfrac{3x^5}{\sqrt{1+x^9}}$.

35. $u = e^x \implies du = e^x\,dx$, so

$$\int e^x \sqrt{1-e^{2x}}\,dx = \int \sqrt{1-u^2}\,du = \tfrac{1}{2}u\sqrt{1-u^2} + \tfrac{1}{2}\sin^{-1}u + C \quad \text{(Formula 30)}$$
$$= \tfrac{1}{2}\left[e^x\sqrt{1-e^{2x}} + \sin^{-1}(e^x)\right] + C.$$

37. $u = x + \tfrac{1}{2} \implies du = dx$, so

$$\int \sqrt{x^2 + x + 1}\,dx = \int \sqrt{\left(x+\tfrac{1}{2}\right)^2 + \tfrac{3}{4}}\,dx = \int \sqrt{u^2 + \left(\tfrac{\sqrt{3}}{2}\right)^2}\,du$$
$$= \tfrac{1}{2}u\sqrt{u^2 + \tfrac{3}{4}} + \tfrac{3}{8}\ln\left|u + \sqrt{u^2 + \tfrac{3}{4}}\right| + C \quad \text{(Formula 21)}$$
$$= \frac{2x+1}{4}\sqrt{x^2 + x + 1} + \tfrac{3}{8}\ln\left|x + \tfrac{1}{2} + \sqrt{x^2 + x + 1}\right| + C$$

39. $f(x) = \sqrt{1+x^4}$, $\Delta x = \dfrac{b-a}{n} = \dfrac{1-0}{10} = \dfrac{1}{10}$

   (a) $T_{10} = \frac{1}{10\cdot 2}\{f(0) + 2[f(0.1) + f(0.2) + \cdots + f(0.9)] + f(1)\} \approx 1.090608$

   (b) $M_{10} = \frac{1}{10}\left[f\left(\tfrac{1}{20}\right) + f\left(\tfrac{3}{20}\right) + f\left(\tfrac{5}{20}\right) + \cdots + f\left(\tfrac{19}{20}\right)\right] \approx 1.088840$

   (c) $S_{10} = \frac{1}{10\cdot 3}\left[f(0) + 4f(0.1) + 2f(0.2) + \cdots + 4f(0.9) + f(1)\right] \approx 1.089429$

f is concave upward, so the Trapezoidal Rule gives us an overestimate, the Midpoint Rule gives an underestimate, and we cannot tell whether Simpson's Rule gives us an overestimate or an underestimate.

41. $f(x) = \left(1 + x^4\right)^{1/2}$, $f'(x) = \tfrac{1}{2}\left(1 + x^4\right)^{-1/2}\left(4x^3\right) = 2x^3\left(1 + x^4\right)^{-1/2}$,

$f''(x) = \left(2x^6 + 6x^2\right)\left(1 + x^4\right)^{-3/2}$. A graph of f'' on $[0, 1]$ shows that it has its maximum at $x = 1$, so $|f''(x)| \leq f''(1) = \sqrt{8}$ on $[0, 1]$. By taking $K = \sqrt{8}$, we find that the error in Exercise 39(a) is

bounded by $\dfrac{K(b-a)^3}{12n^2} = \dfrac{\sqrt{8}}{1200} \approx 0.0024$, and in (b) by about $\tfrac{1}{2}(0.0024) = 0.0012$.

Note: Another way to estimate K is to let $x = 1$ in the factor $2x^6 + 6x^2$ (maximizing the numerator) and let $x = 0$ in the factor $\left(1 + x^4\right)^{-3/2}$ (minimizing the denominator). Doing so gives us $K = 8$ and errors of 0.0067 and 0.003.

43. (a) $f(x) = \sin(\sin x)$. A CAS gives

$$f^{(4)}(x) = \sin(\sin x)\left[\cos^4 x + 7\cos^2 x - 3\right]$$
$$+ \cos(\sin x)\left[6\cos^2 x \sin x + \sin x\right].$$

From the graph, we see that $\left|f^{(4)}(x)\right| < 3.8$ for $x \in [0, \pi]$.

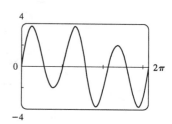

(b) We use Simpson's Rule with $f(x) = \sin(\sin x)$ and $\Delta x = \frac{\pi}{10}$:

$\int_0^\pi f(x)\, dx \approx \frac{\pi}{10 \cdot 3} \left[f(0) + 4f\left(\frac{\pi}{10}\right) + 2f\left(\frac{2\pi}{10}\right) + \cdots + 4f\left(\frac{9\pi}{10}\right) + f(\pi) \right] \approx 1.786721$.

From part (a), we know that $f^{(4)}(x) < 3.8$ on $[0, \pi]$, so we use Theorem 5.8.4 with $K = 3.8$, and estimate the error as $|E_S| \leq \frac{3.8(\pi - 0)^5}{180(10)^4} \approx 0.000646$.

(c) If we want the error to be less than 0.00001, we must have $|E_S| \leq \left(3.8\pi^5\right) / \left(180n^4\right) \leq 0.00001$, so $n^4 \geq \frac{3.8\pi^5}{180(0.00001)} \approx 646{,}041.6 \Longrightarrow n \geq 28.35$. Since n must be even for Simpson's Rule, we must have $n \geq 30$ to ensure the desired accuracy.

45. $\displaystyle\int_0^\infty \frac{dx}{(x+2)^4} = \lim_{t\to\infty} \left[\frac{-1}{3(x+2)^3} \right]_0^t = \lim_{t\to\infty} \left[\frac{1}{3 \cdot 2^3} - \frac{1}{3(t+2)^3} \right] = \frac{1}{24}$

47. $\int_{-\infty}^0 e^{-2x}\, dx = \lim_{t\to-\infty} \int_t^0 e^{-2x}\, dx = \lim_{t\to-\infty} \left[-\frac{1}{2}e^{-2x} \right]_t^0 = \lim_{t\to-\infty} \left(-\frac{1}{2} + \frac{1}{2}e^{-2t} \right) = \infty$. Divergent

49. $u = \ln x \Longrightarrow du = dx/x$, so

$\displaystyle\int_1^e \frac{dx}{x\sqrt{\ln x}} = \lim_{t\to1+} \int_t^e \frac{dx}{x\sqrt{\ln x}} = \lim_{t\to1+} \int_{\ln t}^1 \frac{du}{\sqrt{u}} = \lim_{t\to1+} \left[2\sqrt{u} \right]_{\ln t}^1 = \lim_{t\to1+} \left(2 - 2\sqrt{\ln t} \right) = 2$.

51. $\dfrac{x^3}{x^5 + 2} \leq \dfrac{x^3}{x^5} = \dfrac{1}{x^2}$ for x in $[1, \infty)$. $\displaystyle\int_1^\infty \frac{1}{x^2}\, dx$ is convergent by (5.9.2) with $p = 2 > 1$. Therefore,

$\displaystyle\int_1^\infty \frac{x^3}{x^5 + 2}\, dx$ is convergent by the Comparison Theorem.

53. (a) displacement $= \int_0^5 \left(t^2 - t\right) dt = \left[\frac{1}{3}t^3 - \frac{1}{2}t^2 \right]_0^5 = \frac{125}{3} - \frac{25}{2} = \frac{175}{6} = 29.1\overline{6}$

(b) distance traveled $= \int_0^5 \left| t^2 - t \right| dt = \int_0^5 |t(t-1)|\, dt = \int_0^1 \left(t - t^2\right) dt + \int_1^5 \left(t^2 - t\right) dt$

$= \left[\frac{1}{2}t^2 - \frac{1}{3}t^3 \right]_0^1 + \left[\frac{1}{3}t^3 - \frac{1}{2}t^2 \right]_1^5 = \frac{1}{2} - \frac{1}{3} - 0 + \left(\frac{125}{3} - \frac{25}{2} \right) - \left(\frac{1}{3} - \frac{1}{2} \right) = 29.5$

55. We use Simpson's Rule with $n = 6$ and $\Delta x = 1$:

total increase $= \int_{1988}^{1994} r(t)\, dt \approx S_6$

$= \frac{1}{3} \left[r(1988) + 4r(1989) + 2r(1990) + 4r(1991) + 2r(1992) + 4r(1993) + r(1994) \right]$

$= \frac{1}{3} \left[6.5 + 4(7.7) + 2(9.0) + 4(8.7) + 2(7.4) + 4(5.9) + 4.8 \right] = \frac{1}{3}(133.3) \approx 44.4\%$

57. Both numerator and denominator approach 0 as $a \to 0$, so we use l'Hospital's Rule. (Note that we are differentiating *with respect to* a, since that is the quantity which is changing.) We also use FTC1:

$\displaystyle\lim_{a\to0} T(x, t) = \lim_{a\to0} \frac{C \int_0^a e^{-(x-u)^2/(4kt)}\, du}{a\sqrt{4\pi kt}} \overset{\text{H}}{=} \lim_{a\to0} \frac{Ce^{-(x-a)^2/(4kt)}}{\sqrt{4\pi kt}} = \frac{Ce^{-x^2/(4kt)}}{\sqrt{4\pi kt}}$

59. We differentiate both sides of the given equation using FTC1, and get

$f(x) = e^{2x} + 2xe^{2x} + e^{-x}f(x) \Longrightarrow f(x)(1 - e^{-x}) = e^{2x} + 2xe^{2x} \Longrightarrow f(x) = \dfrac{e^{2x}(1 + 2x)}{1 - e^{-x}}$.

61. Let $u = f(x)$ and $du = f'(x)\, dx$. So

$2 \int_a^b f(x) f'(x)\, dx = 2 \int_{f(a)}^{f(b)} u\, du = \left[u^2 \right]_{f(a)}^{f(b)} = [f(b)]^2 - [f(a)]^2$.

63. By the Fundamental Theorem of Calculus,

$\int_0^\infty f'(x)\, dx = \lim_{t\to\infty} \int_0^t f'(x)\, dx = \lim_{t\to\infty} [f(t) - f(0)] = \lim_{t\to\infty} f(t) - f(0) = 0 - f(0) = -f(0)$.

Focus on Problem Solving

1.

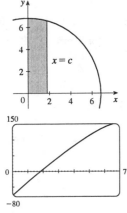

By symmetry, the problem can be reduced to finding the line $x = c$ such that the shaded area is one-third of the area of the quarter-circle. The equation of the circle is $y = \sqrt{49 - x^2}$, so we require that $\int_0^c \sqrt{49 - x^2}\,dx = \frac{1}{3} \cdot \frac{1}{4}\pi\,(7)^2 \iff$
$\left[\frac{1}{2}x\sqrt{49 - x^2} + \frac{49}{2}\sin^{-1}(x/7)\right]_0^c = \frac{49}{12}\pi$ (by Formula 30) $\iff$
$\frac{1}{2}c\sqrt{49 - c^2} + \frac{49}{2}\sin^{-1}(c/7) = \frac{49}{12}\pi$.
This equation would be difficult to solve exactly, so we plot the left-hand side as a function of c, and find that the equation holds for $c \approx 1.85$. So the cuts should be made at distances of about 1.85 inches from the center of the pizza.

3. Differentiating both sides of the equation $x \sin \pi x = \int_0^{x^2} f(t)\,dt$ (using FTC1 and the Chain Rule for the right side) gives $\sin \pi x + \pi x \cos \pi x = 2xf\left(x^2\right)$. Letting $x = 2$ so that $f\left(x^2\right) = f(4)$, we obtain $\sin 2\pi + 2\pi \cos 2\pi = 4f(4)$, so $f(4) = \frac{1}{4}(0 + 2\pi \cdot 1) = \frac{\pi}{2}$.

5. By Part 2 of the Fundamental Theorem of Calculus, $\int_0^1 f'(x)\,dx = f(1) - f(0) = 1 - 0 = 1$.

7. By l'Hospital's Rule and the Fundamental Theorem, using the notation $\exp(y) = e^y$,
$$\lim_{x \to 0} \frac{\int_0^x (1 - \tan 2t)^{1/t}\,dt}{x} \overset{\text{H}}{=} \lim_{x \to 0} \frac{(1 - \tan 2x)^{1/x}}{1} = \exp\left(\lim_{x \to 0} \frac{\ln(1 - \tan 2x)}{x}\right)$$
$$\overset{\text{H}}{=} \exp\left(\lim_{x \to 0} \frac{-2\sec^2 2x}{1 - \tan 2x}\right) = \exp\left(\frac{-2 \cdot 1^2}{1 - 0}\right) = e^{-2}.$$

9. Such a function cannot exist. $f'(x) > 3$ for all x means that f is differentiable (and hence continuous) for all x. So by Part 2 of the Fundamental Theorem, $\int_1^4 f'(x)\,dx = f(4) - f(1) = 7 - (-1) = 8$. However, if $f'(x) > 3$ for all x, then $\int_1^4 f'(x)\,dx \geq 3 \cdot (4 - 1) = 9$ by Comparison Property 3 in Section 5.4.

Another Solution: By the Mean Value Theorem there exists a number $c \in (1, 4)$ such that
$$f'(c) = \frac{f(4) - f(1)}{4 - 1} = \frac{7 - (-1)}{3} = \frac{8}{3} \implies 8 = 3f'(c).$$ But $f'(x) > 3 \implies 3f'(c) > 9$, so such a function cannot exist.

11. $f(x) = 2 + x - x^2 = (-x + 2)(x + 1) = 0 \iff x = 2$ or $x = -1$. $f(x) \geq 0$ for $x \in [-1, 2]$ and $f(x) < 0$ everywhere else. The integral $\int_a^b \left(2 + x - x^2\right)\,dx$ has a maximum on the interval where the integrand is positive, which is $[-1, 2]$. So $a = -1$, $b = 2$. (Any larger interval gives a smaller integral since $f(x) < 0$ outside $[-1, 2]$. Any smaller interval also gives a smaller integral since $f(x) \geq 0$ in $[-1, 2]$.)

13. By FTC1, $\dfrac{d}{dx} \displaystyle\int_0^x \left(\displaystyle\int_1^{\sin t} \sqrt{1+u^4}\, du \right) dt = \displaystyle\int_1^{\sin x} \sqrt{1+u^4}\, du$. Again using FTC1,

$$\frac{d^2}{dx^2} \int_0^x \left(\int_1^{\sin t} \sqrt{1+u^4}\, du \right) dt = \frac{d}{dx} \int_1^{\sin x} \sqrt{1+u^4}\, du = \sqrt{1+\sin^4 x}\, \cos x.$$

15. The given integral represents the difference of the shaded areas, which appears to be 0. It can be calculated by integrating with respect to either x or y, so we find x in terms of y for each curve: $y = \sqrt[3]{1-x^7} \implies x = \sqrt[7]{1-y^3}$ and $y = \sqrt[7]{1-x^3} \implies x = \sqrt[3]{1-y^7}$, so $\displaystyle\int_0^1 \left(\sqrt[3]{1-y^7} - \sqrt[7]{1-y^3} \right) dy = \displaystyle\int_0^1 \left(\sqrt[7]{1-x^3} - \sqrt[3]{1-x^7} \right) dx$. But this equation is of the form $z = -z$. So $\displaystyle\int_0^1 \left(\sqrt[3]{1-x^7} - \sqrt[7]{1-x^3} \right) dx = 0$.

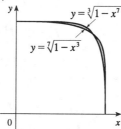

17. In accordance with the hint, we let $I_k = \displaystyle\int_0^1 \left(1-x^2\right)^k dx$, and we find an expression for I_{k+1} in terms of I_k. We integrate I_{k+1} by parts with $u = \left(1-x^2\right)^{k+1}$, $dv = dx \implies du = (k+1)\left(1-x^2\right)^k (-2x)$, $v = x$, and then split the remaining integral into identifiable quantities:

$$I_{k+1} = \left[x\left(1-x^2\right)^{k+1} \right]_0^1 + 2(k+1)\int_0^1 x^2\left(1-x^2\right)^k dx$$

$$= (2k+2)\int_0^1 \left(1-x^2\right)^k \left[1-\left(1-x^2\right)\right] dx = (2k+2)\left(I_k - I_{k+1}\right).$$

So $I_{k+1}\left[1 + (2k+2)\right] = (2k+2)I_k \implies I_{k+1} = \dfrac{2k+2}{2k+3} I_k$.

Now to complete the proof, we use induction: $I_0 = 1 = \dfrac{2^0 \, (0!)^2}{1!}$, so the formula holds for $n = 0$. Now suppose it holds for $n = k$. Then

$$I_{k+1} = \frac{2k+2}{2k+3} I_k = \frac{2k+2}{2k+3}\left[\frac{2^{2k}\,(k!)^2}{(2k+1)!} \right] = \frac{2(k+1)\,2^{2k}\,(k!)^2}{(2k+3)(2k+1)!} = \frac{2(k+1)\,2^{2k}\,(k!)^2}{(2k+3)(2k+1)!} \cdot \frac{2(k+1)}{2k+2}$$

$$= \frac{[2(k+1)]^2\,2^{2k}\,(k!)^2}{(2k+3)(2k+2)(2k+1)!} = \frac{2^{2(k+1)}\,[(k+1)!]^2}{[2(k+1)+1]!}.$$

So by induction, the formula holds for all integers $n \geq 0$.

Chapter 6   Applications of Integration

Section 6.1   More about Areas

1. $A = \int_{-1}^{1} \left[(x^2 + 3) - x \right] dx = 2 \int_0^1 (x^2 + 3)\, dx$ (by Theorem 5.5.6)
  $= 2 \left[\frac{1}{3}x^3 + 3x \right]_0^1 = 2 \left(\frac{1}{3} + 3 \right) = \frac{20}{3}$

3. $A = \int_{-1}^{1} \left[(1 - y^4) - (y^3 - y) \right] dy = 2 \int_0^1 (1 - y^4)\, dx$ (by Theorem 5.5.6)
  $= 2 \left[-\frac{1}{5}y^5 + y \right]_0^1 = 2 \left(-\frac{1}{5} + 1 \right) = \frac{8}{5}$

5. $A = \int_0^1 (x - x^2)\, dx = \left[\frac{1}{2}x^2 - \frac{1}{3}x^3 \right]_0^1$
  $= \frac{1}{2} - \frac{1}{3}$
  $= \frac{1}{6}$

7. $A = \int_0^1 (e^{3x} - e^x)\, dx = \left[\frac{1}{3}e^{3x} - e^x \right]_0^1$
  $= \left(\frac{1}{3}e^3 - e \right) - \left(\frac{1}{3} - 1 \right)$
  $= \frac{1}{3}e^3 - e + \frac{2}{3} \approx 4.64$

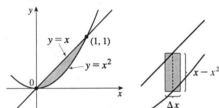

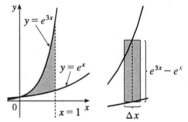

9. $A = \int_{-1}^{1} \left[(x^2 + 3) - 4x^2 \right] dx = 2 \int_0^1 (3 - 3x^2)\, dx = 2 \left[3x - x^3 \right]_0^1 = 2(3 - 1) = 4$

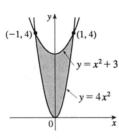

11. $A = \int_{-1}^{3} \left[(2y + 3) - y^2 \right] dy = \left[y^2 + 3y - \frac{1}{3}y^3 \right]_{-1}^{3} = (9 + 9 - 9) - \left(1 - 3 + \frac{1}{3} \right) = \frac{32}{3}$

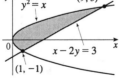

13. $A = \int_{-1}^{1} \left[(1 - y^2) - (y^2 - 1)\right] dy$
$= \int_{-1}^{1} 2 \left(1 - y^2\right) dy$
$= 4 \int_{0}^{1} \left(1 - y^2\right) dy$
$= 4 \left[y - \frac{1}{3}y^3\right]_0^1 = 4 \left(1 - \frac{1}{3}\right) = \frac{8}{3}$

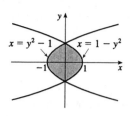

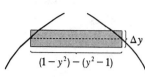

15. $A = \int_{-1}^{1} \left(\frac{2}{x^2 + 1} - x^2\right) dx$
$= 2 \int_{0}^{1} \left(\frac{2}{x^2 + 1} - x^2\right) dx$
$= 2 \left[2 \tan^{-1} x - \frac{1}{3}x^3\right]_0^1$
$= 2 \left(2 \cdot \frac{\pi}{4} - \frac{1}{3}\right) = \pi - \frac{2}{3} \approx 2.47$

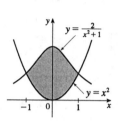

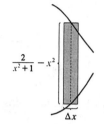

17.

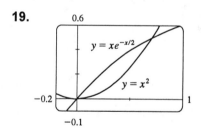

From the graph, we see that the curves intersect at $x \approx \pm 1.02$, with $2 \cos x > x^2$ on $(-1.02, 1.02)$. So the area of the region bounded by the curves is

$A \approx \int_{-1.02}^{1.02} \left(2 \cos x - x^2\right) dx = 2 \int_0^{1.02} \left(2 \cos x - x^2\right) dx$
$= 2 \left[2 \sin x - \frac{1}{3}x^3\right]_0^{1.02} \approx 2.70.$

19.

From the graph, we see that the curves intersect at $x = 0$ and $x \approx 0.70$, with $xe^{-x/2} > x^2$ on $(0, 0.70)$. So the area of the region bounded by the curves is
$A \approx \int_0^{0.70} \left(xe^{-x/2} - x^2\right) dx$
$= \left[4 \left(-\frac{1}{2}x - 1\right) e^{-x/2} - \frac{1}{3}x^3\right]_0^{0.70}$ (Formula 96 with $a = -\frac{1}{2}$)
$= \left[-5.4e^{-0.35} - \frac{1}{3}(0.70)^3\right] - (-4) \approx 0.08$

21. As in Example 4, we approximate the distance between the two cars after ten seconds using Simpson's rule with $\Delta t = 1 \text{ s} = \frac{1}{3600}$ h.
$\int_0^{10} (v_K - v_C) \, dt \approx S_{10}$
$= \frac{1}{3 \cdot 3600} \left[(0 - 0) + 4(22 - 20) + 2(37 - 32) + 4(52 - 46) + 2(61 - 54) + 4(71 - 62)\right.$
$\left. + 2(80 - 69) + 4(86 - 75) + 2(93 - 81) + 4(98 - 86) + (102 - 90)\right]$
$= \frac{1}{10,800} (242) = \frac{121}{5400}$ mi

So after 10 seconds, Kelly's car is about $\frac{121}{5400}$ mi $\left(5280\frac{\text{ft}}{\text{mi}}\right) \approx 118$ ft ahead of Chris's.

23. If x = distance from left end of pool and $w = w(x)$ = width at x, then Simpson's Rule with $n = 8$ and
$\Delta x = 2$ gives
$$\text{Area} = \int_0^{16} w\,dx \approx \tfrac{2}{3}\left[0 + 4(6.2) + 2(7.2) + 4(6.8) + 2(5.6) + 4(5.0) + 2(4.8) + 4(4.8) + 0\right]$$
$$= \tfrac{2}{3}(126.4) \approx 84 \text{ m.}$$

25. $\cos x = \sin 2x = 2\sin x \cos x \iff 2\sin x = 1$ or $\cos x = 0 \iff x = \tfrac{\pi}{6}$ or $\tfrac{\pi}{2}$.

$A = \int_0^{\pi/6} (\cos x - \sin 2x)\,dx + \int_{\pi/6}^{\pi/2} (\sin 2x - \cos x)\,dx$

$= \left[\sin x + \tfrac{1}{2}\cos 2x\right]_0^{\pi/6} + \left[-\tfrac{1}{2}\cos 2x - \sin x\right]_{\pi/6}^{\pi/2}$

$= \tfrac{1}{2} + \tfrac{1}{2}\cdot\tfrac{1}{2} - \left(0 + \tfrac{1}{2}\cdot 1\right)$

$\qquad + \left[-\tfrac{1}{2}\cdot(-1) - 1\right] - \left(-\tfrac{1}{2}\cdot\tfrac{1}{2} - \tfrac{1}{2}\right)$

$= \tfrac{1}{2}$

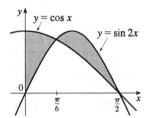

27. By symmetry of the ellipse about the x- and y-axes,

$$A = 4\int_0^a y\,dx = 4\int_{\pi/2}^0 b\sin\theta\,(-a\sin\theta)\,d\theta \qquad \left[\begin{array}{l} x = a\cos\theta = 0 \implies \theta = \tfrac{\pi}{2} \text{ and} \\ x = a\cos\theta = a \implies \theta = 0 \end{array}\right]$$

$$= 4ab\int_0^{\pi/2} \sin^2\theta\,d\theta = 4ab\int_0^{\pi/2} \tfrac{1}{2}(1 - \cos 2\theta)\,d\theta$$

$$= 2ab\left[\theta - \tfrac{1}{2}\sin 2\theta\right]_0^{\pi/2} = 2ab\left(\tfrac{\pi}{2}\right) = \pi ab$$

29.

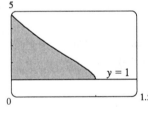

$x = \cos t,\ y = e^t,\ 0 \le t \le \tfrac{\pi}{2}$.

$A = \int_0^1 (y - 1)\,dx = \int_{\pi/2}^0 (e^t - 1)(-\sin t)\,dt$

$= \int_0^{\pi/2} (e^t\sin t - \sin t)\,dt = \tfrac{1}{2}\left(e^{\pi/2} - 1\right)$

31. By symmetry, the area of the region enclosed by the loop is
twice the area above the x-axis inside the loop. The top half of
the loop is described by $x = t^2$, $y = t^3 - 3t$, $-\sqrt{3} \le t \le 0$,
so, using the Substitution Rule with $y = t^3 - 3t$ and
$dx = 2t\,dt$, we find that

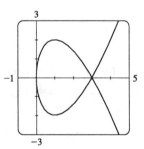

area $= 2\int_0^3 y\,dx = 2\int_0^{-\sqrt{3}} (t^3 - 3t)\,2t\,dt = 4\int_0^{-\sqrt{3}} (t^4 - 3t^2)\,dt$

$= 4\left[\tfrac{1}{5}t^5 - t^3\right]_0^{-\sqrt{3}} = 4\left[\tfrac{1}{5}\left(-3^{1/2}\right)^5 - \left(-3^{1/2}\right)^3\right]$

$= 4\left[\tfrac{1}{5}\left(-9\sqrt{3}\right) - \left(-3\sqrt{3}\right)\right] = \tfrac{24}{5}\sqrt{3} \approx 8.31.$

Section 6.1 More about Areas

33. We first assume that $c > 0$, since c can be replaced by $-c$ in both equations without changing the graphs, and if $c = 0$ the curves do not enclose a region. We see from the graph that the enclosed area lies between $x = -c$ and $x = c$, and by symmetry, it is equal to twice the area under the top half of the graph (whose equation is $y = c^2 - x^2$).

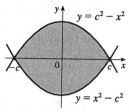

The enclosed area is $2 \int_{-c}^{c} \left(c^2 - x^2\right) dx = 4 \int_{0}^{c} \left(c^2 - x^2\right) dx = 4 \left[c^2 x - \frac{1}{3}x^3\right]_0^c = 4 \left(c^3 - \frac{1}{3}c^3\right) = \frac{8}{3}c^3$, which is equal to 576 when $c = \sqrt[3]{216} = 6$. Note that $c = -6$ is another solution, since the graphs are the same.

35. By the symmetry of the problem, we consider only the first quadrant where $y = x^2 \Longrightarrow x = \sqrt{y}$. We are looking for a number b such that $\int_0^4 x \, dy = 2 \int_0^b x \, dy \Longrightarrow \int_0^4 \sqrt{y} \, dy = 2 \int_0^b \sqrt{y} \, dy \Longrightarrow$ $\frac{2}{3} \left[y^{3/2}\right]_0^4 = \frac{4}{3} \left[y^{3/2}\right]_0^b \Longrightarrow \frac{2}{3}(8 - 0) = \frac{4}{3}\left(b^{3/2} - 0\right) \Longrightarrow b^{3/2} = 4 \Longrightarrow b = 4^{2/3} \approx 2.52$.

37. The area under the graph of f from 0 to t is equal to $\int_0^t f(x) \, dx$, so the requirement is that $\int_0^t f(x) \, dx = t^3$ for all t. We differentiate this equation with respect to t (with the help of FTC1) to get $f(t) = 3t^2$. This function is positive and continuous, as required.

39. The curve and the line will determine a region when they intersect at two or more points. So we solve the equation $x/\left(x^2 + 1\right) = mx \Longrightarrow x = 0$ or $mx^2 + m - 1 = 0 \Longrightarrow x = 0$ or $x^2 = \dfrac{1 - m}{m} \Longrightarrow x = 0$ or $x = \pm\sqrt{\dfrac{1}{m} - 1}$. Note that if $m = 1$, this has only the solution $x = 0$, and no region is determined. But if $1/m - 1 > 0 \Longleftrightarrow 1/m > 1 \Longleftrightarrow 0 < m < 1$, then there are two solutions. [Another way of seeing this is to observe that the slope of the tangent to $y = x/\left(x^2 + 1\right)$ at the origin is $y' = 1$ and therefore we must have $0 < m < 1$.] Note that we cannot just integrate between the positive and negative roots, since the curve and the line cross at the origin. Since mx and $x/\left(x^2 + 1\right)$ are both odd functions, the total area is twice the area between the curves on the interval $\left[0, \sqrt{1/m - 1}\right]$. So the total area enclosed is

$$2 \int_0^{\sqrt{1/m-1}} \left[\frac{x}{x^2 + 1} - mx\right] dx = 2 \left[\frac{1}{2} \ln\left(x^2 + 1\right) - \frac{1}{2}mx^2\right]_0^{\sqrt{1/m-1}}$$
$$= \left[\ln\left(1/m - 1 + 1\right) - m\left(1/m - 1\right)\right] - (\ln 1 - 0)$$
$$= \ln\left(1/m\right) + m - 1 = m - \ln m - 1$$

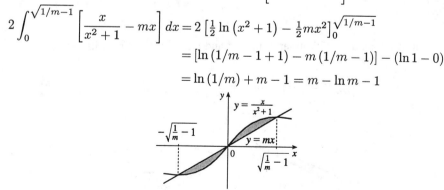

Section 6.2 Volumes

1. A cross-section is circular with radius x^2, so its area is $A(x) = \pi (x^2)^2$.
$$V = \int_0^1 A(x)\,dx = \int_0^1 \pi (x^2)^2\,dx$$
$$= \pi \int_0^1 x^4\,dx = \pi \left[\tfrac{1}{5}x^5\right]_0^1 = \tfrac{\pi}{5}$$

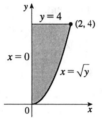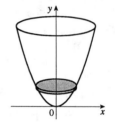

3. A cross-section is circular with radius $\sqrt{y}$, so its area is $A(y) = \pi \left(\sqrt{y}\right)^2$.
$$V = \int_0^4 A(y)\,dy = \int_0^4 \pi \left(\sqrt{y}\right)^2\,dy$$
$$= \pi \int_0^4 y\,dy = \pi \left[\tfrac{1}{2}y^2\right]_0^4 = 8\pi$$

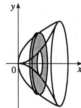

5. A cross-section is an annulus with inner radius x^2 and outer radius $\sqrt{x}$, so its area is $A(x) = \pi \left(\sqrt{x}\right)^2 - \pi (x^2)^2$.
$$V = \int_0^1 A(x)\,dx = \pi \int_0^1 \left[\left(\sqrt{x}\right)^2 - (x^2)^2\right]dx = \pi \int_0^1 (x - x^4)\,dx = \pi \left[\tfrac{1}{2}x^2 - \tfrac{1}{5}x^5\right]_0^1$$
$$= \pi \left(\tfrac{1}{2} - \tfrac{1}{5}\right) = \tfrac{3\pi}{10}$$

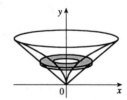

7. A cross-section is an annulus with inner radius y^2 and outer radius $2y$, so its area is $A(y) = \pi (2y)^2 - \pi (y^2)^2$.
$$V = \int_0^2 A(y)\,dy = \pi \int_0^2 \left[(2y)^2 - (y^2)^2\right]dy = \pi \int_0^2 (4y^2 - y^4)\,dy = \pi \left[\tfrac{4}{3}y^3 - \tfrac{1}{5}y^5\right]_0^2$$
$$= \pi \left(\tfrac{32}{3} - \tfrac{32}{5}\right) = \tfrac{64\pi}{15}$$

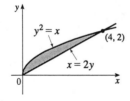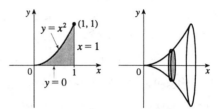

9. A cross-section is an annulus with inner radius $2 - 1$ and outer radius $2 - x^4$, so its area is
$A(x) = \pi (2 - x^4)^2 - \pi (2 - 1)^2$.
$V = \int_{-1}^{1} A(x)\, dx = \pi \int_{-1}^{1} \left[(2 - x^4)^2 - (2 - 1)^2 \right] dx = 2\pi \int_0^1 (3 - 4x^4 + x^8)\, dx$
$= 2\pi \left[3x - \frac{4}{5}x^5 + \frac{1}{9}x^9 \right]_0^1 = 2\pi \left(3 - \frac{4}{5} + \frac{1}{9} \right) = \frac{208}{45}\pi$

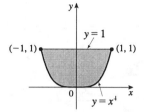

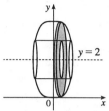

11.

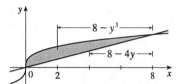

A cross-section is an annulus with inner
radius $8 - 4y$ and outer radius $8 - y^3$, so its
area is $A(y) = \pi (8 - y^3)^2 - \pi (8 - 4y)^2$.
$V = \int_0^2 A(y)\, dy$
$= \pi \int_0^2 \left[(8 - y^3)^2 - (8 - 4y)^2 \right] dy$
$= \pi \int_0^2 (-16y^3 + y^6 + 64y - 16y^2)\, dy$
$= \pi \left[-4y^4 + \frac{1}{7}y^7 + 32y^2 - \frac{16}{3}y^3 \right]_0^2$
$= \pi \left(-64 + \frac{128}{7} + 128 - \frac{128}{3} \right) = \frac{832}{21}\pi$

13.

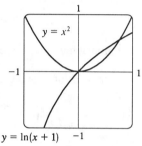

$y = x^2$ and $y = \ln(x + 1)$ intersect at $x = 0$
and at $x \approx 0.747$. $V \approx$
$\pi \int_0^{0.747} \left[[\ln(x + 1)]^2 - (x^2)^2 \right] dx \approx 0.132$

15. We use Simpson's Rule with $n = 10$ and $\Delta x = 1.5$.
$V = \int_0^{15} A(x)\, dx \approx S_{10}$
$= \frac{1.5}{3} \left[0 + 4(18) + 2(58) + 4(79) + 2(94) + 4(106) + 2(117) + 4(128) + 2(63) + 4(39) + 0 \right]$
$= \frac{1}{2} (2144) \approx 1072$ cm^3

17. $V = \pi \int_0^h \left(-\frac{r}{h}y + r \right)^2 dy = \pi \int_0^h \left[\frac{r^2}{h^2}y^2 - \frac{2r^2}{h}y + r^2 \right] dy$
$= \pi \left[\frac{r^2}{3h^2}y^3 - \frac{r^2}{h}y^2 + r^2 y \right]_0^h = \pi \left(\frac{1}{3}r^2 h - r^2 h + r^2 h \right) = \frac{1}{3}\pi r^2 h$

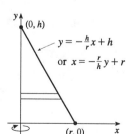

19. $x^2 + y^2 = r^2 \implies x^2 = r^2 - y^2 \implies$

$$V = \pi \int_{r-h}^{r} (r^2 - y^2)\, dy = \pi \left[r^2 y - \frac{y^3}{3} \right]_{r-h}^{r} = \pi \left(\left[r^3 - \frac{r^3}{3} \right] - \left[r^2(r-h) - \frac{(r-h)^3}{3} \right] \right)$$

$$= \pi \left(\tfrac{2}{3} r^3 - \tfrac{1}{3}(r-h)\left[3r^2 - (r-h)^2 \right] \right) = \tfrac{1}{3}\pi \left(2r^3 - (r-h)\left[3r^2 - (r^2 - 2rh + h^2) \right] \right)$$

$$= \tfrac{1}{3}\pi h^2 (3r - h), \text{ or, equivalently, } \pi h^2 \left(r - \frac{h}{3} \right)$$

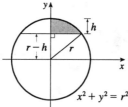

21. For a cross-section at height y, we see from similar triangles that $\dfrac{\alpha/2}{b/2} = \dfrac{h-y}{h}$, so $\alpha = b\left(1 - \dfrac{y}{h}\right)$.

Similarly, for cross-sections having $2b$ as their base and β replacing α, $\beta = 2b\left(1 - \dfrac{y}{h}\right)$. So

$$V = \int_0^h A(y)\, dy = \int_0^h \left[b\left(1 - \frac{y}{h}\right) \right] \left[2b\left(1 - \frac{y}{h}\right) \right] dy$$

$$= \int_0^h 2b^2 \left(1 - \frac{y}{h}\right)^2 dy = 2b^2 \int_0^h \left(1 - \frac{2y}{h} + \frac{y^2}{h^2}\right) dy = 2b^2 \left[y - \frac{y^2}{h} + \frac{y^3}{3h^2} \right]_0^h$$

$$= 2b^2 \left[h - h + \tfrac{1}{3}h \right] = \tfrac{2}{3}b^2 h \quad \left(= \tfrac{1}{3}Bh \text{ where } B \text{ is the area of the base, as with any pyramid.} \right)$$

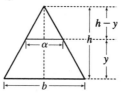

23. A cross-section at height z is a triangle similar to the base, so its

area is $A(z) = \tfrac{1}{2} \cdot 3\left(\dfrac{5-z}{5}\right) \cdot 4\left(\dfrac{5-z}{5}\right) = 6\left(1 - \dfrac{z}{5}\right)^2$, so

$V = \int_0^5 A(z)\, dz = 6\int_0^5 (1 - z/5)^2\, dz = 6\left[(-5)\tfrac{1}{3}\left(1 - \tfrac{1}{5}z\right)^3 \right]_0^5 = -10(-1) = 10 \text{ cm}^3.$

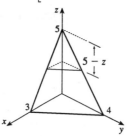

25. If l is a leg of the isosceles right triangle and $2y$ is the hypotenuse, then $l^2 + l^2 = (2y)^2 \Longrightarrow l = \sqrt{2}y$.

$V = \int_{-2}^{2} A(x)\,dx = 2\int_{0}^{2} A(x)\,dx = 2\int_{0}^{2} \frac{1}{2}\left(\sqrt{2}y\right)^2 dx = 2\int_{0}^{2} y^2\,dx = \frac{1}{2}\int_{0}^{2}\left(36 - 9x^2\right) dx$

$= \frac{9}{2}\int_{0}^{2}\left(4 - x^2\right) dx = \frac{9}{2}\left[4x - \frac{1}{3}x^3\right]_{0}^{2} = \frac{9}{2}\left(8 - \frac{8}{3}\right) = 24$

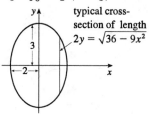

typical cross-
section of length
$2y = \sqrt{36 - 9x^2}$

27.

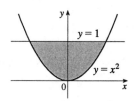

The square has area $A(y) = \left(2\sqrt{y}\right)^2 = 4y$, so

$V = \int_{0}^{1} A(y)\,dy = \int_{0}^{1} 4y\,dy = \left[2y^2\right]_{0}^{1} = 2.$

29. Assume that the base of each isosceles triangle lies in the base of S. Then its
area is $A(x) = \frac{1}{2}bh = \frac{1}{2}\left(1 - \frac{1}{2}x\right)\left(1 - \frac{1}{2}x\right) = \frac{1}{2}\left(1 - \frac{1}{2}x\right)^2$, and the volume is

$V = \int_{0}^{2} A(x)\,dx = \frac{1}{2}\int_{0}^{2}\left(1 - \frac{1}{2}x\right)^2 dx = \frac{1}{2}\left[\frac{2}{3}\left(\frac{1}{2}x - 1\right)^3\right]_{0}^{2} = \frac{1}{3}.$

31. (a) The torus is obtained by rotating the circle $(x - R)^2 + y^2 = r^2$ about the
y-axis. Solving for y, we see that the right half of the circle is given by
$x = R + \sqrt{r^2 - y^2} = f(y)$ and the left half by $x = R - \sqrt{r^2 - y^2} = g(y)$. So
$V = \pi\int_{-r}^{r}\left([f(y)]^2 - [g(y)]^2\right) dy = 2\pi\int_{0}^{r} 4R\sqrt{r^2 - y^2}\,dy = 8\pi R\int_{0}^{r}\sqrt{r^2 - y^2}\,dy.$

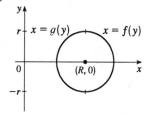

(b) Observe that the integral represents a quarter of the area of a circle with radius r, so
$8\pi R\int_{0}^{r}\sqrt{r^2 - y^2}\,dy = 8\pi R \cdot \frac{1}{4}\left(\pi r^2\right) = 2\pi^2 r^2 R.$

33. (a) Volume$(S_1) = \int_{0}^{h} A(z)\,dz = $ Volume(S_2) since the cross-sectional area $A(z)$ at height z is the
same for both solids.

(b) By Cavalieri's Principle, the volume of the cylinder in the figure is the same as that of a right
circular cylinder with radius r and height h, that is, $\pi r^2 h$.

35. The volume is obtained by rotating the area common to two circles of radius r, as shown. The volume of the right half is

$$V_{\text{right}} = \pi \int_0^{r/2} y^2 \, dx = \pi \int_0^{r/2} \left[r^2 - \left(\tfrac{1}{2}r + x \right)^2 \right] dx = \pi \left[r^2 x - \tfrac{1}{3} \left(\tfrac{1}{2}r + x \right)^3 \right]_0^{r/2}$$

$$= \pi \left[\left(\tfrac{1}{2}r^3 - \tfrac{1}{3}r^3 \right) - \left(0 - \tfrac{1}{24}r^3 \right) \right] = \tfrac{5}{24}\pi r^3.$$

So by symmetry, the total volume is twice this, or $\tfrac{5}{12}\pi r^3$.

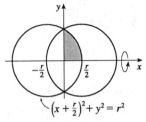

$$\left(x + \tfrac{r}{2} \right)^2 + y^2 = r^2$$

Another Solution: We observe that the volume is the twice the volume of a cap of a sphere, so we can use the formula from Exercise 19 with $h = \tfrac{1}{2}r$: $V = 2 \cdot \tfrac{1}{3}\pi r h^2 (3r - h) = \tfrac{2}{3}\pi \left(\tfrac{1}{2}r \right)^2 (3r - \tfrac{1}{2}r) = \tfrac{5}{12}\pi r^3.$

37. Take the x-axis to be the axis of the cylindrical hole of radius r. A quarter of the cross-section through y, perpendicular to the y-axis, is the rectangle shown. Using Pythagoras twice, we see that the dimensions of this rectangle are $x = \sqrt{R^2 - y^2}$ and $z = \sqrt{r^2 - y^2}$, so $\tfrac{1}{4}A(y) = xz = \sqrt{r^2 - y^2}\sqrt{R^2 - y^2}$, and

$$V = \int_{-r}^{r} A(y) \, dy = \int_{-r}^{r} 4\sqrt{r^2 - y^2}\sqrt{R^2 - y^2} \, dy = 8 \int_0^r \sqrt{r^2 - y^2}\sqrt{R^2 - y^2} \, dy$$

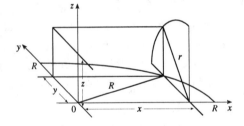

39.

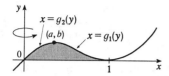

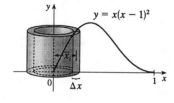

If we were to use the "washer method," we would first have to locate the local maximum point (a, b) of $y = x(x-1)^2$ using the methods of Chapter 4. Then we would have to solve the equation $y = x(x-1)^2$ for x in terms of y to obtain the functions $x = g_1(y)$ and $x = g_2(y)$ shown in the figure above. This step would be difficult because it involves the cubic formula. Finally we would find the volume using

$$V = \pi \int_0^b \left\{ [g_1(y)]^2 - [g_2(y)]^2 \right\} dy.$$

Instead, we use cylindrical shells. As in Example 7, we rotate an approximating rectangle with width Δx about the y-axis, to get a cylindrical shell whose average radius is $\bar{x}_i$ and whose volume is $2\pi \bar{x}_i \left[\bar{x}_i (\bar{x}_i - 1)^2 \right]$.

So the total volume is

$$V = \lim_{n \to \infty} \sum_{i=1}^{n} 2\pi \bar{x}_i \left[\bar{x}_i (\bar{x}_i - 1)^2 \right] \Delta x = \int_0^1 2\pi x \left[x(x-1)^2 \right] dx = 2\pi \int_0^1 \left(x^4 - 2x^3 + x^2 \right) dx$$

$$= 2\pi \left[\frac{x^5}{5} - 2\frac{x^4}{4} + \frac{x^3}{3} \right]_0^1 = \frac{\pi}{15}.$$

41.

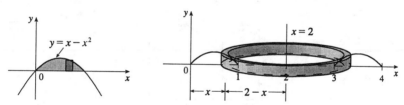

$$V = \int_0^1 (\text{circumference}) \, (\text{height}) \, (\text{thickness}) \, dx$$

$$= \int_0^1 [2\pi (2 - x)] \left(x - x^2 \right) dx$$

$$= 2\pi \int_0^1 \left(x^3 - 3x^2 + 2x \right) dx$$

$$= 2\pi \left[\tfrac{1}{4} x^4 - x^3 + x^2 \right]_0^1 = 2\pi \left(\tfrac{1}{4} \right)$$

$$= \frac{\pi}{2}$$

Section 6.3 Arc Length

1. $y = 2x + 1 \Longrightarrow L = \int_{-1}^{3} \sqrt{1 + \left(\dfrac{dy}{dx}\right)^2}\, dx = \int_{-1}^{3} \sqrt{1 + 2^2}\, dx = \sqrt{5}\,[3 - (-1)] = 4\sqrt{5}.$

The arc length can be calculated using the distance formula, since the curve is a line segment, so

$L = [\text{distance from } (-1, -1) \text{ to } (3, 7)] = \sqrt{[3 - (-1)]^2 + [7 - (-1)]^2} = \sqrt{80} = 4\sqrt{5}.$

3. $x = e^t \cos t,\ y = e^t \sin t,\ 0 \le t \le \pi.$

$\left(\dfrac{dx}{dt}\right)^2 + \left(\dfrac{dy}{dt}\right)^2 = [e^t (\cos t - \sin t)]^2 + [e^t (\sin t + \cos t)]^2 = e^{2t}\left(2\cos^2 t + 2\sin^2 t\right) = 2e^{2t}.$

$L = \int_0^\pi \sqrt{2}\,e^t\, dt = \sqrt{2}\,(e^\pi - 1).$

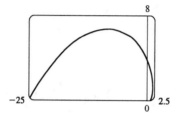

5.

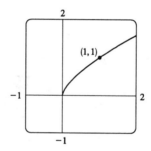

$x = y^{3/2} \Longrightarrow 1 + (dx/dy)^2 = 1 + \left(\tfrac{3}{2}y^{1/2}\right)^2 = 1 + \tfrac{9}{4}y.$

$L = \int_0^1 \sqrt{1 + \tfrac{9}{4}y}\, dy = \tfrac{4}{9} \cdot \tfrac{2}{3} \left[\left(1 + \tfrac{9}{4}y\right)^{3/2}\right]_0^1 = \tfrac{8}{27}\left(\tfrac{13\sqrt{13}}{8} - 1\right) = \tfrac{13\sqrt{13} - 8}{27}.$

7. $y = x^3 \Longrightarrow 1 + (y')^2 = 1 + \left(3x^2\right)^2 = 1 + 9x^4 \Longrightarrow L = \int_0^1 \sqrt{1 + 9x^4}\, dx.$
Let $f(x) = \sqrt{1 + 9x^4}$. Then by Simpson's Rule with $n = 10$,
$L \approx \tfrac{1/10}{3}[f(0) + 4f(0.1) + 2f(0.2) + 4f(0.3) + \cdots + 2f(0.8) + 4f(0.9) + f(1)] \approx 1.548.$

9. $y = \sin x \Longrightarrow 1 + (y')^2 = 1 + \cos^2 x \Longrightarrow L = \int_0^\pi \sqrt{1 + \cos^2 x}\, dx.$ Let $f(x) = \sqrt{1 + \cos^2 x}$. Then
$L \approx \tfrac{\pi/10}{3}\left[f(0) + 4f\left(\tfrac{\pi}{10}\right) + 2f\left(\tfrac{2\pi}{10}\right) + 4f\left(\tfrac{3\pi}{10}\right) + \cdots + 2f\left(\tfrac{8\pi}{10}\right) + 4f\left(\tfrac{9\pi}{10}\right) + f(\pi)\right] \approx 3.820.$

11. (a)

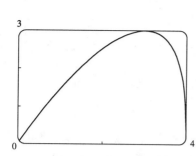

(b)

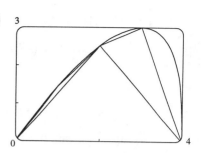

Let $f(x) = y = x\sqrt[3]{4-x}$. The polygon with one side is just the line segment joining the points $(0, f(0)) = (0, 0)$ and $(4, f(4)) = (4, 0)$, and its length is 4. The polygon with two sides joins the points $(0, 0)$, $(2, f(2)) = (2, 2\sqrt[3]{2})$ and $(4, 0)$. Its length is

$$\sqrt{(2-0)^2 + (2\sqrt[3]{2} - 0)^2} + \sqrt{(4-2)^2 + (0 - 2\sqrt[3]{2})^2} = 2\sqrt{4 + 2^{8/3}} \approx 6.43. \text{ Similarly, the}$$

inscribed polygon with four sides joins the points $(0, 0)$, $(1, \sqrt[3]{3})$, $(2, 2\sqrt[3]{2})$, $(3, 3)$, and $(4, 0)$, so its

length is $\sqrt{1 + (\sqrt[3]{3})^2} + \sqrt{1 + (2\sqrt[3]{2} - \sqrt[3]{3})^2} + \sqrt{1 + (3 - 2\sqrt[3]{2})^2} + \sqrt{1 + 9} \approx 7.50$.

(c) Using the arc length formula with $\dfrac{dy}{dx} = x\left[\dfrac{1}{3}(4-x)^{-2/3}(-1)\right] + \sqrt[3]{4-x} = \dfrac{12 - 4x}{3(4-x)^{2/3}}$, the

length of the curve is $L = \displaystyle\int_0^4 \sqrt{1 + \left(\dfrac{dy}{dx}\right)^2}\, dx = \int_0^4 \sqrt{1 + \left[\dfrac{12 - 4x}{3(4-x)^{2/3}}\right]^2}\, dx$.

(d) According to a CAS, the length of the curve is $L \approx 7.7988$. The actual value is larger than any of the approximations in part (b). This is always true, since any approximating straight line between two points on the curve is shorter than the length of the curve between the two points.

13. $x = t^3 \implies dx/dt = 3t^2$ and $y = t^4 \implies dy/dt = 4t^3 \implies$
$L = \int_0^1 \sqrt{9t^4 + 16t^6}\, dt = \int_0^1 t^2\sqrt{9 + 16t^2}\, dt = \dfrac{205}{128} - \dfrac{81\ln 3}{512} \approx 1.428$.

15. $y = \ln(\cos x) \implies y' = \dfrac{1}{\cos x}(-\sin x) = -\tan x \implies 1 + (y')^2 = 1 + \tan^2 x = \sec^2 x$.
So $L = \int_0^{\pi/4} \sec x\, dx = [\ln(\sec x + \tan x)]_0^{\pi/4} = \ln(\sqrt{2} + 1) \approx 0.881$.

17. The sine wave has amplitude 1 and period 14, since it goes through two periods in a distance of 28 in., so its equation is $y = 1\sin\left(\dfrac{2\pi}{14}x\right) = \sin\left(\dfrac{\pi}{7}x\right)$. The width w of the flat metal sheet needed to make the panel is the arc length of the sine curve from $x = 0$ to $x = 28$.
We set up the integral to evaluate w using the arc length formula with $\dfrac{dy}{dx} = \dfrac{\pi}{7}\cos\left(\dfrac{\pi}{7}x\right)$:
$L = \int_0^{28} \sqrt{1 + \left[\dfrac{\pi}{7}\cos\left(\dfrac{\pi}{7}x\right)\right]^2}\, dx = 2\int_0^{14} \sqrt{1 + \left[\dfrac{\pi}{7}\cos\left(\dfrac{\pi}{7}x\right)\right]^2}\, dx$. This integral would be very difficult to evaluate exactly, so we use a CAS, and find that $L \approx 29.36$ inches.

19. $x = a \sin \theta, y = b \cos \theta, 0 \le \theta \le 2\pi$.

$$\left(\frac{dx}{d\theta}\right)^2 + \left(\frac{dy}{d\theta}\right)^2 = (a \cos \theta)^2 + (-b \sin \theta)^2 = a^2 \cos^2 \theta + b^2 \sin^2 \theta = a^2 \left(1 - \sin^2 \theta\right) + b^2 \sin^2 \theta$$

$$= a^2 - \left(a^2 - b^2\right) \sin^2 \theta = a^2 - c^2 \sin^2 \theta = a^2 \left(1 - \frac{c^2}{a^2} \sin^2 \theta\right)$$

$$= a^2 \left(1 - e^2 \sin^2 \theta\right)$$

So $L = 4 \int_0^{\pi/2} \sqrt{a^2 \left(1 - e^2 \sin^2 \theta\right)} \, d\theta$ (by symmetry) $= 4a \int_0^{\pi/2} \sqrt{1 - e^2 \sin^2 \theta} \, d\theta$.

21. (a) Notice that $0 \le t \le 2\pi$ does not give the complete curve because $x(0) \ne x(2\pi)$. In fact, we must take $t \in [0, 4\pi]$ in order to obtain the complete curve, since the first term in each of the parametric equations has period 2π and the second has period $\frac{4\pi}{11}$, and the least common integer multiple of these two numbers is 4π.

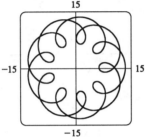

(b) For $x = 11 \cos t - 4 \cos(11t/2)$ and $y = 11 \sin t - 4 \sin(11t/2)$, we can show that $L = 11 \int_0^{4\pi} \sqrt{5 - 4 \cos(9t/2)} \, dt \approx 294$.

Section 6.4 Average Value of a Function

1. $f_{ave} = \frac{1}{3-0} \int_0^3 \left(x^2 - 2x\right) dx = \frac{1}{3} \left[\frac{1}{3}x^3 - x^2\right]_0^3 = \frac{1}{3}(9-9) = 0$

3. $f_{ave} = \frac{1}{2-0} \int_0^2 e^x dx = \frac{1}{2} [e^x]_0^2 = \frac{1}{2}\left(e^2 - 1\right) \approx 3.19$

5. (a) $f_{ave} = \frac{1}{2-0} \int_0^2 \left(4 - x^2\right) dx = \frac{1}{2} \left[4x - \frac{1}{3}x^3\right]_0^2$

$= \frac{1}{2}\left(8 - \frac{8}{3}\right) = \frac{8}{3}$

(b) $f_{ave} = f(c) \iff \frac{8}{3} = 4 - c^2 \iff c^2 = \frac{4}{3} \iff$

$c = \frac{2}{\sqrt{3}} \approx 1.15$

(c)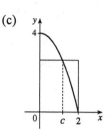

7. (a) $f_{ave} = \frac{1}{2-0} \int_0^2 \left(x^3 - x + 1\right) dx = \frac{1}{2} \left[\frac{1}{4}x^4 - \frac{1}{2}x^2 + x\right]_0^2$

$= \frac{1}{2}(4 - 2 + 2) = 2$

(b) From the graph, $f(x) = 2$ at $x \approx 1.32$.

(c)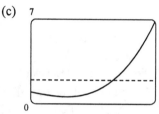

9. Since f is continuous on $[1,3]$, by the Mean Value Theorem for Integrals there exists a number c in $[1,3]$ such that $\int_1^3 f(x)\, dx = f(c)(3-1) \implies 8 = 2f(c)$; that is, there is a number c such that $f(c) = \frac{8}{2} = 4$.

11. $T_{ave} = \frac{1}{12} \int_0^{12} \left[50 + 14\sin\frac{1}{12}\pi t\right] dt = \frac{1}{12} \left[50t - 14 \cdot \frac{12}{\pi}\cos\frac{1}{12}\pi t\right]_0^{12}$

$= \frac{1}{12} \left[50 \cdot 12 + 14 \cdot \frac{12}{\pi} + 14 \cdot \frac{12}{\pi}\right] = \left(50 + \frac{28}{\pi}\right) \,^\circ\text{F} \approx 59^\circ\text{F}$

13. $\rho_{ave} = \frac{1}{8} \int_0^8 \frac{12}{\sqrt{x+1}}\, dx = \frac{3}{2} \int_0^8 (x+1)^{-1/2}\, dx = \left[3\sqrt{x+1}\right]_0^8 = 9 - 3 = 6\text{ kg/m}$

15. $V_{ave} = \frac{1}{5} \int_0^5 V(t)\, dt = \frac{1}{5} \int_0^5 \frac{5}{4\pi}\left[1 - \cos\left(\frac{2}{5}\pi t\right)\right] dt = \frac{1}{4\pi} \int_0^5 \left[1 - \cos\left(\frac{2}{5}\pi t\right)\right] dt$

$= \frac{1}{4\pi} \left[t - \frac{5}{2\pi}\sin\left(\frac{2}{5}\pi t\right)\right]_0^5 = \frac{1}{4\pi}[(5-0) - 0] = \frac{5}{4\pi} \approx 0.4\text{ L}$

17. Let $F(x) = \int_a^x f(t)\, dt$ for x in $[a,b]$. Then F is continuous on $[a,b]$ and differentiable on (a,b), so by the Mean Value Theorem there is a number c in (a,b) such that $F(b) - F(a) = F'(c)(b-a)$. But $F'(x) = f(x)$ by the Fundamental Theorem of Calculus. Therefore, $\int_a^b f(t)\, dt - 0 = f(c)(b-a)$.

Section 6.5 Applications to Physics and Engineering

1. $W = \int_a^b f(x)\, dx = \int_0^{10} (5x^2 + 1)\, dx = \left[\frac{5}{3}x^3 + x\right]_0^{10} = \frac{5000}{3} + 10 = \frac{5030}{3}$ ft-lb.

3. $10 = f(x) = kx = \frac{1}{3}k$ (4 inches $= \frac{1}{3}$ foot), so $k = 30$ lb/ft and $f(x) = 30x$. Now 6 inches $= \frac{1}{2}$ foot,

so $W = \int_0^{1/2} 30x\, dx = \left[15x^2\right]_0^{1/2} = \frac{15}{4}$ ft-lb.

5. (a) If $\int_0^{0.12} kx\, dx = 2$ J, then $2 = \left[\frac{1}{2}kx^2\right]_0^{0.12} = \frac{1}{2}k(0.0144) = 0.0072k$ and

$k = \frac{2}{0.0072} = \frac{2500}{9} \approx 277.78$. Thus, the work needed to stretch the spring from 35 cm to 40 cm is

$\int_{0.05}^{0.10} \frac{2500}{9}x\, dx = \left[\frac{1250}{9}x^2\right]_{1/20}^{1/10} = \frac{1250}{9}\left(\frac{1}{100} - \frac{1}{400}\right) = \frac{25}{24} \approx 1.04$ J.

(b) $f(x) = kx$, so $30 = \frac{2500}{9}x$ and $x = \frac{270}{2500}$ m $= 10.8$ cm

Note: In Exercises 7–11, n is the number of subintervals of length Δx, and x_i^* is a sample point in the ith subinterval $[x_{i-1}, x_i]$.

7. First notice that the exact height of the building does not matter (as long as it is more than 50 ft).
The portion of the rope from x ft to $(x + \Delta x)$ ft below the top of the building weighs $\frac{1}{2}\Delta x$ lb and must be lifted x_i^* ft, so its contribution to the total work is $\frac{1}{2}x_i^*\Delta x$ ft-lb. The total work is

$$W = \lim_{n \to \infty} \sum_{i=1}^{n} \frac{1}{2}x_i^*\Delta x = \int_0^{50} \frac{1}{2}x\, dx = \left[\frac{1}{4}x^2\right]_0^{50} = \frac{2500}{4} = 625 \text{ ft-lb.}$$

9. The work needed to lift the cable is $\lim\limits_{n \to \infty} \sum\limits_{i=1}^{n} 2x_i^*\Delta x = \int_0^{500} 2x\, dx = \left[x^2\right]_0^{500} = 250{,}000$ ft-lb. The
work needed to lift the coal is 800 lb $\cdot 500$ ft $= 400{,}000$ ft-lb. Thus, the total work required is
$250{,}000 + 400{,}000 = 650{,}000$ ft-lb.

11. A "slice" of water Δx m thick and lying at a depth of x_i^* m (where $0 \le x_i^* \le \frac{1}{2}$) has
volume $2\Delta x$ m^3, a mass of $2000\Delta x$ kg, weighs about $(9.8)(2000\Delta x) = 19{,}600\Delta x$ N,
and thus requires about $19{,}600x_i^*\Delta x$ J of work for its removal. So

$$W = \lim_{n \to \infty} \sum_{i=1}^{n} 19{,}600x_i^*\Delta x = \int_0^{1/2} 19{,}600x\, dx = \left[9800x^2\right]_0^{1/2} = 2450 \text{ J.}$$

13. (a) A rectangular "slice" of water Δx m thick and lying x ft above the bottom has width x ft and
volume $8x\Delta x$ m^3. It weighs about $(9.8 \times 10^3)(8x\Delta x)$ N, and must be lifted $(5 - x)$ m by the
pump, so the work needed is about $(9.8 \times 10^3)(5 - x)(8x\Delta x)$J. The total work required is

$W \approx \int_0^3 (9.8 \times 10^3)(5 - x)\,8x\, dx = (9.8 \times 10^3)\int_0^3 (40x - 8x^2)\, dx$

$= (9.8 \times 10^3)\left[20x^2 - \frac{8}{3}x^3\right]_0^3 = (9.8 \times 10^3)(180 - 72) = (9.8 \times 10^3)(108)$

$= 1058.4 \times 10^3 \approx 1.06 \times 10^6$ J.

(b) If only 4.7×10^5 J of work is done, then only the water above a certain level
(call it h) will be pumped out. So we use the same formula as in part (a), except
that the work is fixed, and we are trying to find the lower limit of integration:
$4.7 \times 10^5 \approx \int_h^3 \left(9.8 \times 10^3\right)(5-x)\,8x\,dx = \left(9.8 \times 10^3\right)\left[20x^2 - \frac{8}{3}x^3\right]_h^3 \iff$
$\frac{4.7}{9.8} \times 10^2 \approx 48 = (180 - 72) - \left(20h^2 - \frac{8}{3}h^3\right) \iff 2h^3 - 15h^2 + 45 = 0.$

To find the solution of this equation, we plot
$2h^3 - 15h^2 + 45$ between $h = 0$ and $h = 3$. We see
that the equation is satisfied for $h \approx 2.0$. So the depth
of water remaining in the tank is about 2.0 m.

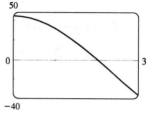

15. $V = \pi r^2 x$, so V is a function of x and P can also be regarded as a function of x. If $V_1 = \pi r^2 x_1$ and
$V_2 = \pi r^2 x_2$, then
$$W = \int_{x_1}^{x_2} F(x)\,dx = \int_{x_1}^{x_2} \pi r^2 P\left(V(x)\right)\,dx = \int_{x_1}^{x_2} P\left(V(x)\right)\,dV(x) \quad \left[\begin{array}{l} \text{Put } V(x) = \pi r^2 x, \\ \text{so } dV(x) = \pi r^2 \, dx \end{array}\right]$$
$$= \int_{V_1}^{V_2} P(V)\,dV \text{ by the Substitution Rule.}$$

17. (a) $W = \displaystyle\int_a^b F(r)\,dr = \int_a^b G\frac{m_1 m_2}{r^2}\,dr = Gm_1 m_2 \left[\frac{-1}{r}\right]_a^b = Gm_1 m_2 \left(\frac{1}{a} - \frac{1}{b}\right)$

(b) By part (a), $W = GMm \left(\dfrac{1}{R} - \dfrac{1}{R + 1{,}000{,}000}\right)$ where $M = $ mass of earth in kg,

$R = $ radius of earth in m, $m = $ mass of satellite in kg. (Note that 1000 km $= 1{,}000{,}000$ m.) Thus,

$$W = \left(6.67 \times 10^{-11}\right)\left(5.98 \times 10^{24}\right)(1000) \times \left(\frac{1}{6.37 \times 10^6} - \frac{1}{7.37 \times 10^6}\right) \approx 8.50 \times 10^9 \text{ J.}$$

Note: In Exercises 19 and 21, n is the number of subintervals of length Δx and x_i^* is a sample point in the
ith subinterval $[x_{i-1}, x_i]$.

19. In the middle of the figure in the text, draw a vertical x-axis that increases in the downward direction. The
area of the ith rectangular strip is $2\sqrt{100 - x_i^*}\,\Delta x$ and the pressure on the strip is $\rho g x_i^*$ [$\rho = 1000 \text{ kg/m}^3$
and $g = 9.8 \text{ m/s}^2$]. Thus, the hydrostatic force on the ith strip is the product $\rho g x_i^* 2\sqrt{100 - x_i^*}\,\Delta x$.

$F = \displaystyle\lim_{n \to \infty} \sum_{i=1}^n \rho g x_i^* 2\sqrt{100 - x_i^*}\,\Delta x = \int_0^{10} \rho g x \cdot 2\sqrt{100 - x^2}\,dx = 9.8 \times 10^3 \int_0^{10} \sqrt{100 - x^2}\,2x\,dx$

$= 9.8 \times 10^3 \int_{100}^0 u^{1/2}\,(-du) \quad \text{(put } u = 100 - x^2\text{)} \quad = 9.8 \times 10^3 \int_0^{100} u^{1/2}\,du$

$= 9.8 \times 10^3 \left[\frac{2}{3}u^{3/2}\right]_0^{100} = \frac{2}{3} \cdot 9.8 \times 10^6 \approx 6.5 \times 10^6 \text{ N}$

21. Place an x-axis as in Exercise 19. Using similar triangles, $\dfrac{4 \text{ ft wide}}{6 \text{ ft high}} = \dfrac{w \text{ ft wide}}{x_i^* \text{ ft high}}$, so $w = \frac{4}{6}x_i^*$ and the

area of the ith rectangular strip is $\frac{4}{6}x_i^*\Delta x$. The pressure on the strip is $\delta\,(x_i^* - 2)$ $[\delta = \rho g = 62.5 \text{ lb/ft}^3]$
and the hydrostatic force is $\delta\,(x_i^* - 2)\frac{4}{6}x_i^*\Delta x$.

$$F = \lim_{n\to\infty} \sum_{i=1}^{n} \Delta\,(x_i^* - 2)\tfrac{4}{6}x_i^*\Delta x = \int_2^6 \delta\,(x-2)\tfrac{2}{3}x\,dx = \tfrac{2}{3}\delta \int_2^6 (x^2 - 2x)\,dx = \tfrac{2}{3}\delta\left[\tfrac{1}{3}x^3 - x^2\right]_2^6$$

$$= \tfrac{2}{3}\delta\left[36 - \left(-\tfrac{4}{3}\right)\right] = \tfrac{224}{9}\delta \approx 1.56 \times 10^3 \text{ lb}$$

23. Assume that the pool is filled with water.

(a) $F = \int_0^3 \delta x 20\,dx = 20\delta\left[\tfrac{1}{2}x^2\right]_0^3 = 20\delta \cdot \tfrac{9}{2} = 90\delta \approx 5625 \text{ lb} \approx 5.63 \times 10^3 \text{ lb}$

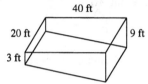

(b) $F = \int_0^9 \delta x 20\,dx = 20\delta\left[\tfrac{1}{2}x^2\right]_0^9 = 810\delta \approx 50{,}625 \text{ lb} \approx 5.06 \times 10^4 \text{ lb}.$

(c) For the first 3 ft, the length of the side is constant at 40 ft. For $3 < x \le 9$, we can use similar
triangles to find the length a: $\dfrac{a}{40} = \dfrac{9-x}{6} \Longrightarrow a = 40 \cdot \dfrac{9-x}{6}.$

$$F = \int_0^3 \delta x 40\,dx + \int_3^9 \delta x\,(40)\tfrac{9-x}{6}\,dx = 40\delta\left[\tfrac{1}{2}x^2\right]_0^3 + \tfrac{20}{3}\delta \int_3^9 (9x - x^2)\,dx$$

$$= 180\delta + \tfrac{20}{3}\delta\left[\tfrac{9}{2}x^2 - \tfrac{1}{3}x^3\right]_3^9 = 180\delta + \tfrac{20}{3}\delta\left[\left(\tfrac{729}{2} - 243\right) - \left(\tfrac{81}{2} - 9\right)\right]$$

$$= 780\delta \approx 48{,}750 \text{ lb} \approx 4.88 \times 10^4 \text{ lb}$$

(d) For any right triangle with hypotenuse on the bottom,

$$\csc\theta = \frac{\Delta x}{\text{hypotenuse}} \Longrightarrow$$

$$\text{hypotenuse} = \Delta x \csc\theta = \Delta x\frac{\sqrt{40^2 + 6^2}}{6} = \frac{\sqrt{409}}{3}\Delta x.$$

$$F = \int_3^9 \delta x 20\frac{\sqrt{409}}{3}\,dx = \tfrac{1}{3}\,(20\sqrt{409})\,\delta\left[\tfrac{1}{2}x^2\right]_3^9$$
$$= \tfrac{1}{3}\cdot 10\sqrt{409}\delta\,(81 - 9)$$
$$\approx 303{,}356 \text{ lb} \approx 3.03 \times 10^5 \text{ lb}$$

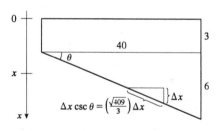

25. $m_1 = 4,\ m_2 = 8;\ P_1\,(-1,2),\ P_2\,(2,4).\ m = \sum_{i=1}^{2} m_i = m_1 + m_2 = 12.$

$M_x = \sum_{i=1}^{2} m_i y_i = 4\cdot 2 + 8\cdot 4 = 40;\ M_y = \sum_{i=1}^{2} m_i x_i = 4\cdot(-1) + 8\cdot 2 = 12;\ \overline{x} = M_y/m = 1$ and

$\overline{y} = M_x/m = \tfrac{10}{3}$, so the center of mass of the system is $(\overline{x}, \overline{y}) = \left(1, \tfrac{10}{3}\right).$

27. $A = \int_0^1 e^x \, dx = [e^x]_0^1 = e - 1,$

$\bar{x} = \dfrac{1}{A} \int_0^1 x e^x \, dx = \dfrac{1}{e-1} [x e^x - e^x]_0^1$ (integration by parts) $= \dfrac{1}{e-1} [0 - (-1)] = \dfrac{1}{e-1},$

$\bar{y} = \dfrac{1}{A} \int_0^1 \tfrac{1}{2} (e^x)^2 \, dx = \dfrac{1}{e-1} \cdot \dfrac{1}{4} [e^{2x}]_0^1 = \dfrac{1}{4(e-1)} (e^2 - 1) = \dfrac{e+1}{4}.$

$(\bar{x}, \bar{y}) = \left(\dfrac{1}{e-1}, \dfrac{e+1}{4} \right) \approx (0.58, 0.93).$

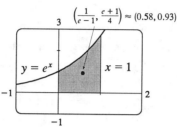

29. By symmetry, $\bar{x} = 0$. $A = 2 \int_0^{\pi/4} \cos 2x \, dx = [\sin 2x]_0^{\pi/4} = 1.$

$\bar{y} = \dfrac{1}{A} \int_{-\pi/4}^{\pi/4} \dfrac{1}{2} \cos^2 2x \, dx = \int_0^{\pi/4} \cos^2 2x \, dx = \dfrac{1}{2} \int_0^{\pi/4} (1 + \cos 4x) \, dx = \dfrac{1}{2} \left[x + \dfrac{1}{4} \sin 4x \right]_0^{\pi/4}$

$= \tfrac{1}{2} \left(\tfrac{\pi}{4} + \tfrac{1}{4} \cdot 0 \right) = \tfrac{\pi}{8}. \; (\bar{x}, \bar{y}) = \left(0, \tfrac{\pi}{8} \right).$

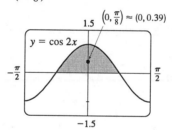

31. By symmetry, $M_y = 0$ and $\bar{x} = 0$. $A = \tfrac{1}{2} bh = \tfrac{1}{2} \cdot 2 \cdot 2 = 2.$

$M_x = 2\rho \int_0^1 \tfrac{1}{2} (2 - 2x)^2 \, dx = \left(2 \cdot 1 \cdot \tfrac{1}{2} \cdot 2^2 \right) \int_0^1 (1 - x)^2 \, dx = 4 \left[-\tfrac{1}{3} (1 - x)^3 \right]_0^1 = 4 \cdot \tfrac{1}{3} = \tfrac{4}{3}.$

$\bar{y} = \dfrac{1}{\rho A} M_x = \dfrac{1}{1 \cdot 2} \cdot \dfrac{4}{3} = \dfrac{2}{3}. \; (\bar{x}, \bar{y}) = \left(0, \tfrac{2}{3} \right).$

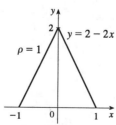

33. (a)

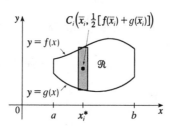

Suppose the region lies between two curves $y = f(x)$ and $y = g(x)$ where
$f(x) \geq g(x)$, as illustrated in the figure. Use n subintervals determined by points x_i with
$a = x_0 < x_1 < \cdots < x_n = b$ and choose $x_i^* = \overline{x}_i$ to be the midpoint of the ith subinterval;
that is, $\overline{x}_i = \frac{1}{2}(x_{i-1} + x_i)$. Then the centroid of the ith approximating rectangle R_i is
its center $C_i = \left(\overline{x}_i, \frac{1}{2}[f(\overline{x}_i) + g(\overline{x}_i)]\right)$. Its area is $[f(\overline{x}_i) - g(\overline{x}_i)]\Delta x$, so its mass is
$\rho[f(\overline{x}_i) - g(\overline{x}_i)]\Delta x$. Thus, $M_y(R_i) = \rho[f(\overline{x}_i) - g(\overline{x}_i)]\Delta x \cdot \overline{x}_i = \rho\overline{x}_i[f(\overline{x}_i) - g(\overline{x}_i)]\Delta x$
and $M_x(R_i) = \rho[f(\overline{x}_i) - g(\overline{x}_i)]\Delta x \cdot \frac{1}{2}[f(\overline{x}_i) + g(\overline{x}_i)] = \rho \cdot \frac{1}{2}\left\{[f(\overline{x}_i)]^2 - [g(\overline{x}_i)]^2\right\}\Delta x$.
Summing over i and taking the limit as $n \to \infty$, we get

$$M_y = \lim_{n\to\infty} \sum_{i=1}^{n} \rho\overline{x}_i[f(\overline{x}_i) - g(\overline{x}_i)]\Delta x = \rho\int_a^b x[f(x) - g(x)]\,dx \text{ and}$$

$$M_x = \lim_{n\to\infty} \sum_{i=1}^{n} \rho \cdot \frac{1}{2}\left[f(\overline{x}_i)^2 - g(\overline{x}_i)^2\right]\Delta x = \rho\int_a^b \frac{1}{2}\left\{[f(x)]^2 - [g(x)]^2\right\}\,dx.$$

Thus, $\overline{x} = \dfrac{M_y}{m} = \dfrac{M_y}{\rho A} = \dfrac{1}{A}\int_a^b x[f(x) - g(x)]\,dx$ and

$$\overline{y} = \frac{M_x}{m} = \frac{M_x}{\rho A} = \frac{1}{A}\int_a^b \frac{1}{2}\left\{[f(x)]^2 - [g(x)]^2\right\}\,dx.$$

(b)

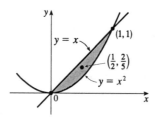

The region is sketched in the figure. We take $f(x) = x$, $g(x) = x^2$, $a = 0$, and $b = 1$ in Formulas 11.
First we note that the area of the region is $A = \int_0^1 (x - x^2)\,dx = \left[\frac{1}{2}x^2 - \frac{1}{3}x^3\right]_0^1 = \frac{1}{6}$. Therefore,

$$\overline{x} = \frac{1}{A}\int_0^1 x[f(x) - g(x)]\,dx = \frac{1}{1/6}\int_0^1 x(x - x^2) = 6\int_0^1 (x^2 - x^3)\,dx = 6\left[\frac{1}{3}x^3 - \frac{1}{4}x^4\right]_0^1$$

$$= \tfrac{1}{2} \text{ and}$$

$$\overline{y} = \frac{1}{A}\int_0^1 \frac{1}{2}\left\{[f(x)]^2 - [g(x)]^2\right\}\,dx = \frac{1}{1/6}\int_0^1 \frac{1}{2}(x^2 - x^4)\,dx = 3\left[\frac{1}{3}x^3 - \frac{1}{5}x^5\right]_0^1 = \tfrac{2}{5}.$$

The centroid is $\left(\frac{1}{2}, \frac{2}{5}\right)$.

Section 6.6 Applications to Economics and Biology

1. $C(2000) = C(0) + \int_0^{2000} C'(x)\,dx = 1{,}500{,}000 + \int_0^{2000} \left(0.006x^2 - 1.5x + 8\right) dx$

$= 1{,}500{,}000 + \left[0.002x^3 - 0.75x^2 + 8x\right]_0^{2000} = \$14{,}516{,}000$

3. $C(5000) - C(3000) = \int_{3000}^{5000} \left(140 - 0.5x + 0.012x^2\right) dx = \left[140x - 0.25x^2 + 0.004x^3\right]_{3000}^{5000}$

$= 494{,}450{,}000 - 106{,}170{,}000 = \$388{,}280{,}000$

5. $p(x) = 20 = \frac{1000}{x+20} \implies x + 20 = 50 \implies x = 30.$

Consumer surplus $= \int_0^{30} [p(x) - 20]\,dx = \int_0^{30} \left(\frac{1000}{x+20} - 20\right) dx = \left[1000\ln(x+20) - 20x\right]_0^{30}$

$= 1000\ln\left(\frac{50}{20}\right) - 600 \approx \$316.29.$

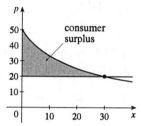

7. $P = p_S(x) = 10 = 5 + \frac{1}{10}\sqrt{x} \implies 50 = \sqrt{x} \implies x = 2500.$

Producer surplus $= \int_0^{2500} [P - p_S(x)]\,dx = \int_0^{2500} \left(10 - 5 - \frac{1}{10}\sqrt{x}\right) dx = \left[5x - \frac{1}{15}x^{3/2}\right]_0^{2500}$

$\approx \$4166.67$

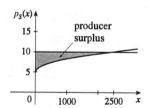

9. The demand function is linear, with slope $\frac{-10}{100}$ and $p(1000) = 450$. So its equation is

$p - 450 = -\frac{1}{10}(x - 1000)$ or $p = -\frac{1}{10}x + 550.$ A selling price of $\$400 \implies 400 = -\frac{1}{10}x + 550 \implies$

$x = 1500.$ Consumer surplus $= \int_0^{1500} \left(550 - \frac{1}{10}x - 400\right) dx = \left[150x - \frac{1}{20}x^2\right]_0^{1500} = \$112{,}500$

11. $F = \dfrac{\pi P R^4}{8\eta\ell} = \dfrac{\pi(4000)(0.008)^4}{8(0.027)(2)} \approx 1.19 \times 10^{-4}\ \text{cm}^3/\text{s}$

13. $\int_0^{12} c(t)\,dt = \int_0^{12} \frac{1}{4}t(12 - t)\,dt = \left[\frac{3}{2}t^2 - \frac{1}{12}t^3\right]_0^{12} = (216 - 144) = 72\ \text{mg} \cdot \text{s/L}.$ Therefore,

$F = A/72 = \frac{8}{72} = \frac{1}{9}\ \text{L/s} = \frac{60}{9}\ \text{L/min}.$

Section 6.7 Probability

1. (a) $\int_{100}^{200} f(t)\,dt$ is the probability that a randomly chosen battery will have a lifetime of between 100 and 200 hours.

(b) $\int_{200}^{\infty} f(t)\,dt$ is the probability that a randomly chosen battery will have a lifetime of at least 200 hours.

3. (a) In general, we must satisfy the two conditions that are mentioned before Example 1 — namely, (1) $f(x) \geq 0$ for all x, and (2) $\int_{-\infty}^{\infty} f(x)\,dx = 1$. Clearly, condition (1) is satisfied. For condition (2), we see that $\int_{-\infty}^{\infty} f(x)\,dx = \int_{0}^{10} 0.1\,dx = \left[\frac{1}{10}x\right]_{0}^{10} = 1$. Thus, $f(x)$ is a probability density function.

(b) Since all the numbers between 0 and 10 are equally likely to be selected, we expect the mean to be halfway between the endpoints of the interval; that is, $x = 5$.
$\mu = \int_{-\infty}^{\infty} xf(x)\,dx = \int_{0}^{10} x(0.1)\,dx = \left[\frac{1}{20}x^2\right]_{0}^{10} = \frac{100}{20} = 5$, as expected.

5. We need to find m so that $\int_{m}^{\infty} f(t)\,dt = \frac{1}{2} \implies \lim_{x\to\infty} \int_{m}^{x} \frac{1}{5}e^{-t/5}dt = \frac{1}{2} \implies$
$\lim_{x\to\infty} \left[\frac{1}{5}(-5)e^{-t/5}\right]_{m}^{x} = \frac{1}{2} \implies (-1)\left(0 - e^{-m/5}\right) = \frac{1}{2} \implies e^{-m/5} = \frac{1}{2} \implies -m/5 = \ln\frac{1}{2} \implies$
$m = -5\ln\frac{1}{2} = 5\ln 2 \approx 3.47$ min.

7. We use an exponential density function with $\mu = 2.5$ min.

(a) $P(X > 4) = \int_{4}^{\infty} f(t)\,dt = \lim_{x\to\infty} \int_{4}^{x} \frac{1}{2.5}e^{-t/2.5}\,dt = \lim_{x\to\infty} \left[-e^{-t/2.5}\right]_{4}^{x} = 0 + e^{-4/2.5} \approx 0.202$

(b) $P(0 \leq X \leq 2) = \int_{0}^{2} f(t)\,dt = \left[-e^{-t/2.5}\right]_{0}^{2} = -e^{-2/2.5} + 1 \approx 0.551$

(c) We need to find a value a so that $P(X \geq a) = 0.02$, or, equivalently, $P(0 \leq X \leq a) = 0.98 \iff$
$\int_{0}^{a} f(t)\,dt = 0.98 \iff \left[-e^{-t/2.5}\right]_{0}^{a} = 0.98 \implies -e^{-a/2.5} + 1 = 0.98 \iff e^{-a/2.5} = 0.02 \implies$

$-a/2.5 = \ln 0.02 \implies a = -2.5\ln\frac{1}{50} = 2.5\ln 50 \approx 9.78$ min ≈ 10 min. The ad should say that if you aren't served within 10 minutes, you get a free hamburger.

9. $P\left(X \geq 10\right) = \int_{10}^{\infty} \frac{1}{4.2\sqrt{2\pi}} \exp\left(-\frac{(x-9.4)^2}{2 \cdot 4.2^2}\right) dx$. To avoid the improper integral we approximate

it by the integral from 10 to 100. Thus,

$$P\left(X \geq 10\right) \approx \int_{10}^{100} \frac{1}{4.2\sqrt{2\pi}} \exp\left(-\frac{(x-9.4)^2}{2 \cdot 4.2^2}\right) dx$$

$$\approx 0.443$$

(using a calculator or computer to estimate the integral), so 44.3% of the households throw out at least 10 lb of paper a week.

11. $P\left(\mu - 2\sigma \leq X \leq \mu + 2\sigma\right) = \int_{\mu - 2\sigma}^{\mu + 2\sigma} \frac{1}{\sigma\sqrt{2\pi}} \exp\left(-\frac{(x-\mu)^2}{2\sigma^2}\right) dx$. Substituting $t = \frac{x-\mu}{\sigma}$ and

$dt = \frac{1}{\sigma} dx$ gives us

$$\int_{-2}^{2} \frac{1}{\sigma\sqrt{2\pi}} e^{-t^2/2} \left(\sigma \, dt\right) = \frac{1}{\sqrt{2\pi}} \int_{-2}^{2} e^{-t^2/2} \, dt$$

$$\approx 0.9545$$

13. If $h = L$, then

$$P = \frac{\text{area under } y = L \sin \theta}{\text{area of rectangle}}$$

$$= \frac{\int_0^\pi L \sin \theta \, d\theta}{\pi L}$$

$$= \frac{[-\cos \theta]_0^\pi}{\pi}$$

$$= \frac{-(-1) + 1}{\pi}$$

$$= \frac{2}{\pi}$$

If $h = \frac{L}{2}$, then

$$P = \frac{\text{area under } y = \frac{1}{2}L \sin \theta}{\text{area of rectangle}}$$

$$= \frac{\int_0^\pi \frac{1}{2}L \sin \theta \, d\theta}{\pi L}$$

$$= \frac{[-\cos \theta]_0^\pi}{2\pi}$$

$$= \frac{2}{2\pi}$$

$$= \frac{1}{\pi}$$

Chapter 6 Review

Concept Check

1. (a) See Section 6.1, Figure 2 and Equations 6.1.1 and 6.1.2.

(b) Instead of using "top minus bottom" and integrating from left to right, we use "right minus left" and integrate from bottom to top. See Figures 9 and 10 in Section 6.1.

2. The numerical value of the area represents the number of meters that Sue is ahead of Kathy after 1 minute.

3. See the discussion in Section 6.2, near Figures 2 and 3, ending in the Definition of Volume.

4. (a) The length of a curve is defined to be the limit of the lengths of the inscribed polygons, as described near Figure 3 in Section 6.3.

(b) See Equation 6.3.1.

(c) See Equations 6.3.2 and 6.3.3.

5. (a) See the boxed equation preceding Example 1 in Section 6.4.

(b) The Mean Value Theorem for Integrals says that there is a number c at which the value of f is exactly equal to the average value of the function, that is, $f(c) = f_{ave}$. For a geometric interpretation of the Mean Value Theorem for Integrals , see Figure 2 in Section 6.4 and the discussion which accompanies it.

6. $\int_0^6 f(x)\,dx$ represents the amount of work done. Its units are newton-meters, or joules.

7. (a) The center of mass is the point at which the plate balances horizontally.

(b) See Equations 6.5.11.

8. See Figure 3 in Section 6.6, and the discussion which precedes it.

9. (a) See the definition before Figure 6 in Section 6.6.

(b) See the discussion after Figure 6 in Section 6.6.

10. (a) $\int_0^{100} f(x)\,dx$ represents the probability that the weight of a randomly chosen female college student is less than 100 pounds.

(b) $\mu = \int_{-\infty}^{\infty} xf(x)\,dx = \int_0^{\infty} xf(x)\,dx$

Review

Exercises

1. $A = \int_0^6 \left[(12x - 2x^2) - (x^2 - 6x)\right] dx = \int_0^6 \left(18x - 3x^2\right) dx = \left[9x^2 - x^3\right]_0^6 = 9 \cdot 36 - 216 = 108$

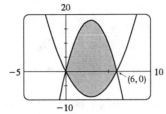

3. (a) Using the Midpoint Rule on $[0, 1]$ with $f(x) = \tan(x^2)$ and $n = 4$, we estimate

$A = \int_0^1 \tan(x^2)\, dx \approx \frac{1}{4}\left[\tan\left(\left(\frac{1}{8}\right)^2\right) + \tan\left(\left(\frac{3}{8}\right)^2\right) + \tan\left(\left(\frac{5}{8}\right)^2\right) + \tan\left(\left(\frac{7}{8}\right)^2\right)\right]$

$\approx \frac{1}{4}(1.53) \approx 0.38.$

(b) Using the Midpoint Rule on $[0, 1]$ with $f(x) = \pi \tan^2(x^2)$ (for disks) and $n = 4$, we estimate

$V = \int_0^1 f(x)\, dx \approx \frac{1}{4}\pi\left[\tan^2\left(\left(\frac{1}{8}\right)^2\right) + \tan^2\left(\left(\frac{3}{8}\right)^2\right) + \tan^2\left(\left(\frac{5}{8}\right)^2\right) + \tan^2\left(\left(\frac{7}{8}\right)^2\right)\right]$

$\approx \frac{\pi}{4}(1.114) \approx 0.87.$

5. (a) $V = \int_0^1 \pi\left[(x)^2 - (x^2)^2\right] dx = \int_0^1 \pi\left(x^2 - x^4\right) dx = \pi\left[\frac{1}{3}x^3 - \frac{1}{5}x^5\right]_0^1 = \pi\left[\frac{1}{3} - \frac{1}{5}\right] = \frac{2\pi}{15}$

(b) $V = \int_0^1 \pi\left[(\sqrt{y})^2 - y^2\right] dy = \int_0^1 \pi\left(y - y^2\right) dy = \pi\left[\frac{1}{2}y^2 - \frac{1}{3}y^3\right]_0^1 = \pi\left[\frac{1}{2} - \frac{1}{3}\right] = \frac{\pi}{6}$

(c) $V = \int_0^1 \pi\left[(2 - x^2)^2 - (2 - x)^2\right] dx = \int_0^1 \pi\left(x^4 - 5x^2 + 4x\right) dx = \pi\left[\frac{1}{5}x^5 - \frac{5}{3}x^3 + 2x^2\right]_0^1$

$= \pi\left[\frac{1}{5} - \frac{5}{3} + 2\right] = \frac{8\pi}{15}$

7. We graph the curve $x = t^3 - 3t$, $y = t^2 + t + 1$ in the parameter interval $-2.2 \le t \le 1.5$.

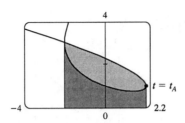

By zooming in, we estimate the coordinates of the point of intersection to be $(-2, 3)$. In fact this is exact, since both $t = -2$ and $t = 1$ give the point $(-2, 3)$. Now as in Example 4 in Section 6.1, we can evaluate the area of the loop simply by integrating $y\,dx$ between these two t-values, since this integral represents the area under the upper part of the loop for t between the first t-value and t_A, minus the area under the bottom part between t_A and t_B, plus the area under the top part between t_B and the final t-value. So since $dx = (3t^2 - 3)\,dt$, the area of the loop is

$A = \int_{-2}^{1} (t^2 + t + 1)(3t^2 - 3)\,dt = \int_{-2}^{1} (3t^4 + 3t^3 - 3t - 3)\,dt = \left[\frac{3}{5}t^5 + \frac{3}{4}t^4 - \frac{3}{2}t^2 - 3t \right]_{-2}^{1}$
$= \left(\frac{3}{5} + \frac{3}{4} - \frac{3}{2} - 3 \right) - \left[-\frac{96}{5} + 12 - 6 - (-6) \right] = \frac{81}{20}$.

9. Take the base to be the disk $x^2 + y^2 \le 9$. Then $V = \int_{-3}^{3} A(x)\,dx$, where $A(x_0)$ is the area of the isosceles right triangle whose hypotenuse lies along the line $x = x_0$ in the xy-plane. $A(x) = \frac{1}{2}\left(\sqrt{2}\sqrt{9 - x^2} \right)^2 = 9 - x^2$, so $V = 2\int_{0}^{3} A(x)\,dx = 2\int_{0}^{3} (9 - x^2)\,dx = 2\left[9x - \frac{1}{3}x^3 \right]_{0}^{3} = 2(27 - 9) = 36$.

11. Equilateral triangles with sides measuring $\frac{1}{4}x$ meters have height $\frac{1}{4}x \sin 60° = \frac{\sqrt{3}}{8}x$. Therefore, $A(x) = \frac{1}{2} \cdot \frac{1}{4}x \cdot \frac{\sqrt{3}}{8}x = \frac{\sqrt{3}}{64}x^2$.
$V = \int_{0}^{20} A(x)\,dx = \frac{\sqrt{3}}{64}\int_{0}^{20} x^2\,dx = \frac{\sqrt{3}}{64}\left[\frac{1}{3}x^3 \right]_{0}^{20} = \frac{8000\sqrt{3}}{64 \cdot 3} = \frac{125\sqrt{3}}{3} \text{ m}^3$

13. $x = 3t^2$, $y = 2t^3$, $0 \le t \le 2$.
$L = \int_{0}^{2} \sqrt{(dx/dt)^2 + (dy/dt)^2}\,dt = \int_{0}^{2} \sqrt{(6t)^2 + (6t^2)^2}\,dt = 6\int_{0}^{2} t\sqrt{1 + t^2}\,dt = 2\left[(1 + t^2)^{3/2} \right]_{0}^{2}$
$= 2(5\sqrt{5} - 1)$

15. $f(x) = kx \Longrightarrow 30\,\text{N} = k(15 - 12)\,\text{cm} \Longrightarrow k = 10\,\text{N/cm} = 1000\,\text{N/m}$. $20\,\text{cm} - 12\,\text{cm} = 0.08\,\text{m} \Longrightarrow$
$W = \int_{0}^{0.08} kx\,dx = 1000\int_{0}^{0.08} x\,dx = 500\left[x^2 \right]_{0}^{0.08} = 500(0.08)^2 = 3.2\,\text{N-m} = 3.2\,\text{J}$.

17. (a) The parabola has equation $y = ax^2$ with vertex at the origin and passing through $(4, 4)$.
$4 = a \cdot 4^2 \implies a = \frac{1}{4} \implies y = \frac{1}{4}x^2 \implies x^2 = 4y \implies x = 2\sqrt{y}$. (See the figure in Exercise 18.)
Each circular disk has radius $2\sqrt{y}$ and is moved $4 - y$ ft.
$W = \int_0^4 \pi \left(2\sqrt{y}\right)^2 62.5 \left(4 - y\right) dy = 250\pi \int_0^4 y \left(4 - y\right) dy = 250\pi \left[2y^2 - \frac{1}{3}y^3\right]_0^4$
$= 250\pi \left(32 - \frac{64}{3}\right) = \frac{8000\pi}{3} \approx 8377.6$ ft-lb

(b) In part (a) we knew the final water level (0) but not the amount of
work done. Here we use the same equation, except with the work
fixed, and the lower limit of integration (that is, the final water
level — call it h) unknown: $W = 4000 \iff$

$250\pi \left[2y^2 - \frac{1}{3}y^3\right]_h^4 = 4000 \iff$

$\frac{16}{\pi} = \left[\left(32 - \frac{64}{3}\right) - \left(2h^2 - \frac{1}{3}h^3\right)\right] \iff h^3 - 6h^2 + 32 - \frac{48}{\pi} = 0.$
We plot the graph of the function $f(h) = h^3 - 6h^2 + 32 - \frac{48}{\pi}$ on
the interval $[0, 4]$ to see where it is 0. From the graph, $f(h) = 0$
for $h \approx 2.06$. So the depth of water remaining is about 2.06 ft.

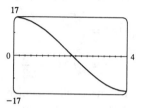

19. As in Example 4 of Section 6.5,
$F = \int_0^2 \rho g x \left(5 - x\right) dx = \rho g \left[\frac{5}{2}x^2 - \frac{1}{3}x^3\right]_0^2 = \rho g \left(10 - \frac{8}{3}\right) = \frac{22}{3}\delta \approx \frac{22}{3} \cdot 62.5 \approx 458.3$ lb.

21. $x = 100 \implies P = 2000 - 0.1\left(100\right) - 0.01\left(100\right)^2 = 1890$
Consumer surplus $= \int_0^{100} \left[p(x) - P\right] dx = \int_0^{100} \left(2000 - 0.1x - 0.01x^2 - 1890\right) dx$
$= \left[110x - 0.05x^2 - \frac{0.01}{3}x^3\right]_0^{100} = 11{,}000 - 500 - \frac{10{,}000}{3} \approx \$7166.67$

23. $\lim_{h \to 0} f_{\text{ave}} = \lim_{h \to 0} \frac{1}{h} \int_x^{x+h} f(t)\, dt = \lim_{h \to 0} \frac{F(x + h) - F(x)}{h}$, where $F(x) = \int_a^x f(t)\, dt$.

But we recognize this limit as being $F'(x)$ by the definition of a derivative. Therefore,
$\lim_{h \to 0} f_{\text{ave}} = F'(x) = f(x)$ by FTC1.

25. (a) The probability density function is $f(t) = \begin{cases} 0 & \text{if } t < 0 \\ \frac{1}{8}e^{-t/8} & \text{if } t \geq 0 \end{cases}$

$P\left(0 \leq X \leq 3\right) = \int_0^3 \frac{1}{8}e^{-t/8}\, dt = \left[-e^{-t/8}\right]_0^3 = -e^{-3/8} + 1 \approx 0.3127$

(b) $P\left(X > 10\right) = \int_{10}^{\infty} \frac{1}{8}e^{-t/8}\, dt = \lim_{x \to \infty} \left[-e^{-t/8}\right]_{10}^x = \lim_{x \to \infty} \left(-e^{-x/8} + e^{-10/8}\right) = 0 + e^{-5/4}$
≈ 0.2865

(c) We need to find m such that $P\left(X \geq m\right) = \frac{1}{2} \implies \int_m^{\infty} \frac{1}{8}e^{-t/8}\, dt = \frac{1}{2} \implies$
$\lim_{x \to \infty} \left[-e^{-t/8}\right]_m^x = \frac{1}{2} \implies \lim_{x \to \infty} \left(-e^{-x/8} + e^{-m/8}\right) = \frac{1}{2} \implies e^{-m/8} = \frac{1}{2} \implies -m/8 = \ln\frac{1}{2} \implies$
$m = -8\ln\frac{1}{2} = 8\ln 2 \approx 5.55$ minutes.

Focus on Problem Solving

1. The volume generated from $x = 0$ to $x = b$ is $\int_0^b \pi \, [f(x)]^2 \, dx$. Hence, we are given that $b^2 = \int_0^b \pi \, [f(x)]^2 \, dx$ for all $b > 0$. Differentiating both sides of this equation using the Fundamental Theorem of Calculus gives $2b = \pi \, [f(b)]^2 \implies f(b) = \sqrt{\dfrac{2b}{\pi}}$, since f is positive. Therefore,

$$f(x) = \sqrt{\frac{2x}{\pi}}.$$

3. (a) $V = \pi h^2 \left(r - \dfrac{h}{3} \right) = \frac{1}{3}\pi h^2 (3r - h)$. See the solution to Exercise 6.2.19.

(b) The smaller segment has height $h = 1 - x$ and so by part (a) its volume is
$V = \frac{1}{3}\pi (1-x)^2 [3(1) - (1-x)] = \frac{1}{3}\pi (x-1)^2 (x+2)$. This volume must be $\frac{1}{3}$ of
the total volume of the sphere, which is $\frac{4}{3}\pi (1)^3$. So $\frac{1}{3}\pi (x-1)^2 (x+2) = \frac{1}{3}\left(\frac{4}{3}\pi\right) \implies$
$(x^2 - 2x + 1)(x+2) = \frac{4}{3} \implies x^3 - 3x + 2 = \frac{4}{3} \implies 3x^3 - 9x + 2 = 0$. Using Newton's method
with $f(x) = 3x^3 - 9x + 2$, $f'(x) = 9x^2 - 9$, we get $x_{n+1} = x_n - \dfrac{3x_n^3 - 9x_n + 2}{9x_n^2 - 9}$. Taking
$x_1 = 0$, we get $x_2 \approx 0.2222$, and $x_3 \approx 0.2261 \approx x_4$, so, correct to four decimal places, $x \approx 0.2261$.

(c) With $r = 0.5$ and $s = 0.75$, the equation $x^3 - 3rx^2 + 4r^3 s = 0$ becomes
$x^3 - 3(0.5)x^2 + 4(0.5)^3 (0.75) = 0 \implies x^3 - \frac{3}{2}x^2 + 4\left(\frac{1}{8}\right)\frac{3}{4} = 0 \implies 8x^3 - 12x^2 + 3 = 0$.
We use Newton's method with $f(x) = 8x^3 - 12x^2 + 3$, $f'(x) = 24x^2 - 24x$, so
$x_{n+1} = x_n - \dfrac{8x_n^3 - 12x_n^2 + 3}{24x_n^2 - 24x_n}$. Take $x_1 = 0.5$. Then $x_2 \approx 0.6667$, and $x_3 \approx 0.6736 \approx x_4$. So to
four decimal places the depth is 0.6736 m.

(d) (i) From part (a) with $r = 5$ in., the volume of water in the bowl is

$V = \frac{1}{3}\pi h^2 (3r - h) = \frac{1}{3}\pi h^2 (15 - h) = 5\pi h^2 - \frac{1}{3}\pi h^3$. We are given that $\dfrac{dV}{dt} = 0.2$ m^3/s and

we want to find $\dfrac{dh}{dt}$ when $h = 3$. Now $\dfrac{dV}{dt} = 10\pi h \dfrac{dh}{dt} - \pi h^2 \dfrac{dh}{dt}$, so $\dfrac{dh}{dt} = \dfrac{0.2}{\pi(10h - h^2)}$.

When $h = 3$, we have $\dfrac{dh}{dt} = \dfrac{0.2}{\pi(10 \cdot 3 - 3^2)} = \dfrac{1}{105\pi} \approx 0.003$ in/s.

(iii) From part (a), the volume of water required to fill the bowl from the instant that the water
is 4 in. deep is $V = \frac{1}{2} \cdot \frac{4}{3}\pi (5)^3 - \frac{1}{3}\pi (4)^2 (15 - 4) = \frac{2}{3} \cdot 125\pi - \frac{16}{3} \cdot 11\pi = \frac{74}{3}\pi$.
To find the time required to fill the bowl we divide this volume by the rate:
Time $= \dfrac{74\pi/3}{0.2} = \dfrac{370\pi}{3} \approx 387$ s ≈ 6.5 min

5. We are given that the rate of change of the volume of water is $\dfrac{dV}{dt} = -kA(x)$, where k is some positive constant and $A(x)$ is the area of the surface when the water has depth x. Now we are concerned with the rate of change of the depth of the water with respect to time, that is, $\dfrac{dx}{dt}$. But by the Chain Rule, $\dfrac{dV}{dt} = \dfrac{dV}{dx}\dfrac{dx}{dt}$, so the first equation can be written $\dfrac{dV}{dx}\dfrac{dx}{dt} = -kA(x)$ (★). Also, we know that the total volume of water up to a depth x is $V(x) = \int_0^x A(s)\,ds$, where $A(s)$ is the area of a cross-section of the water at a depth s. Differentiating this equation with respect to x, we get $dV/dx = A(x)$. Substituting this into equation ★, we get $A(x)(dx/dt) = -kA(x) \Longrightarrow dx/dt = -k$, a constant.

7. (a) Choose a vertical x-axis pointing downward with its origin at the surface. In order to calculate the pressure at depth z, consider n subintervals of the interval $[0, z]$ by points x_i and choose a point $x_i^* \in [x_{i-1}, x_i]$ for each i. The thin layer of water lying between depth x_{i-1} and depth x_i has a density of approximately $\rho(x_i^*)$, so the weight of a piece of that layer with unit cross-sectional area would be $\rho(x_i^*)g\Delta x$. The total weight of a column of water extending from the surface to depth z (with unit cross-sectional area) would be approximately $\sum_{i=1}^{n}\rho(x_i^*)g\Delta x$. The estimate becomes exact if we take the limit as $n \to \infty$; weight (or force) per unit area at depth z is $W = \lim_{n\to\infty}\sum_{i=1}^{n}\rho(x_i^*)g\Delta x$. In other words, $P(z) = \int_0^z \rho(x)g\,dx$. More generally, if we make no assumptions about the location of the origin, then $P(z) = P_0 + \int_0^z \rho(x)g\,dx$, where P_0 is the pressure at $x = 0$. Differentiating, we get $dP/dz = \rho(z)g$.

(b)

$$F = \int_{-r}^{r} P(L + x)\cdot 2\sqrt{r^2 - x^2}\,dx = \int_{-r}^{r}\left(P_0 + \int_0^{L+x}\rho_0 e^{z/H}g\,dz\right)\cdot 2\sqrt{r^2 - x^2}\,dx$$

$$= P_0\int_{-r}^{r} 2\sqrt{r^2 - x^2}\,dx + \rho_0 gH\int_{-r}^{r}\left(e^{(L+x)/H} - 1\right)\cdot 2\sqrt{r^2 - x^2}\,dx$$

$$= (P_0 - \rho_0 gH)\int_{-r}^{r} 2\sqrt{r^2 - x^2}\,dx + \rho_0 gH\int_{-r}^{r} e^{(L+x)/H}\cdot 2\sqrt{r^2 - x^2}\,dx$$

$$= (P_0 - \rho_0 gH)\left(\pi r^2\right) + \rho_0 gH e^{L/H}\int_{-r}^{r} e^{x/H}\cdot 2\sqrt{r^2 - x^2}\,dx$$

9.

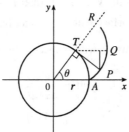

The coordinates of T are $(x_1, y_1) = (r\cos\theta, r\sin\theta)$. Since TP was unwound from arc TA, TP has length $r\theta$. Also $\angle PTQ = \angle PTR - \angle QTR = \frac{1}{2}\pi - \theta$, so P has coordinates

$$
\begin{aligned}
x &= x_1 + |TP|\cos\angle PTQ = r\cos\theta + r\theta\cos\left(\tfrac{1}{2}\pi - \theta\right)\\
&= r\left(\cos\theta + \theta\sin\theta\right),\\
y &= y_1 + |TP|\sin\angle PTQ = r\sin\theta - r\theta\sin\left(\tfrac{1}{2}\pi - \theta\right)\\
&= r\left(\sin\theta - \theta\cos\theta\right)
\end{aligned}
$$

11. $x = \displaystyle\int_1^t \frac{\cos u}{u}\,du$, $y = \displaystyle\int_1^t \frac{\sin u}{u}\,du$, so by FTC1, we have $\dfrac{dx}{dt} = \dfrac{\cos t}{t}$ and $\dfrac{dy}{dt} = \dfrac{\sin t}{t}$. Vertical tangent lines occur when $dx/dt = 0 \iff \cos t = 0 \iff t = \frac{\pi}{2} + n\pi$. The parameter value corresponding to the origin, $(x, y) = (0, 0)$, is $t = 1$, so the nearest vertical tangent occurs when $t = \frac{\pi}{2}$. Therefore, the arc length between these points is

$$
\begin{aligned}
L &= \int_1^{\pi/2} \sqrt{\left(\frac{dx}{dt}\right)^2 + \left(\frac{dy}{dt}\right)^2}\,dt\\
&= \int_1^{\pi/2} \sqrt{\frac{\cos^2 t}{t^2} + \frac{\sin^2 t}{t^2}}\,dt\\
&= \int_1^{\pi/2} \frac{dt}{t}\\
&= [\ln t]_1^{\pi/2}\\
&= \ln \tfrac{\pi}{2}
\end{aligned}
$$

Chapter 7   Differential Equations

Section 7.1   Modeling with Differential Equations

1. $y = 2 + e^{-x^3} \Longrightarrow y' = -3x^2 e^{-x^3}$.

LHS $= y' + 3x^2 y = -3x^2 e^{-x^3} + 3x^2 \left(2 + e^{-x^3}\right) = -3x^2 e^{-x^3} + 6x^2 + 3x^2 e^{-x^3} = 6x^2$

$\qquad$ = RHS

3. (a) $y = \sin kt \Longrightarrow y' = k \cos kt \Longrightarrow y'' = -k^2 \sin kt$. $y'' + 9y = 0 \Longrightarrow -k^2 \sin kt + 9 \sin kt = 0 \Longrightarrow$
$\qquad \left(9 - k^2\right) \sin kt = 0 \quad$ (for all t) $\quad \Longrightarrow k = \pm 3$

$\quad$ **(b)** $y = A \sin kt + B \cos kt \Longrightarrow y' = Ak \cos kt - Bk \sin kt \Longrightarrow y'' = -Ak^2 \sin kt - Bk^2 \cos kt \Longrightarrow$
$\qquad y'' + 9y = -Ak^2 \sin kt - Bk^2 \cos kt + 9 \left(A \sin kt + B \cos kt\right)$
$\qquad \qquad = \left(9 - k^2\right) A \sin kt + \left(9 - k^2\right) B \cos kt = 0.$
$\qquad$ The last equation is true for all values of A and B if $k = \pm 3$.

5. (a) $y = e^t \Longrightarrow y' = e^t \Longrightarrow y'' = e^t$. LHS $= y'' + 2y' + y = e^t + 2e^t + e^t = 4e^t \neq 0$, so $y = e^t$ is not
$\qquad$ a solution of the differential equation.

$\quad$ **(b)** $y = e^{-t} \Longrightarrow y' = -e^{-t} \Longrightarrow y'' = e^{-t}$. LHS $= y'' + 2y' + y = e^{-t} - 2e^{-t} + e^{-t} = 0 = $ RHS, so
$\qquad y = e^{-t}$ is a solution.

$\quad$ **(c)** $y = te^{-t} \Longrightarrow y' = e^{-t} \left(1 - t\right) \Longrightarrow y'' = e^{-t} \left(t - 2\right)$.
$\qquad$ LHS $= y'' + 2y' + y = e^{-t} \left(t - 2\right) + 2e^{-t} \left(1 - t\right) + te^{-t} = e^{-t} \left[(t - 2) + 2(1 - t) + t\right]$
$\qquad \qquad = e^{-t} \left(0\right) = 0 = $ RHS, so $y = te^{-t}$ is a solution.

$\quad$ **(d)** $y = t^2 e^{-t} \Longrightarrow y' = te^{-t} \left(2 - t\right) \Longrightarrow y'' = e^{-t} \left(t^2 - 4t + 2\right)$.
$\qquad$ LHS $= y'' + 2y' + y = e^{-t} \left(t^2 - 4t + 2\right) + 2te^{-t} \left(2 - t\right) + t^2 e^{-t}$
$\qquad \qquad = e^{-t} \left[(t^2 - 4t + 2) + 2t(2 - t) + t^2\right] = e^{-t} \left(2\right) \neq 0,$
$\qquad$ so $y = t^2 e^{-t}$ is not a solution.

7. (a) Since the derivative y' is always negative (or 0), the function y must be decreasing (or have a
$\qquad$ horizontal tangent) on any interval on which it is defined.

$\quad$ **(b)** $y = \dfrac{1}{x + C} \Longrightarrow y' = -\dfrac{1}{(x + C)^2}$. LHS $= y' = -\dfrac{1}{(x + C)^2} = -\left(\dfrac{1}{x + C}\right)^2 = -y^2 = $ RHS

$\quad$ **(c)** $y = 0$ is a solution of $y' = -y^2$.

$\quad$ **(d)** $y\left(0\right) = \dfrac{1}{0 + C} = \dfrac{1}{C}$ and $y\left(0\right) = 0.5 \Longrightarrow C = 2$, so $y = \dfrac{1}{x + 2}$.

9. (a) $\dfrac{dP}{dt} = 1.2P\left(1 - \dfrac{P}{4200}\right)$. $\dfrac{dP}{dt} > 0 \Longrightarrow 1 - \dfrac{P}{4200} > 0 \Longrightarrow P < 4200 \Longrightarrow$ the population is

increasing for $P < 4200$ (assuming that $P \geq 0$).

(b) $\dfrac{dP}{dt} < 0 \Longrightarrow P > 4200$

(c) $\dfrac{dP}{dt} = 0 \Longrightarrow P = 4200$ or $P = 0$

11. (a) P increases most rapidly at the beginning, since there are usually many simple, easily-learned sub-skills associated with learning a skill. As t increases, we would expect dP/dt to remain positive, but decrease. This is because as time progresses, the only points left to learn are the more difficult ones.

(b) dP/dt is always positive, so the level of performance is increasing. As P gets close to M, dP/dt gets close to 0, that is, the performance levels off, as explained in part (a).

(c)

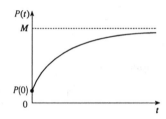

Section 7.2 Direction Fields

1. (a)

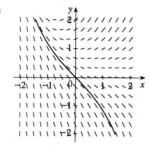

(b)

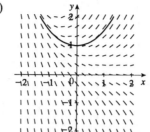

(c)

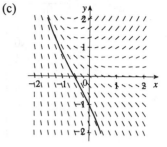

3. $y' = y - 1$. The slopes at each point are independent of x, so the slopes are the same along each line parallel to the x-axis. Thus, IV is the direction field for this equation. Note that for $y = 1$, $y' = 0$.

5. $y' = y^2 - x^2 = 0 \implies y = \pm x$. There are horizontal tangents on these lines only in graph III, so this equation corresponds to direction field III.

7. (a)

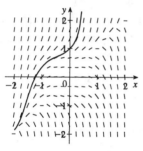

(b)

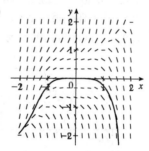

(c)

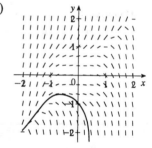

9. $y' = x - y$

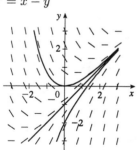

11.

x	y	$y' = y^2$
0	0	0
0	1	1
0	−1	1
1	0	0
−1	0	0
1	−1	1
1	1	1
1	2	4
1	−2	4
−1	2	4
−1	−2	4

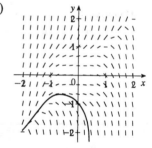

The solution curve through $(0, 1)$

13.

x	y	$y' = x^2 + y^2$
0	0	0
0	1	1
1	0	1
1	1	2
−1	1	2
0	2	4
2	0	4
2	2	8
2	1	5
−2	−1	5
1	2	5

The solution curve through $(0,0)$

15. In Maple, we can use either `directionfield` (in Maple's share library) or `plots[fieldplot]` to plot the direction field. To plot the solution, we can either use the initial-value option in `directionfield`, or actually solve the equation. In Mathematica, we use `PlotVectorField` for the direction field, and the `Plot[Evaluate[...]]` construction to plot the solution, which is $y = e^{(1-\cos 2x)/2}$. In Derive, use `Direction_Field` (in utility file `ODE_APPR`) to plot the direction field. Then use `DSOLVE1(-y*SIN(2*x),1,x,y,0,1)` (in utility file `ODE1`) to solve the equation. Simplify each result.

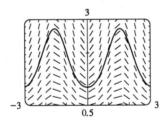

17.

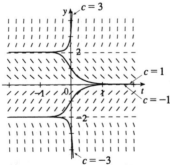

$L = \lim\limits_{t \to \infty} y\,(t)$ exists for $-2 \le c \le 2$; $L = \pm 2$ for $c = \pm 2$ and $L = 0$ for $-2 < c < 2$. For other values of c, L does not exist.

19. (a) $R\dfrac{dQ}{dt} + \dfrac{1}{C}Q = E(t)$ becomes $5Q' + \dfrac{1}{0.05}Q = 60$ or $Q' + 4Q = 12$.

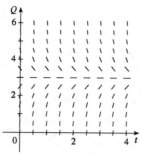

(b) From the graph in part (a), it appears that the limiting value of the charge Q is about 3.

(c) If $Q' = 0$, then $4Q = 12 \Longrightarrow Q = 3$ is an equilibrium solution.

(d)

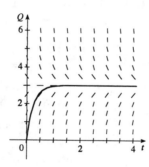

Section 7.3 Euler's Method

1. (a) $y' = F(x, y) = y$ and $y(0) = 1 \Longrightarrow x_0 = 0$, $y_0 = 1$.

(i) $h = 0.4$ and $y_1 = y_0 + hF(x_0, y_0) \Longrightarrow y_1 = 1 + 0.4 \cdot 1 = 1.4$. $x_1 = x_0 + h = 0 + 0.4 = 0.4$,
so $y_1 = y(0.4) = 1.4$.

(ii) $h = 0.2 \Longrightarrow x_1 = 0.2$ and $x_2 = 0.4$, so we need to find y_2.
$$y_1 = y_0 + hF(x_0, y_0) = 1 + 0.2y_0 = 1 + 0.2 \cdot 1 = 1.2,$$
$$y_2 = y_1 + hF(x_1, y_1) = 1.2 + 0.2y_1 = 1.2 + 0.2 \cdot 1.2 = 1.44.$$

(iii) $h = 0.1 \Longrightarrow x_4 = 0.4$, so we need to find y_4.
$$y_1 = y_0 + hF(x_0, y_0) = 1 + 0.1y_0 = 1 + 0.1 \cdot 1 = 1.1,$$
$$y_2 = y_1 + hF(x_1, y_1) = 1.1 + 0.1y_1 = 1.1 + 0.1 \cdot 1.1 = 1.21,$$
$$y_3 = y_2 + hF(x_2, y_2) = 1.21 + 0.1y_2 = 1.21 + 0.1 \cdot 1.21 = 1.331,$$
$$y_4 = y_3 + hF(x_3, y_3) = 1.331 + 0.1y_3 = 1.331 + 0.1 \cdot 1.331 = 1.4641.$$

(b)

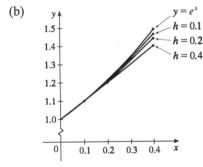

We see that the estimates are underestimates since they are all below the graph of $y = e^x$.

(c) (i) For $h = 0.4$:
(exact value) − (approximate value)
$$= e^{0.4} - 1.4 \approx 0.0918$$

(ii) For $h = 0.2$:
(exact value) − (approximate value)
$$= e^{0.4} - 1.44 \approx 0.0518$$

(iii) For $h = 0.1$:
(exact value) − (approximate value)
$$= e^{0.4} - 1.4641 \approx 0.0277$$

Each time the step size is halved, the error estimate also appears to be halved (approximately).

3. $h = 0.5$, $x_0 = 1$, $y_0 = 2$, and $F(x, y) = 1 + 3x - 2y$. So
$$y_n = y_{n-1} + hF(x_{n-1}, y_{n-1}) = y_{n-1} + 0.5(1 + 3x_{n-1} - 2y_{n-1}) = 0.5 + 1.5x_{n-1}.$$ Thus,
$y_1 = 0.5 + 1.5 \cdot 1 = 2$, $y_2 = 0.5 + 1.5 \cdot 1.5 = 2.75$, $y_3 = 0.5 + 1.5 \cdot 2 = 3.5$, $y_4 = 0.5 + 1.5 \cdot 2.5 = 4.25$.

5. $h = 0.1$, $x_0 = 0$, $y_0 = 1$, and $F(x, y) = x^2 + y^2$. We need to find y_5, because $x_5 = 0.5$. So
$y_n = y_{n-1} + 0.1(x_{n-1}^2 + y_{n-1}^2)$. $y_1 = 1 + 0.1(0^2 + 1^2) = 1.1$, $y_2 = 1.1 + 0.1(0.1^2 + 1.1^2) = 1.222$,
$y_3 = 1.222 + 0.1(0.2^2 + 1.222^2) \approx 1.37533$, $y_4 = 1.37533 + 0.1(0.3^2 + 1.37533^2) \approx 1.57348$,
$y_5 = 1.57348 + 0.1(0.4^2 + 1.57348^2) \approx 1.8371 \approx y(0.5)$.

7. (a) $dy/dx + 3x^2y = 6x^2 \implies y' = 6x^2 - 3x^2y$. Store this expression in Y_1 and use the following simple program to evaluate $y(1)$ for each part, using $H = h = 1$ and $N = 1$ for part (i), $H = 0.1$ and $N = 10$ for part (ii), and so forth.

$$h \to \text{H: } 0 \to \text{X: } 3 \to \text{Y:}$$
$$\text{For(I, 1, N):}$$
$$\quad Y + HY_1 \to \text{Y:}$$
$$\quad X + H \to \text{X:}$$
$$\text{End(loop):}$$
$$\text{Display Y.}$$

(i) $H = 1, N = 1 \implies y(1) = 3$

(ii) $H = 0.1, N = 10 \implies y(1) \approx 2.3928$

(iii) $H = 0.01, N = 100 \implies y(1) \approx 2.3701$

(iv) $H = 0.001, N = 1000 \implies y(1) \approx 2.3681$

(b) $y = 2 + e^{-x^3} \implies y' = -3x^2e^{-x^3}$.
LHS $= y' + 3x^2y = -3x^2e^{-x^3} + 3x^2\left(2 + e^{-x^3}\right) = -3x^2e^{-x^3} + 6x^2 + 3x^2e^{-x^3} = 6x^2 =$ RHS.
$y(0) = 2 + e^{-0} = 2 + 1 = 3$.

(c) (i) For $h = 1$: (exact value) $-$ (approximate value) $= 2 + e^{-1} - 3 \approx -0.6321$

(ii) For $h = 0.1$: (exact value) $-$ (approximate value) $= 2 + e^{-1} - 2.3928 \approx -0.0249$

(iii) For $h = 0.01$: (exact value) $-$ (approximate value) $= 2 + e^{-1} - 2.3701 \approx -0.0022$

(iv) For $h = 0.001$: (exact value) $-$ (approximate value) $= 2 + e^{-1} - 2.3681 \approx -0.0002$

In (ii)–(iv), it seems that when the step size is divided by 10, the error estimate is also divided by 10 (approximately).

9. Substituting $R = 5$, $C = 0.05$, and $E(t) = 60$ in $R\dfrac{dQ}{dt} = \dfrac{1}{C}Q = E(t)$ gives us $Q' = 12 - 4Q$.
$Q(0) = 0$, so $t_0 = 0$ and $Q_0 = 0$.
$Q_1 = Q_0 + hF(t_0, Q_0) = 0 + 0.1(12 - 4 \cdot 0) = 1.2$,
$Q_2 = Q_1 + hF(t_1, Q_1) = 1.2 + 0.1(12 - 4 \cdot 1.2) = 1.92$,
$Q_3 = Q_2 + hF(t_2, Q_2) = 1.92 + 0.1(12 - 4 \cdot 1.92) = 2.352$,
$Q_4 = Q_3 + hF(t_3, Q_3) = 2.352 + 0.1(12 - 4 \cdot 2.352) = 2.6112$,
$Q_5 = Q_4 + hF(t_4, Q_4) = 2.6112 + 0.1(12 - 4 \cdot 2.6112) = 2.76672$.
Thus, $Q_5 = Q(0.5) \approx 2.77$ C.

Section 7.4   Separable Equations

1. $\dfrac{dy}{dx} = y^2 \Longrightarrow \dfrac{dy}{y^2} = dx \ (y \neq 0) \Longrightarrow \displaystyle\int \dfrac{dy}{y^2} = \int dx \Longrightarrow -\dfrac{1}{y} = x + C \Longrightarrow -y = \dfrac{1}{x+C} \Longrightarrow$

$y = \dfrac{-1}{x+C}$, and $y = 0$ is also a solution.

3. $yy' = x \Longrightarrow \displaystyle\int y\,dy = \int x\,dx \Longrightarrow \dfrac{y^2}{2} = \dfrac{x^2}{2} + C_1 \Longrightarrow y^2 = x^2 + 2C_1 \Longrightarrow x^2 - y^2 = C$ (where

$C = -2C_1$).   This represents a family of hyperbolas.

5. $\dfrac{du}{dt} = e^{u+2t} = e^u e^{2t} \Longrightarrow \displaystyle\int e^{-u}\,du = \int e^{2t}\,dt \Longrightarrow -e^{-u} = \tfrac{1}{2}e^{2t} + C_1 \Longrightarrow$

$e^{-u} = -\tfrac{1}{2}e^{2t} + C$ (where $C = -C_1$ and the right-hand side is positive, since $e^{-u} > 0$) $\Longrightarrow$

$-u = \ln\left(C - \tfrac{1}{2}e^{2t}\right) \Longrightarrow u = -\ln\left(C - \tfrac{1}{2}e^{2t}\right)$

7. $\dfrac{dy}{dx} = y^2 + 1,\ y(1) = 0.\ \displaystyle\int \dfrac{dy}{y^2+1} = \int dx \Longleftrightarrow \tan^{-1}y = x + C.\ y = 0$ when $x = 1$, so

$1 + C = \tan^{-1}0 = 0 \Longrightarrow C = -1$. Thus, $\tan^{-1}y = x - 1$ and $y = \tan(x-1)$.

9. $xe^{-t}\dfrac{dx}{dt} = t,\ x(0) = 1.\ \int x\,dx = \int te^t\,dt \Longrightarrow \tfrac{1}{2}x^2 = (t-1)e^t + C.\ x(0) = 1$, so

$\tfrac{1}{2} = (0-1)e^0 + C$ and $C = \tfrac{3}{2}$. Thus, $x^2 = 2(t-1)e^t + 3 \Longrightarrow x = \sqrt{2(t-1)e^t + 3}$.

11. $\dfrac{du}{dt} = \dfrac{2t+1}{2(u-1)},\ u(0) = -1.\ \displaystyle\int 2(u-1)\,du = \int (2t+1)\,dt \Longrightarrow u^2 - 2u = t^2 + t + C.\ u(0) = -1$

so $(-1)^2 - 2(-1) = 0^2 + 0 + C \Longrightarrow C = 3$. Thus, $u^2 - 2u = t^2 + t + 3 \Longrightarrow u^2 - 2u - (t^2 + t + 3) = 0$.

Using the quadratic formula gives $u = \dfrac{2 \pm \sqrt{4 + 4(t^2 + t + 3)}}{2} = 1 - \sqrt{t^2 + t + 4}$ for $u(0) = -1$.

13. $\dfrac{dy}{dx} = 4x^3y,\ y(0) = 7.\ \dfrac{dy}{y} = 4x^3\,dx$ (if $y \neq 0$) $\Longrightarrow \displaystyle\int \dfrac{dy}{y} = \int 4x^3\,dx \Longrightarrow \ln|y| = x^4 + C \Longrightarrow$

$e^{\ln|y|} = e^{x^4+C} \Longrightarrow |y| = e^{x^4}e^C \Longrightarrow y = Ae^{x^4};\ y(0) = 7 \Longrightarrow A = 7 \Longrightarrow y = 7e^{x^4}$.

15. $y' = y\sin x,\ y(0) = 1.\ \displaystyle\int \dfrac{dy}{y} = \int \sin x\,dx \Longleftrightarrow \ln|y| = -\cos x + C \Longrightarrow |y| = e^{-\cos x + C} \Longrightarrow$

$y(x) = Ae^{-\cos x}.\ y(0) = Ae^{-1} = 1 \Longleftrightarrow A = e^1$, so $y = e \cdot e^{-\cos x} = e^{1-\cos x}$.

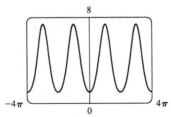

17. $\dfrac{dy}{dx} = \dfrac{\sin x}{\sin y}$, $y(0) = \dfrac{\pi}{2}$. So $\int \sin y\, dy = \int \sin x\, dx \Longleftrightarrow -\cos y = -\cos x + C \Longleftrightarrow \cos y = \cos x - C$.

From the initial condition, we need $\cos\frac{\pi}{2} = \cos 0 - C \Longrightarrow 0 = 1 - C \Longrightarrow C = 1$, so the solution is $\cos y = \cos x - 1$. Note that we cannot take $\cos^{-1}$ of both sides, since that would unnecessarily restrict the solution to the case where $-1 \le \cos x - 1 \Longleftrightarrow 0 \le \cos x$, as $\cos^{-1}$ is defined only on $[-1, 1]$. Instead we plot the graph using Maple's `plots[implicitplot]` or Mathematica's `Plot[Evaluate[···]]`.

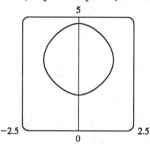

19. (a) In Maple, we can use either `directionfield` (in Maple's share library) or `plots[fieldplot]` to plot the direction field. To plot the solution, we can either use the initial-value option in `directionfield`, or actually solve the equation. In Mathematica, we use `PlotVectorField` for the direction field, and the `Plot[Evaluate[···]]` construction to plot the solution, which is $y = \pm\sqrt{2(x+c)}$.

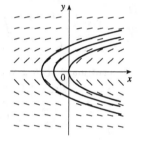

(b) $y' = 1/y \Longrightarrow y\,dy = dx$, so $\frac{1}{2}y^2 = x + c$ or $y = \pm\sqrt{2(x+c)}$.

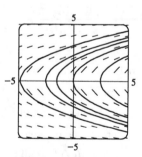

21. The curves $y = kx^2$ form a family of parabolas with axis the y-axis. Differentiating gives $y' = 2kx$, but $k = y/x^2$, so $y' = 2y/x$. Thus, the slope of the tangent line at any point (x, y) on one of the parabolas is $y' = 2y/x$, so the orthogonal trajectories must satisfy $y' = -x/(2y) \Longleftrightarrow$ $2y\,dy = -x\,dx \Longleftrightarrow y^2 = -x^2/2 + c_1 \Longleftrightarrow x^2 + 2y^2 = c$. This is a family of ellipses.

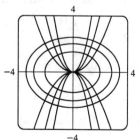

23. Differentiating $y = (x + k)^{-1}$ gives $y' = -\dfrac{1}{(x + k)^2}$, but

$k = \dfrac{1}{y} - x$, so $y' = -\dfrac{1}{(1/y)^2} = -y^2$. Thus, the orthogonal

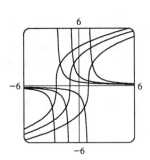

trajectories must satisfy $y' = -\dfrac{1}{-y^2} = \dfrac{1}{y^2} \iff$

$y^2\, dy = dx \iff \dfrac{y^3}{3} = x + c$ or $y = [3\,(x + c)]^{1/3}$

25. From Exercise 7.2.19, $\dfrac{dQ}{dt} = 12 - 4Q \iff \displaystyle\int \dfrac{dQ}{12 - 4Q} = \int dt \iff -\tfrac{1}{4}\ln|12 - 4Q| = t + C \iff$

$\ln|12 - 4Q| = -4t - 4C \iff |12 - 4Q| = e^{-4t - 4C} \iff 12 - 4Q = Ke^{-4t} \quad (K = \pm e^{-4C}) \iff$

$4Q = 12 - Ke^{-4t} \iff Q = 3 - Ae^{-4t} \quad (A = K/3).\ Q(0) = 0 \iff 0 = 3 - A \iff A = 3 \iff$

$Q(t) = 3 - 3e^{-4t}.$ As $t \to \infty$, $Q(t) \to 3 - 0 = 3$ (the limiting value).

27. $\dfrac{dP}{dt} = k\,(M - P) \iff \displaystyle\int \dfrac{dP}{P - M} = \int -k\, dt \iff \ln|P - M| = -kt + C \iff$

$|P - M| = e^{-kt + C} \iff P - M = Ae^{-kt} \quad (A = \pm e^C) \iff P = M + Ae^{-kt}.$ If we assume

that performance is at level 0 when $t = 0$, then $P(0) = 0 \iff 0 = M + A \iff A = -M \iff$

$P(t) = M - Me^{-kt}.\ \displaystyle\lim_{t \to \infty} P(t) = M - M \cdot 0 = M.$

29. (a) $\dfrac{dC}{dt} = r - kC \implies \dfrac{dC}{dt} = -(kC - r) \implies \displaystyle\int \dfrac{dC}{kC - r} = \int -dt \implies$

$(1/k)\ln|kC - r| = -t + M_1 \implies \ln|kC - r| = -kt + M_2 \implies |kC - r| = e^{-kt + M_2} \implies$

$kC - r = M_3 e^{-kt} \implies kC = M_3 e^{-kt} + r \implies C(t) = M_4 e^{-kt} + r/k.\ C(0) = C_0 \implies$

$C_0 = M_4 + r/k \implies M_4 = C_0 - r/k \implies C(t) = (C_0 - r/k)\, e^{-kt} + r/k.$

(b) If $C_0 < r/k$, then $C_0 - r/k < 0$ and the formula for $C(t)$ shows that $C(t)$ increases and

$\displaystyle\lim_{t \to \infty} C(t) = r/k.$ As t increases, the formula for $C(t)$ shows how the role of C_0 steadily

diminishes as that of r/k increases.

31. (a) Let $y(t)$ be the amount of salt (in kg) after t minutes. Then $y(0) = 15$. The amount of liquid in the

tank is 1000 L at all times, so the concentration at time t (in minutes) is $y(t)/1000$ kg/L and

$$\dfrac{dy}{dt} = -\left[\dfrac{y(t)\ \text{kg}}{1000\ \text{L}}\right]\left(10\dfrac{\text{L}}{\text{min}}\right) = -\dfrac{y(t)}{100}\dfrac{\text{kg}}{\text{min}}.\ \int \dfrac{dy}{y} = -\dfrac{1}{100}\int dt \implies \ln y = -\dfrac{t}{100} + C,$$

and $y(0) = 15 \implies \ln 15 = C$, so $\ln y = \ln 15 - \dfrac{t}{100}.$ It follows that $\ln\left(\dfrac{y}{15}\right) = -\dfrac{t}{100}$ and

$\dfrac{y}{15} = e^{-t/100}$, so $y = 15e^{-t/100}$ kg.

(b) After 20 minutes, $y = 15e^{-20/100} = 15e^{-0.2} \approx 12.3$ kg.

33. Assume that the raindrop begins at rest, so that $v(0) = 0$. $dm/dt = km$ and $(mv)' = gm \Longrightarrow$

$m'v + mv' = gm \Longrightarrow (km)v + mv' = gm \Longrightarrow v' = g - kv \Longrightarrow \displaystyle\int \frac{dv}{g - kv} = \int dt \Longrightarrow$

$-(1/k)\ln|g - kv| = t + C \Longrightarrow g - kv = Ae^{-kt}$. $v(0) = 0 \Longrightarrow A = g$. So $v = (g/k)\left(1 - e^{-kt}\right)$.

Since $k > 0$, as $t \to \infty$, $e^{-kt} \to 0$ and therefore, $\displaystyle\lim_{t\to\infty} v(t) = g/k$.

35. (a) The rate of growth of the area is jointly proportional to $\sqrt{A(t)}$ and $M - A(t)$; that is, the rate is proportional to the product of those two quantities. So for some constant k, $dA/dt = k\sqrt{A}(M - A)$. We are interested in the maximum of the function dA/dt (when the tissue grows the fastest), so we differentiate, using the Chain Rule and then substituting for dA/dt from the differential equation:

$$\frac{d}{dt}\left(\frac{dA}{dt}\right) = k\left[\tfrac{1}{2}A^{-1/2}(M - A)\frac{dA}{dt} + \sqrt{A}(-1)\frac{dA}{dt}\right] = \tfrac{1}{2}kA^{-1/2}\frac{dA}{dt}[(M - A) - 2A]$$

$$= \tfrac{1}{2}k^2(M - A)(M - 3A).$$

This is 0 when $M - A = 0$ [this situation never actually occurs, since the graph of $A(t)$ is asymptotic to the line $y = M$, as in the logistic model] and when $M - 3A = 0 \Longleftrightarrow A(t) = M/3$. This represents a maximum by the First Derivative Test, since $\dfrac{d}{dt}\left(\dfrac{dA}{dt}\right)$ goes from positive to negative when $A(t) = M/3$.

(b) From the CAS, we get $A(t) = M\left(\dfrac{Ce^{\sqrt{M}kt} - 1}{Ce^{\sqrt{M}kt} + 1}\right)^2$. To get C in terms of the initial area

A_0 and the maximum area M, we substitute $t = 0$ and $A = A_0$: $A_0 = M\left(\dfrac{C - 1}{C + 1}\right)^2 \Longleftrightarrow$

$(C + 1)\sqrt{A_0} = (C - 1)\sqrt{M} \Longleftrightarrow C\sqrt{A_0} + \sqrt{A_0} = C\sqrt{M} - \sqrt{M} \Longleftrightarrow$

$\sqrt{A_0} + \sqrt{M} = C\sqrt{M} - C\sqrt{A_0} \Longleftrightarrow C = \dfrac{\sqrt{M} + \sqrt{A_0}}{\sqrt{M} - \sqrt{A_0}}$. (Notice that if $A_0 = 0$, then $C = 1$.)

37. (a) We have $V(t) = \pi r^2 y(t) \Longrightarrow \dfrac{dV}{dy} = \pi r^2 = 4\pi$ where $\dfrac{dV}{dt} = \dfrac{dV}{dy}\dfrac{dy}{dt}$. Thus, $\dfrac{dV}{dt} = -a\sqrt{2gy} \Longrightarrow$

$\dfrac{dV}{dy}\dfrac{dy}{dt} = -\pi\left(\tfrac{1}{12}\right)^2\sqrt{2 \cdot 32y} \Longrightarrow 4\pi\dfrac{dy}{dt} = -\pi\tfrac{8}{144}\sqrt{y} \Longrightarrow \dfrac{dy}{dt} = -\tfrac{1}{72}\sqrt{y}$.

(b) $\dfrac{dy}{dt} = -\tfrac{1}{72}\sqrt{y} \Longrightarrow y^{-1/2}\,dy = -\tfrac{1}{72}\,dt \Longrightarrow 2\sqrt{y} = -\tfrac{1}{72}t + C$. $y(0) = 6 \Longrightarrow 2\sqrt{6} = 0 + C \Longrightarrow$

$C = 2\sqrt{6} \Longrightarrow y = \left(-\tfrac{1}{144}t + \sqrt{6}\right)^2$.

(c) We want to find t when $y = 0$, so we set $y = 0 = \left(-\tfrac{1}{144}t + \sqrt{6}\right)^2 \Longrightarrow t = 144\sqrt{6} \approx 5$ min 53 s.

Section 7.5 Exponential Growth and Decay

1. The relative growth rate is $\dfrac{1}{P}\dfrac{dP}{dt} = 0.7944$, so $\dfrac{dP}{dt} = 0.7944P$ and, by Theorem 2,

   $P(t) = P(0)e^{0.7944t} = 2e^{0.7944t}$. Thus, $P(6) = 2e^{0.7944(6)} \approx 234.99$ or about 235 members.

3. (a) By Theorem 2, $y(t) = y(0)e^{kt} = 500e^{kt}$. $y(3) = 500e^{3k} = 8000 \Longrightarrow e^{3k} = 16 \Longrightarrow$

   $3k = \ln 16 \Longrightarrow k = (\ln 16)/3$. So $y(t) = 500e^{(\ln 16)t/3} = 500 \cdot 16^{t/3}$

   (b) $y(4) = 500 \cdot 16^{4/3} \approx 20{,}159$

   (c) $y(t) = 500 \cdot 16^{t/3} = 30{,}000 \Longrightarrow 16^{t/3} = 60 \Longrightarrow \frac{1}{3}t\ln 16 = \ln 60 \Longrightarrow$

   $t = 3(\ln 60)/(\ln 16) \approx 4.4$ h

5. (a) Let the population (in millions) in the year t be $P(t)$. Since the initial time is the year 1750, we
   substitute $t - 1750$ for t in Theorem 2, so the exponential model gives $P(t) = P(1750)e^{k(t-1750)}$.
   Then $P(1800) = 906 = 728e^{k(1800-1750)} \Longrightarrow \ln \frac{906}{728} = k(50) \Longrightarrow k = \frac{1}{50}\ln\frac{906}{728} \approx 0.0043748$.
   So with this model, we have $P(1900) = 728e^{150(0.0043748)} \approx 1403$ million, and
   $P(1950) \approx 728e^{200(0.0043748)} \approx 1746$ million. Both of these estimates are much too low.

   (b) In this case, the exponential model gives $P(t) = P(1850)e^{k(t-1850)} \Longrightarrow$
   $P(1900) = 1608 = 1171e^{k(1900-1850)} \Longrightarrow \ln\frac{1608}{1171} = k(50) \Longrightarrow k = \frac{1}{50}\ln\frac{1608}{1171} \approx 0.006343$. So
   with this model, we estimate $P(1950) \approx 1171e^{100(0.006343)} \approx 2208$ million. This is still too low,
   but closer than the estimate of $P(1950)$ in part (a).

   (c) The exponential model gives $P(t) = P(1900)e^{k(t-1900)} \Longrightarrow$
   $P(1950) = 2517 = 1608e^{k(1950-1900)} \Longrightarrow \ln\frac{2517}{1608} = k(50) \Longrightarrow k = \frac{1}{50}\ln\frac{2517}{1608} \approx 0.008962$.
   With this model, we estimate $P(1992) \approx 1608e^{0.008962(1992-1900)} \approx 3667$ million. This is much
   too low. The discrepancy is explained by the fact that the world birth rate (average yearly number of
   births per person) is about the same as always, whereas the mortality rate (especially the infant
   mortality rate) is much lower, owing mostly to advances in medical science and to the wars in the
   first part of the twentieth century. The exponential model assumes, among other things, that the
   birth and mortality rates will remain constant.

7. (a) If $y = [N_2O_5]$ then by Theorem 2, $\dfrac{dy}{dt} = -0.0005y \Longrightarrow y(t) = y(0)e^{-0.0005t} = Ce^{-0.0005t}$.

   (b) $y(t) = Ce^{-0.0005t} = 0.9C \Longrightarrow e^{-0.0005t} = 0.9 \Longrightarrow -0.0005t = \ln 0.9 \Longrightarrow$
   $t = -2000\ln 0.9 \approx 211$ s

9. (a) If $y(t)$ is the mass remaining after t days, then $y(t) = y(0)e^{kt} = 50e^{kt}$.
   $y(0.00014) = 50e^{0.00014k} = 25 \Longrightarrow e^{0.00014k} = \frac{1}{2} \Longrightarrow k = -(\ln 2)/0.00014 \Longrightarrow$
   $y(t) = 50e^{-(\ln 2)t/0.00014} = 50 \cdot 2^{-t/0.00014}$

   (b) $y(0.01) = 50 \cdot 2^{-0.01/0.00014} \approx 1.57 \times 10^{-20}$ mg

   (c) $50e^{-(\ln 2)t/0.00014} = 40 \Longrightarrow -(\ln 2)t/0.00014 = \ln 0.8 \Longrightarrow t = -0.00014\dfrac{\ln 0.8}{\ln 2} \approx 4.5 \times 10^{-5}$ s

11. Let $y(t)$ be the level of radioactivity. Thus, $y(t) = y(0)e^{-kt}$ and k is determined by using the

half-life: $y(5730) = \frac{1}{2}y(0) \Longrightarrow \frac{1}{2} = e^{-5730k} \Longrightarrow k = -\frac{\ln\frac{1}{2}}{5730} = \frac{\ln 2}{5730}$. If 74% of the ^{14}C

remains, then we know that $y(t) = 0.74y(0) \Longrightarrow 0.74 = e^{-t(\ln 2)/5730} \Longrightarrow \ln 0.74 = -\frac{t\ln 2}{5730} \Longrightarrow$

$t = -\frac{5730(\ln 0.74)}{\ln 2} \approx 2489 \approx 2500$ years.

13. (a) If $y = u - 75$, $u(0) = 185 \Longrightarrow y(0) = 185 - 75 = 110$, and the initial-value problem is

$dy/dt = ky$ with $y(0) = 110$. So the solution is $y(t) = 110e^{kt}$.

(b) $y(30) = 110e^{30k} = 150 - 75 \Longrightarrow e^{30k} = \frac{75}{110} = \frac{15}{22} \Longrightarrow k = \frac{1}{30}\ln\frac{15}{22}$, so $y(t) = 110e^{\frac{1}{30}t\ln\left(\frac{15}{22}\right)}$

and $y(45) = 110e^{\frac{45}{30}\ln\left(\frac{15}{22}\right)} \approx 62°$F. Thus, $u(45) \approx 62 + 75 = 137°$F.

(c) $u(t) = 100 \Longrightarrow y(t) = 25.$ $y(t) = 110e^{\frac{1}{30}t\ln\left(\frac{15}{22}\right)} = 25 \Longrightarrow e^{\frac{1}{30}t\ln\left(\frac{15}{22}\right)} = \frac{25}{110} \Longrightarrow$

$\frac{1}{30}t\ln\frac{15}{22} = \ln\frac{25}{110} \Longrightarrow t = \frac{30\ln\frac{25}{110}}{\ln\frac{15}{22}} \approx 116$ min.

15. (a) Let $P(h)$ be the pressure at altitude h. Then $dP/dh = kP \Longrightarrow P(h) = P(0)e^{kh} = 101.3e^{kh}$.

$P(1000) = 101.3e^{1000k} = 87.14 \Longrightarrow 1000k = \ln\left(\frac{87.14}{101.3}\right) \Longrightarrow P(h) = 101.3\,e^{\frac{1}{1000}h\ln\left(\frac{87.14}{101.3}\right)}$, so

$P(3000) = 101.3e^{3\ln\left(\frac{87.14}{101.3}\right)} \approx 64.5$ kPa.

(b) $P(6187) = 101.3\,e^{\frac{6187}{1000}\ln\left(\frac{87.14}{101.3}\right)} \approx 39.9$ kPa

17. (a) Using $A = A_0\left(1 + \dfrac{r}{n}\right)^{nt}$ with $A_0 = 3000$, $r = 0.05$, and $t = 5$, we have:

  (i) Annually: $n = 1$; $A = 3000(1.05)^5 = \$3828.84$

  (ii) Semiannually: $n = 2$; $A = 3000\left(1 + \frac{0.05}{2}\right)^{10} = \$3840.25$

  (iii) Monthly: $n = 12$; $A = 3000\left(1 + \frac{0.05}{12}\right)^{60} = \$3850.08$

  (iv) Weekly: $n = 365$; $A = 3000\left(1 + \frac{0.05}{52}\right)^{5\cdot52} = \$3851.61$

  (v) Daily: $n = 365$; $A = 3000\left(1 + \frac{0.05}{365}\right)^{5\cdot365} = \$3852.01$

  (vi) Continuously: $A = 3000e^{(0.05)5} = \$3852.08$

(b) $dA/dt = 0.05A$ and $A(0) = 3000$.

19. (a) $\dfrac{dP}{dt} = kP - m = k\left(P - \dfrac{m}{k}\right)$. Let $y = P - \dfrac{m}{k}$, so the equation becomes $\dfrac{dy}{dt} = ky$. The solution

is $y = y_0e^{kt} \Longrightarrow P - \dfrac{m}{k} = \left(P_0 - \dfrac{m}{k}\right)e^{kt} \Longrightarrow P(t) = \dfrac{m}{k} + \left(P_0 - \dfrac{m}{k}\right)e^{kt}$.

(b) There will be an exponential expansion $\Longleftrightarrow P_0 - \dfrac{m}{k} > 0 \Longleftrightarrow m < kP_0$.

(c) The population will be constant if $P_0 - \dfrac{m}{k} = 0 \Longleftrightarrow m = kP_0$. It will decline if $P_0 - \dfrac{m}{k} < 0 \Longleftrightarrow$

$m > kP_0$.

(d) $P_0 = 8{,}000{,}000$, $k = \alpha - \beta = 0.016$, $m = 210{,}000 \Longrightarrow m > kP_0\ (= 128{,}000)$, so by part (c), the

population was declining.

Section 7.6 The Logistic Equation

1. (a) $dP/dt = 0.05P - 0.0005P^2 = 0.05P\,(1 - 0.01P) = 0.05P\,(1 - P/100)$. Comparing to Equation 1, $dP/dt = kP\,(1 - P/K)$, we see that the carrying capacity is $K = 100$ and the value of k is 0.05.

(b) The slopes close to 0 occur where P is near 100. The largest slopes appear to be on the line $P = 50$. The solutions are increasing for $0 < P_0 < 100$ and decreasing for $P_0 > 100$.

(c)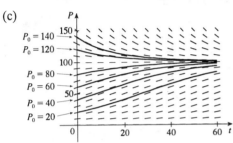

All of the solutions approach $P = 100$ as t increases. As in part (b), the solutions differ since for $0 < P_0 < 100$ they are increasing, and for $P_0 > 100$ they are decreasing. Also, some have an IP and some don't. It appears that the solutions which have $P_0 = 20$ and $P_0 = 40$ have inflection points at $P = 50$.

(d) The equilibrium solutions are $P = 0$ (trivial solution) and $P = 100$. The increasing solutions move away from $P = 0$ and all nonzero solutions approach $P = 100$ as $t \to \infty$.

3. (a) $\dfrac{dy}{dt} = ky\left(1 - \dfrac{y}{K}\right) \Longrightarrow y\,(t) = \dfrac{K}{1 + Ae^{-kt}}$ with $A = \dfrac{K - y\,(0)}{y\,(0)}$. With $K = 8 \times 10^7, k = 0.71$, and $y\,(0) = 2 \times 10^7$, we get the model $y\,(t) = \dfrac{8 \times 10^7}{1 + 3e^{-0.71t}}$, so $y\,(1) = \dfrac{8 \times 10^7}{1 + 3e^{-0.71}} \approx 3.23 \times 10^7$ kg.

(b) $y\,(t) = 4 \times 10^7 \Longrightarrow \dfrac{8 \times 10^7}{1 + 3e^{-0.71t}} = 4 \times 10^7 \Longrightarrow 2 = 1 + 3e^{-0.71t} \Longrightarrow e^{-0.71t} = \tfrac{1}{3} \Longrightarrow$

$-0.71t = \ln\tfrac{1}{3} \Longrightarrow t = \dfrac{\ln 3}{0.71} \approx 1.55$ yr.

5. (a) We will assume that the difference in the birth and death rates is 20 million/yr. Let $t = 0$ correspond to the year 1990 and use a unit of 1 million for all calculations. $k \approx \dfrac{1}{P}\dfrac{dP}{dt} = \tfrac{1}{5300}\,(20) = \tfrac{1}{265}$, so

$$\dfrac{dP}{dt} = kP\left(1 - \dfrac{P}{K}\right) = \dfrac{1}{265}P\left(1 - \dfrac{P}{100{,}000}\right).$$

(b) $A = \dfrac{K - P_0}{P_0} = \dfrac{100{,}000 - 5300}{5300} = \dfrac{947}{53} \approx 17.8679$. $P\,(t) = \dfrac{K}{1 + Ae^{-kt}} = \dfrac{100{,}000}{1 + \frac{947}{53}e^{-(1/265)t}}$,

so $P\,(10) \approx 5492.6$ (or 5.5 billion), $P\,(110) \approx 7813.8$, and $P\,(510) \approx 27{,}718.3$.

(c) If $K = 50{,}000$, then $P\,(t) = \dfrac{50{,}000}{1 + \frac{447}{53}e^{-(1/265)t}}$. So $P\,(10) \approx 5481.5$, $P\,(110) \approx 7611.8$, and $P\,(510) \approx 22{,}412.6$

7. (a) Our assumption is that $\dfrac{dy}{dt} = ky\,(1-y)$, where y is the fraction of the population that has heard the rumor.

(b) Using the logistic Equation (1), $\dfrac{dP}{dt} = kP\left(1 - \dfrac{P}{K}\right)$, we substitute $y = \dfrac{P}{K}$, $P = Ky$, and

$\dfrac{dP}{dt} = K\dfrac{dy}{dt}$, to obtain $K\dfrac{dy}{dt} = k\,(Ky)\,(1-y) \iff \dfrac{dy}{dt} = ky\,(1-y)$, our equation in part (a).

Now the solution to (1) is $P\,(t) = \dfrac{K}{1 + Ae^{-kt}}$, where $A = \dfrac{K - P_0}{P_0}$. We use the same substitution

to obtain $Ky = \dfrac{K}{1 + \dfrac{K - Ky_0}{Ky_0}e^{-kt}} \implies y = \dfrac{y_0}{y_0 + (1 - y_0)\,e^{-kt}}$.

Alternatively, we could use the same steps as outlined in "The Analytic Solution", following Example 2.

(c) Let t be the number of hours since 8 A.M. Then $y_0 = y\,(0) = \frac{80}{1000} = 0.08$ and $y\,(4) = \frac{1}{2}$,

so $\dfrac{1}{2} = y\,(4) = \dfrac{0.08}{0.08 + 0.92e^{-4k}}$. Thus, $0.08 + 0.92e^{-4k} = 0.16$, $e^{-4k} = \frac{0.08}{0.92} = \frac{2}{23}$, and

$e^{-k} = \left(\frac{2}{23}\right)^{1/4}$, so $y = \dfrac{0.08}{0.08 + 0.92\,(2/23)^{t/4}} = \dfrac{2}{2 + 23\,(2/23)^{t/4}}$. Solving this equation for t, we

get $2y + 23y\left(\dfrac{2}{23}\right)^{t/4} = \dfrac{2 - 2y}{23y} \implies \left(\dfrac{2}{23}\right)^{t/4} = \dfrac{2}{23}\cdot\dfrac{1-y}{y} \implies \left(\dfrac{2}{23}\right)^{t/4-1} = \dfrac{1-y}{y}$. It

follows that $\dfrac{t}{4} - 1 = \dfrac{\ln\left[(1-y)/y\right]}{\ln\frac{2}{23}}$, so $t = 4\left[1 + \dfrac{\ln\left((1-y)/y\right)}{\ln\frac{2}{23}}\right]$. When $y = 0.9$, $\dfrac{1-y}{y} = \dfrac{1}{9}$,

so $t = 4\left(1 - \dfrac{\ln 9}{\ln\frac{2}{23}}\right) \approx 7.6$ h or 7 h 36 min. Thus, 90% of the population will have heard the rumor by 3:36 P.M..

9. (a) $\dfrac{dP}{dt} = k\,(P)\left(1 - \dfrac{P}{K}\right) \implies$

$\dfrac{d^2P}{dt^2} = k\left[P\left(-\dfrac{1}{K}\dfrac{dP}{dt}\right) + \left(1 - \dfrac{P}{K}\right)\dfrac{dP}{dt}\right] = k\dfrac{dP}{dt}\left(-\dfrac{P}{K} + 1 - \dfrac{P}{K}\right)$

$= k\left[kP\left(1 - \dfrac{P}{K}\right)\right]\left[1 - \dfrac{2P}{K}\right] = k^2P\left(1 - \dfrac{P}{K}\right)\left(1 - \dfrac{2P}{K}\right)$

(b) P grows fastest when P' has a maximum, that is, when $P'' = 0$. From part (a), $P'' = 0 \iff P = 0$, $P = K$, or $P = K/2$. Since $0 < P < K$, we see that $P'' = 0 \iff P = K/2$.

11. (a) The term -15 represents a harvesting of fish at a constant rate — in this case, 15 fish/week. This is the rate at which fish are caught.

(b)

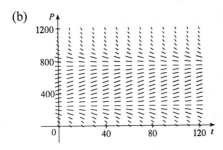

(c) From the graph in part (b), it appears that $P(t) = 250$ and $P(t) = 750$ are the equilibrium solutions. We confirm this analytically by solving the equation $dP/dt = 0$ as follows: $0.08P\left(1 - \dfrac{P}{1000}\right) - 15 = 0 \Longrightarrow$

$0.08P - 0.00008P^2 - 15 = 0 \Longrightarrow$

$-0.00008\left(P^2 - 1000P + 187{,}500\right) = 0 \Longrightarrow$

$(P - 250)(P - 750) = 0 \Longrightarrow P = 250 \text{ or } 750.$

(d)

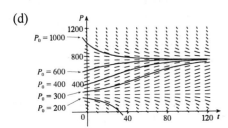

For $0 < P_0 < 250$, $P(t)$ decreases to 0. For $P_0 = 250$, $P(t)$ remains constant. For $250 < P_0 < 750$, $P(t)$ increases and approaches 750. For $P_0 = 750$, $P(t)$ remains constant. For $P_0 > 750$, $P(t)$ decreases and approaches 750.

(e) $\dfrac{dP}{dt} = 0.08P\left(1 - \dfrac{P}{1000}\right) - 15 \iff$

$-\dfrac{100{,}000}{8} \cdot \dfrac{dP}{dt} = \left(0.08P - 0.00008P^2 - 15\right) \cdot \left(-\dfrac{100{,}000}{8}\right) \iff$

$-12{,}500\dfrac{dP}{dt} = P^2 - 1000P + 187{,}500 \iff \dfrac{dP}{(P-250)(P-750)} = -\dfrac{1}{12{,}500}\,dt \iff$

$\displaystyle\int\left(\dfrac{-1/500}{P-250} + \dfrac{1/500}{P-750}\right)dP = -\dfrac{1}{12{,}500}\,dt \iff \int\left(\dfrac{1}{P-250} - \dfrac{1}{P-750}\right)dP = \dfrac{1}{25}\,dt \iff$

$\ln|P-250| - \ln|P-750| = \dfrac{1}{25}t + C \iff \ln\left|\dfrac{P-250}{P-750}\right| = \dfrac{1}{25}t + C \iff$

$\left|\dfrac{P-250}{P-750}\right| = e^{t/25+C} = ke^{t/25} \iff \dfrac{P-250}{P-750} = ke^{t/25} \iff P-250 = Pke^{t/25} - 750ke^{t/25} \iff$

$P - Pke^{t/25} = 250 - 750ke^{t/25} \iff P(t) = \dfrac{250 - 750ke^{t/25}}{1 - ke^{t/25}}$. If $t = 0$ and $P = 200$, then

$200 = \dfrac{250 - 750k}{1 - k} \iff 200 - 200k = 250 - 750k \iff$

$550k = 50 \iff k = \dfrac{1}{11}$. Similarly, if $t = 0$ and $P = 300$, then

$k = -\dfrac{1}{9}$. Simplifying P with these two values of k gives us

$P(t) = \dfrac{250\left(3e^{t/25} - 11\right)}{e^{t/25} - 11}$ and $P(t) = \dfrac{750\left(e^{t/25} + 3\right)}{e^{t/25} + 9}$.

P 1200

0   120
                      t

13. (a) $\dfrac{dP}{dt} = (kP)\left(1 - \dfrac{P}{K}\right)\left(1 - \dfrac{m}{P}\right).$

If $m < P < K$, then $dP/dt = (+)(+)(+) = + \Longrightarrow P$ is increasing.

If $0 < P < m$, then $dP/dt = (+)(+)(-) = - \Longrightarrow P$ is decreasing.

(b)

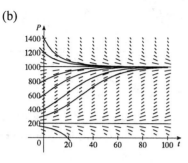

$k = 0.08$, $K = 1000$, and $m = 200 \Longrightarrow$

$$\dfrac{dP}{dt} = 0.08P\left(1 - \dfrac{P}{1000}\right)\left(1 - \dfrac{200}{P}\right).$$

For $0 < P_0 < 200$, the population dies out. For $P_0 = 200$, the population is steady.

For $200 < P_0 < 1000$, the population increases and approaches 1000.

For $P_0 > 1000$, the population decreases and approaches 1000.

The equilibrium solutions are $P(t) = 200$ and $P(t) = 1000$.

(c) $\dfrac{dP}{dt} = kP\left(1 - \dfrac{P}{K}\right)\left(1 - \dfrac{m}{P}\right) = kP\left(\dfrac{K-P}{K}\right)\left(\dfrac{P-m}{P}\right) = \dfrac{k}{K}(K-P)(P-m) \Longleftrightarrow$

$$\int \dfrac{dP}{(K-P)(P-m)} = \int \dfrac{k}{K}\, dt.$$

By partial fractions, $\dfrac{1}{(K-P)(P-m)} = \dfrac{A}{K-P} + \dfrac{B}{P-m}$, so $A(P-m) + B(K-P) = 1.$

If $P = m$, $B = \dfrac{1}{K-m}$; if $P = K$, $A = \dfrac{1}{K-m}$,

so $\dfrac{1}{K-m}\int\left(\dfrac{1}{K-P} + \dfrac{1}{P-m}\right) dP = \int \dfrac{k}{K}\, dt \Longrightarrow$

$\dfrac{1}{K-m}\left(-\ln|K-P| + \ln|P-m|\right) = \dfrac{k}{K}t + M.$

But $m < P < K$, so $\dfrac{1}{K-m}\ln\dfrac{P-m}{K-P} = \dfrac{k}{K}t + M \Longrightarrow$

$\ln\dfrac{P-m}{K-P} = (K-m)\dfrac{k}{K}t + M_1 \Longleftrightarrow \dfrac{P-m}{K-P} = De^{(K-m)(k/K)t}$ $(D = e^{M_1}).$ Let

$t = 0$: $\dfrac{P_0 - m}{K - P_0} = D.$ So $\dfrac{P-m}{K-P} = \dfrac{P_0 - m}{K - P_0}e^{(K-m)(k/K)t}.$ Solving for P, we get

$$P(t) = \dfrac{m(K-P_0) + K(P_0 - m)e^{(K-m)(k/K)t}}{K - P_0 + (P_0 - m)e^{(K-m)(k/K)t}}.$$

(d) If $P_0 < m$, then $P_0 - m < 0$. Let $N(t)$ be the numerator of the expression for $P(t)$ in part (c).

Then $N(0) = P_0(K-m) > 0$, and $P_0 - m < 0 \Longleftrightarrow \lim\limits_{t\to\infty} K(P_0 - m)e^{(K-m)(k/K)t} = -\infty \Longrightarrow$

$\lim\limits_{t\to\infty} N(t) = -\infty.$ Since N is continuous, there is a number t such that $N(t) = 0$ and thus $P(t) = 0$. So the species will become extinct.

15. (a) $\dfrac{dP}{dt} = kP \cos{(rt - \phi)} \implies \dfrac{dP}{P} = k \cos{(rt - \phi)} \, dt \implies \displaystyle\int \dfrac{dP}{P} = k \int \cos{(rt - \phi)} \, dt \implies$

$\ln P = \dfrac{k}{r} \sin{(rt - \phi)} + C$. (Since this is a growth model, $P > 0$ and we can write $\ln P$ instead

of $\ln |P|$.) Since $P(0) = P_0$, we obtain $\ln P_0 = \dfrac{k}{r} \sin{(-\phi)} + C = -\dfrac{k}{r} \sin{\phi} + C \implies$

$C = \ln P_0 + \dfrac{k}{r} \sin{\phi}$. Thus, $\ln P = \dfrac{k}{r} \sin{(rt - \phi)} + \ln P_0 + \dfrac{k}{r} \sin{\phi}$, which we can rewrite as

$\ln \dfrac{P}{P_0} = \dfrac{k}{r} [\sin{(rt - \phi)} + \sin{\phi}]$ or, after exponentiation, $P(t) = P_0 e^{(k/r)[\sin(rt - \phi) + \sin\phi]}$.

(b) As k increases, the amplitude increases, but the minimum value stays the same.

As r increases, the amplitude and the period decrease.

A change in ϕ produces slight adjustments in the phase shift and amplitude.

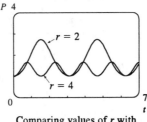

Comparing values of k with $P_0 = 1$, $r = 2$, and $\phi = \pi/2$

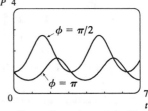

Comparing values of r with $P_0 = 1$, $k = 1$, and $\phi = \pi/2$

Comparing values of ϕ with $P_0 = 1$, $k = 1$, and $r = 2$

$P(t)$ oscillates between $P_0 e^{(k/r)(1 + \sin \phi)}$ and $P_0 e^{(k/r)(-1 + \sin \phi)}$ (the extreme values are attained

when $rt - \phi$ is an odd multiple of $\frac{\pi}{2}$), so $\lim\limits_{t \to \infty} P(t)$ does not exist.

Section 7.7 Predator-Prey Systems

1. (a) $dx/dt = -0.05x + 0.0001xy$. If $y = 0$, we have $dx/dt = -0.05x$, which indicates that in the absence of y, x declines at a rate proportional to itself. So x represents the predator population and y represents the prey population. The growth of the prey population, $0.1y$ (from $dy/dt = 0.1y - 0.005xy$), is restricted only by encounters with predators (the term $-0.005xy$). The predator population increases only through the term $0.0001xy$, that is, by encounters with the prey and not through additional food sources.

(b) $dy/dt = -0.015y + 0.00008xy$. If $x = 0$, we have $dy/dt = -0.015y$, which indicates that in the absence of x, y would decline at a rate proportional to itself. So y represents the predator population and x represents the prey population. The growth of the prey population, $0.2x$ (from $dx/dt = 0.2x - 0.0002x^2 - 0.006xy = 0.2x(1 - 0.001x) - 0.006xy$), is restricted by a carrying capacity of 1000 [from the term $1 - 0.001x = 1 - x/1000$] and by encounters with predators (the term $-0.006xy$). The predator population increases only through the term $0.00008xy$, that is, by encounters with the prey and not through additional food sources.

3. (a) At $t = 0$, there are about 300 rabbits and 100 foxes. At $t = t_1$, the number of foxes reaches a minimum of about 20 while the number of rabbits is about 1000. At $t = t_2$, the number of rabbits reaches a maximum of about 2400, while the number of foxes rebounds to 100. At $t = t_3$, the number of rabbits has decreased to about 1000 and the number of foxes has reached a maximum of about 315. As t increases, the number of foxes decreases dramatically to 100, and the number of rabbits decreases to 300 (the initial populations), and the cycle starts again.

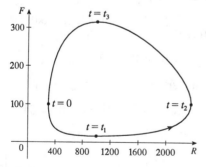

(b)

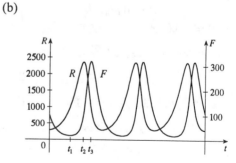

Chapter 7   Differential Equations

5.

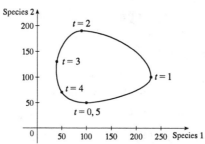

7. $\dfrac{dW}{dR} = \dfrac{-0.02W + 0.00002RW}{0.08R - 0.001RW} \iff (0.08 - 0.001W)\,R\,dW = (-0.02 + 0.00002R)\,W\,dR \iff$

$\dfrac{0.08 - 0.001W}{W}\,dW = \dfrac{-0.02 + 0.00002R}{R}\,dR \iff$

$\int (0.08/W - 0.001)\,dW = \int (-0.02/R + 0.00002)\,dR \iff$

$0.08 \ln|W| - 0.001W = -0.02 \ln|R| + 0.00002R + K \iff$

$0.08 \ln W + 0.02 \ln R = 0.001W + 0.00002R + K \iff$

$\ln\left(W^{0.08}R^{0.02}\right) = 0.00002R + 0.001W + K \iff W^{0.08}R^{0.02} = e^{0.00002R + 0.001W + K} \iff$

$R^{0.02}W^{0.08} = Ce^{0.00002R}e^{0.001W} \iff$

$\dfrac{R^{0.02}W^{0.08}}{e^{0.00002R}e^{0.001W}} = C.$ In general, if $\dfrac{dy}{dx} = \dfrac{-ry + bxy}{kx - axy}$, then $C = \dfrac{x^r y^k}{e^{bx}e^{ay}}$.

9. (a) Letting $W = 0$ gives us $dR/dt = 0.08R\,(1 - 0.0002R)$. $dR/dt = 0 \iff R = 0$ or 5000. Since $dR/dt > 0$ for $0 < R < 5000$, we would expect the rabbit population to *increase* to 5000 for these values of R. Since $dR/dt < 0$ for $R > 5000$, we would expect the rabbit population to *decrease* to 5000 for these values of R. Hence, in the absence of wolves, we would expect the rabbit population to stabilize at 5000.

(b) R and W are constant $\implies R' = 0$ and $W' = 0 \implies$

$\left\{ \begin{array}{l} 0 = 0.08R\,(1 - 0.0002R) - 0.001RW \\ 0 = -0.02W + 0.00002RW \end{array} \right\} \implies \left\{ \begin{array}{l} 0 = R\,[0.08\,(1 - 0.0002R) - 0.001W] \\ 0 = W\,(-0.02 + 0.00002R) \end{array} \right.$

The second equation is true if $W = 0$ or $R = \frac{0.02}{0.00002} = 1000$. If $W = 0$ in the first

equation, then either $R = 0$ or $R = \frac{1}{0.0002} = 5000$ [as in Part (a)]. If $R = 1000$, then

$0 = 1000\,[0.08\,(1 - 0.0002 \cdot 1000) - 0.001W] \iff 0 = 80\,(1 - 0.2) - W \iff W = 64.$

Case (i): $W = 0$, $R = 0$: both populations are zero   (d)

Case (ii): $W = 0$, $R = 5000$: see part (a)

Case (iii): $R = 1000$, $W = 64$: the predator/prey interaction balances and the populations are stable.

(c) The populations of wolves and rabbits fluctuate around 64 and 1000, respectively, and eventually stabilize at those values.

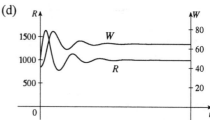

Chapter 7 Review

Concept Check

1. (a) A differential equation is an equation that contains an unknown function and one or more of its derivatives.

(b) The order of a differential equation is the order of the highest derivative that occurs in the equation.

(c) An initial condition is a condition of the form $y(t_0) = y_0$.

2. $y' = x^2 + y^2 \geq 0$ for all x and y. $y' = 0$ only at the origin, so there is a horizontal tangent at $(0,0)$, but nowhere else. The graph of the solution is increasing on every interval.

3. See the paragraph preceding Example 1 in Section 7.2.

4. See the third paragraph in Section 7.3.

5. A separable equation is a first-order differential equation in which the expression for dy/dx can be factored as a function of x times a function of y, that is, $dy/dx = g(x) f(y)$. We can solve the equation by integrating both sides of the equation $dy/f(y) = g(x)\,dx$ and solving for y.

6. (a) $dy/dt = ky$

(b) The equation in part (a) is an appropriate model for population growth, assuming that there is enough room and nutrition to support the growth.

(c) If $y(0) = y_0$, then the solution is $y(t) = y_0 e^{kt}$.

7. (a) $dP/dt = kP(1 - P/K)$, where K is the carrying capacity.

(b) The equation in part (a) is an appropriate model for population growth, assuming that the population grows at a rate proportional to the size of the population in the beginning, but eventually levels off and approaches its carrying capacity because of limited resources.

8. (a) $dF/dt = kF - aFS$ and $dS/dt = -rS + bFS$.

(b) In the absence of sharks, an ample food supply would support exponential growth of the fish population, that is, $dF/dt = kF$, where k is a positive constant. In the absence of fish, we assume that the shark population would decline at a rate proportional to itself, that is, $dS/dt = -rS$, where r is a positive constant.

True-False Quiz

1. False. $y = 0$ is a solution of $y' = -y^4$, but $y = 0$ is not a decreasing function. (All non-trivial solutions are decreasing, however.)

3. False. $x + y$ cannot be written in the form $g(x) f(y)$.

5. True. By comparing $\dfrac{dy}{dt} = 2y \left(1 - \dfrac{y}{5}\right)$ with the logistic differential equation (7.6.1), we see that the carrying capacity is 5, that is, $\lim\limits_{t \to \infty} y = 5$.

Exercises

1. (a)

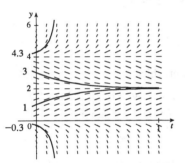

(b) $\lim\limits_{t \to \infty} y(t)$ appears to be finite for $0 \le c \le 4$.

In fact $\lim\limits_{t \to \infty} y(t) = 4$ for $c = 4$,

$\lim\limits_{t \to \infty} y(t) = 2$ for $0 < c < 4$, and

$\lim\limits_{t \to \infty} y(t) = 0$ for $c = 0$. The equilibrium

solutions are $y(t) = 0$, $y(t) = 2$, and

$y(t) = 4$.

3. (a)

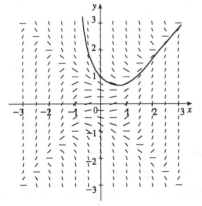

We estimate that when $x = 0.3$, $y = 0.8$, so $y(0.3) \approx 0.8$.

(b) $h = 0.1$, $x_0 = 0$, $y_0 = 1$ and $F(x, y) = x^2 - y^2$. So $y_n = y_{n-1} + 0.1 \left(x_{n-1}^2 - y_{n-1}^2\right)$. Thus,

$y_1 = 1 + 0.1 \left(0^2 - 1^2\right) = 0.9$,

$y_2 = 0.9 + 0.1 \left(0.1^2 - 0.9^2\right) = 0.82$,

$y_3 = 0.82 + 0.1 \left(0.2^2 - 0.82^2\right) =$

$0.75676 \approx y(0.3)$.

(c) The centers of the horizontal line segments of the direction field are located on the lines $y = x$ and $y = -x$. When a solution curve crosses one of these lines, it has a local maximum or minimum.

5. $y' = 2 + 2x^2 + y + x^2 y = (2 + y)\left(1 + x^2\right) \implies \dfrac{dy}{2 + y} = \left(1 + x^2\right) dx \implies$

$\ln|2 + y| = x + \dfrac{x^3}{3} + c_1 \implies 2 + y = ke^{x + x^3/3}$ and the solution is $y(x) = ke^{x + x^3/3} - 2$.

7. $xyy' = \ln x \Longrightarrow y\,dy = \dfrac{\ln x}{x}\,dx \Longrightarrow \int y\,dy = \displaystyle\int \dfrac{\ln x}{x}\,dx$ (Make the substitution $u = \ln x$; then

$du = dx/x$.) So $\int y\,dy = \int u\,du \Longrightarrow \frac{1}{2}y^2 = \frac{1}{2}u^2 + C \Longrightarrow \frac{1}{2}y^2 = \frac{1}{2}\left(\ln x\right)^2 + C.\ y(1) = 2 \Longrightarrow$

$\frac{1}{2}2^2 = \frac{1}{2}\left(\ln 1\right)^2 + C = C \Longleftrightarrow C = 2.$ Therefore, $\frac{1}{2}y^2 = \frac{1}{2}\left(\ln x\right)^2 + 2$, or $y = \sqrt{\left(\ln x\right)^2 + 4}$.

The negative square root is inadmissible, since $y(1) > 0$.

9. The curves $kx^2 + y^2 = 1$ form a family of ellipses for $k > 0$, a family of hyperbolas for $k < 0$, and two

parallel lines $y = \pm 1$ for $k = 0$. Solving $kx^2 + y^2 = 1$ for k gives $k = \dfrac{1 - y^2}{x^2}$. Differentiating

gives $2kx + 2yy' = 0 \Longleftrightarrow y' = -\dfrac{kx}{y} = -\left(1 - y^2\right)\dfrac{x}{yx^2} = \dfrac{y^2 - 1}{xy}$. Thus, for $k \neq 0$ the orthogonal

trajectories must satisfy $y' = -\dfrac{xy}{y^2 - 1} \Longrightarrow \dfrac{y^2 - 1}{y}\,dy = -x\,dx \Longrightarrow \dfrac{y^2}{2} - \ln|y| = \dfrac{-x^2}{2} + c$. For

$k = 0$, the orthogonal trajectories are given by $x = c_1$ for c_1 an arbitrary constant.

11. (a) $y(t) = y(0)\,e^{kt} = 1000e^{kt} \Longrightarrow y(2) = 1000e^{2k} = 9000 \Longrightarrow e^{2k} = 9 \Longrightarrow 2k = \ln 9 \Longrightarrow$

$\quad k = \frac{1}{2}\ln 9 = \ln 3 \Longrightarrow y(t) = 1000e^{(\ln 3)t} = 1000 \cdot 3^t$

(b) $y(3) = 1000 \cdot 3^3 = 27{,}000$

(c) $1000 \cdot 3^t = 2000 \Longrightarrow 3^t = 2 \Longrightarrow t\ln 3 = \ln 2 \Longrightarrow t = (\ln 2)/(\ln 3) \approx 0.63\ \text{h}$

13. (a) $C'(t) = -kC(t) \Longrightarrow C(t) = C(0)\,e^{-kt}$ by Theorem 7.5.2. But $C(0) = C_0$, so $C(t) = C_0 e^{-kt}$.

(b) $C(30) = \frac{1}{2}C_0$ since the concentration is reduced by half. Thus, $\frac{1}{2}C_0 = C_0 e^{-30k} \Longrightarrow$

$\quad \ln\frac{1}{2} = -30k \Longrightarrow k = -\frac{1}{30}\ln\frac{1}{2} = \frac{1}{30}\ln 2.$ Since 10% of the original concentration remains if 90%

$\quad$ is eliminated, we want the value of t such that $C(t) = \frac{1}{10}C_0$. Therefore, $\frac{1}{10}C_0 = C_0 e^{-t(\ln 2)/30} \Longrightarrow$

$\quad \ln 0.1 = -t(\ln 2)/30 \Longrightarrow t = -\frac{30}{\ln 2}\ln 0.1 \approx 100\ \text{h}.$

15. (a) $\dfrac{dL}{dt} \propto L_\infty - L \Longrightarrow \dfrac{dL}{dt} = k(L_\infty - L) \Longrightarrow \displaystyle\int \dfrac{dL}{L_\infty - L} = \int k\,dt \Longrightarrow -\ln|L_\infty - L| = kt + C \Longrightarrow$

$\quad \ln|L_\infty - L| = -kt - C \Longrightarrow |L_\infty - L| = e^{-kt-C} \Longrightarrow L_\infty - L = Ae^{-kt} \Longrightarrow L = L_\infty - Ae^{-kt}.$

$\quad$ At $t = 0,\ L = L(0) = L_\infty - A \Longrightarrow A = L_\infty - L(0) \Longrightarrow L(t) = L_\infty - [L_\infty - L(0)]\,e^{-kt}$

(b) $L_\infty = 53\ \text{cm},\ L(0) = 10\ \text{cm, and } k = 0.2 \Longrightarrow L(t) = 53 - (53 - 10)\,e^{-0.2t} = 53 - 43e^{-0.2t}.$

17. Let P be the population and I be the number of infected people. The rate of spread dI/dt is jointly proportional to I and to $P - I$, so for some constant k, $dI/dt = kI\,(P - I) \Longrightarrow$
$I = \dfrac{I_0 P}{I_0 + (P - I_0)\,e^{-kPt}}$ (from the discussion of logistic growth in Section 7.6).

Now, measuring t in days, we substitute $t = 7$, $P = 5000$, $I_0 = 160$ and $I\,(7) = 1200$ to find k:

$1200 = \dfrac{160 \cdot 5000}{160 + (5000 - 160)\,e^{-5000 \cdot 7 \cdot k}} \Longleftrightarrow k \approx 0.00006448$. So, putting $I = 5000 \times 80\% = 4000$,

we solve for t: $4000 = \dfrac{160 \cdot 5000}{160 + (5000 - 160)\,e^{-0.00006448 \cdot 5000 \cdot t}} \Longleftrightarrow 160 + 4840e^{-0.3224t} = 200 \Longleftrightarrow$

$-0.3224t = \ln\frac{40}{4840} \Longleftrightarrow t \approx 14.9$. So it takes about 15 days for 80% of the population to be infected.

19. (a) $dx/dt = 0.4x\,(1 - 0.000005x) - 0.002xy$, $dy/dt = -0.2y + 0.000008xy$. If $y = 0$, then $dx/dt = 0.4x\,(1 - 0.000005x)$, so $dx/dt = 0 \Longleftrightarrow x = 0$ or $x = 200{,}000$, which shows that the insect population increases logistically with a carrying capacity of 200,000. Since $dx/dt > 0$ for $0 < x < 200{,}000$ and $dx/dt < 0$ for $x > 200{,}000$, we expect the insect population to stabilize at 200,000.

(b) x and y are constant $\Longrightarrow x' = 0$ and $y' = 0 \Longrightarrow$

$\begin{cases} 0 = 0.4x\,(1 - 0.000005x) - 0.002xy \\ 0 = -0.2y + 0.000008xy \end{cases} \Longrightarrow \begin{cases} 0 = 0.4x\,[(1 - 0.000005x) - 0.005y] \\ 0 = y\,(-0.2 + 0.000008x) \end{cases}$

The second equation is true if $y = 0$ or $x = \frac{0.2}{0.000008} = 25{,}000$. If $y = 0$ in the

first equation, then either $x = 0$ or $x = \frac{1}{0.000005} = 200{,}000$. If $x = 25{,}000$, then

$0 = 0.4\,(25{,}000)\,[(1 - 0.000005 \cdot 25{,}000) - 0.005y] \Longrightarrow 0 = 10{,}000\,[(1 - 0.125) - 0.005y] \Longrightarrow$
$0 = 8750 - 50y \Longrightarrow y = 175$.

Case (i): $y = 0$, $x = 0$: zero populations

Case (ii): $y = 0$, $x = 200{,}000$: in the absence of birds, the insect population is constantly 200,000.

Case (iii): $x = 25{,}000$, $y = 175$: the predator/prey interaction balances and the populations are stable.

(c) The populations of the birds and insects fluctuate around 175 and 25,000, respectively, and eventually stabilize at those values.

(d)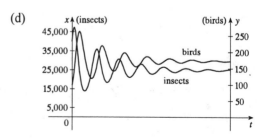

Focus on Problem Solving

1. We use the Fundamental Theorem of Calculus to differentiate the given equation:

$$[f(x)]^2 = 100 + \int_0^x \left\{ [f(t)]^2 + [f'(t)]^2 \right\} dt \implies 2f(x)f'(x) = [f(x)]^2 + [f'(x)]^2 \implies$$

$$[f(x)]^2 + [f'(x)]^2 - 2f(x)f'(x) = 0 \implies [f(x) - f'(x)]^2 = 0 \iff f(x) = f'(x). \text{ We can solve}$$

this as a separable equation, or else use Theorem 7.5.2 with $k = 1$, which says that the solutions are $f(x) = Ce^x$. Now $[f(0)]^2 = 100$, so $f(0) = C = \pm 10$, and hence $f(x) = \pm 10e^x$ are the only functions satisfying the given equation.

3.
$$f'(x) = \lim_{h \to 0} \frac{f(x+h) - f(x)}{h} = \lim_{h \to 0} \frac{f(x)[f(h) - 1]}{h} \quad [\text{since } f(x+h) = f(x)f(h)]$$

$$= f(x) \lim_{h \to 0} \frac{f(h) - 1}{h} = f(x) \lim_{h \to 0} \frac{f(h) - f(0)}{h - 0} = f(x)f'(0) = f(x)$$

Therefore, $f'(x) = f(x)$ for all x and from Theorem 7.5.2 we get $f(x) = Ae^x$.
Now $f(0) = 1 \implies A = 1 \implies f(x) = e^x$.

5. (a) We are given that $V = \frac{1}{3}\pi r^2 h$, $dV/dt = 60{,}000\pi$ ft^3/h, and $r = 1.5h = \frac{3}{2}h$.

So $V = \frac{1}{3}\pi \left(\frac{3}{2}h\right)^2 h = \frac{3}{4}\pi h^3 \implies \dfrac{dV}{dt} = \frac{3}{4}\pi \cdot 3h^2 \dfrac{dh}{dt} = \frac{9}{4}\pi h^2 \dfrac{dh}{dt}$. Therefore,

$$\frac{dh}{dt} = \frac{4\,(dV/dt)}{9\pi h^2} = \frac{240{,}000\pi}{9\pi h^2} = \frac{80{,}000}{3h^2} \; (\bigstar) \implies \int 3h^2 dh = \int 80{,}000\,dt \implies h^3 = 80{,}000t + C.$$

When $t = 0$, $h = 60$. Thus, $C = 60^3 = 216{,}000$, so $h^3 = 80{,}000t + 216{,}000$. Let $h = 100$. Then $100^3 = 1{,}000{,}000 = 80{,}000t + 216{,}000 \implies 80{,}000t = 784{,}000 \implies t = 9.8$, so the time required is 9.8 hours.

(b) The floor area of the silo is $F = \pi \cdot 200^2 = 40{,}000\pi$ ft^2, and the area of the base of the pile is $A = \pi r^2 = \pi \left(\frac{3}{2}h\right)^2 = \frac{9\pi}{4}h^2$. So the area of the floor which is not covered when $h = 60$ is

$$F - A = 40{,}000\pi - 8100\pi = 31{,}900\pi \approx 100{,}217 \text{ ft}^2. \text{ Now } A = \frac{9\pi}{4}h^2 \implies \frac{dA}{dt} = \frac{9\pi}{4} \cdot 2h\frac{dh}{dt},$$

and from $(\bigstar)$ in part (a) we know that when $h = 60$, $\dfrac{dh}{dt} = \dfrac{80{,}000}{3(60)^2} = \dfrac{200}{27} \dfrac{\text{ft}}{\text{h}}$. Therefore,

$$dA/dt = \tfrac{9\pi}{4}(2)(60)\left(\tfrac{200}{27}\right) = 2000\pi \approx 6283 \text{ ft}^2/\text{h}.$$

(c) At $h = 90$ ft, $\dfrac{dV}{dt} = 60{,}000\pi - 20{,}000\pi = 40{,}000\pi$ ft^3/h. From $(\bigstar)$ in part (a),

$$\frac{dh}{dt} = \frac{4\,(dV/dt)}{9\pi h^2} = \frac{4\,(40{,}000\pi)}{9\pi h^2} = \frac{160{,}000}{9h^2} \implies \int 9h^2 dh = \int 160{,}000\,dt \implies$$

$3h^3 = 160{,}000t + C$. When $t = 0$, $h = 90$; therefore, $C = 3 \cdot 729{,}000 = 2{,}187{,}000$. So $3h^3 = 160{,}000t + 2{,}187{,}000$. At the top, $h = 100 \implies 3(100)^3 = 160{,}000t + 2{,}187{,}000 \implies$ $t = \frac{813{,}000}{160{,}000} \approx 5.1$. The pile reaches the top after about 5.1 h.

7. (a) While running from $(L, 0)$ to (x, y), the dog travels a distance

$$s = \int_x^L \sqrt{1 + (dy/dx)^2}\, dx = -\int_L^x \sqrt{1 + (dy/dx)^2}\, dx, \text{ so } \frac{ds}{dx} = -\sqrt{1 + (dy/dx)^2}. \text{ The dog}$$

and rabbit run at the same speed, so the rabbit's position when the dog has traveled a distance s is

$(0, s)$. Since the dog runs straight for the rabbit, $\dfrac{dy}{dx} = \dfrac{s - y}{0 - x}$ (see the figure).

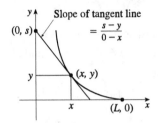

Thus, $s = y - x\dfrac{dy}{dx} \implies \dfrac{ds}{dx} = \dfrac{dy}{dx} - \left(x\dfrac{d^2y}{dx^2} + 1\dfrac{dy}{dx}\right) = -x\dfrac{d^2y}{dx^2}$. Equating the two expressions

for $\dfrac{ds}{dx}$ gives us $x\dfrac{d^2y}{dx^2} = \sqrt{1 + \left(\dfrac{dy}{dx}\right)^2}$, as claimed.

(b) Letting $z = \dfrac{dy}{dx}$, we obtain the differential equation $x\dfrac{dz}{dx} = \sqrt{1 + z^2}$, or $\dfrac{dz}{\sqrt{1 + z^2}} = \dfrac{dx}{x}$.

Integrating: $\ln x = \int \dfrac{dz}{\sqrt{1 + z^2}} = \ln\left|z + \sqrt{1 + z^2}\right| + C$ (by Formula 25).

When $x = L$, $z = dy/dx = 0$, so $\ln L = \ln 1 + C$. Therefore, $C = \ln L$, so

$\ln x = \ln\left(\sqrt{1 + z^2} + z\right) + \ln L = \ln\left(L\left(\sqrt{1 + z^2} + z\right)\right) \implies x = L\left(\sqrt{1 + z^2} + z\right) \implies$

$\sqrt{1 + z^2} = \dfrac{x}{L} - z \implies 1 + z^2 = \left(\dfrac{x}{L}\right)^2 - \dfrac{2xz}{L} + z^2 \implies \left(\dfrac{x}{L}\right)^2 - 2z\left(\dfrac{x}{L}\right) - 1 = 0 \implies$

$z = \dfrac{(x/L)^2 - 1}{2(x/L)} = \dfrac{x^2 - L^2}{2Lx} = \dfrac{x}{2L} - \dfrac{L}{2}\dfrac{1}{x}$ (for $x > 0$). Since $z = \dfrac{dy}{dx}$, $y = \dfrac{x^2}{4L} - \dfrac{L}{2}\ln x + C_1$.

Since $y = 0$ when $x = L$, $0 = \dfrac{L}{4} - \dfrac{L}{2}\ln L + C_1 \implies C_1 = \dfrac{L}{2}\ln L - \dfrac{L}{4}$. Thus,

$$y = \dfrac{x^2}{4L} - \dfrac{L}{2}\ln x + \dfrac{L}{2}\ln L - \dfrac{L}{4} = \dfrac{x^2 - L^2}{4L} - \dfrac{L}{2}\ln\left(\dfrac{x}{L}\right).$$

(c) As $x \to 0^+$, $y \to \infty$, so the dog never catches the rabbit.

Chapter 8 Infinite Sequences and Series

Section 8.1 Sequences

1. (a) A sequence is an ordered list of numbers. It can also be defined as a function whose domain is the set of positive integers.

(b) The terms a_n approach 8 as n becomes large. In fact, we can make a_n as close to 8 as we like by taking n sufficiently large.

(c) The terms a_n become large as n becomes large.

3. The first six terms of $a_n = \dfrac{n}{2n+1}$ are: $\dfrac{1}{3}, \dfrac{2}{5}, \dfrac{3}{7}, \dfrac{4}{9}, \dfrac{5}{11}, \dfrac{6}{13}$. It appears that the sequence is approaching $\dfrac{1}{2}$. $\displaystyle\lim_{n\to\infty} \dfrac{n}{2n+1} = \lim_{n\to\infty} \dfrac{1}{2+1/n} = \dfrac{1}{2}$

5. The numerators are all 1 and the denominators are powers of 2, so $a_n = \dfrac{1}{2^n}$.

7. In each term, the numerator is 2 greater than the number of the term, and the denominator is the square of the number that is 3 greater than the term number. So $a_n = \dfrac{n+2}{(n+3)^2}$.

9. $\displaystyle\lim_{n\to\infty} \dfrac{1}{5^n} = \lim_{n\to\infty} \left(\dfrac{1}{5}\right)^n = 0$ by Equation 6 with $r = \dfrac{1}{5}$. Convergent

11. $\displaystyle\lim_{n\to\infty} \dfrac{n^2-1}{n^2+1} = \lim_{n\to\infty} \dfrac{1-1/n^2}{1+1/n^2} = 1$. Convergent

13. $\{a_n\}$ diverges since $\dfrac{n^2}{n+1} = \dfrac{n}{1+1/n} \to \infty$ as $n \to \infty$.

15. $\{a_n\} = \left\{\cos\frac{\pi}{2}, \cos\pi, \cos\frac{3\pi}{2}, \cos 2\pi, \cos\frac{5\pi}{2}, \cos 3\pi, \cos\frac{7\pi}{2}, \cos 4\pi, \dots\right\}$
$= \{0, -1, 0, 1, 0, -1, 0, 1, \dots\}$
This sequence oscillates among 0, -1, and 1 and so diverges.

17. $a_n = \dfrac{\pi^n}{3^n} = \left(\dfrac{\pi}{3}\right)^n$, so $\{a_n\}$ diverges by Equation 6 with $r = \frac{\pi}{3} > 1$.

19. $\displaystyle\lim_{x\to\infty} \dfrac{\ln(x^2)}{x} \overset{\text{H}}{=} \lim_{x\to\infty} \dfrac{2\ln x}{x} \overset{\text{H}}{=} \lim_{x\to\infty} \dfrac{2/x}{1} = 0$, so by Theorem 2, $\left\{\dfrac{\ln(n^2)}{n}\right\}$ converges to 0.

21. $b_n = \sqrt{n+2} - \sqrt{n} = (\sqrt{n+2} - \sqrt{n}) \dfrac{\sqrt{n+2}+\sqrt{n}}{\sqrt{n+2}+\sqrt{n}} = \dfrac{2}{\sqrt{n+2}+\sqrt{n}} < \dfrac{2}{2\sqrt{n}} = \dfrac{1}{\sqrt{n}} \to 0$ as
$n \to \infty$. So by the Squeeze Theorem with $a_n = 0$ and $c_n = 1/\sqrt{n}$, $\{\sqrt{n+2} - \sqrt{n}\}$ converges to 0.

23. $\displaystyle\lim_{x\to\infty} \dfrac{x}{2^x} \overset{\text{H}}{=} \lim_{x\to\infty} \dfrac{1}{(\ln 2)\, 2^x} = 0$, so by Theorem 2, $\{n2^{-n}\}$ converges to 0.

25. $0 \le \dfrac{\cos^2 n}{2^n} \le \dfrac{1}{2^n}$ [since $0 \le \cos^2 n \le 1$], so since $\lim\limits_{n \to \infty} \dfrac{1}{2^n} = 0$, $\left\{ \dfrac{\cos^2 n}{2^n} \right\}$ converges to 0 by the

Squeeze Theorem.

27.

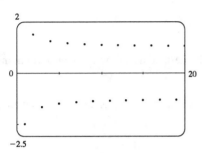

From the graph, we see that the sequence

$\left\{ (-1)^n \dfrac{n+1}{n} \right\}$ is divergent, since it

oscillates between 1 and -1 (approximately).

29.

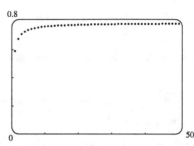

From the graph, it appears that the sequence
converges to about 0.78.

$\lim\limits_{n \to \infty} \dfrac{2n}{2n+1} = \lim\limits_{n \to \infty} \dfrac{2}{2 + 1/n} = 1$, so

$\lim\limits_{n \to \infty} \arctan\left(\dfrac{2n}{2n+1} \right) = \arctan 1 = \dfrac{\pi}{4}$.

31.

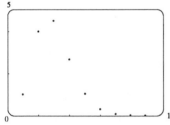

From the graph, it appears that the sequence converges to 0.

$0 < a_n = \dfrac{n^3}{n!} = \dfrac{n}{n} \cdot \dfrac{n}{(n-1)} \cdot \dfrac{n}{(n-2)} \cdot \dfrac{1}{(n-3)} \cdots \cdots \dfrac{1}{3} \cdot \dfrac{1}{2} \cdot \dfrac{1}{1}$

$\le \dfrac{n^2}{(n-1)(n-2)(n-3)}$ (for $n \ge 4$)

$= \dfrac{1/n}{(1 - 1/n)(1 - 2/n)(1 - 3/n)} \to 0$ as $n \to \infty$

So by the Squeeze Theorem, $\{ n^3/n! \}$ converges to 0.

33. (a) $a_1 = 1$, $a_2 = 4 - a_1 = 4 - 1 = 3$, $a_3 = 4 - a_2 = 4 - 3 = 1$, $a_4 = 4 - a_3 = 4 - 1 = 3$,

$a_5 = 4 - a_4 = 4 - 3 = 1$. Since the terms of the sequence alternate between 1 and 3, the sequence
is divergent.

(b) $a_1 = 2$, $a_2 = 4 - a_1 = 4 - 2 = 2$, $a_3 = 4 - a_2 = 4 - 2 = 2$. Since all of the terms are 2,

$\lim\limits_{n \to \infty} a_n = 2$ and hence, the sequence is convergent.

35. (a) Let a_n be the number of rabbit pairs in the nth month. Clearly $a_1 = 1 = a_2$. In the nth month, each
pair that is 2 or more months old (that is, a_{n-2} pairs) will have a pair of children to add to the a_{n-1}
pairs already present. Thus, $a_n = a_{n-1} + a_{n-2}$, so that $\{ a_n \} = \{ f_n \}$, the Fibonacci sequence.

(b) $a_n = \dfrac{f_{n+1}}{f_n} \implies a_{n-1} = \dfrac{f_n}{f_{n-1}} = \dfrac{f_{n-1} + f_{n-2}}{f_{n-1}} = 1 + \dfrac{f_{n-2}}{f_{n-1}} = 1 + \dfrac{1}{f_{n-1}/f_{n-2}} = 1 + \dfrac{1}{a_{n-2}}$. If

$L = \lim\limits_{n \to \infty} a_n$, then $L = \lim\limits_{n \to \infty} a_{n-1}$ and $L = \lim\limits_{n \to \infty} a_{n-2}$, so L must satisfy $L = 1 + 1/L \implies$

$L^2 - L - 1 = 0 \implies L = \dfrac{1 + \sqrt{5}}{2}$ (since L must be positive).

37. $3(n+1) + 5 > 3n + 5$ so $\dfrac{1}{3(n+1)+5} < \dfrac{1}{3n+5} \iff a_{n+1} < a_n$, so $\{a_n\}$ is decreasing.

39. $\left\{\dfrac{n-2}{n+2}\right\}$ is increasing since $a_n < a_{n+1} \iff \dfrac{n-2}{n+2} < \dfrac{(n+1)-2}{(n+1)+2} \iff$

$(n-2)(n+3) < (n+2)(n-1) \iff n^2 + n - 6 < n^2 + n - 2 \iff -6 < -2$, which is true.

41. Since $\{a_n\}$ is a decreasing sequence, $a_n > a_{n+1}$ for all $n \geq 1$. Because all of its terms lie between 5 and 8, $\{a_n\}$ is a bounded sequence. By the Monotonic Sequence Theorem, $\{a_n\}$ is convergent, that is, $\{a_n\}$ has a limit L. L must be less than 8 since $\{a_n\}$ is decreasing, so $5 \leq L < 8$.

43. We show by induction that $\{a_n\}$ is increasing and bounded above by 3.

Let P_n be the proposition that $a_{n+1} > a_n$ and $0 < a_n < 3$. Clearly P_1 is true. Assume that P_n is true. Then $a_{n+1} > a_n \implies 1/a_{n+1} < 1/a_n \implies -1/a_{n+1} > -1/a_n$.

Now $a_{n+2} = 3 - 1/a_{n+1} > 3 - 1/a_n = a_{n+1} \iff P_{n+1}$. This proves that $\{a_n\}$ is increasing and bounded above by 3, so $1 = a_1 < a_n < 3$, that is, $\{a_n\}$ is bounded, and hence convergent by the Monotonic Sequence Theorem. If $L = \lim\limits_{n\to\infty} a_n$, then $\lim\limits_{n\to\infty} a_{n+1} = L$ also, so L must satisfy

$L = 3 - 1/L \implies L^2 - 3L + 1 = 0 \implies L = \frac{3 \pm \sqrt{5}}{2}$. But $L > 1$, so $L = \frac{3+\sqrt{5}}{2}$.

45. $(0.8)^n < 0.000001 \implies \ln(0.8)^n < \ln(0.000001) \implies n\ln(0.8) < \ln(0.000001) \implies$

$n > \dfrac{\ln(0.000001)}{\ln(0.8)} \implies n > 61.9$, so n must be at least 62 to satisfy the given inequality.

47. (a) First we show that $a > a_1 > b_1 > b$.

$a_1 - b_1 = \dfrac{a+b}{2} - \sqrt{ab} = \frac{1}{2}\left(a - 2\sqrt{ab} + b\right) = \frac{1}{2}\left(\sqrt{a} - \sqrt{b}\right)^2 > 0$ (since $a > b$) $\implies a_1 > b_1$.

Also $a - a_1 = a - \frac{1}{2}(a+b) = \frac{1}{2}(a-b) > 0$ and $b - b_1 = b - \sqrt{ab} = \sqrt{b}\left(\sqrt{b} - \sqrt{a}\right) < 0$, so $a > a_1 > b_1 > b$. In the same way we can show that $a_1 > a_2 > b_2 > b_1$ and so the given assertion is true for $n = 1$. Suppose it is true for $n = k$, that is, $a_k > a_{k+1} > b_{k+1} > b_k$. Then

$a_{k+2} - b_{k+2} = \frac{1}{2}(a_{k+1} + b_{k+1}) - \sqrt{a_{k+1}b_{k+1}} = \frac{1}{2}\left(a_{k+1} - 2\sqrt{a_{k+1}b_{k+1}} + b_{k+1}\right)$

$= \frac{1}{2}\left(\sqrt{a_{k+1}} - \sqrt{b_{k+1}}\right)^2 > 0,$

$a_{k+1} - a_{k+2} = a_{k+1} - \frac{1}{2}(a_{k+1} + b_{k+1}) = \frac{1}{2}(a_{k+1} - b_{k+1}) > 0$, and

$b_{k+1} - b_{k+2} = b_{k+1} - \sqrt{a_{k+1}b_{k+1}} = \sqrt{b_{k+1}}\left(\sqrt{b_{k+1}} - \sqrt{a_{k+1}}\right) < 0 \implies$

$a_{k+1} > a_{k+2} > b_{k+2} > b_{k+1}$, so the assertion is true for $n = k+1$. Thus, it is true for all n by mathematical induction.

(b) From part (a) we have $a > a_n > a_{n+1} > b_{n+1} > b_n > b$, which shows that both sequences, $\{a_n\}$ and $\{b_n\}$, are monotonic and bounded. So they are both convergent by the Monotonic Sequence Theorem.

(c) Let $\lim\limits_{n\to\infty} a_n = \alpha$ and $\lim\limits_{n\to\infty} b_n = \beta$. Then $\lim\limits_{n\to\infty} a_{n+1} = \lim\limits_{n\to\infty} \dfrac{a_n + b_n}{2} \implies \alpha = \dfrac{\alpha + \beta}{2} \implies$

$2\alpha = \alpha + \beta \implies \alpha = \beta$.

Section 8.2 Series

1. (a) A sequence is an ordered list of numbers whereas a series is the *sum* of a list of numbers.

(b) A series is convergent if the sequence of partial sums is a convergent sequence. A series is divergent if it is not convergent.

3.

n	s_n
1	3.33333
2	4.44444
3	4.81481
4	4.93827
5	4.97942
6	4.99314
7	4.99771
8	4.99924
9	4.99975
10	4.99992
11	4.99997
12	4.99999

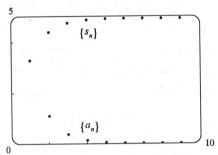

From the graph, it seems that the series converges. In fact, it is a geometric series with $a = \frac{10}{3}$ and $r = \frac{1}{3}$, so its sum is

$$\sum_{n=1}^{\infty} \frac{10}{3^n} = \frac{10/3}{1 - 1/3} = 5.$$ Note that the dot corresponding to $n = 1$ is part of both $\{a_n\}$ and $\{s_n\}$.

5.

n	s_n
1	0.50000
2	1.16667
3	1.91667
4	2.71667
5	3.55000
6	4.40714
7	5.28214
8	6.17103
9	7.07103
10	7.98012

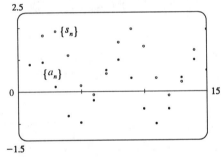

The series diverges, since its terms do not approach 0.

276

7.

n	s_n
1	0.64645
2	0.80755
3	0.87500
4	0.91056
5	0.93196
6	0.94601
7	0.95581
8	0.96296
9	0.96838
10	0.97259

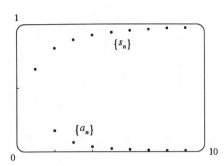

From the graph, it seems that the series converges to 1. To find the sum, we write

$$s_n = \sum_{i=1}^{n} \left(\frac{1}{i^{1.5}} - \frac{1}{(i+1)^{1.5}} \right) = \left(1 - \frac{1}{2^{1.5}}\right) + \left(\frac{1}{2^{1.5}} - \frac{1}{3^{1.5}}\right)$$

$$+ \left(\frac{1}{3^{1.5}} - \frac{1}{4^{1.5}}\right) + \cdots + \left(\frac{1}{n^{1.5}} - \frac{1}{(n+1)^{1.5}}\right) = 1 - \frac{1}{(n+1)^{1.5}}.$$

So the sum is $\lim_{n\to\infty} s_n = 1$.

9. (a) $\lim_{n\to\infty} a_n = \lim_{n\to\infty} \frac{2n}{3n+1} = \frac{2}{3}$, so the *sequence* $\{a_n\}$ is convergent by (8.1.1).

(b) Since $\lim_{n\to\infty} a_n = \frac{2}{3} \neq 0$, the *series* $\sum_{n=1}^{\infty} a_n$ is divergent by the Test for Divergence (7).

11. $4 + \frac{8}{5} + \frac{16}{25} + \frac{32}{125} + \cdots$ is a geometric series with $a = 4$ and $r = \frac{2}{5}$. Since $|r| = \frac{2}{5} < 1$, the series converges to $\frac{4}{1-2/5} = \frac{4}{3/5} = \frac{20}{3}$.

13. $\sum_{n=1}^{\infty} \frac{(-3)^{n-1}}{4^n} = \frac{1}{4}\sum_{n=1}^{\infty} \left(-\frac{3}{4}\right)^{n-1}$. The latter series is geometric with $a = 1$ and $r = -\frac{3}{4}$. Since $|r| = \frac{3}{4} < 1$, it converges to $\frac{1}{1-(-3/4)} = \frac{4}{7}$. Thus, the given series converges to $\left(\frac{1}{4}\right)\left(\frac{4}{7}\right) = \frac{1}{7}$.

15. For $\sum_{n=1}^{\infty} 3^{-n}8^{n+1} = \sum_{n=1}^{\infty} 8\left(\frac{8}{3}\right)^n$, $a = \frac{64}{3}$ and $r = \frac{8}{3} > 1$, so the series diverges.

17. $\sum_{n=1}^{\infty} [2(0.1)^n + (0.2)^n] = 2\sum_{n=1}^{\infty}(0.1)^n + \sum_{n=1}^{\infty}(0.2)^n$. These are convergent geometric series and so by Theorem 8, the sum is also convergent. $2\left(\frac{0.1}{1-0.1}\right) + \frac{0.2}{1-0.2} = \frac{2}{9} + \frac{1}{4} = \frac{17}{36}$

19. $\lim_{n\to\infty} a_n = \lim_{n\to\infty} \frac{n}{\sqrt{1+n^2}} = \lim_{n\to\infty} \frac{1}{\sqrt{1+1/n^2}} = 1 \neq 0$, so the series diverges by the Test for Divergence.

21. Converges. $s_n = \displaystyle\sum_{i=1}^{n} \frac{1}{i(i+2)} = \sum_{i=1}^{n}\left(\frac{1/2}{i} - \frac{1/2}{i+2}\right)$ (using partial fractions) $= \frac{1}{2}\displaystyle\sum_{i=1}^{n}\left(\frac{1}{i} - \frac{1}{i+2}\right)$.

The latter sum is

$$\left(1 - \frac{1}{3}\right) + \left(\frac{1}{2} - \frac{1}{4}\right) + \left(\frac{1}{3} - \frac{1}{5}\right) + \cdots + \left(\frac{1}{n-1} - \frac{1}{n+1}\right) + \left(\frac{1}{n} - \frac{1}{n+2}\right)$$

$$= 1 + \frac{1}{2} - \frac{1}{n+1} - \frac{1}{n+2} \quad \text{(telescoping series)}.$$

Thus, $\displaystyle\sum_{n=1}^{\infty} \frac{1}{n(n+2)} = \frac{1}{2}\lim_{n\to\infty}\left(1 + \frac{1}{2} - \frac{1}{n+1} - \frac{1}{n+2}\right) = \frac{1}{2}\left(1 + \frac{1}{2}\right) = \frac{3}{4}$.

23. Converges. $\displaystyle\sum_{n=1}^{\infty} \frac{3^n + 2^n}{6^n} = \sum_{n=1}^{\infty}\left(\frac{3^n}{6^n} + \frac{2^n}{6^n}\right) = \sum_{n=1}^{\infty}\left[\left(\tfrac{1}{2}\right)^n + \left(\tfrac{1}{3}\right)^n\right] = \frac{1/2}{1-1/2} + \frac{1/3}{1-1/3} = 1 + \frac{1}{2} = \frac{3}{2}$

25. Converges. $s_n = \left(\sin 1 - \sin\frac{1}{2}\right) + \left(\sin\frac{1}{2} - \sin\frac{1}{3}\right) + \cdots + \left(\sin\frac{1}{n} - \sin\frac{1}{n+1}\right) = \sin 1 - \sin\frac{1}{n+1}$,

so $\displaystyle\sum_{n=1}^{\infty}\left(\sin\frac{1}{n} - \sin\frac{1}{n+1}\right) = \lim_{n\to\infty} s_n = \sin 1 - \sin 0 = \sin 1$.

27. $\displaystyle\lim_{n\to\infty} a_n = \lim_{n\to\infty} \arctan n = \frac{\pi}{2} \neq 0$, so the series diverges by the Test for Divergence.

29. $0.\overline{5} = 0.5 + 0.05 + 0.005 + \cdots = \frac{0.5}{1-0.1} = \frac{5}{9}$

31. $0.\overline{307} = 0.307 + 0.000307 + 0.000000307 + \cdots = \frac{0.307}{1-0.001} = \frac{307}{999}$

33. $\displaystyle\sum_{n=0}^{\infty}(x-3)^n$ is a geometric series with $r = x - 3$, so it converges whenever $|x - 3| < 1 \iff$

$-1 < x - 3 < 1 \iff 2 < x < 4$, and the sum is $\dfrac{1}{1-(x-3)} = \dfrac{1}{4-x}$.

35. $\displaystyle\sum_{n=0}^{\infty}\left(\frac{1}{x}\right)^n$ is geometric with $r = \dfrac{1}{x}$, so it converges whenever $\left|\dfrac{1}{x}\right| < 1 \iff |x| > 1 \iff x > 1$ or

$x < -1$, and the sum is $\dfrac{1}{1-1/x} = \dfrac{x}{x-1}$.

37. After defining f, We use `convert(f,parfrac);` in Maple, `Apart` in Mathematica,

or `Expand Rational` and `Simplify` in Derive to find that the general term is

$\dfrac{1}{(4n+1)(4n-3)} = -\dfrac{1/4}{4n+1} + \dfrac{1/4}{4n-3}$. So the nth partial sum is

$$s_n = \sum_{k=1}^{n}\left(-\frac{1/4}{4k+1} + \frac{1/4}{4k-3}\right) = \frac{1}{4}\left(\frac{1}{4k-3} - \frac{1}{4k+1}\right)$$

$$= \frac{1}{4}\left[\left(1 - \frac{1}{5}\right) + \left(\frac{1}{5} - \frac{1}{9}\right) + \left(\frac{1}{9} - \frac{1}{13}\right) + \cdots + \left(\frac{1}{4n-3} - \frac{1}{4n+1}\right)\right] = \frac{1}{4}\left(1 - \frac{1}{4n+1}\right)$$

The series converges to $\displaystyle\lim_{n\to\infty} s_n = \frac{1}{4}$. This can be confirmed by directly computing the sum using

`sum(f,1..infinity);` (in Maple), `Sum[f,{n,1,Infinity}]` (in Mathematica), or `Calculus Sum`

(from 1 to ∞) and `Simplify` (in Derive).

39. For $n = 1$, $a_1 = 0$ since $s_1 = 0$. For $n > 1$,

$$a_n = s_n - s_{n-1} = \frac{n-1}{n+1} - \frac{(n-1)-1}{(n-1)+1}$$

$$= \frac{(n-1)\,n - (n+1)\,(n-2)}{(n+1)\,n}$$

$$= \frac{2}{n\,(n+1)}$$

Also,

$$\sum_{n=1}^{\infty} a_n = \lim_{n\to\infty} s_n$$

$$= \lim_{n\to\infty} \frac{1 - 1/n}{1 + 1/n}$$

$$= 1$$

41. (a) The first step in the chain occurs when the local government spends D dollars. The people who receive it spend a fraction c of those D dollars, that is, Dc dollars. Those who receive the Dc dollars spend a fraction c of it, that is, Dc^2 dollars. Continuing in this way, we see that the total spending after n transactions is

$$S_n = D + Dc + Dc^2 + \cdots + Dc^{n-1}$$

$$= \frac{D\,(1 - c^n)}{1 - c} \text{ by (3).}$$

(b)

$$\lim_{n\to\infty} S_n = \lim_{n\to\infty} \frac{D\,(1 - c^n)}{1 - c} = \frac{D}{1 - c} \lim_{n\to\infty} (1 - c^n)$$

$$= \frac{D}{1 - c} \quad (\text{since } 0 < c < 1 \implies \lim_{n\to\infty} c^n = 0)$$

$$= \frac{D}{s} \ (\text{since } c + s = 1)$$

$$= kD \ (\text{since } k = 1/s)$$

If $c = 0.8$, then $s = 1 - c = 0.2$ and the multiplier is $k = 1/s = 5$.

43. $\displaystyle\sum_{n=2}^{\infty} (1 + c)^{-n}$ is a geometric series with $a = (1 + c)^{-2}$ and $r = (1 + c)^{-1}$, so the series converges

when $\left|(1 + c)^{-1}\right| < 1 \implies |1 + c| > 1 \implies 1 + c > 1$ or $1 + c < -1 \implies c > 0$ or $c < -2$. We calculate

the sum of the series and set it equal to 2: $\dfrac{(1 + c)^{-2}}{1 - (1 + c)^{-1}} = 2 \iff \left(\dfrac{1}{1 + c}\right)^2 = 2 - 2\left(\dfrac{1}{1 + c}\right) \iff$

$1 = 2\,(1 + c)^2 - 2\,(1 + c) = 0 \iff 2c^2 + 2c - 1 = 0 \iff c = \dfrac{-2 \pm \sqrt{12}}{4} = \dfrac{\pm\sqrt{3} - 1}{2}$. However, the

negative root is inadmissible because $-2 < \dfrac{-\sqrt{3} - 1}{2} < 0$. So $c = \dfrac{\sqrt{3} - 1}{2}$.

45.

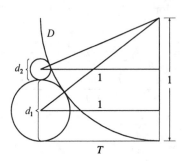

Let d_n be the diameter of C_n. We draw lines from the centers of the C_i to the center of D (or C), and using the Pythagorean Theorem, we can write $1^2 + \left(1 - \frac{1}{2}d_1\right)^2 = \left(1 + \frac{1}{2}d_1\right)^2 \Longleftrightarrow$

$1 = \left(1 + \frac{1}{2}d_1\right)^2 - \left(1 - \frac{1}{2}d_1\right)^2 = 2d_1$ (difference of squares) $\Longrightarrow d_1 = \frac{1}{2}$. Similarly,

$1 = \left(1 + \frac{1}{2}d_2\right)^2 - \left(1 - d_1 - \frac{1}{2}d_2\right)^2 = 2d_2 + 2d_1 - d_1^2 - d_1 d_2 = (2 - d_1)(d_1 + d_2) \Longleftrightarrow$

$d_2 = \dfrac{1}{2 - d_1} - d_1 = \dfrac{(1 - d_1)^2}{2 - d_1}, 1 = \left(1 + \frac{1}{2}d_3\right)^2 - \left(1 - d_1 - d_2 - \frac{1}{2}d_3\right)^2 \Longleftrightarrow d_3 = \dfrac{[1 - (d_1 + d_2)]^2}{2 - (d_1 + d_2)}$,

and in general, $d_{n+1} = \dfrac{\left(1 - \sum\limits_{i=1}^{n} d_i\right)^2}{2 - \sum\limits_{i=1}^{n} d_i}$. If we actually calculate d_2 and d_3 from the formulas above, we

find that they are $\dfrac{1}{6} = \dfrac{1}{2 \cdot 3}$ and $\dfrac{1}{12} = \dfrac{1}{3 \cdot 4}$ respectively, so we suspect that in general, $d_n = \dfrac{1}{n(n+1)}$.

To prove this, we use induction: assume that for all $k \le n$, $d_k = \dfrac{1}{k(k+1)} = \dfrac{1}{k} - \dfrac{1}{k+1}$. Then

$\sum\limits_{i=1}^{n} d_i = 1 - \dfrac{1}{n+1} = \dfrac{n}{n+1}$ (telescoping sum). Substituting this into our formula for d_{n+1}, we get

$d_{n+1} = \dfrac{\left[1 - \dfrac{n}{n+1}\right]^2}{2 - \left(\dfrac{n}{n+1}\right)} = \dfrac{\dfrac{1}{(n+1)^2}}{\dfrac{n+2}{n+1}} = \dfrac{1}{(n+1)(n+2)}$, and the induction is complete.

Now, we observe that the partial sums $\sum\limits_{i=1}^{n} d_i$ of the diameters of the circles approach 1 as $n \to \infty$; that

is, $\sum\limits_{n=1}^{\infty} a_n = \sum\limits_{n=1}^{\infty} \dfrac{1}{n(n+1)} = 1$, which is what we wanted to prove.

47. The series $1 - 1 + 1 - 1 + 1 - 1 + \cdots$ diverges (geometric series with $r = -1$) so we cannot say
$0 = 1 - 1 + 1 - 1 + 1 - 1 + \cdots$.

49. Suppose on the contrary that $\sum(a_n + b_n)$ converges. Then by Theorem 8(iii), so would
$\sum[(a_n + b_n) - a_n] = \sum b_n$, a contradiction.

51. The partial sums $\{s_n\}$ form an increasing sequence, since $s_n - s_{n-1} = a_n > 0$ for all n. Also, the sequence $\{s_n\}$ is bounded since $s_n \leq 1000$ for all n. So by Theorem 10.1.7 , the sequence of partial sums converges, that is, the series $\sum a_n$ is convergent.

53. (a) At the first step, only the interval $\left(\frac{1}{3}, \frac{2}{3}\right)$ (length $\frac{1}{3}$) is removed. At the second step, we remove the intervals $\left(\frac{1}{9}, \frac{2}{9}\right)$ and $\left(\frac{7}{9}, \frac{8}{9}\right)$, which have a total length of $2 \cdot \left(\frac{1}{3}\right)^2$. At the third step, we remove 2^2 intervals, each of length $\left(\frac{1}{3}\right)^3$. In general, at the nth step we remove 2^{n-1} intervals, each of length $\left(\frac{1}{3}\right)^n$, for a length of $2^{n-1} \cdot \left(\frac{1}{3}\right)^n = \frac{1}{3} \left(\frac{2}{3}\right)^{n-1}$. Thus, the total length of all removed intervals is

$$\sum_{n=1}^{\infty} \frac{1}{3} \left(\frac{2}{3}\right)^{n-1} = \frac{1/3}{1 - 2/3} = 1 \text{ (geometric series with } a = \frac{1}{3} \text{ and } r = \frac{2}{3}). \text{ Notice that at the } n\text{th step,}$$

the leftmost interval that is removed is $\left(\left(\frac{1}{3}\right)^n, \left(\frac{2}{3}\right)^n\right)$, so we never remove 0, and 0 is in the Cantor set. Also, the rightmost interval removed is $\left(1 - \left(\frac{2}{3}\right)^n, 1 - \left(\frac{1}{3}\right)^n\right)$, so 1 is never removed. Some other numbers in the Cantor set are $\frac{1}{3}, \frac{2}{3}, \frac{1}{9}, \frac{2}{9}, \frac{7}{9},$ and $\frac{8}{9}$.

(b) The area removed at the first step is $\frac{1}{9}$; at the second step, $8 \cdot \left(\frac{1}{9}\right)^2$; at the third step, $(8)^2 \cdot \left(\frac{1}{9}\right)^3$. In general, the area removed at the nth step is $(8)^{n-1} \left(\frac{1}{9}\right)^n = \frac{1}{9} \left(\frac{8}{9}\right)^{n-1}$, so the total area of all removed squares is $\sum_{n=1}^{\infty} \frac{1}{9} \left(\frac{8}{9}\right)^{n-1} = \frac{1/9}{1 - 8/9} = 1$.

55. (a) $\displaystyle\sum_{n=1}^{\infty} \frac{n}{(n+1)!} \Longrightarrow s_1 = \frac{1}{1 \cdot 2} = \frac{1}{2}, s_2 = \frac{1}{2} + \frac{2}{1 \cdot 2 \cdot 3} = \frac{5}{6}, s_3 = \frac{5}{6} + \frac{3}{1 \cdot 2 \cdot 3 \cdot 4} = \frac{23}{24},$

$s_4 = \frac{23}{24} + \frac{4}{1 \cdot 2 \cdot 3 \cdot 4 \cdot 5} = \frac{119}{120}$. The denominators are $(n+1)!$, so a guess would be

$s_n = \frac{(n+1)! - 1}{(n+1)!}$.

(b) For $n = 1$, $s_1 = \frac{1}{2} = \frac{2! - 1}{2!}$, so the formula holds for $n = 1$. Assume $s_k = \frac{(k+1)! - 1}{(k+1)!}$. Then

$$s_{k+1} = \frac{(k+1)! - 1}{(k+1)!} + \frac{k+1}{(k+2)!} = \frac{(k+1)! - 1}{(k+1)!} + \frac{k+1}{(k+1)!\,(k+2)}$$

$$= \frac{(k+2)! - (k+2) + k + 1}{(k+2)!} = \frac{(k+2)! - 1}{(k+2)!}$$

Thus, the formula is true for $n = k + 1$. So by induction, the guess is correct.

(c) $\displaystyle\lim_{n \to \infty} s_n = \lim_{n \to \infty} \frac{(n+1)! - 1}{(n+1)!} = \lim_{n \to \infty} \left[1 - \frac{1}{(n+1)!}\right] = 1$ and so $\displaystyle\sum_{n=0}^{\infty} \frac{n}{(n+1)!} = 1$.

Section 8.3 The Integral and Comparison Tests; Estimating Sums

1.

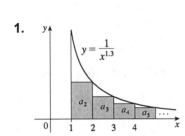

The picture shows that $a_2 = \dfrac{1}{2^{1.3}} < \displaystyle\int_1^2 \dfrac{1}{x^{1.3}}\,dx,$

$a_3 = \dfrac{1}{3^{1.3}} < \displaystyle\int_2^3 \dfrac{1}{x^{1.3}}\,dx,$ and so on, so

$\displaystyle\sum_{n=2}^{\infty} \dfrac{1}{n^{1.3}} < \int_1^{\infty} \dfrac{1}{x^{1.3}}\,dx.$ The integral converges by (5.9.2)

with $p = 1.3 > 1$, so the series converges.

3. (a) We cannot say anything about $\sum a_n$. If $a_n > b_n$ for all n and $\sum b_n$ is convergent, then $\sum a_n$ could be convergent or divergent.

(b) If $a_n < b_n$ for all n, then $\sum a_n$ is convergent. [This is part (a) of the Comparison Test.]

5. $\displaystyle\sum_{n=1}^{\infty} n^b$ is a p-series with $p = -b$. $\displaystyle\sum_{n=1}^{\infty} b^n$ is a geometric series. By (1), the p-series is convergent if $p > 1$.

In this case, $\displaystyle\sum_{n=1}^{\infty} n^b = \sum_{n=1}^{\infty} \left(1/n^{-b}\right)$, so $-b > 1 \iff b < -1$ are the values for which the series converge.

A geometric series $\displaystyle\sum_{n=1}^{\infty} ar^{n-1}$ converges if $|r| < 1$, so $\displaystyle\sum_{n=1}^{\infty} b^n$ converges if $|b| < 1 \iff -1 < b < 1$.

7. $f(x) = xe^{-x^2}$ is continuous and positive on $[1, \infty)$, and since $f'(x) = e^{-x^2}\left(1 - 2x^2\right) < 0$ for $x > 1$, f is decreasing as well. Thus, we can use the Integral Test.

$\displaystyle\int_1^{\infty} xe^{-x^2}\,dx = \lim_{t\to\infty}\left[-\tfrac{1}{2}e^{-x^2}\right]_1^t = 0 - \left(-\dfrac{e^{-1}}{2}\right) = \dfrac{1}{2e}.$ Since the integral converges, the series converges.

9. $f(x) = \dfrac{x}{x^2 + 1}$ is continuous and positive on $[1, \infty)$, and since $f'(x) = \dfrac{1 - x^2}{(x^2 + 1)^2} < 0$ for $x > 1$, f

is also decreasing. Using the Integral Test, $\displaystyle\int_1^{\infty} \dfrac{x}{x^2 + 1}\,dx = \lim_{t\to\infty}\left[\dfrac{\ln\left(x^2 + 1\right)}{2}\right]_1^t = \infty,$ so the series diverges.

11. $f(x) = \dfrac{1}{x \ln x}$ is continuous and positive on $[2, \infty)$, and also decreasing

since $f'(x) = -\dfrac{1 + \ln x}{x^2 (\ln x)^2} < 0$ for $x > 2$, so we can use the Integral Test.

$\displaystyle\int_2^{\infty} \dfrac{1}{x \ln x}\,dx = \lim_{t\to\infty}\left[\ln\left(\ln x\right)\right]_2^t = \lim_{t\to\infty}\left[\ln\left(\ln t\right) - \ln\left(\ln 2\right)\right] = \infty,$ so the series diverges.

13. $\dfrac{1}{n^3 + n^2} < \dfrac{1}{n^3}$ since $n^3 + n^2 > n^3$ for all n, and since $\displaystyle\sum_{n=1}^{\infty} \dfrac{1}{n^3}$ is a convergent p-series ($p = 3 > 1$),

$\displaystyle\sum_{n=1}^{\infty} \dfrac{1}{n^3 + n^2}$ also converges by the Comparison Test [part (a)].

15. $\dfrac{1 + 5^n}{4^n} > \dfrac{5^n}{4^n} = \left(\dfrac{5}{4}\right)^n$. $\displaystyle\sum_{n=0}^{\infty} \left(\dfrac{5}{4}\right)^n$ is a divergent geometric series ($|r| = \frac{5}{4} > 1$), so $\displaystyle\sum_{n=0}^{\infty} \dfrac{1 + 5^n}{4^n}$

diverges by the Comparison Test.

17. $\dfrac{3}{n(n+3)} < \dfrac{3}{n^2}$. $\displaystyle\sum_{n=1}^{\infty} \dfrac{3}{n^2} = 3\displaystyle\sum_{n=1}^{\infty} \dfrac{1}{n^2}$ is a convergent p-series ($p = 2 > 1$), so $\displaystyle\sum_{n=1}^{\infty} \dfrac{3}{n(n+3)}$ converges

by the Comparison Test.

19. Use the Limit Comparison Test with $a_n = \dfrac{1}{1 + \sqrt{n}}$ and $b_n = \dfrac{1}{\sqrt{n}}$: $\displaystyle\lim_{n\to\infty} \dfrac{a_n}{b_n} = \lim_{n\to\infty} \dfrac{\sqrt{n}}{1 + \sqrt{n}} = 1 > 0$.

Since $\displaystyle\sum_{n=1}^{\infty} \dfrac{1}{\sqrt{n}}$ is a divergent p-series ($p = \frac{1}{2} \leq 1$), $\displaystyle\sum_{n=1}^{\infty} \dfrac{1}{1 + \sqrt{n}}$ also diverges.

21. Use the Limit Comparison Test with $a_n = \sin(1/n)$ and $b_n = 1/n$:

$\displaystyle\lim_{n\to\infty} \dfrac{a_n}{b_n} = \lim_{n\to\infty} \dfrac{\sin(1/n)}{1/n} = \lim_{\theta\to 0} \dfrac{\sin\theta}{\theta} = 1 > 0$. Since $\displaystyle\sum_{n=1}^{\infty} b_n$ is the divergent harmonic series,

$\displaystyle\sum_{n=1}^{\infty} \sin(1/n)$ also diverges.

23. We have already shown (in Exercise 11) that when $p = 1$ the series $\displaystyle\sum_{n=2}^{\infty} \dfrac{1}{n(\ln n)^p}$ diverges,

so assume $p \neq 1$. $f(x) = 1/[x(\ln x)^p]$ is continuous and positive on $[2, \infty)$, and

$f'(x) = -\dfrac{p + \ln x}{x^2(\ln x)^{p+1}} < 0$ if $x > e^{-p}$, so that f is eventually decreasing and we can use the Integral

Test. $\displaystyle\int_2^{\infty} \dfrac{1}{x(\ln x)^p}\,dx = \lim_{t\to\infty} \left[\dfrac{(\ln x)^{1-p}}{1-p}\right]_2^t$ (for $p \neq 1$) $= \displaystyle\lim_{t\to\infty} \left[\dfrac{(\ln t)^{1-p}}{1-p}\right] - \dfrac{(\ln 2)^{1-p}}{1-p}$.

This limit exists whenever $1 - p < 0 \iff p > 1$, so the series converges for $p > 1$.

25. (a) $f(x) = 1/x^2$ is positive and continuous and $f'(x) = -2/x^3$ is negative for $x > 1$, and

so the Integral Test applies. $\displaystyle\sum_{n=1}^{\infty} \dfrac{1}{n^2} \approx s_{10} = \dfrac{1}{1^2} + \dfrac{1}{2^2} + \dfrac{1}{3^2} + \cdots + \dfrac{1}{10^2} \approx 1.549768$.

$R_{10} \leq \displaystyle\int_{10}^{\infty} \dfrac{1}{x^2}\,dx = \lim_{t\to\infty} \left[\dfrac{-1}{x}\right]_{10}^t = \lim_{t\to\infty} \left(-\dfrac{1}{t} + \dfrac{1}{10}\right) = \dfrac{1}{10}$, so the error is at most 0.1.

(b) $s_{10} + \displaystyle\int_{11}^{\infty} (1/x^2)\,dx \leq s \leq s_{10} + \int_{10}^{\infty} (1/x^2)\,dx \implies s_{10} + \frac{1}{11} \leq s \leq s_{10} + \frac{1}{10} \implies$

$1.549768 + 0.090909 = 1.640677 \leq s \leq 1.549768 + 0.1 = 1.649768$, so we get $s \approx 1.64522$ (the

average of 1.640677 and 1.649768) with error ≤ 0.005 (the maximum of $1.649768 - 1.64522$ and

$1.64522 - 1.640677$, rounded up).

(c) $R_n \leq \displaystyle\int_n^{\infty} (1/x^2)\,dx = 1/n$. So $R_n < 0.001$ if $1/n < \frac{1}{1000} \iff n > 1000$.

27. $f(x) = x^{-3/2}$ is positive and continuous and $f'(x) = -\frac{3}{2}x^{-5/2}$ is negative for $x > 1$, so the Integral Test applies. From the end of Example 7, we see that the error is at most half the length of the interval. From (4), the interval is $\left(s_n + \int_{n+1}^{\infty} f(x)\,dx, s_n + \int_n^{\infty} f(x)\,dx\right)$, so its length is $\int_n^{\infty} f(x)\,dx - \int_{n+1}^{\infty} f(x)\,dx$. Thus, we need n such that

$$0.01 > \frac{1}{2}\left(\int_n^{\infty} x^{-3/2}\,dx - \int_{n+1}^{\infty} x^{-3/2}\,dx\right) = \frac{1}{2}\left(\lim_{t\to\infty}\left[\frac{-2}{\sqrt{x}}\right]_n^t - \lim_{t\to\infty}\left[\frac{-2}{\sqrt{x}}\right]_{n+1}^t\right)$$

$$= \frac{1}{\sqrt{n}} - \frac{1}{\sqrt{n+1}} \iff n > 13.08$$

Again from the end of Example 7, we approximate s by the midpoint of this interval. In general, the midpoint is $\frac{1}{2}\left[\left(s_n + \int_{n+1}^{\infty} f(x)\,dx\right) + \left(s_n + \int_n^{\infty} f(x)\,dx\right)\right] =$ $s_n + \frac{1}{2}\left(\int_{n+1}^{\infty} f(x)\,dx + \int_n^{\infty} f(x)\,dx\right)$. So using $n = 14$, we have

$s \approx s_{14} + \frac{1}{2}\left(\int_{14}^{\infty} x^{-3/2}\,dx + \int_{15}^{\infty} x^{-3/2}\,dx\right) = 2.0872 + \frac{1}{\sqrt{14}} + \frac{1}{\sqrt{15}} \approx 2.6127$. Any larger value of n will also work. For instance, $s \approx s_{30} + \frac{1}{\sqrt{30}} + \frac{1}{\sqrt{31}} \approx 2.6124$.

29. $\displaystyle\sum_{n=1}^{10} \frac{1}{n^4 + n^2} = \frac{1}{2} + \frac{1}{20} + \frac{1}{90} + \cdots + \frac{1}{10{,}100} \approx 0.567975$.

Now $\dfrac{1}{n^4 + n^2} < \dfrac{1}{n^4}$, so using the reasoning and notation of Example 7, the error is

$$R_{10} \le T_{10} = \sum_{n=11}^{\infty} \frac{1}{n^4} \le \int_{10}^{\infty} \frac{dx}{x^4} = \lim_{t\to\infty}\left[-\frac{x^{-3}}{3}\right]_{10}^t = \frac{1}{3000} = 0.000\overline{3}.$$

31. (a) From the figure, $a_2 + a_3 + \cdots + a_n \le \int_1^n f(x)\,dx$, so with

$$f(x) = \frac{1}{x}, \quad \frac{1}{2} + \frac{1}{3} + \frac{1}{4} + \cdots + \frac{1}{n} \le \int_1^n \frac{1}{x}\,dx = \ln n.$$

Thus, $s_n = 1 + \dfrac{1}{2} + \dfrac{1}{3} + \dfrac{1}{4} + \cdots + \dfrac{1}{n} \le 1 + \ln n.$

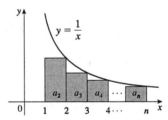

(b) By part (a), $s_{10^6} \le 1 + \ln 10^6 \approx 14.82 < 15$ and $s_{10^9} \le 1 + \ln 10^9 \approx 21.72 < 22$.

33. Since $\dfrac{d_n}{10^n} \le \dfrac{9}{10^n}$ for each n, and since $\displaystyle\sum_{n=1}^{\infty} \frac{9}{10^n}$ is a convergent geometric series ($|r| = \frac{1}{10} < 1$),

$0.d_1 d_2 d_3 \ldots = \displaystyle\sum_{n=1}^{\infty} \frac{d_n}{10^n}$ will always converge by the Comparison Test.

35. Yes. Since $\sum a_n$ converges, its terms approach 0 as $n \to \infty$, so $\displaystyle\lim_{n\to\infty} \frac{\sin a_n}{a_n} = 1$ by Theorem 3.4.2. Thus, $\sum \sin a_n$ converges by the Limit Comparison Test.

Section 8.4 Other Convergence Tests

1. (a) An alternating series is a series whose terms are alternately positive and negative.

(b) An alternating series $\sum\limits_{n=1}^{\infty} (-1)^{n-1} b_n$ converges if $0 < b_{n+1} \leq b_n$ for all n and $\lim\limits_{n\to\infty} b_n = 0$. (This is the Alternating Series Test.)

(c) The error involved in using the partial sum s_n as an approximation to the total sum s is the remainder $R_n = s - s_n$ and the size of the error is smaller than b_{n+1}, that is, $|R_n| \leq b_{n+1}$. (This is the Alternating Series Estimation Theorem.)

3. $\sum\limits_{n=1}^{\infty} (-1)^{n-1} \dfrac{3}{n+4}$. $b_n = \dfrac{3}{n+4} > 0$ and $b_{n+1} \leq b_n$ for all n; $\lim\limits_{n\to\infty} b_n = 0$, so the series converges by the Alternating Series Test.

5. $\sum\limits_{n=1}^{\infty} (-1)^{n+1} \dfrac{n}{5n+1}$. $\lim\limits_{n\to\infty} \dfrac{n}{5n+1} = \dfrac{1}{5}$, so $\lim\limits_{n\to\infty} (-1)^{n+1} \dfrac{n}{5n+1}$ does not exist and the series diverges by the Test for Divergence.

7. $\sum\limits_{n=1}^{\infty} (-1)^n \dfrac{n}{n^2+1}$. $b_n = \dfrac{n}{n^2+1} > 0$ for all n. $b_{n+1} \leq b_n \iff \dfrac{n+1}{(n+1)^2+1} \leq \dfrac{n}{n^2+1} \iff$

$(n+1)(n^2+1) \leq \left[(n+1)^2 + 1\right] n \iff n^3 + n^2 + n + 1 \leq n^3 + 2n^2 + 2n \iff 0 \leq n^2 + n - 1$,

which is true for all $n \geq 1$. Also, $\lim\limits_{n\to\infty} b_n = \lim\limits_{n\to\infty} \dfrac{n}{n^2+1} = \lim\limits_{n\to\infty} \dfrac{1/n}{1+1/n^2} = 0$. Therefore, the series converges by the Alternating Series Test.

9. $\sum\limits_{n=1}^{\infty} \dfrac{(-1)^{n-1}}{n} = 1 - \dfrac{1}{2} + \dfrac{1}{3} - \dfrac{1}{4} + \cdots + \dfrac{1}{49} - \dfrac{1}{50} + \dfrac{1}{51} - \dfrac{1}{52} + \cdots$. The 50th partial sum of this series is an underestimate, since $\sum\limits_{n=1}^{\infty} \dfrac{(-1)^{n-1}}{n} = s_{50} + \left(\dfrac{1}{51} - \dfrac{1}{52}\right) + \left(\dfrac{1}{53} - \dfrac{1}{54}\right) + \cdots$, and the terms in parentheses are all positive. The result can be seen geometrically in Figure 1.

11. If $p > 0$, $\dfrac{1}{(n+1)^p} \leq \dfrac{1}{n^p}$ and $\lim\limits_{n\to\infty} \dfrac{1}{n^p} = 0$, so the series converges by the Alternating Series Test.

If $p \leq 0$, $\lim\limits_{n\to\infty} \dfrac{(-1)^{n-1}}{n^p}$ does not exist, so the series diverges by the Test for Divergence.

Thus, $\sum\limits_{n=1}^{\infty} \dfrac{(-1)^{n-1}}{n^p}$ converges $\iff p > 0$.

13. $b_7 = 2^7/7! \approx 0.025 > 0.001$ and $b_8 = 2^8/8! \approx 0.006 < 0.01$, so by the Alternating Series Estimation Theorem, $n = 7$. (That is, since the 8th term is less than the desired error, we need to add the first 7 terms to get the sum to the desired accuracy.)

15.

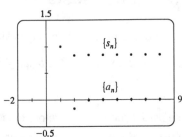

The graph gives us an estimate for the sum of the series $\displaystyle\sum_{n=1}^{\infty} \frac{(-1)^{n-1}}{(2n-1)!}$ of 0.84.

$$b_5 = \frac{1}{(2 \cdot 5 - 1)!} = \frac{1}{362{,}880} \approx 0.000003 < 0.00001, \text{ so } \sum_{n=1}^{\infty} \frac{(-1)^{n-1}}{(2n-1)!} \approx \sum_{n=1}^{4} \frac{(-1)^{n-1}}{(2n-1)!} \approx 0.8415.$$

17. $b_6 = \dfrac{1}{2^6 6!} = \dfrac{1}{46{,}080} \approx 0.000022 < 0.00001, \text{ so } \displaystyle\sum_{n=0}^{\infty} \frac{(-1)^n}{2^n n!} \approx \sum_{n=0}^{5} \frac{(-1)^n}{2^n n!} \approx 0.6065.$

19. Using the Ratio Test, $\displaystyle\lim_{n\to\infty} \left| \frac{a_{n+1}}{a_n} \right| = \lim_{n\to\infty} \left| \frac{(-3)^{n+1} \big/ (n+1)^3}{(-3)^n / n^3} \right| = 3 \lim_{n\to\infty} \left(\frac{n}{n+1} \right)^3 = 3 > 1$, so

the series diverges.

21. $\displaystyle\sum_{n=1}^{\infty} \frac{1}{2n+1}$ diverges (use the Integral Test or the Limit Comparison Test with $b_n = 1/n$).

23. $\left| \dfrac{\sin 2n}{n^2} \right| \le \dfrac{1}{n^2}$ and $\displaystyle\sum_{n=1}^{\infty} \frac{1}{n^2}$ converges (p-series, $p = 2 > 1$), so $\displaystyle\sum_{n=1}^{\infty} \frac{\sin 2n}{n^2}$ converges absolutely by the

Comparison Test.

25. $\displaystyle\lim_{n\to\infty} \left| \frac{a_{n+1}}{a_n} \right| = \lim_{n\to\infty} \frac{(n+2)\,5^{n+1} \big/ \left[(n+1)\,3^{2(n+1)} \right]}{(n+1)\,5^n \big/ (n 3^{2n})} = \lim_{n\to\infty} \frac{5n\,(n+2)}{9\,(n+1)^2} = \frac{5}{9} < 1$, so the series

converges absolutely by the Ratio Test.

27. $\displaystyle\lim_{n\to\infty} \left| \frac{a_{n+1}}{a_n} \right| = \lim_{n\to\infty} \frac{(n+3)! \big/ \left[(n+1)! 10^{n+1} \right]}{(n+2)! / (n! 10^n)} = \frac{1}{10} \lim_{n\to\infty} \frac{n+3}{n+1} = \frac{1}{10} < 1$, so the series

converges absolutely by the Ratio Test.

29. By the recursive definition, $\displaystyle\lim_{n\to\infty} \left| \frac{a_{n+1}}{a_n} \right| = \lim_{n\to\infty} \left| \frac{5n+1}{4n+3} \right| = \frac{5}{4} > 1$, so the series diverges by the

Ratio Test.

31. (a) $\lim\limits_{n\to\infty} \left| \dfrac{1 \big/ (n+1)^3}{1/n^3} \right| = \lim\limits_{n\to\infty} \dfrac{n^3}{(n+1)^3} = \lim\limits_{n\to\infty} \dfrac{1}{(1+1/n)^3} = 1$. Inconclusive.

(b) $\lim\limits_{n\to\infty} \left| \dfrac{(n+1)}{2^{n+1}} \cdot \dfrac{2^n}{n} \right| = \lim\limits_{n\to\infty} \dfrac{n+1}{2n} = \lim\limits_{n\to\infty} \left(\dfrac{1}{2} + \dfrac{1}{2n} \right) = \dfrac{1}{2}$. Conclusive (convergent).

(c) $\lim\limits_{n\to\infty} \left| \dfrac{(-3)^n}{\sqrt{n+1}} \cdot \dfrac{\sqrt{n}}{(-3)^{n-1}} \right| = 3 \lim\limits_{n\to\infty} \sqrt{\dfrac{n}{n+1}} = 3 \lim\limits_{n\to\infty} \sqrt{\dfrac{1}{1+1/n}} = 3$. Conclusive (divergent).

(d) $\lim\limits_{n\to\infty} \left| \dfrac{\sqrt{n+1}}{1+(n+1)^2} \cdot \dfrac{1+n^2}{\sqrt{n}} \right| = \lim\limits_{n\to\infty} \left[\sqrt{1+\dfrac{1}{n}} \cdot \dfrac{1/n^2+1}{1/n^2+(1+1/n)^2} \right] = 1$. Inconclusive.

33. (a) $\lim\limits_{n\to\infty} \left| \dfrac{a_{n+1}}{a_n} \right| = \lim\limits_{n\to\infty} \dfrac{|x|^{n+1}/(n+1)!}{|x|^n/n!} = |x| \lim\limits_{n\to\infty} \dfrac{1}{n+1} = 0$, so by the Ratio Test the series converges for all x.

(b) Since the series of part (a) always converges, we must have $\lim\limits_{n\to\infty} \dfrac{x^n}{n!} = 0$ by Theorem 8.2.6.

35. (a) $s_5 = \sum\limits_{n=1}^{5} \dfrac{1}{n2^n} = \dfrac{1}{2} + \dfrac{1}{8} + \dfrac{1}{24} + \dfrac{1}{64} + \dfrac{1}{160} = \dfrac{661}{960} \approx 0.68854$. Now the ratios

$r_n = \dfrac{a_{n+1}}{a_n} = \dfrac{n2^n}{(n+1)\,2^{n+1}} = \dfrac{n}{2(n+1)}$ form an increasing sequence, since

$r_{n+1} - r_n = \dfrac{n+1}{2(n+2)} - \dfrac{n}{2(n+1)} = \dfrac{(n+1)^2 - n(n+2)}{2(n+1)(n+2)} = \dfrac{1}{2(n+1)(n+2)} > 0$. So by

Exercise 34(b), the error is less than $\dfrac{a_6}{1 - \lim\limits_{n\to\infty} r_n} = \dfrac{1/(6 \cdot 2^6)}{1 - 1/2} = \dfrac{1}{192} \approx 0.00521$.

(b) The error in using s_n as an approximation to the sum is $R_n = \dfrac{a_{n+1}}{1 - \frac{1}{2}} = \dfrac{2}{(n+1)\,2^{n+1}}$. We want

$R_n < 0.00005 \iff \dfrac{1}{(n+1)\,2^n} < 0.00005 \iff (n+1)\,2^n > 20{,}000$. To find such an n we can

use trial and error or a graph. We calculate $(11+1)\,2^{11} = 24{,}576$, so $s_{11} = \sum\limits_{n=1}^{11} \dfrac{1}{n2^n} \approx 0.693109$

is within 0.00005 of the actual sum.

Section 8.5 Power Series

Note: "R" stands for "radius of convergence" and "I" stands for "interval of convergence" in this section.

1. A power series is a series of the form $\sum_{n=0}^{\infty} c_n x^n = c_0 + c_1 x + c_2 x^2 + c_3 x^3 + \cdots$, where x is a variable and the c_n's are constants called the coefficients of the series.

   More generally, a series of the form $\sum_{n=0}^{\infty} c_n (x-a)^n = c_0 + c_1 (x-a) + c_2 (x-a)^2 + \cdots$ is called a power series in $(x-a)$ or a power series centered at a or a power series about a.

3. (a) We are given that the power series $\sum_{n=0}^{\infty} c_n x^n$ is convergent for $x = 4$. So by Theorem 3 it must converge for at least $-4 < x \le 4$. In particular, it converges when $x = -2$, that is, $\sum_{n=0}^{\infty} c_n (-2)^n$ is convergent.

   (b) It does not follow that $\sum_{n=0}^{\infty} c_n (-4)^n$ is necessarily convergent. [See the comments after Theorem 3 about convergence at the endpoint of an interval. An example is $c_n = (-1)^n / (n 4^n)$.]

5. If $a_n = \dfrac{x^n}{n+2}$, then $\lim\limits_{n \to \infty} \left| \dfrac{a_{n+1}}{a_n} \right| = \lim\limits_{n \to \infty} \left| \dfrac{x^{n+1}}{n+3} \cdot \dfrac{n+2}{x^n} \right| = |x| \lim\limits_{n \to \infty} \dfrac{n+2}{n+3} = |x| < 1$ for convergence (by the Ratio Test), and $R = 1$. When $x = 1$, the series is $\sum_{n=0}^{\infty} \dfrac{1}{n+2}$, which diverges (Integral Test or Comparison Test), and when $x = -1$, it is $\sum_{n=0}^{\infty} \dfrac{(-1)^n}{n+2}$, which converges (Alternating Series Test), so $I = [-1, 1)$.

7. If $a_n = \dfrac{x^n}{n!}$, then $\lim\limits_{n \to \infty} \left| \dfrac{a_{n+1}}{a_n} \right| = \lim\limits_{n \to \infty} \left| \dfrac{x^{n+1}/(n+1)!}{x^n/n!} \right| = |x| \lim\limits_{n \to \infty} \dfrac{1}{n+1} = 0 < 1$ for all x. So, by the Ratio Test, $R = \infty$, and $I = (-\infty, \infty)$.

9. If $a_n = \dfrac{(-1)^n x^n}{n 2^n}$, then $\lim\limits_{n \to \infty} \left| \dfrac{a_{n+1}}{a_n} \right| = \lim\limits_{n \to \infty} \left| \dfrac{x^{n+1}/[(n+1) 2^{n+1}]}{x^n/(n 2^n)} \right| = \left| \dfrac{x}{2} \right| \lim\limits_{n \to \infty} \dfrac{n}{n+1} = \left| \dfrac{x}{2} \right| < 1$ for convergence, so $|x| < 2$ and $R = 2$. When $x = 2$, $\sum_{n=1}^{\infty} \dfrac{(-1)^n x^n}{n 2^n} = \sum_{n=1}^{\infty} \dfrac{(-1)^n}{n}$, which converges by the Alternating Series Test. When $x = -2$, $\sum_{n=1}^{\infty} \dfrac{(-1)^n x^n}{n 2^n} = \sum_{n=1}^{\infty} \dfrac{1}{n}$, which diverges (harmonic series), so $I = (-2, 2]$.

11. If $a_n = \dfrac{n}{4^n}(2x-1)^n$, then

$$\left|\frac{a_{n+1}}{a_n}\right| = \left|\frac{(n+1)(2x-1)^{n+1}}{4^{n+1}} \cdot \frac{4^n}{n(2x-1)^n}\right| = \left|\frac{2x-1}{4}\left(1+\frac{1}{n}\right)\right| \to \tfrac{1}{2}\left|x-\tfrac{1}{2}\right| \text{ as } n \to \infty. \text{ For }$$

convergence, $\tfrac{1}{2}\left|x-\tfrac{1}{2}\right| < 1 \Longrightarrow \left|x-\tfrac{1}{2}\right| < 2 \Longrightarrow R = 2$ and $-2 < x - \tfrac{1}{2} < 2 \Longrightarrow -\tfrac{3}{2} < x < \tfrac{5}{2}$. If

$x = -\tfrac{3}{2}$, the series becomes $\displaystyle\sum_{n=0}^{\infty} \frac{n}{4^n}(-4)^n = \sum_{n=0}^{\infty}(-1)^n n$, which is divergent by the Test for

Divergence. If $x = \tfrac{5}{2}$, the series is $\displaystyle\sum_{n=0}^{\infty} \frac{n}{4^n}4^n = \sum_{n=0}^{\infty} n$, also divergent by the Test for Divergence. So

$I = \left(-\tfrac{3}{2}, \tfrac{5}{2}\right)$.

13. If $a_n = \dfrac{(-1)^n(x-1)^n}{\sqrt{n}}$, then

$$\lim_{n\to\infty}\left|\frac{a_{n+1}}{a_n}\right| = \lim_{n\to\infty}\left|\frac{(x-1)^{n+1}}{\sqrt{n+1}} \cdot \frac{\sqrt{n}}{(x-1)^n}\right| = |x-1| \lim_{n\to\infty}\sqrt{\frac{n}{n+1}}$$

$$= |x-1| < 1 \text{ for convergence, or } 0 < x < 2, \text{ and } R = 1.$$

When $x = 0$, $\displaystyle\sum_{n=1}^{\infty} \frac{(-1)^n(x-1)^n}{\sqrt{n}} = \sum_{n=1}^{\infty} \frac{1}{\sqrt{n}}$ which is a divergent p-series ($p = \tfrac{1}{2} \le 1$). When $x = 2$,

the series is $\displaystyle\sum_{n=1}^{\infty} \frac{(-1)^n}{\sqrt{n}}$, which converges by the Alternating Series Test. So $I = (0, 2]$.

15. If $a_n = \dfrac{2^n(x-3)^n}{n+3}$, then

$$\lim_{n\to\infty}\left|\frac{a_{n+1}}{a_n}\right| = \lim_{n\to\infty}\left|\frac{2^{n+1}(x-3)^{n+1}}{n+4} \cdot \frac{n+3}{2^n(x-3)^n}\right| = 2\,|x-3|\lim_{n\to\infty}\frac{n+3}{n+4}$$

$$= 2\,|x-3| < 1 \text{ for convergence, or}$$

$|x-3| < \tfrac{1}{2} \Longleftrightarrow \tfrac{5}{2} < x < \tfrac{7}{2}$, and $R = \tfrac{1}{2}$. When $x = \tfrac{5}{2}$, $\displaystyle\sum_{n=0}^{\infty} \frac{2^n(x-3)^n}{n+3} = \sum_{n=0}^{\infty} \frac{(-1)^n}{n+3}$ which

converges by the Alternating Series Test. When $x = \tfrac{7}{2}$, $\displaystyle\sum_{n=0}^{\infty} \frac{2^n(x-3)^n}{n+3} = \sum_{n=0}^{\infty} \frac{1}{n+3}$, similar to the

harmonic series, which diverges. So $I = \left[\tfrac{5}{2}, \tfrac{7}{2}\right)$.

17. If $a_n = n!\,(2x-1)^n$, then

$$\lim_{n\to\infty}\left|\frac{a_{n+1}}{a_n}\right| = \lim_{n\to\infty}\left|\frac{(n+1)!\,(2x-1)^{n+1}}{n!\,(2x-1)^n}\right| = \lim_{n\to\infty}(n+1)\,|2x-1| \to \infty \text{ as } n \to \infty \text{ for all } x \ne \tfrac{1}{2}.$$

Since the series diverges for all $x \ne \tfrac{1}{2}$, $R = 0$ and $I = \left\{\tfrac{1}{2}\right\}$.

19. If $a_n = \dfrac{(n!)^k}{(kn)!} x^n$, then

$$\lim_{n\to\infty} \left| \frac{a_{n+1}}{a_n} \right| = \lim_{n\to\infty} \frac{[(n+1)!]^k (kn)!}{(n!)^k [k(n+1)]!} |x|$$

$$= \lim_{n\to\infty} \frac{(n+1)^k}{(kn+k)(kn+k-1)\cdots(kn+2)(kn+1)} |x|$$

$$= \lim_{n\to\infty} \left[\frac{(n+1)}{(kn+1)} \frac{(n+1)}{(kn+2)} \cdots \frac{(n+1)}{(kn+k)} \right] |x|$$

$$= \lim_{n\to\infty} \left[\frac{n+1}{kn+1} \right] \lim_{n\to\infty} \left[\frac{n+1}{kn+2} \right] \cdots \lim_{n\to\infty} \left[\frac{n+1}{kn+k} \right] |x|$$

$$= \left(\frac{1}{k} \right)^k |x| < 1 \iff$$

$|x| < k^k$ for convergence, and the radius of convergence is $R = k^k$.

21. (a) If $a_n = \dfrac{(-1)^n x^{2n+1}}{n!(n+1)!2^{2n+1}}$, then $\lim_{n\to\infty} \left| \dfrac{a_{n+1}}{a_n} \right| = \left(\dfrac{x}{2} \right)^2 \lim_{n\to\infty} \dfrac{1}{(n+1)(n+2)} = 0$ for all x. So $J_1(x)$ converges for all x; the domain is $(-\infty, \infty)$.

(b), (c) The initial terms of $J_1(x)$ up to $n = 5$ are $a_0 = \dfrac{x}{2}$, $a_1 = -\dfrac{x^3}{16}$, $a_2 = \dfrac{x^5}{384}$, $a_3 = -\dfrac{x^7}{18,432}$, $a_4 = \dfrac{x^9}{1,474,560}$, and $a_5 = -\dfrac{x^{11}}{176,947,200}$.

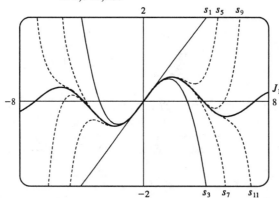

The partial sums seem to approximate $J_1(x)$ well near the origin, but as $|x|$ increases, we need to take a large number of terms to get a good approximation.

23.

$$s_{2n-1} = 1 + 2x + x^2 + 2x^3 + \cdots + x^{2n-2} + 2x^{2n-1}$$
$$= (1 + 2x)\left(1 + x^2 + x^4 + \cdots + x^{2n-2}\right)$$
$$= (1 + 2x)\,\frac{1 - x^{2n}}{1 - x^2}\ \text{[by (8.2.3) with } r = x^2\text{]}$$
$$\rightarrow \frac{1 + 2x}{1 - x^2}\ \text{as } n \rightarrow \infty\ \text{[by (8.2.4)], when } |x| < 1.$$

Also, $s_{2n} = s_{2n-1} + x^{2n} \rightarrow \dfrac{1 + 2x}{1 - x^2}$, since $x^{2n} \rightarrow 0$ for $|x| < 1$. Therefore, $s_n \rightarrow \dfrac{1 + 2x}{1 - x^2}$, since

s_{2n} and s_{2n-1} both approach $\dfrac{1 + 2x}{1 - x^2}$ as $n \rightarrow \infty$. Thus, the interval of convergence is $(-1, 1)$ and

$f(x) = \dfrac{1 + 2x}{1 - x^2}$.

25. $\sum (c_n + d_n)\, x^n = \sum c_n x^n + \sum d_n x^n$ on the interval $(-2, 2)$, since both series converge there. So the radius of convergence must be at least 2. Now since $\sum c_n x^n$ has $R = 2$, it must diverge either at $x = -2$ or at $x = 2$. So by Exercise 8.2.49, $\sum (c_n + d_n)\, x^n$ diverges either at $x = -2$ or at $x = 2$, and so its radius of convergence is 2.

Section 8.6 Representations of Functions as Power Series

Note: "R" stands for "radius of convergence" and "I" stands for "interval of convergence" in this section.

1. If $f(x) = \sum\limits_{n=0}^{\infty} c_n x^n$ has radius of convergence 10, then $f'(x) = \sum\limits_{n=1}^{\infty} n c_n x^{n-1}$ also has radius of convergence 10 by Theorem 2.

3. $f(x) = \dfrac{1}{1+x} = \dfrac{1}{1-(-x)} = \sum\limits_{n=0}^{\infty}(-x)^n = \sum\limits_{n=0}^{\infty}(-1)^n\, x^n$ with $|-x| < 1 \iff |x| < 1$, so $R = 1$ and $I = (-1,1)$.

5. $f(x) = \dfrac{1}{1+4x^2} = \sum\limits_{n=0}^{\infty}(-1)^n \left(4x^2\right)^n$ (substituting $4x^2$ for x in the series from Exercise 3)

$= \sum\limits_{n=0}^{\infty}(-1)^n\, 4^n x^{2n}$, with $\left|4x^2\right| < 1 \iff x^2 < \tfrac{1}{4} \iff |x| < \tfrac{1}{2}$, so $R = \tfrac{1}{2}$ and $I = \left(-\tfrac{1}{2}, \tfrac{1}{2}\right)$.

7. $f(x) = \dfrac{x}{x-3} = 1 + \dfrac{3}{x-3} = 1 - \dfrac{1}{1-x/3} = 1 - \sum\limits_{n=0}^{\infty}\left(\dfrac{x}{3}\right)^n = -\sum\limits_{n=1}^{\infty}\left(\dfrac{x}{3}\right)^n$ (since the 0th term of

the series is 1). For convergence, $\dfrac{|x|}{3} < 1 \iff |x| < 3$, so $R = 3$ and $I = (-3,3)$.

Another Method: $\dfrac{x}{x-3} = -\dfrac{x}{3(1-x/3)} = -\dfrac{x}{3}\sum\limits_{n=0}^{\infty}\left(\dfrac{x}{3}\right)^n = -\sum\limits_{n=0}^{\infty}\dfrac{x^{n+1}}{3^{n+1}} = -\sum\limits_{n=1}^{\infty}\dfrac{x^n}{3^n}$

9. $f(x) = \dfrac{1}{(1+x)^2} = -\dfrac{d}{dx}\left(\dfrac{1}{1+x}\right) = -\dfrac{d}{dx}\left(\sum\limits_{n=0}^{\infty}(-1)^n\, x^n\right)$ (from Exercise 3)

$= \sum\limits_{n=1}^{\infty}(-1)^{n+1}\, n x^{n-1}$ [from Theorem 2(a)] $= \sum\limits_{n=0}^{\infty}(-1)^n\, (n+1)\, x^n$ with $R = 1$.

11. $f(x) = \dfrac{1}{(1+x)^3} = -\dfrac{1}{2}\dfrac{d}{dx}\left[\dfrac{1}{(1+x)^2}\right] = -\dfrac{1}{2}\dfrac{d}{dx}\left[\sum\limits_{n=0}^{\infty}(-1)^n\,(n+1)\,x^n\right]$ (from Exercise 9)

$= -\dfrac{1}{2}\sum\limits_{n=1}^{\infty}(-1)^n\,(n+1)\,n x^{n-1} = \dfrac{1}{2}\sum\limits_{n=0}^{\infty}(-1)^n\,(n+2)\,(n+1)\,x^n$ with $R = 1$.

13. $f(x) = \ln(5-x) = -\displaystyle\int \dfrac{dx}{5-x} = -\dfrac{1}{5}\int \dfrac{dx}{1-x/5}$

$= -\dfrac{1}{5}\displaystyle\int \left[\sum\limits_{n=0}^{\infty}\left(\dfrac{x}{5}\right)^n\right] dx = C - \dfrac{1}{5}\sum\limits_{n=0}^{\infty}\dfrac{x^{n+1}}{5^n\,(n+1)} = C - \sum\limits_{n=1}^{\infty}\dfrac{x^n}{n5^n}$

Putting $x = 0$, we get $C = \ln 5$. The series converges for $|x/5| < 1 \iff |x| < 5$, so $R = 5$.

15. $f(x) = \ln(3+x) = \int \dfrac{dx}{3+x} = \dfrac{1}{3}\int \dfrac{dx}{1+x/3} = \dfrac{1}{3}\int \displaystyle\sum_{n=0}^{\infty}(-1)^n\left(\dfrac{x}{3}\right)^n dx$ (from Exercise 3)

$= C + \dfrac{1}{3}\displaystyle\sum_{n=0}^{\infty}\dfrac{(-1/3)^n}{n+1}x^{n+1} = \ln 3 + \dfrac{1}{3}\displaystyle\sum_{n=1}^{\infty}\dfrac{(-1/3)^{n-1}}{n}x^n$ $[C = f(0) = \ln 3]$

$= \ln 3 + \displaystyle\sum_{n=1}^{\infty}\dfrac{(-1)^{n-1}}{n3^n}x^n$ with $R = 3$.

The terms of the series are $a_0 = \ln 3$, $a_1 = \dfrac{x}{3}$, $a_2 = -\dfrac{x^2}{18}$, $a_3 = \dfrac{x^3}{81}$, $a_4 = -\dfrac{x^4}{324}$, $a_5 = \dfrac{x^5}{1215}, \ldots$

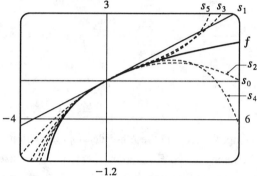

As n increases, $s_n(x)$ approximates f better on the interval of convergence, which is $(-3, 3)$.

17. $f(x) = \ln\left(\dfrac{1+x}{1-x}\right) = \ln(1+x) - \ln(1-x) = \displaystyle\int \dfrac{dx}{1+x} + \int \dfrac{dx}{1-x}$

$= \displaystyle\int \left[\sum_{n=0}^{\infty}(-1)^n x^n + \sum_{n=0}^{\infty}x_n\right]dx = \int \sum_{n=0}^{\infty}2x^{2n}\,dx = \sum_{n=0}^{\infty}\dfrac{2x^{2n+1}}{2n+1} + C$

But $f(0) = \ln\frac{1}{1} = 0$, so $C = 0$ and we have $f(x) = \displaystyle\sum_{n=0}^{\infty}\dfrac{2x^{2n+1}}{2n+1}$ with $R = 1$. If $x = \pm 1$, then

$f(x) = \pm 2\displaystyle\sum_{n=0}^{\infty}\dfrac{1}{2n+1}$, which both diverge by the Limit Comparison Test with $b_n = \dfrac{1}{n}$.

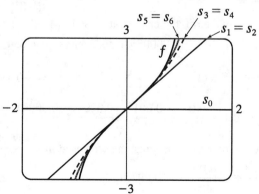

As n increases, $s_n(x)$ approximates f better on the interval of convergence, which is $(-1, 1)$.

19. $\displaystyle\int \frac{dx}{1+x^4} = \int \sum_{n=0}^{\infty} (-1)^n \, x^{4n} \, dx = C + \sum_{n=0}^{\infty} \frac{(-1)^n \, x^{4n+1}}{4n+1}$ with $R = 1$.

21. By Example 7, $\displaystyle\arctan x = \sum_{n=0}^{\infty} (-1)^n \frac{x^{2n+1}}{2n+1}$, so

$$\int \frac{\arctan x}{x} \, dx = \int \sum_{n=0}^{\infty} (-1)^n \frac{x^{2n}}{2n+1} \, dx$$

$$= C + \sum_{n=0}^{\infty} (-1)^n \frac{x^{2n+1}}{(2n+1)^2} \text{ with } R = 1.$$

23. We use the representation $\displaystyle\int \frac{dx}{1+x^4} = C + \sum_{n=0}^{\infty} \frac{(-1)^n \, x^{4n+1}}{4n+1}$ from Exercise 19 with $C = 0$. So

$$\int_0^{0.2} \frac{dx}{1+x^4} = \left[x - \frac{x^5}{5} + \frac{x^9}{9} - \frac{x^{13}}{13} + \cdots \right]_0^{0.2}$$

$$= 0.2 - \frac{0.2^5}{5} + \frac{0.2^9}{9} - \frac{0.2^{13}}{13} + \cdots$$

Since the series is alternating, the error in the nth-order approximation is less than the first neglected term, by The Alternating Series Estimation Theorem. If we use only the first two terms of the series, then the error is at most $0.2^9/9 \approx 5.7 \times 10^{-8}$. So, to six decimal places,

$$\int_0^{0.2} \frac{dx}{1+x^4} \approx 0.2 - \frac{0.2^5}{5} = 0.199936.$$

25. We substitute x^4 for x in Example 7, and find that

$$\int x^2 \tan^{-1}(x^4) \, dx = \int x^2 \sum_{n=0}^{\infty} (-1)^n \frac{(x^4)^{2n+1}}{2n+1} \, dx$$

$$= \int \sum_{n=0}^{\infty} (-1)^n \frac{x^{8n+6}}{2n+1} \, dx = C + \sum_{n=0}^{\infty} (-1)^n \frac{x^{8n+7}}{(2n+1)(8n+7)}$$

So $\displaystyle\int_0^{1/3} x^2 \tan^{-1}(x^4) \, dx = \left[\frac{x^7}{7} - \frac{x^{15}}{45} + \cdots \right]_0^{1/3} = \frac{1}{7 \cdot 3^7} - \frac{1}{45 \cdot 3^{15}} + \cdots$. The series is alternating, so if we use only one term, the error is at most $1/(45 \cdot 3^{15}) \approx 1.5 \times 10^{-9}$. So $\displaystyle\int_0^{1/3} x^2 \tan^{-1}(x^4) \, dx \approx 1/(7 \cdot 3^7) \approx 0.000065$ to six decimal places.

27. Using the result of Example 6, $\displaystyle\ln(1-x) = -\sum_{n=1}^{\infty} \frac{x^n}{n}$, with $x = -0.1$, we have

$$\ln 1.1 = \ln[1 - (-0.1)] = 0.1 - \frac{0.01}{2} + \frac{0.001}{3} - \frac{0.0001}{4} + \frac{0.00001}{5} - \cdots. \text{ The series is}$$

alternating, so if we use only the first four terms, the error is at most $\dfrac{0.00001}{5} = 0.000002$. So

$$\ln 1.1 \approx 0.1 - \frac{0.01}{2} + \frac{0.001}{3} - \frac{0.0001}{4} \approx 0.09531.$$

29. (a) $J_0(x) = \displaystyle\sum_{n=0}^{\infty} \frac{(-1)^n \, x^{2n}}{2^{2n} \, (n!)^2}$, $J_0'(x) = \displaystyle\sum_{n=1}^{\infty} \frac{(-1)^n \, 2n x^{2n-1}}{2^{2n} \, (n!)^2}$, and

$$J_0''(x) = \sum_{n=1}^{\infty} \frac{(-1)^n \, 2n \, (2n-1) \, x^{2n-2}}{2^{2n} \, (n!)^2}, \text{ so}$$

$$x^2 J_0''(x) + x J_0'(x) + x^2 J_0(x) = \sum_{n=1}^{\infty} \frac{(-1)^n \, 2n \, (2n-1) \, x^{2n}}{2^{2n} \, (n!)^2} + \sum_{n=1}^{\infty} \frac{(-1)^n \, 2n x^{2n}}{2^{2n} \, (n!)^2} + \sum_{n=0}^{\infty} \frac{(-1)^n \, x^{2n+2}}{2^{2n} \, (n!)^2}$$

$$= \sum_{n=1}^{\infty} \frac{(-1)^n \, 2n \, (2n-1) \, x^{2n}}{2^{2n} \, (n!)^2} + \sum_{n=1}^{\infty} \frac{(-1)^n \, 2n x^{2n}}{2^{2n} \, (n!)^2} + \sum_{n=1}^{\infty} \frac{(-1)^{n-1} \, x^{2n}}{2^{2n-2} \, [(n-1)!]^2}$$

$$= \sum_{n=1}^{\infty} (-1)^n \left[\frac{2n \, (2n-1) + 2n - 2^2 n^2}{2^{2n} \, (n!)^2} \right] x^{2n} = \sum_{n=1}^{\infty} (-1)^n \left[\frac{4n^2 - 2n + 2n - 4n^2}{2^{2n} \, (n!)^2} \right] x^{2n} = 0$$

(b) $\displaystyle\int_0^1 J_0(x)\, dx = \int_0^1 \left[\sum_{n=0}^{\infty} \frac{(-1)^n \, x^{2n}}{2^{2n} \, (n!)^2} \right] dx = \int_0^1 \left(1 - \frac{x^2}{4} + \frac{x^4}{64} - \frac{x^6}{2304} + \cdots \right) dx$

$$= \left[x - \frac{x^3}{3 \cdot 4} + \frac{x^5}{5 \cdot 64} - \frac{x^7}{7 \cdot 2304} + \cdots \right]_0^1 = 1 - \frac{1}{12} + \frac{1}{320} - \frac{1}{16{,}128} + \cdots$$

Since $\frac{1}{16{,}128} \approx 0.000062$, it follows from The Alternating Series Estimation Theorem that, correct to three decimal places, $\int_0^1 J_0(x)\, dx \approx 1 - \frac{1}{12} + \frac{1}{320} \approx 0.920$.

31. (a) $f(x) = \displaystyle\sum_{n=0}^{\infty} \frac{x^n}{n!} \implies f'(x) = \sum_{n=1}^{\infty} \frac{n x^{n-1}}{n!} = \sum_{n=1}^{\infty} \frac{x^{n-1}}{(n-1)!} = \sum_{n=0}^{\infty} \frac{x^n}{n!} = f(x)$.

(b) By Theorem 7.5.2, the only solution to the differential equation $\dfrac{df(x)}{dx} = f(x)$ is $f(x) = Ke^x$, but $f(0) = 1$, so $K = 1$ and $f(x) = e^x$.

Or: We could solve the equation $\dfrac{df(x)}{dx} = f(x)$ as a separable differential equation.

33. If $a_n = \dfrac{x^n}{n^2}$, then by the Ratio Test, $\displaystyle\lim_{n\to\infty} \left| \frac{a_{n+1}}{a_n} \right| = |x| \lim_{n\to\infty} \left(\frac{n}{n+1} \right)^2 = |x| < 1$ for convergence, so

$R = 1$. When $x = \pm 1$, $\displaystyle\sum_{n=1}^{\infty} \left| \frac{x^n}{n^2} \right| = \sum_{n=1}^{\infty} \frac{1}{n^2}$ which is a convergent p-series ($p = 2 > 1$), so the interval

of convergence for f is $[-1, 1]$. By Theorem 2, the radii of convergence of f' and f'' are both 1, so we

need only check the endpoints. $f(x) = \displaystyle\sum_{n=1}^{\infty} \frac{x^n}{n^2} \implies f'(x) = \sum_{n=1}^{\infty} \frac{n x^{n-1}}{n^2} = \sum_{n=0}^{\infty} \frac{x^n}{n+1}$, and this series

diverges for $x = 1$ (harmonic series) and converges for $x = -1$ (Alternating Series Test), so the interval

of convergence is $[-1, 1)$. $f''(x) = \displaystyle\sum_{n=1}^{\infty} \frac{n x^{n-1}}{n+1}$ diverges at both 1 and -1 (Test for Divergence) since

$\displaystyle\lim_{n\to\infty} \frac{n}{n+1} = 1 \neq 0$, so its interval of convergence is $(-1, 1)$.

Section 8.7 Taylor and Maclaurin Series

1. Using Theorem 5 with $\sum\limits_{n=0}^{\infty} b_n (x-5)^n$, $b_n = \dfrac{f^{(n)}(a)}{n!}$, so $b_8 = \dfrac{f^{(8)}(5)}{8!}$.

3.

n	$f^{(n)}(x)$	$f^{(n)}(0)$
0	$\cos x$	1
1	$-\sin x$	0
2	$-\cos x$	-1
3	$\sin x$	0
4	$\cos x$	1
...	...	...

$$\cos x = f(0) + f'(0)\,x + \frac{f''(0)}{2!}x^2 + \frac{f^{(3)}(0)}{3!}x^3 + \frac{f^{(4)}(0)}{4!}x^4 + \cdots$$

$$= 1 - \frac{x^2}{2!} + \frac{x^4}{4!} - \cdots = \sum_{n=0}^{\infty} \frac{(-1)^n\, x^{2n}}{(2n)!}$$

If $a_n = \dfrac{(-1)^n\, x^{2n}}{(2n)!}$, then $\lim\limits_{n\to\infty}\left|\dfrac{a_{n+1}}{a_n}\right| = x^2 \lim\limits_{n\to\infty} \dfrac{1}{(2n+2)(2n+1)} = 0 < 1$ for all x.
So $R = \infty$ (Ratio Test).

5.

n	$f^{(n)}(x)$	$f^{(n)}(0)$
0	$(1+x)^{-2}$	1
1	$-2(1+x)^{-3}$	-2
2	$2 \cdot 3(1+x)^{-4}$	$2 \cdot 3$
3	$-2 \cdot 3 \cdot 4(1+x)^{-5}$	$-2 \cdot 3 \cdot 4$
4	$2 \cdot 3 \cdot 4 \cdot 5(1+x)^{-6}$	$2 \cdot 3 \cdot 4 \cdot 5$
...	...	...

So $f^{(n)}(0) = (-1)^n (n+1)!$ and

$$\frac{1}{(1+x)^2} = \sum_{n=0}^{\infty} \frac{(-1)^n (n+1)!}{n!} x^n = \sum_{n=0}^{\infty} (-1)^n (n+1)\, x^n$$

If $a_n = (-1)^n (n+1) x^n$, then $\lim\limits_{n\to\infty}\left|\dfrac{a_{n+1}}{a_n}\right| = |x| < 1$ for convergence, so $R = 1$.

7. Clearly, $f^{(n)}(x) = e^x$, so $f^{(n)}(3) = e^3$ and $e^x = \sum\limits_{n=0}^{\infty} \dfrac{e^3}{n!}(x-3)^n$. If $a_n = \dfrac{e^3}{n!}(x-3)^n$, then

$$\lim_{n\to\infty}\left|\frac{a_{n+1}}{a_n}\right| = \lim_{n\to\infty} \frac{|x-3|}{n+1} = 0 \text{ for all } x, \text{ so } R = \infty.$$

9.

n	$f^{(n)}(x)$	$f^{(n)}(1)$
0	x^{-1}	1
1	$-x^{-2}$	-1
2	$2x^{-3}$	2
3	$-3\cdot 2x^{-4}$	$-3\cdot 2$
4	$4\cdot 3\cdot 2x^{-5}$	$4\cdot 3\cdot 2$
...	...	...

So $f^{(n)}(1) = (-1)^n\, n!$, and

$$\frac{1}{x} = \sum_{n=0}^{\infty} \frac{(-1)^n\, n!}{n!}(x-1)^n = \sum_{n=0}^{\infty} (-1)^n (x-1)^n.$$ If

$a_n = (-1)^n (x-1)^n$ then $\displaystyle\lim_{n\to\infty}\left|\frac{a_{n+1}}{a_n}\right| = |x-1| < 1$ for

convergence, so $0 < x < 2$ and $R = 1$.

11.

n	$f^{(n)}(x)$	$f^{(n)}\left(\frac{\pi}{4}\right)$
0	$\sin x$	$\sqrt{2}/2$
1	$\cos x$	$\sqrt{2}/2$
2	$-\sin x$	$-\sqrt{2}/2$
3	$-\cos x$	$-\sqrt{2}/2$
4	$\sin x$	$\sqrt{2}/2$
...	...	...

$$\sin x = f\left(\tfrac{\pi}{4}\right) + f'\left(\tfrac{\pi}{4}\right)\left(x - \tfrac{\pi}{4}\right) + \frac{f''\left(\tfrac{\pi}{4}\right)}{2!}\left(x - \tfrac{\pi}{4}\right)^2$$

$$+ \frac{f^{(3)}\left(\tfrac{\pi}{4}\right)}{3!}\left(x - \tfrac{\pi}{4}\right)^3 + \frac{f^{(4)}\left(\tfrac{\pi}{4}\right)}{4!}\left(x - \tfrac{\pi}{4}\right)^4 + \cdots$$

$$= \frac{\sqrt{2}}{2}\left[1 + \left(x - \tfrac{\pi}{4}\right) - \tfrac{1}{2!}\left(x - \tfrac{\pi}{4}\right)^2 \right.$$

$$\left. - \tfrac{1}{3!}\left(x - \tfrac{\pi}{4}\right)^3 + \tfrac{1}{4!}\left(x - \tfrac{\pi}{4}\right)^4 + \cdots\right]$$

$$= \frac{\sqrt{2}}{2}\sum_{n=0}^{\infty}\frac{(-1)^{n(n-1)/2}\left(x - \tfrac{\pi}{4}\right)^n}{n!}$$

If $a_n = \dfrac{(-1)^{n(n-1)/2}\left(x - \tfrac{\pi}{4}\right)^n}{n!}$, then $\displaystyle\lim_{n\to\infty}\left|\frac{a_{n+1}}{a_n}\right| = \lim_{n\to\infty}\frac{\left|x - \tfrac{\pi}{4}\right|}{n+1} = 0 < 1$ for all x, so $R = \infty$.

13. If $f(x) = \cos x$, then by Formula 9 with $a = 0$, $|R_n(x)| \le \dfrac{\left|f^{(n+1)}(x)\right|}{(n+1)!}|x|^{n+1}$. But

$f^{(n+1)}(x) = \pm\sin x$ or $\pm\cos x$. In each case, $\left|f^{(n+1)}(x)\right| \le 1$, so $|R_n(x)| \le \dfrac{1}{(n+1)!}|x|^{n+1} \to 0$

as $n \to \infty$ by Equation 10. So $\displaystyle\lim_{n\to\infty} R_n(x) = 0$ and, by Theorem 8, the series in Exercise 3 represents $\cos x$ for all x.

15. $e^x = \displaystyle\sum_{n=0}^{\infty} \frac{x^n}{n!} \implies e^{3x} = \sum_{n=0}^{\infty}\frac{(3x)^n}{n!} = \sum_{n=0}^{\infty}\frac{3^n x^n}{n!}$, with $R = \infty$.

17. $\cos x = \displaystyle\sum_{n=0}^{\infty}\frac{(-1)^n x^{2n}}{(2n)!} \implies x^2\cos x = x^2\sum_{n=0}^{\infty}\frac{(-1)^n x^{2n}}{(2n)!} = \sum_{n=0}^{\infty}\frac{(-1)^n x^{2n+2}}{(2n)!}$, $R = \infty$

19. $\sin x = \displaystyle\sum_{n=0}^{\infty}\frac{(-1)^n x^{2n+1}}{(2n+1)!} \implies x\sin\left(\tfrac{x}{2}\right) = x\sum_{n=0}^{\infty}\frac{(-1)^n (x/2)^{2n+1}}{(2n+1)!} = \sum_{n=0}^{\infty}\frac{(-1)^n x^{2n+2}}{(2n+1)!\,2^{2n+1}}$, with
$R = \infty$.

21. $\sin^2 x = \tfrac{1}{2}[1 - \cos 2x] = \dfrac{1}{2}\left[1 - \displaystyle\sum_{n=0}^{\infty}\frac{(-1)^n (2x)^{2n}}{(2n)!}\right] = 2^{-1}\left[1 - 1 - \sum_{n=1}^{\infty}\frac{(-1)^n (2x)^{2n}}{(2n)!}\right]$

$$= \sum_{n=1}^{\infty}\frac{(-1)^{n+1}\,2^{2n-1} x^{2n}}{(2n)!}, \text{ with } R = \infty.$$

23.

n	$f^{(n)}(x)$	$f^{(n)}(0)$
0	$(1+x)^{1/2}$	1
1	$\frac{1}{2}(1+x)^{-1/2}$	$\frac{1}{2}$
2	$-\frac{1}{4}(1+x)^{-3/2}$	$-\frac{1}{4}$
3	$\frac{3}{8}(1+x)^{-5/2}$	$\frac{3}{8}$
4	$-\frac{15}{16}(1+x)^{-7/2}$	$-\frac{15}{16}$
...	...	...

So $f^{(n)}(0) = \dfrac{(-1)^{n-1} \, 1 \cdot 3 \cdot 5 \cdots (2n-3)}{2^n}$ for $n \geq 2$, and

$$\sqrt{1+x} = 1 + \frac{x}{2} + \sum_{n=2}^{\infty} \frac{(-1)^{n-1} \, 1 \cdot 3 \cdot 5 \cdots (2n-3)}{2^n n!} x^n$$

If $a_n = \dfrac{(-1)^{n-1} \, 1 \cdot 3 \cdot 5 \cdots (2n-3)}{2^n n!} x^n$, then

$$\lim_{n \to \infty} \left| \frac{a_{n+1}}{a_n} \right| = \frac{|x|}{2} \lim_{n \to \infty} \frac{2n-1}{n+1}$$

$$= |x| < 1 \text{ for convergence, so } R = 1.$$

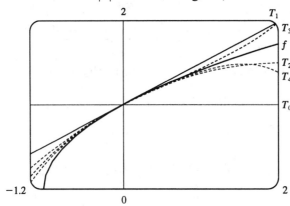

25. $f(x) = (1+x)^{-3} = -\dfrac{1}{2}\dfrac{d}{dx}\left[\dfrac{1}{(1+x)^2}\right] = -\dfrac{1}{2}\dfrac{d}{dx}\left[\displaystyle\sum_{n=0}^{\infty}(-1)^n(n+1)x^n\right]$ (from Exercise 5)

$\qquad = -\dfrac{1}{2}\displaystyle\sum_{n=1}^{\infty}(-1)^n n(n+1)x^{n-1} = \displaystyle\sum_{n=1}^{\infty}\dfrac{(-1)^{n+1}n(n+1)x^{n-1}}{2}$

$\qquad = \displaystyle\sum_{n=0}^{\infty}\dfrac{(-1)^n(n+1)(n+2)x^n}{2}$, with $R = 1$ as in Exercise 5.

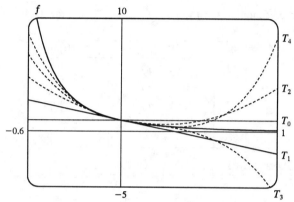

27. $\ln(1+x) = \displaystyle\int\dfrac{dx}{1+x} = \int\displaystyle\sum_{n=0}^{\infty}(-1)^n x^n\,dx = C + \displaystyle\sum_{n=0}^{\infty}(-1)^n\dfrac{x^{n+1}}{n+1} = \displaystyle\sum_{n=1}^{\infty}\dfrac{(-1)^{n-1}x^n}{n}$

with $C = 0$ and $R = 1$, so $\ln(1.1) = \displaystyle\sum_{n=1}^{\infty}\dfrac{(-1)^{n-1}(0.1)^n}{n}$.   This is an alternating series with

$b_5 = \dfrac{(0.1)^5}{5} = 0.000002$, so to five decimal places, $\ln(1.1) \approx \displaystyle\sum_{n=1}^{4}\dfrac{(-1)^{n-1}(0.1)^n}{n} \approx 0.09531$.

29. $\displaystyle\int \sin(x^2)\,dx = \int\displaystyle\sum_{n=0}^{\infty}(-1)^n\dfrac{(x^2)^{2n+1}}{(2n+1)!}\,dx = \int\displaystyle\sum_{n=0}^{\infty}\dfrac{(-1)^n x^{4n+2}}{(2n+1)!}\,dx$

$\qquad = C + \displaystyle\sum_{n=0}^{\infty}\dfrac{(-1)^n x^{4n+3}}{(4n+3)(2n+1)!}$

31. Using the series from Exercise 23 and substituting x^3 for x, we get

$\displaystyle\int\sqrt{x^3+1}\,dx = \int\left[1 + \dfrac{x^3}{2} + \displaystyle\sum_{n=2}^{\infty}\dfrac{(-1)^{n-1}1\cdot 3\cdot 5\cdot\,\cdots\,\cdot(2n-3)}{2^n n!}x^{3n}\right]dx$

$\qquad = C + x + \dfrac{x^4}{8} + \displaystyle\sum_{n=2}^{\infty}\dfrac{(-1)^{n-1}1\cdot 3\cdot 5\cdot\,\cdots\,\cdot(2n-3)}{2^n n!(3n+1)}x^{3n+1}.$

33. Using our series from Exercise 29, we get $\displaystyle\int_0^1 \sin\left(x^2\right) dx = \sum_{n=0}^{\infty}\left[\frac{(-1)^n\, x^{4n+3}}{(4n+3)\,(2n+1)!}\right]_0^1 =$

$\displaystyle\sum_{n=0}^{\infty}\frac{(-1)^n}{(4n+3)\,(2n+1)!}$ and $|c_3| = \dfrac{1}{75{,}600} < 0.000014$, so by the Alternating Series Estimation

Theorem, we have $\displaystyle\sum_{n=0}^{2}\frac{(-1)^n}{(4n+3)\,(2n+1)!} \approx \frac{1}{3} - \frac{1}{42} + \frac{1}{1320} \approx 0.310$ (correct to three decimal places).

35. We first find a series representation for $f(x) = (1+x)^{-1/2}$, and then substitute.

n	$f^{(n)}(x)$	$f^{(n)}(0)$
0	$(1+x)^{-1/2}$	1
1	$-\frac{1}{2}(1+x)^{-3/2}$	$-\frac{1}{2}$
2	$\frac{3}{4}(1+x)^{-5/2}$	$\frac{3}{4}$
3	$-\frac{15}{8}(1+x)^{-7/2}$	$-\frac{15}{8}$
$\ldots$	$\ldots$	$\ldots$

$\dfrac{1}{\sqrt{1+x}} = 1 - \dfrac{x}{2} + \dfrac{3}{4}\left(\dfrac{x^2}{2!}\right) - \dfrac{15}{8}\left(\dfrac{x^3}{3!}\right) + \cdots \implies \dfrac{1}{\sqrt{1+x^3}} = 1 - \dfrac{1}{2}x^3 + \dfrac{3}{8}x^6 - \dfrac{5}{16}x^9 + \cdots \implies$

$\displaystyle\int_0^{0.1}\frac{dx}{\sqrt{1+x^3}} = \left[x - \frac{1}{8}x^4 + \frac{3}{56}x^7 - \frac{1}{32}x^{10} + \cdots\right]_0^{0.1} \approx (0.1) - \frac{1}{8}(0.1)^4$, by the Alternating Series

Estimation Theorem, since $\frac{3}{56}(0.1)^7 \approx 0.0000000054 < 10^{-8}$, which is the maximum desired error.

Therefore, $\displaystyle\int_0^{0.1}\frac{dx}{\sqrt{1+x^3}} \approx 0.09998750$.

37. $\displaystyle\lim_{x\to 0}\frac{x - \tan^{-1}x}{x^3} = \lim_{x\to 0}\frac{x - \left(x - \frac{1}{3}x^3 + \frac{1}{5}x^5 - \frac{1}{7}x^7 + \cdots\right)}{x^3} = \lim_{x\to 0}\frac{\frac{1}{3}x^3 - \frac{1}{5}x^5 + \frac{1}{7}x^7 - \cdots}{x^3}$

$\displaystyle = \lim_{x\to 0}\left(\frac{1}{3} - \frac{1}{5}x^2 + \frac{1}{7}x^4 - \cdots\right) = \frac{1}{3}$

since power series are continuous functions.

39. $\displaystyle\lim_{x\to 0}\frac{\sin x - x + \frac{1}{6}x^3}{x^5} = \lim_{x\to 0}\frac{\left(x - \frac{1}{3!}x^3 + \frac{1}{5!}x^5 - \frac{1}{7!}x^7 + \cdots\right) - x + \frac{1}{6}x^3}{x^5}$

$\displaystyle = \lim_{x\to 0}\frac{\frac{1}{5!}x^5 - \frac{1}{7!}x^7 + \cdots}{x^5} = \lim_{x\to 0}\left(\frac{1}{5!} - \frac{x^2}{7!} + \frac{x^4}{9!} - \cdots\right) = \frac{1}{5!} = \frac{1}{120}$

since power series are continuous functions.

41. As in Example 8(a), we have $e^{-x^2} = 1 - \dfrac{x^2}{1!} + \dfrac{x^4}{2!} - \dfrac{x^6}{3!} + \cdots$ and we

know that $\cos x = 1 - \dfrac{x^2}{2!} + \dfrac{x^4}{4!} - \cdots$ from Equation 16. Therefore,

$e^{-x^2}\cos x = \left(1 - x^2 + \frac{1}{2}x^4 - \cdots\right)\left(1 - \frac{1}{2}x^2 + \frac{1}{24}x^4 - \cdots\right)$. Writing only the terms with degree ≤ 4,

we get $e^{-x^2}\cos x = 1 - \frac{1}{2}x^2 + \frac{1}{24}x^4 - x^2 + \frac{1}{2}x^4 + \frac{1}{2}x^4 + \cdots = 1 - \frac{3}{2}x^2 + \frac{25}{24}x^4 + \cdots$.

43.

$$
1 + x + \tfrac{1}{2}x^2 + \tfrac{1}{6}x^3 + \cdots \overline{\smash{\big)}\,}
$$

$$
\begin{array}{r}
-x + \tfrac{1}{2}x^2 - \tfrac{1}{3}x^3 + \cdots \\
\hline
-x - \tfrac{1}{2}x^2 - \tfrac{1}{3}x^3 - \cdots \\
-x - x^2 - \tfrac{1}{2}x^3 - \cdots \\
\hline
\tfrac{1}{2}x^2 + \tfrac{1}{6}x^3 - \cdots \\
\tfrac{1}{2}x^2 + \tfrac{1}{2}x^3 + \cdots \\
\hline
-\tfrac{1}{3}x^3 + \cdots \\
-\tfrac{1}{3}x^3 + \cdots \\
\hline
\cdots
\end{array}
$$

From Example 6 in Section 8.6, we have $\ln(1-x) = -x - \tfrac{1}{2}x^2 - \tfrac{1}{3}x^3 - \cdots$, $|x| < 1$.

Therefore, $y = \dfrac{\ln(1-x)}{e^x} = \dfrac{-x - \tfrac{1}{2}x^2 - \tfrac{1}{3}x^3 - \cdots}{1 + x + \tfrac{1}{2}x^2 + \tfrac{1}{6}x^3 + \cdots}$. So by the long division above,

$$
\frac{\ln(1-x)}{e^x} = -x + \frac{x^2}{2} - \frac{x^3}{3} + \cdots, \ |x| < 1.
$$

45. $\displaystyle\sum_{n=0}^{\infty} (-1)^n \frac{x^{4n}}{n!} = \sum_{n=0}^{\infty} \frac{\left(-x^4\right)^n}{n!} = e^{-x^4}$, by (11).

47. $\displaystyle\sum_{n=0}^{\infty} \frac{(-1)^n \, \pi^{2n+1}}{4^{2n+1} \, (2n+1)!} = \sum_{n=0}^{\infty} \frac{(-1)^n \left(\frac{\pi}{4}\right)^{2n+1}}{(2n+1)!} = \sin\frac{\pi}{4} = \frac{1}{\sqrt{2}}$, by (15).

49. $\displaystyle\sum_{n=0}^{\infty} \frac{x^{n+1}}{(n+1)!} = \frac{x}{1!} + \frac{x^2}{2!} + \frac{x^3}{3!} + \cdots = \left(1 + \frac{x}{1!} + \frac{x^2}{2!} + \frac{x^3}{3!} + \cdots\right) - 1 = e^x - 1$, by (11).

Section 8.8 The Binomial Series

1. The general binomial series in (2) is

$$(1+x)^k = \sum_{n=0}^{\infty} \binom{k}{n} x^n$$

$$= 1 + kx + \frac{k(k-1)}{2!}x^2 + \frac{k(k-1)(k-2)}{3!}x^3 + \cdots$$

$$(1+x)^{1/2} = \sum_{n=0}^{\infty} \binom{\frac{1}{2}}{n} x^n = 1 + \binom{1}{2}x + \frac{\left(\frac{1}{2}\right)\left(-\frac{1}{2}\right)}{2!}x^2 + \frac{\left(\frac{1}{2}\right)\left(-\frac{1}{2}\right)\left(-\frac{3}{2}\right)}{3!}x^3 + \cdots$$

$$= 1 + \frac{x}{2} - \frac{x^2}{2^2 \cdot 2!} + \frac{1 \cdot 3 \cdot x^3}{2^3 \cdot 3!} - \frac{1 \cdot 3 \cdot 5 \cdot x^4}{2^4 \cdot 4!} + \cdots$$

$$= 1 + \frac{x}{2} + \sum_{n=2}^{\infty} \frac{(-1)^{n-1} 1 \cdot 3 \cdot 5 \cdots \cdot (2n-3) x^n}{2^n \cdot n!}, \; R = 1$$

3. $[1+(2x)]^{-4} = 1 + (-4)(2x) + \dfrac{(-4)(-5)}{2!}(2x)^2 + \dfrac{(-4)(-5)(-6)}{3!}(2x)^3 + \cdots$

$$= 1 + \sum_{n=1}^{\infty} \frac{(-1)^n 2^n \cdot 4 \cdot 5 \cdot 6 \cdots \cdot (n+3)}{n!} x^n$$

$$= 1 + \sum_{n=1}^{\infty} \frac{(-1)^n 2^n \cdot 2 \cdot 3 \cdot 4 \cdot 5 \cdot 6 \cdots \cdot (n+1)(n+2)(n+3)}{2 \cdot 3 \cdot n!} x^n$$

$$= \sum_{n=0}^{\infty} (-1)^n \frac{2^n (n+1)(n+2)(n+3)}{6} x^n, \; |2x| < 1 \iff |x| < \tfrac{1}{2}, \text{ so } R = \tfrac{1}{2}.$$

5. $[1+(-x^4)]^{1/4} = 1 + \left(\frac{1}{4}\right)(-x^4) + \dfrac{\left(\frac{1}{4}\right)\left(-\frac{3}{4}\right)}{2!}(-x^4)^2 + \dfrac{\left(\frac{1}{4}\right)\left(-\frac{3}{4}\right)\left(-\frac{7}{4}\right)}{3!}(-x^4)^3 + \cdots$

$$= 1 - \frac{x^4}{4} - \sum_{n=2}^{\infty} \frac{3 \cdot 7 \cdot 11 \cdots \cdot (4n-5)}{4^n \cdot n!} x^{4n}, \text{ with } R = 1.$$

7. $\dfrac{1}{\sqrt[3]{8+x}} = (8+x)^{-1/3} = 8^{-1/3}\left(1+\dfrac{x}{8}\right)^{-1/3} = \dfrac{1}{2}\left(1+\dfrac{x}{8}\right)^{-1/3}$

$$= \frac{1}{2}\left[1 + \left(-\frac{1}{3}\right)\left(\frac{x}{8}\right) + \frac{\left(-\frac{1}{3}\right)\left(-\frac{4}{3}\right)}{2!}\left(\frac{x}{8}\right)^2 + \cdots\right]$$

$$= \frac{1}{2}\left[1 + \sum_{n=1}^{\infty} \frac{(-1)^n 1 \cdot 4 \cdot 7 \cdots \cdot (3n-2)}{3^n \cdot n! \, 8^n} x^n\right]$$

and $\left|\dfrac{x}{8}\right| < 1 |x| < 8$, so $R = 8$.

The three Taylor polynomials are $T_1(x) = \frac{1}{2} - \frac{1}{48}x$, $T_2(x) = \frac{1}{2} - \frac{1}{48}x + \frac{1}{576}x^2$, and $T_3(x) = \frac{1}{2} - \frac{1}{48}x + \frac{1}{576}x^2 - \frac{4 \cdot 7}{2 \cdot 27 \cdot 6 \cdot 512}x^3 = \frac{1}{2} - \frac{1}{48}x + \frac{1}{576}x^2 - \frac{7}{41,472}x^3$.

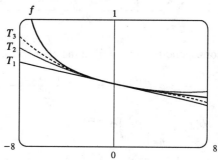

9. (a)

$$[1 + (-x^2)]^{-1/2} = 1 + \left(-\frac{1}{2}\right)(-x^2) + \frac{\left(-\frac{1}{2}\right)\left(-\frac{3}{2}\right)}{2!}(-x^2)^2$$

$$+ \frac{\left(-\frac{1}{2}\right)\left(-\frac{3}{2}\right)\left(-\frac{5}{2}\right)}{3!}(-x^2)^3 + \cdots$$

$$= 1 + \sum_{n=1}^{\infty} \frac{1 \cdot 3 \cdot 5 \cdots \cdots (2n-1)}{2^n \cdot n!}x^{2n}$$

(b)

$$\sin^{-1}x = \int \frac{1}{\sqrt{1-x^2}}\,dx = C + x + \sum_{n=1}^{\infty} \frac{1 \cdot 3 \cdot 5 \cdots \cdots (2n-1)}{(2n+1)\,2^n \cdot n!}x^{2n+1}$$

$$= x + \sum_{n=1}^{\infty} \frac{1 \cdot 3 \cdot 5 \cdots \cdots (2n-1)}{(2n+1)\,2^n \cdot n!}x^{2n+1}$$

since $0 = \sin^{-1} 0 = C$.

11. (a)

$$[1 + (-x)]^{-2} = 1 + (-2)(-x) + \frac{(-2)(-3)}{2!}(-x)^2 + \frac{(-2)(-3)(-4)}{3!}(-x)^3 + \cdots$$

$$= 1 + 2x + 3x^2 + 4x^3 + \cdots$$

$$= \sum_{n=0}^{\infty} (n+1)x^n$$

so $\dfrac{x}{(1-x)^2} = \displaystyle\sum_{n=0}^{\infty}(n+1)x^{n+1} = \sum_{n=1}^{\infty} nx^n$.

(b) With $x = \frac{1}{2}$ in part (a), we have $\displaystyle\sum_{n=1}^{\infty}\frac{n}{2^n} = \frac{\frac{1}{2}}{\left(1-\frac{1}{2}\right)^2} = 2$.

13. (a)
$$(1+x^2)^{1/2} = 1 + \left(\frac{1}{2}\right)x^2 + \frac{\left(\frac{1}{2}\right)\left(-\frac{1}{2}\right)}{2!}(x^2)^2 + \frac{\left(\frac{1}{2}\right)\left(-\frac{1}{2}\right)\left(-\frac{3}{2}\right)}{3!}(x^2)^3 + \cdots$$

$$= 1 + \frac{x^2}{2} + \sum_{n=2}^{\infty} \frac{(-1)^{n-1}\, 1 \cdot 3 \cdot 5 \cdot \cdots \cdot (2n-3)}{2^n \cdot n!} x^{2n}$$

(b) The coefficient of x^{10} (corresponding to $n = 5$) in the above Maclaurin series is $\dfrac{f^{(10)}(0)}{10!}$, so

$$\frac{f^{(10)}(0)}{10!} = \frac{(-1)^4 \cdot 1 \cdot 3 \cdot 5 \cdot 7}{2^5 \cdot 5!} \implies f^{(10)}(0) = 10!\left(\frac{1 \cdot 3 \cdot 5 \cdot 7}{2^5 \cdot 5!}\right) = 99{,}225.$$

15. (a) $g(x) = \displaystyle\sum_{n=0}^{\infty} \binom{k}{n} x^n \implies g'(x) = \sum_{n=1}^{\infty} \binom{k}{n} n x^{n-1}$, so

$$(1+x)\, g'(x) = (1+x) \sum_{n=1}^{\infty} \binom{k}{n} n x^{n-1} = \sum_{n=1}^{\infty} \binom{k}{n} n x^{n-1} + \sum_{n=1}^{\infty} \binom{k}{n} n x^{n}$$

$$= \sum_{n=0}^{\infty} \binom{k}{n+1}(n+1) x^n + \sum_{n=0}^{\infty} \binom{k}{n} n x^n \qquad \left[\begin{array}{c}\text{Replace } n \text{ with } n+1 \\ \text{in the first series}\end{array}\right]$$

$$= \sum_{n=0}^{\infty} (n+1)\frac{k(k-1)(k-2)\cdots(k-n+1)(k-n)}{(n+1)!} x^n$$

$$+ \sum_{n=0}^{\infty}\left[(n)\frac{k(k-1)(k-2)\cdots(k-n+1)}{n!}\right] x^n$$

$$= \sum_{n=0}^{\infty} \frac{(n+1)\,k(k-1)(k-2)\cdots(k-n+1)}{(n+1)!}\,[(k-n)+n]\,x^n$$

$$= k\sum_{n=0}^{\infty} \frac{k(k-1)(k-2)\cdots(k-n+1)}{n!} x^n = k\sum_{n=0}^{\infty} \binom{k}{n} x^n = kg(x)$$

Thus, $g'(x) = \dfrac{kg(x)}{1+x}$.

(b) $h(x) = (1+x)^{-k}\, g(x) \implies$

$$h'(x) = -k(1+x)^{-k-1} g(x) + (1+x)^{-k} g'(x) \qquad \text{[Product Rule]}$$

$$= -k(1+x)^{-k-1} g(x) + (1+x)^{-k} \frac{kg(x)}{1+x} \qquad \text{[from part (a)]}$$

$$= -k(1+x)^{-k-1} g(x) + k(1+x)^{-k-1} g(x) = 0$$

(c) From part (b) we see that $h(x)$ must be constant for $x \in (-1, 1)$, so $h(x) = h(0) = 1$ for $x \in (-1, 1)$. Thus, $h(x) = 1 = (1+x)^{-k} g(x) \iff g(x) = (1+x)^k$ for $x \in (-1, 1)$.

Section 8.9 Applications of Taylor Polynomials

1. (a)

n	$f^{(n)}(x)$	$f^{(n)}(0)$	$T_n(x)$
0	$\cos x$	1	1
1	$-\sin x$	0	1
2	$-\cos x$	-1	$1 - \frac{1}{2}x^2$
3	$\sin x$	0	$1 - \frac{1}{2}x^2$
4	$\cos x$	1	$1 - \frac{1}{2}x^2 + \frac{1}{24}x^4$
5	$-\sin x$	0	$1 - \frac{1}{2}x^2 + \frac{1}{24}x^4$
6	$-\cos x$	-1	$1 - \frac{1}{2}x^2 + \frac{1}{24}x^4 - \frac{1}{720}x^6$

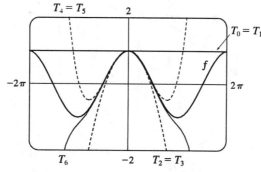

(b)

x	f	$T_0 = T_1$	$T_2 = T_3$	$T_4 = T_5$	T_6
$\frac{\pi}{4}$	0.7071	1	0.6916	0.7074	0.7071
$\frac{\pi}{2}$	0	1	-0.2337	0.0200	-0.0009
π	-1	1	-3.9348	0.1239	-1.2114

(c) As n increases, $T_n(x)$ is a good approximation to $f(x)$ on a larger and larger interval.

3.

n	$f^{(n)}(x)$	$f^{(n)}\left(\frac{\pi}{6}\right)$
0	$\sin x$	$\frac{1}{2}$
1	$\cos x$	$\frac{\sqrt{3}}{2}$
2	$-\sin x$	$-\frac{1}{2}$
3	$-\cos x$	$-\frac{\sqrt{3}}{2}$

$$T_3(x) = \sum_{n=0}^{3} \frac{f^{(n)}\left(\frac{\pi}{6}\right)}{n!}\left(x - \frac{\pi}{6}\right)^n$$
$$= \frac{1}{2} + \frac{\sqrt{3}}{2}\left(x - \frac{\pi}{6}\right) - \frac{1}{4}\left(x - \frac{\pi}{6}\right)^2 - \frac{\sqrt{3}}{12}\left(x - \frac{\pi}{6}\right)^3$$

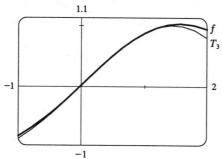

5.

n	$f^{(n)}(x)$	$f^{(n)}(0)$
0	$\tan x$	0
1	$\sec^2 x$	1
2	$2\sec^2 x \tan x$	0
3	$4\sec^2 x \tan^2 x + 2\sec^4 x$	2
4	$8\sec^2 x \tan^3 x$ $+16\sec^4 x \tan x$	0

$$T_4(x) = \sum_{n=0}^{4} \frac{f^{(n)}(0)}{n!} x^n = x + \frac{2x^3}{3!} = x + \frac{x^3}{3}$$

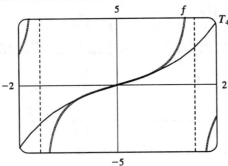

7.

n	$f^{(n)}(x)$	$f^{(n)}\left(\frac{\pi}{3}\right)$
0	$\sec x$	2
1	$\sec x \tan x$	$2\sqrt{3}$
2	$\sec x \tan^2 x + \sec^3 x$	14
3	$\sec x \tan^3 x$ $+5\sec^3 x \tan x$	$46\sqrt{3}$

$$T_3(x) = \sum_{n=0}^{3} \frac{f^{(n)}\left(\frac{\pi}{3}\right)}{n!}\left(x - \frac{\pi}{3}\right)^n$$
$$= 2 + 2\sqrt{3}\left(x - \frac{\pi}{3}\right) + 7\left(x - \frac{\pi}{3}\right)^2$$
$$+ \frac{23\sqrt{3}}{3}\left(x - \frac{\pi}{3}\right)^3$$

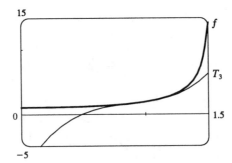

9. In Maple, we can find the Taylor polynomials by the following method: first
define `f:=sec(x);` and then set `T2:=convert(taylor(f,x=0,3),polynom);`,
`T4:=convert(taylor(f,x=0,5),polynom);`, etc. (The third argument in the `taylor` function is
one more than the degree of the desired polynomial).   We must `convert` to the type `polynom` because
the output of the `taylor` function contains an error term which we do not want.   In Mathematica,
we use `Tn:=Normal[Series[f,{x,0,n}]]`, with `n=2, 4`, etc.   Note that in Mathematica, the
"degree" argument is the same as the degree of the desired polynomial.   In Derive, author $\sec x$, then
enter `Calculus,Taylor,8,0`; and then simplify the expression.   The eighth Taylor polynomial is
$T_8(x) = 1 + \frac{1}{2}x^2 + \frac{5}{24}x^4 + \frac{61}{720}x^6 + \frac{277}{8064}x^8$.

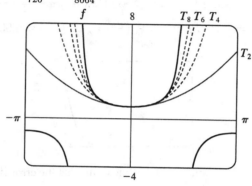

11.

$$f(x) = (1+x)^{1/2}$$
$$f'(x) = \tfrac{1}{2}(1+x)^{-1/2}$$
$$f''(x) = -\tfrac{1}{4}(1+x)^{-3/2} = -\frac{1}{4(1+x)^{3/2}}$$

$$f(0) = 1$$
$$f'(0) = \tfrac{1}{2}$$

(a) $(1+x)^{1/2} \approx T_1(x) = 1 + \frac{1}{2}x$

(b) By Taylor's Inequality, the remainder is $|R_1(x)| \le \dfrac{M}{2!}|x|^2$, where $|f''(x)| \le M$.
Now $0 \le x \le 0.1 \Longrightarrow 0 \le x^2 \le 0.01$, and letting $x = 0$ gives $M = 0.25$, so
$|R_1(x)| \le \frac{0.25}{2}(0.01) = 0.00125$.

(c)

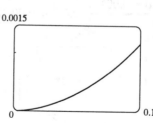

From the graph of $|R_1(x)| = \left|\sqrt{1+x} - \left(1 + \frac{1}{2}x\right)\right|$, it seems that the error is at most 0.0012 on
$[0, 0.1]$.

13.

$$f(x) = \sin x \qquad f\left(\tfrac{\pi}{4}\right) = \tfrac{\sqrt{2}}{2}$$
$$f'(x) = \cos x \qquad f'\left(\tfrac{\pi}{4}\right) = \tfrac{\sqrt{2}}{2}$$
$$f''(x) = -\sin x \qquad f''\left(\tfrac{\pi}{4}\right) = -\tfrac{\sqrt{2}}{2}$$
$$f'''(x) = -\cos x \qquad f'''\left(\tfrac{\pi}{4}\right) = -\tfrac{\sqrt{2}}{2}$$
$$f^{(4)}(x) = \sin x \qquad f^{(4)}\left(\tfrac{\pi}{4}\right) = \tfrac{\sqrt{2}}{2}$$
$$f^{(5)}(x) = \cos x \qquad f^{(5)}\left(\tfrac{\pi}{4}\right) = \tfrac{\sqrt{2}}{2}$$
$$f^{(6)}(x) = -\sin x$$

(a) $\sin x \approx T_5(x) = \tfrac{\sqrt{2}}{2} + \tfrac{\sqrt{2}}{2}\left(x - \tfrac{\pi}{4}\right) - \tfrac{\sqrt{2}}{4}\left(x - \tfrac{\pi}{4}\right)^2 - \tfrac{\sqrt{2}}{12}\left(x - \tfrac{\pi}{4}\right)^3 + \tfrac{\sqrt{2}}{48}\left(x - \tfrac{\pi}{4}\right)^4 + \tfrac{\sqrt{2}}{240}\left(x - \tfrac{\pi}{4}\right)^5$

(b) $|R_5(x)| \leq \dfrac{M}{6!}\left|x - \tfrac{\pi}{4}\right|^6$, where $\left|f^{(6)}(x)\right| \leq M$. Now $0 \leq x \leq \tfrac{\pi}{2} \Longrightarrow \left(x - \tfrac{\pi}{4}\right)^6 \leq \left(\tfrac{\pi}{4}\right)^6$, and

letting $x = \tfrac{\pi}{2}$ gives $M = 1$, so $|R_5(x)| \leq \tfrac{1}{6!}\left(\tfrac{\pi}{4}\right)^6 = \tfrac{1}{720}\left(\tfrac{\pi}{4}\right)^6 \approx 0.00033$.

(c)

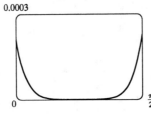

0.0003

From the graph of $|R_5(x)| = |\sin x - T_5(x)|$, it seems that the error is less than 0.00026 on $\left[0, \tfrac{\pi}{2}\right]$.

15.

$$f(x) = e^{x^2} \qquad\qquad f(0) = 1$$
$$f'(x) = e^{x^2}(2x) \qquad\qquad f'(0) = 0$$
$$f''(x) = e^{x^2}(2 + 4x^2) \qquad\qquad f''(0) = 2$$
$$f'''(x) = e^{x^2}(12x + 8x^3) \qquad\qquad f'''(0) = 0$$
$$f^{(4)}(x) = e^{x^2}(12 + 48x^2 + 16x^4)$$

(a) $e^{x^2} \approx T_3(x) = 1 + \tfrac{2}{2!}x^2 = 1 + x^2$

(b) $|R_3(x)| \leq \dfrac{M}{4!}|x|^4$, where $\left|f^{(4)}(x)\right| \leq M$. Now $0 \leq x \leq 0.1 \Longrightarrow x^4 \leq (0.1)^4$, and letting $x = 0.1$

gives $|R_3(x)| \leq \dfrac{e^{0.01}(12 + 0.48 + 0.0016)}{24}(0.1)^4 < 0.00006$.

(c)

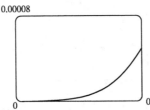

0.00008

0.1

From the graph of $|R_3(x)| = \left|e^{x^2} - (1 + x^2)\right|$, it appears that the error is less than 0.000051 on $[0, 0.1]$.

17. From Exercise 3, $\sin x = \frac{1}{2} + \frac{\sqrt{3}}{2}\left(x - \frac{\pi}{6}\right) - \frac{1}{4}\left(x - \frac{\pi}{6}\right)^2 - \frac{\sqrt{3}}{12}\left(x - \frac{\pi}{6}\right)^3 + R_3(x)$, where

$R_3(x) \leq \frac{M}{4!}\left|x - \frac{\pi}{6}\right|^4$ with $\left|f^{(4)}(x)\right| = |\sin x| \leq M = 1$. Now $35° = \left(\frac{\pi}{6} + \frac{\pi}{36}\right)$ radians,

so the error is $\left|R_3\left(\frac{\pi}{36}\right)\right| \leq \dfrac{\left(\frac{\pi}{36}\right)^4}{4!} < 0.000003$. Therefore, to five decimal places,

$\sin 35° \approx \frac{1}{2} + \frac{\sqrt{3}}{2}\left(\frac{\pi}{36}\right) - \frac{1}{4}\left(\frac{\pi}{36}\right)^2 - \frac{\sqrt{3}}{12}\left(\frac{\pi}{36}\right)^3 \approx 0.57358$.

19. All derivatives of e^x are e^x, so $|R_n(x)| \leq \dfrac{e^x}{(n+1)!}|x|^{n+1}$, where $0 < x < 0.1$. Letting $x = 0.1$,

$R_n(0.1) \leq \dfrac{e^{0.1}}{(n+1)!}(0.1)^{n+1} < 0.00001$, and by trial and error we find that $n = 3$ satisfies this

inequality since $R_3(0.1) < 0.0000046$. Thus, by adding the three terms of the Maclaurin series for e^x

corresponding to $n = 0, 1$, and 2, we can estimate $e^{0.1}$ to within 0.00001.

21. $\sin x = x - \frac{1}{3!}x^3 + \frac{1}{5!}x^5 - \cdots$. By the Alternating Series Estimation Theorem, the error in the

approximation $\sin x = x - \frac{1}{3!}x^3$ is less than $\left|\frac{1}{5!}x^5\right| < 0.01 \iff |x^5| < 120(0.01) \iff$

$|x| < (1.2)^{1/5} \approx 1.037$.

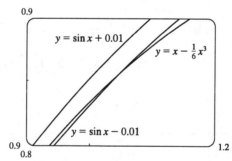

The curves intersect at $x \approx 1.043$, so the graph confirms our estimate. Since both the sine function and
the given approximation are odd functions, we only need to check the estimate for $x > 0$.

23. Let $s(t)$ be the position function of the car, and for convenience set $s(0) = 0$. The velocity of the

car is $v(t) = s'(t)$ and the acceleration is $a(t) = s''(t)$, so the second degree Taylor polynomial is

$T_2(t) = s(0) + v(0)t + \dfrac{a(0)}{2}t^2 = 20t + t^2$. We estimate the distance travelled during the next

second to be $s(1) \approx T_2(1) = 20 + 1 = 21$ m. The function $T_2(t)$ would not be accurate over a full
minute, since the car could not possibly maintain an acceleration of 2 m/s^2 for that long (if it did, its
final speed would be 140 m/s ≈ 315 mi/h!)

25. $E = \dfrac{q}{D^2} - \dfrac{q}{(D+d)^2} = \dfrac{q}{D^2} - \dfrac{q}{D^2\,(1+d/D)^2} = \dfrac{q}{D^2}\left[1 - \left(1 + \dfrac{d}{D}\right)^{-2}\right].$

We use the Binomial Series to expand $(1 + d/D)^{-2}$:

$$E = \frac{q}{D^2}\left[1 - \left(1 - 2\left(\frac{d}{D}\right) + \frac{2\cdot 3}{2!}\left(\frac{d}{D}\right)^2 - \frac{2\cdot 3\cdot 4}{3!}\left(\frac{d}{D}\right)^3 + \cdots\right)\right]$$

$$= \frac{q}{D^2}\left[2\left(\frac{d}{D}\right) - 3\left(\frac{d}{D}\right)^2 + 4\left(\frac{d}{D}\right)^3 - \cdots\right] \approx 2qd\cdot\frac{1}{D^3}$$

when D is much larger than d, that is, when P is far away from the dipole.

27. Using $f(x) = T_n(x) + R_n(x)$ with $n = 1$ and $x = r$, we have $f(r) = T_1(r) + R_1(r)$, where T_1 is the first-degree Taylor polynomial of f at a. Because $a = x_n$, $f(r) = f(x_n) + f'(x_n)(r - x_n) + R_1(r)$. But r is a root of f, so $f(r) = 0$ and we have $0 = f(x_n) + f'(x_n)(r - x_n) + R_1(r)$. Taking the first two terms to the left side and dividing by $f'(x_n)$, we have $f'(x_n)(x_n - r) - f(x_n) = R_1(r) \Longrightarrow$

$x_n - r - \dfrac{f(x_n)}{f'(x_n)} = \dfrac{R_1(r)}{f'(x_n)}$. By the formula for Newton's method, the left side of the

preceding equation is $x_{n+1} - r$, so $|x_{n+1} - r| = \left|\dfrac{R_1(r)}{f'(x_n)}\right|$. Taylor's Inequality gives us

$|R_1(r)| \le \dfrac{|f''(r)|}{2!}\,|r - x_n|^2$. Combining this inequality with the facts $|f''(x)| \le M$ and $|f'(x)| \ge K$

gives us $|x_{n+1} - r| \le \dfrac{M}{2K}\,|x_n - r|^2$.

Section 8.10 Using Series to Solve Differential Equations

1. Let $y(x) = \sum\limits_{n=0}^{\infty} c_n x^n$. Thus, $y'(x) = \sum\limits_{n=1}^{\infty} n c_n x^{n-1}$, which can be written as $y' = \sum\limits_{n=0}^{\infty} (n+1) c_{n+1} x^n$

by replacing n with $n+1$. $y' = 6y \Longrightarrow y' - 6y = 0 \Longrightarrow \sum\limits_{n=0}^{\infty} \left[(n+1) c_{n+1} - 6c_n \right] x^n = 0.$ Hence,

the recurrence relation is $c_{n+1} = \dfrac{6c_n}{n+1}$, $n = 0, 1, 2, \ldots.$ Then $c_1 = 6c_0$, $c_2 = \dfrac{6c_1}{2} = \dfrac{6^2 c_0}{2}$,

$c_3 = \dfrac{6c_2}{3} = \dfrac{6^3 c_0}{2 \cdot 3}, \ldots, c_n = \dfrac{6c_{n-1}}{n} = \dfrac{6^n c_0}{n!}.$

Thus, the solution is

$$y(x) = \sum_{n=0}^{\infty} c_0 \frac{6^n}{n!} x^n = \sum_{n=0}^{\infty} \left[c_0 \frac{(6x)^n}{n!} \right]$$
$$= c_0 e^{6x}$$

3. Assuming $y(x) = \sum\limits_{n=0}^{\infty} c_n x^n$, we have $y'(x) = \sum\limits_{n=1}^{\infty} n c_n x^{n-1} = \sum\limits_{n=0}^{\infty} (n+1) c_{n+1} x^n$ and

$-x^2 y = -\sum\limits_{n=0}^{\infty} c_n x^{n+2} = -\sum\limits_{n=2}^{\infty} c_{n-2} x^n.$ Hence, the equation $y' = x^2 y$ becomes

$$\sum_{n=0}^{\infty} (n+1) c_{n+1} x^n - \sum_{n=2}^{\infty} c_{n-2} x^n = 0$$

or

$$c_1 + 2c_2 x + \sum_{n=2}^{\infty} \left[(n+1) c_{n+1} - c_{n-2} \right] x^n = 0$$

Equating coefficients gives $c_1 = c_2 = 0$ and $c_{n+1} = \dfrac{c_{n-2}}{n+1}$ for $n = 2, 3, \ldots.$ But $c_1 = 0$, so

$c_4 = 0$ and $c_7 = 0$ and in general $c_{3n+1} = 0$. Similarly $c_2 = 0$ so $c_{3n+2} = 0$. Finally $c_3 = \dfrac{c_0}{3}$,

$c_6 = \dfrac{c_3}{6} = \dfrac{c_0}{6 \cdot 3} = \dfrac{c_0}{3^2 \cdot 2!}$, $c_9 = \dfrac{c_6}{9} = \dfrac{c_0}{9 \cdot 6 \cdot 3} = \dfrac{c_0}{3^3 \cdot 3!}, \ldots,$ and $c_{3n} = \dfrac{c_0}{3^n \cdot n!}.$ Thus, the solution is

$$y(x) = \sum_{n=0}^{\infty} c_n x^n = \sum_{n=0}^{\infty} c_{3n} x^{3n}$$
$$= c_0 \sum_{n=0}^{\infty} \frac{\left(x^3/3 \right)^n}{n!}$$
$$= c_0 e^{x^3/3}$$

5. Assuming $y(x) = \sum_{n=0}^{\infty} c_n x^n$, $y''(x) = \sum_{n=2}^{\infty} n(n-1) c_n x^{n-2} = \sum_{n=0}^{\infty} (n+2)(n+1) c_{n+2} x^n$,

$3xy'(x) = 3x \sum_{n=0}^{\infty} n c_n x^{n-1} = \sum_{n=0}^{\infty} 3n c_n x^n$ and the equation $y'' + 3xy' + 3y = 0$

becomes $\sum_{n=0}^{\infty} (n+2)(n+1) c_{n+2} x^n + \sum_{n=0}^{\infty} 3n c_n x^n + \sum_{n=0}^{\infty} 3 c_n x^n = 0 \iff$

$\sum_{n=0}^{\infty} [(n+2)(n+1) c_{n+2} + 3(n+1) c_n] x^n = 0.$ Thus, the recurrence relation is

$c_{n+2} = -\dfrac{3 c_n}{n+2}$ for $n = 0, 1, 2, \ldots.$ Given c_0 and c_1, $c_2 = -\dfrac{3 c_0}{2}$, $c_4 = -\dfrac{3 c_2}{4} = (-1)^2 \dfrac{3^2 c_0}{2^2 \cdot 2!}$,

$c_6 = -\dfrac{3 c_4}{6} = (-1)^3 \dfrac{3^3 c_0}{2^3 \cdot 3!}, \ldots, c_{2n} = (-1)^n \dfrac{3^n c_0}{2^n n!} = (-1)^n \left(\dfrac{3}{2}\right)^n \dfrac{c_0}{n!}.$ Also,

$c_3 = -\dfrac{3 c_1}{3}, c_5 = -\dfrac{3 c_3}{5} = (-1)^2 \dfrac{3^2 c_1}{5 \cdot 3}, c_7 = -\dfrac{3 c_5}{7} = (-1)^3 \dfrac{3^3 c_1}{7 \cdot 5 \cdot 3}, \ldots,$

$c_{2n+1} = (-1)^n \dfrac{3^n c_1}{(2n+1)(2n-1) \cdots \cdots 5 \cdot 3} = (-1)^n \dfrac{3^n c_1 2^n n!}{(2n+1)!} = (-1)^n \dfrac{6^n n! c_1}{(2n+1)!}$ because

$(2n+1)(2n-1) \cdots \cdots 5 \cdot 3 = \dfrac{(2n+1)!}{2^n \cdot n!}.$ Thus, the solution is

$$y(x) = \sum_{n=0}^{\infty} c_{2n} x^{2n} + \sum_{n=0}^{\infty} c_{2n+1} x^{2n+1}$$

$$= c_0 \sum_{n=0}^{\infty} (-1)^n \dfrac{(3x^2/2)^n}{n!} + c_1 \sum_{n=0}^{\infty} \left[(-1)^n \dfrac{6^n n! \, x^{2n+1}}{(2n+1)!}\right]$$

$$= c_0 e^{-3x^2/2} + c_1 \sum_{n=0}^{\infty} \left[(-1)^n \dfrac{6^n n! \, x^{2n+1}}{(2n+1)!}\right]$$

7. Let $y(x) = \sum_{n=0}^{\infty} c_n x^n$. Then $y''(x) = \sum_{n=0}^{\infty} (n+2)(n+1) c_{n+2} x^n$, $-xy'(x) = -\sum_{n=0}^{\infty} n c_n x^n$ and

the equation $y'' - xy' - y = 0$ becomes $\sum_{n=0}^{\infty} [(n+2)(n+1) c_{n+2} - (n+1) c_n] x^n = 0.$ Thus,

the recurrence relation is $c_{n+2} = \dfrac{c_n}{n+2}$ for $n = 0, 1, 2, \ldots.$ But $c_0 = y(0) = 1$, so $c_2 = \dfrac{1}{2}$,

$c_4 = \dfrac{c_2}{4} = \dfrac{1}{2 \cdot 4}$, $c_6 = \dfrac{c_4}{6} = \dfrac{1}{2 \cdot 4 \cdot 6}, \ldots, c_{2n} = \dfrac{1}{2^n n!}.$ Also, $c_1 = y'(0) = 0$ and by the

recurrence relation $c_{2n+1} = 0$ for $n = 0, 1, 2, \ldots.$ Thus, the solution to the initial-value problem is

$$y(x) = \sum_{n=0}^{\infty} c_n x^n = \sum_{n=0}^{\infty} \dfrac{x^{2n}}{2^n n!} = \sum_{n=0}^{\infty} \dfrac{(x^2/2)^n}{n!} = e^{x^2/2}.$$

9. Assuming that $y(x) = \sum\limits_{n=0}^{\infty} c_n x^n$, we have

$$y''(x) = \sum_{n=2}^{\infty} n(n-1) c_n x^{n-2} = \sum_{n=-1}^{\infty} (n+3)(n+2) c_{n+3} x^{n+1} \qquad \text{[replace } n \text{ with } n+3]$$

$$= 2c_2 + \sum_{n=0}^{\infty} (n+3)(n+2) c_{n+3} x^{n+1},$$

$$x^2 y' = x^2 \sum_{n=1}^{\infty} n c_n x^{n-1} = \sum_{n=0}^{\infty} n c_n x^{n+1}, \quad xy = x \sum_{n=0}^{\infty} c_n x^n = \sum_{n=0}^{\infty} c_n x^{n+1}, \text{ and the equation}$$

$y'' + x^2 y' + xy = 0$ becomes $2c_2 + \sum\limits_{n=0}^{\infty} [(n+3)(n+2) c_{n+3} + (n+1) c_n] x^{n+1} = 0$. So $c_2 = 0$ and

the recurrence relation is $c_{n+3} = -\dfrac{(n+1) c_n}{(n+3)(n+2)}$, $n = 0, 1, 2, \ldots$. But $c_0 = y(0) = 0 = c_2$ and by

the recurrence relation, $c_{3n} = c_{3n+2} = 0$ for $n = 0, 1, 2, \ldots$. Also, $c_1 = y'(0) = 1$ so

$$c_4 = -\frac{2}{4 \cdot 3}, \quad c_7 = -\frac{5c_4}{7 \cdot 6} = (-1)^2 \frac{2 \cdot 5}{7 \cdot 6 \cdot 4 \cdot 3} = (-1)^2 \frac{2^2 5^2}{7!},$$

$$\ldots, c_{3n+1} = (-1)^n \frac{2^2 5^2 \cdot \cdots \cdot (3n-1)^2}{(3n+1)!}. \quad \text{Thus, the solution is}$$

$$y(x) = \sum_{n=0}^{\infty} c_n x^n = x + \sum_{n=1}^{\infty} \left[(-1)^n \frac{2^2 5^2 \cdot \cdots \cdot (3n-1)^2 \, x^{3n+1}}{(3n+1)!} \right].$$

Chapter 8 Review

Concept Check

1. (a) See Definition 8.1.1.

   (b) See Definition 8.2.2.

   (c) The terms of the sequence $\{a_n\}$ approach 3 as n becomes large.

   (d) By adding sufficiently many terms of the series, we can make the partial sums as close to 3 as we like.

2. (a) See the definition on page 566.

   (b) A sequence is monotonic if it is either increasing or decreasing.

   (c) By Theorem 8.1.7, every bounded, monotonic sequence is convergent.

3. (a) See (4) in Section 8.2.

   (b) See (1) in Section 8.3.

4. If $\sum a_n = 3$, then $\lim\limits_{n \to \infty} a_n = 0$ and $\lim\limits_{n \to \infty} s_n = 3$.

5. (a) See the Test for Divergence on page 575.

   (b) See the Integral Test on page 581.

   (c) See the Comparison Test on page 583.

   (d) See the Limit Comparison Test on page 584.

   (e) See the Alternating Series Test on page 589.

   (f) See the Ratio Test on page 594.

6. (a) See the definition on page 592.

   (b) By (8.4.1), it is convergent.

7. (a) Use either (3) or (4) in Section 8.3.

   (b) See Example 8 in Section 8.3.

   (c) By adding terms until you reach the desired accuracy given by the Alternating Series Estimation Theorem on page 591.

8. (a) $\displaystyle\sum_{n=0}^{\infty} c_n (x-a)^n$

(b) Given the power series $\displaystyle\sum_{n=0}^{\infty} c_n (x-a)^n$, the radius of convergence is:

    (i):   0 if the series converges only when $x = a$

    (ii): ∞ if the series converges for all x, or

    (iii):   a positive number R such that the series converges if $|x-a| < R$ and diverges if $|x-a| > R$.

(c) The interval of convergence of a power series is the interval that consists of all values of x for which the series converges. Corresponding to the cases in part (b), the interval of convergence is: (i) the single point $\{a\}$, (ii) all real numbers, that is, the real number line $(-\infty, \infty)$, or (iii) an interval with endpoints $a - R$ and $a + R$ which can contain neither, either, or both of the endpoints. In this case, we must test the series for convergence at each endpoint to determine the interval of convergence.

9. (a), (b) See Theorem 2 on page 604.

10. (a) $\displaystyle T_n(x) = \sum_{i=0}^{n} \frac{f^{(i)}(a)}{i!} (x-a)^i$ (b) $\displaystyle\sum_{n=0}^{\infty} \frac{f^{(n)}(a)}{n!} (x-a)^n$

(c) $\displaystyle\sum_{n=0}^{\infty} \frac{f^{(n)}(0)}{n!} x^n$ $[a = 0$ in part (b)$]$

(d) See Theorem 8 on page 611.

(e) See (9) on page 611.

11. (a) – (e) See page 616.

12. See (2) on page 621 for the expansion. The radius of convergence for the binomial series is 1.

True-False Quiz

1. False. See Note 2 after Theorem 8.2.6.

3. False. For example, take $c_n = (-1)^n / (n6^n)$.

5. False, since $\displaystyle\lim_{n\to\infty} \left| \frac{a_{n+1}}{a_n} \right| = \lim_{n\to\infty} \left| \frac{n^3}{(n+1)^3} \right| = \lim_{n\to\infty} \frac{1}{(1+1/n)^3} = 1.$

7. False. See the note after Example 4 in Section 8.3.

9. True. See (6) in Section 8.1.

11. True. By Theorem 8.7.5 the coefficient of x^3 is $\dfrac{f'''(0)}{3!} = \dfrac{1}{3} \Longrightarrow f'''(0) = 2.$
    Or: Use Theorem 8.6.2 to differentiate f three times.

13. False. For example, let $a_n = b_n = (-1)^n$. Then $\{a_n\}$ and $\{b_n\}$ are divergent, but $a_n b_n = 1$, so $\{a_n b_n\}$ is convergent.

15. True by Theorem 8.4.1. $\left[\sum (-1)^n a_n \text{ is absolutely convergent and hence convergent.} \right]$

Exercises

1. $\lim\limits_{n\to\infty} a_n = \lim\limits_{n\to\infty}\dfrac{n}{2n+5} = \lim\limits_{n\to\infty}\dfrac{1}{2+5/n} = \dfrac{1}{2}$. Since the limit exists, the sequence is convergent.

3. $\{2n+5\}$ is divergent since $2n+5 \to \infty$ as $n \to \infty$.

5. $\{\sin n\}$ is divergent since $\lim\limits_{n\to\infty}\sin n$ does not exist.

7. $\left\{\left(1+\dfrac{3}{n}\right)^{4n}\right\}$ is convergent. Let $y = \left(1+\dfrac{3}{x}\right)^{4x}$. Then

$$\lim_{x\to\infty}\ln y = \lim_{x\to\infty} 4x\ln\left(1+\dfrac{3}{x}\right) = \lim_{x\to\infty}\dfrac{\ln(1+3/x)}{1/(4x)} \overset{\text{H}}{=} \lim_{x\to\infty}\dfrac{\dfrac{1}{1+3/x}\left(-\dfrac{3}{x^2}\right)}{-1/(4x^2)} = \lim_{x\to\infty}\dfrac{12}{1+3/x}$$
$$= 12$$

so $\lim\limits_{x\to\infty} y = \lim\limits_{n\to\infty}\left(1+\dfrac{3}{n}\right)^{4n} = e^{12}$.

9. Use the Limit Comparison Test with $a_n = \dfrac{n^2}{n^3+1}$ and $b_n = \dfrac{1}{n}$. Then

$$\lim_{n\to\infty}\dfrac{a_n}{b_n} = \lim_{n\to\infty}\dfrac{n^2/(n^3+1)}{1/n} = \lim_{n\to\infty}\dfrac{1}{1+1/n^3} = 1 > 0. \text{ Since } \sum_{n=1}^{\infty}\dfrac{1}{n} \text{ (the harmonic series)}$$

diverges, $\displaystyle\sum_{n=1}^{\infty}\dfrac{n^2}{n^3+1}$ also diverges.

11. This is an alternating series with $a_n = \dfrac{1}{n^{1/4}}$, $a_n > 0$ for all n, and $a_n > a_{n+1}$.

$\lim\limits_{n\to\infty} a_n = \lim\limits_{n\to\infty}\dfrac{1}{n^{1/4}} = 0$, so the series converges by the Alternating Series Test.

13. $\lim\limits_{n\to\infty}\left|\dfrac{a_{n+1}}{a_n}\right| = \lim\limits_{n\to\infty}\dfrac{4^{n+1}}{(n+1)\,3^{n+1}}\cdot\dfrac{n3^n}{4^n} = \dfrac{4}{3}\lim\limits_{n\to\infty}\dfrac{n}{n+1} = \dfrac{4}{3} > 1$ so the series diverges by the

Ratio Test.

15. $\left|\dfrac{\sin n}{1+n^2}\right| \le \dfrac{1}{1+n^2} < \dfrac{1}{n^2}$ and since $\displaystyle\sum_{n=1}^{\infty}\dfrac{1}{n^2}$ converges (p-series with $p = 2 > 1$), so does $\displaystyle\sum_{n=1}^{\infty}\left|\dfrac{\sin n}{1+n^2}\right|$

by the Comparison Test, and so does $\displaystyle\sum_{n=1}^{\infty}\dfrac{\sin n}{1+n^2}$ by Theorem 8.4.1.

17. $\lim\limits_{n\to\infty}\left|\dfrac{a_{n+1}}{a_n}\right| = \lim\limits_{n\to\infty}\dfrac{1\cdot3\cdot5\cdots(2n-1)(2n+1)}{5^{n+1}(n+1)!}\cdot\dfrac{5^n n!}{1\cdot3\cdot5\cdots(2n-1)} = \lim\limits_{n\to\infty}\dfrac{2n+1}{5(n+1)}$
$= \tfrac{2}{5} < 1$, so the series converges by the Ratio Test.

19. Convergent geometric series. $\displaystyle\sum_{n=1}^{\infty}\dfrac{2^{2n+1}}{5^n} = \sum_{n=1}^{\infty}\dfrac{(2^2)^n\cdot2^1}{5^n} = 2\sum_{n=1}^{\infty}\dfrac{4^n}{5^n} = 2\left(\dfrac{\frac{4}{5}}{1-\frac{4}{5}}\right) = 8.$

Review

21. $\displaystyle\sum_{n=1}^{\infty} \left[\tan^{-1}(n+1) - \tan^{-1} n\right] = \lim_{n\to\infty} \left[(\tan^{-1} 2 - \tan^{-1} 1) + (\tan^{-1} 3 - \tan^{-1} 2) + \cdots\right.$

$$+ \left. (\tan^{-1}(n+1) - \tan^{-1} n)\right]$$

$$= \lim_{n\to\infty} \left[\tan^{-1}(n+1) - \tan^{-1} 1\right] = \frac{\pi}{2} - \frac{\pi}{4} = \frac{\pi}{4}$$

23. $1.2 + 0.0\overline{345} = \dfrac{12}{10} + \dfrac{345/10{,}000}{1 - 1/1000} = \dfrac{12}{10} + \dfrac{345}{9990} = \dfrac{4111}{3330}$

25. $\displaystyle\sum_{n=1}^{\infty} \frac{(-1)^{n+1}}{n^5} = 1 - \frac{1}{32} + \frac{1}{243} - \frac{1}{1024} + \frac{1}{3125} - \frac{1}{7776} + \frac{1}{16{,}807} - \frac{1}{32{,}768} + \cdots .$ Since

$\dfrac{1}{32{,}768} < 0.000031,$ $\displaystyle\sum_{n=1}^{\infty} \frac{(-1)^{n+1}}{n^5} \approx \sum_{n=1}^{7} \frac{(-1)^{n+1}}{n^5} \approx 0.9721.$

27. $\displaystyle\sum_{n=1}^{\infty} \frac{1}{2 + 5^n} \approx \sum_{n=1}^{8} \frac{1}{2 + 5^n} \approx 0.18976224.$ To estimate the error, note that $\dfrac{1}{2 + 5^n} < \dfrac{1}{5^n}$, so the

remainder term is $R_8 = \displaystyle\sum_{n=9}^{\infty} \frac{1}{2 + 5^n} < \sum_{n=9}^{\infty} \frac{1}{5^n} = \frac{1/5^9}{1 - 1/5} = 6.4 \times 10^{-7}$ (geometric series with

$a = \frac{1}{5^9}$ and $r = \frac{1}{5}$).

29. Use the Limit Comparison Test. $\displaystyle\lim_{n\to\infty} \left| \frac{\left(\dfrac{n+1}{n}\right) a_n}{a_n} \right| = \lim_{n\to\infty} \frac{n+1}{n} = \lim_{n\to\infty} \left(1 + \frac{1}{n}\right) = 1 > 0.$

Since $\sum |a_n|$ is convergent, so is $\displaystyle\sum \left| \left(\frac{n+1}{n}\right) a_n \right|$ by the Limit Comparison Test.

31. $\displaystyle\lim_{n\to\infty} \left| \frac{a_{n+1}}{a_n} \right| = \lim_{n\to\infty} \left| \frac{x^{n+1}}{3^{n+1}(n+1)^3} \cdot \frac{3^n n^3}{x^n} \right| = \frac{|x|}{3} \lim_{n\to\infty} \left(\frac{n}{n+1} \right)^3 = \frac{|x|}{3} < 1$ for convergence

(Ratio Test) $\Longrightarrow |x| < 3$ and the radius of convergence is 3. When $x = \pm 3,$ $\displaystyle\sum_{n=1}^{\infty} |a_n| = \sum_{n=1}^{\infty} \frac{1}{n^3}$, which

is a convergent p-series $(p = 3 > 1)$, so the interval of convergence is $[-3, 3]$.

33. $\displaystyle\lim_{n\to\infty} \left| \frac{a_{n+1}}{a_n} \right| = \lim_{n\to\infty} \left| \frac{2^{n+1}(x-3)^{n+1}}{\sqrt{n+4}} \cdot \frac{\sqrt{n+3}}{2^n (x-3)^n} \right| = 2|x-3| \lim_{n\to\infty} \sqrt{\frac{n+3}{n+4}} = 2|x-3| <$

$1 \Longleftrightarrow |x-3| < \frac{1}{2},$ so the radius of convergence is $\frac{1}{2}$. For $x = \frac{7}{2},$ the series becomes

$\displaystyle\sum_{n=0}^{\infty} \frac{1}{\sqrt{n+3}} = \sum_{n=3}^{\infty} \frac{1}{n^{1/2}}$, which diverges $(p = \frac{1}{2} \le 1)$, but for $x = \frac{5}{2},$ we get $\displaystyle\sum_{n=0}^{\infty} \frac{(-1)^n}{\sqrt{n+3}}$, which is a

convergent alternating series, so the interval of convergence is $\left[\frac{5}{2}, \frac{7}{2}\right)$.

35.

$$f(x) = \sin x \qquad f\left(\tfrac{\pi}{6}\right) = \tfrac{1}{2}$$
$$f'(x) = \cos x \qquad f'\left(\tfrac{\pi}{6}\right) = \tfrac{\sqrt{3}}{2}$$
$$f''(x) = -\sin x \qquad f''\left(\tfrac{\pi}{6}\right) = -\tfrac{1}{2}$$
$$f'''(x) = -\cos x \qquad f'''\left(\tfrac{\pi}{6}\right) = -\tfrac{\sqrt{3}}{2}$$
$$f^{(4)}(x) = \sin x \qquad f^{(4)}\left(\tfrac{\pi}{6}\right) = \tfrac{1}{2}$$

$$\cdots \qquad\qquad \cdots$$

$$f^{(2n)}\left(\tfrac{\pi}{6}\right) = (-1)^n \cdot \tfrac{1}{2} \text{ and } f^{(2n+1)}\left(\tfrac{\pi}{6}\right) = (-1)^n \cdot \tfrac{\sqrt{3}}{2}.$$

$$\sin x = \sum_{n=0}^{\infty} \frac{f^{(n)}\left(\tfrac{\pi}{6}\right)}{n!}\left(x - \frac{\pi}{6}\right)^n$$

$$= \sum_{n=0}^{\infty} \frac{(-1)^n}{2(2n)!}\left(x - \frac{\pi}{6}\right)^{2n} + \sum_{n=0}^{\infty} \frac{(-1)^n \sqrt{3}}{2(2n+1)!}\left(x - \frac{\pi}{6}\right)^{2n+1}$$

37. $\dfrac{1}{1+x} = \dfrac{1}{1-(-x)} = \displaystyle\sum_{n=0}^{\infty} (-1)^n x^n$ for $|x| < 1 \Longrightarrow \dfrac{x^2}{1+x} = \displaystyle\sum_{n=0}^{\infty} (-1)^n x^{n+2}$ with $R = 1$.

39. $\dfrac{1}{1-x} = \displaystyle\sum_{n=0}^{\infty} x^n$ for $|x| < 1 \Longrightarrow \ln(1-x) = -\displaystyle\int \frac{dx}{1-x} = -\int \sum_{n=0}^{\infty} x^n \, dx = C - \sum_{n=0}^{\infty} \frac{x^{n+1}}{n+1}.$

$\ln(1-0) = C - 0 \Longrightarrow C = 0 \Longrightarrow \ln(1-x) = -\displaystyle\sum_{n=0}^{\infty} \frac{x^{n+1}}{n+1} = \sum_{n=1}^{\infty} \frac{-x^n}{n}$ with $R = 1$.

41. $\sin x = \displaystyle\sum_{n=0}^{\infty} \frac{(-1)^n x^{2n+1}}{(2n+1)!} \Longrightarrow \sin(x^4) = \sum_{n=0}^{\infty} \frac{(-1)^n (x^4)^{2n+1}}{(2n+1)!} = \sum_{n=0}^{\infty} \frac{(-1)^n x^{8n+4}}{(2n+1)!}$ for all x, so the

radius of convergence is ∞.

43. $f(x) = 1/\sqrt[4]{16-x} = (16-x)^{-1/4} = \tfrac{1}{2}\left(1 - \tfrac{1}{16}x\right)^{-1/4}$

$$= \tfrac{1}{2}\left[1 + \left(-\tfrac{1}{4}\right)\left(-\frac{x}{16}\right) + \frac{\left(-\tfrac{1}{4}\right)\left(-\tfrac{5}{4}\right)}{2!}\left(-\frac{x}{16}\right)^2 + \cdots\right]$$

$$= \frac{1}{2} + \sum_{n=1}^{\infty} \frac{1 \cdot 5 \cdot 9 \cdot \cdots \cdot (4n-3)}{2 \cdot 4^n \cdot n! \cdot 16^n} x^n = \frac{1}{2} + \sum_{n=1}^{\infty} \frac{1 \cdot 5 \cdot 9 \cdot \cdots \cdot (4n-3)}{2^{6n+1} \, n!} x^n$$

for $\left|-\dfrac{x}{16}\right| < 1 \Longrightarrow R = 16$.

45. $e^x = \displaystyle\sum_{n=0}^{\infty} \frac{x^n}{n!}$ so $\dfrac{e^x}{x} = \dfrac{1}{x} + \displaystyle\sum_{n=1}^{\infty} \frac{x^{n-1}}{n!}$ and $\displaystyle\int \frac{e^x}{x}\, dx = C + \ln|x| + \sum_{n=1}^{\infty} \frac{x^n}{n \cdot n!}$

47. (a)

$$f(x) = x^{1/2} \qquad\qquad f(1) = 1$$
$$f'(x) = \tfrac{1}{2}x^{-1/2} \qquad\quad f'(1) = \tfrac{1}{2}$$
$$f''(x) = -\tfrac{1}{4}x^{-3/2} \qquad f''(1) = -\tfrac{1}{4}$$
$$f'''(x) = \tfrac{3}{8}x^{-5/2} \qquad\quad f'''(1) = \tfrac{3}{8}$$
$$f^{(4)}(x) = -\tfrac{15}{16}x^{-7/2}$$

$$\sqrt{x} \approx T_3(x) = 1 + \frac{1/2}{1!}(x-1) - \frac{1/4}{2!}(x-1)^2 + \frac{3/8}{3!}(x-1)^3$$
$$= 1 + \tfrac{1}{2}(x-1) - \tfrac{1}{8}(x-1)^2 + \tfrac{1}{16}(x-1)^3$$

(b)

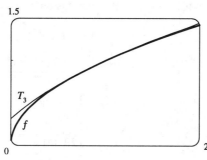

(c) $|R_3(x)| \leq \dfrac{M}{4!}|x-1|^4$, where $|f^{(4)}(x)| \leq M$ with $f^{(4)}(x) = -\tfrac{15}{16}x^{-7/2}$. Now

$0.9 \leq x \leq 1.1 \Longrightarrow (x-1)^4 \leq (0.1)^4$, and letting $x = 0.9$ gives $M = \dfrac{15}{16(0.9)^{7/2}}$, so

$$|R_3(x)| \leq \frac{15}{16(0.9)^{7/2}\,4!}(0.1)^4 \approx 0.000005648.$$

(d)

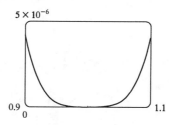

From the graph of $|R_3(x)| = |\sqrt{x} - T_3(x)|$, it appears that the error is less than 5×10^{-6} on $[0.9, 1.1]$.

49. $e^x = \sum\limits_{n=0}^{\infty} \dfrac{x^n}{n!} \Longrightarrow e^{-1/x^2} = \sum\limits_{n=0}^{\infty} \dfrac{\left(-1/x^2\right)^n}{n!} = 1 - \dfrac{1}{x^2} + \dfrac{1}{2x^4} - \cdots \Longrightarrow$

$x^2 \left(1 - e^{-1/x^2}\right) = x^2 \left(\dfrac{1}{x^2} - \dfrac{1}{2x^4} + \cdots\right) = 1 - \dfrac{1}{2x^2} + \cdots \rightarrow 1$ as $x \rightarrow \infty$.

51. Let $y(x) = \sum\limits_{n=0}^{\infty} c_n x^n$. Then $y''(x) = \sum\limits_{n=2}^{\infty} n(n-1) c_n x^{n-2} = \sum\limits_{n=0}^{\infty} (n+2)(n+1) c_{n+2} x^n$,

$xy' = x \sum\limits_{n=1}^{\infty} n c_n x^{n-1} = \sum\limits_{n=0}^{\infty} n c_n x^n$, and the equation $y'' + xy' + y = 0$ becomes

$\sum\limits_{n=0}^{\infty} \left[(n+2)(n+1) c_{n+2} + (n+1) c_n\right] x^n = 0$. Thus, the recurrence relation is $c_{n+2} = -\, c_n / (n+2)$

for $n = 0, 1, 2, \ldots$. But $c_0 = y(0) = 0$, so $c_{2n} = 0$ for $n = 0, 1, 2, \ldots$. Also, $c_1 = y'(0) = 1$, so

$c_3 = -\dfrac{1}{3}$, $c_5 = \dfrac{(-1)^2}{3 \cdot 5} = \dfrac{(-1)^2 \, 2^2 2!}{5!}$, $c_7 = \dfrac{(-1)^3}{3 \cdot 5 \cdot 7} = \dfrac{(-1)^3 \, 2^3 3!}{7!}$, $\ldots$, $c_{2n+1} = \dfrac{(-1)^n \, 2^n n!}{(2n+1)!}$ for

$n = 0, 1, 2, \ldots$. Note that $2^n n! = (2 \cdot 1) \cdot (2 \cdot 2) \cdot (2 \cdot 3) \cdots \cdots (2 \cdot n)$. Thus, the solution to the

initial-value problem is $y(x) = \sum\limits_{n=0}^{\infty} c_n x^n = \sum\limits_{n=0}^{\infty} \dfrac{(-1)^n \, 2^n n! \, x^{2n+1}}{(2n+1)!}$.

Focus on Problem Solving

1. It would be far too much work to compute 15 derivatives of f. The key idea is to remember that $f^{(n)}(0)$ occurs in the coefficient of x^n in the Maclaurin series of f. We start with the Maclaurin series for sin:

$\sin x = x - \dfrac{x^3}{3!} + \dfrac{x^5}{5!} - \cdots$. Then $\sin(x^3) = x^3 - \dfrac{x^9}{3!} + \dfrac{x^{15}}{5!} - \cdots$ and so the coefficient of x^{15} is

$\dfrac{f^{(15)}(0)}{15!} = \dfrac{1}{5!}$. Therefore, $f^{(15)}(0) = \dfrac{15!}{5!} = 6 \cdot 7 \cdot 8 \cdot 9 \cdot 10 \cdot 11 \cdot 12 \cdot 13 \cdot 14 \cdot 15 = 10{,}897{,}286{,}400$.

3. (a) From Formula 14a in Appendix C, with $x = y = \theta$, we get $\tan 2\theta = \dfrac{2\tan\theta}{1 - \tan^2\theta}$, so

$\cot 2\theta = \dfrac{1 - \tan^2\theta}{2\tan\theta} \implies 2\cot 2\theta = \dfrac{1 - \tan^2\theta}{\tan\theta} = \cot\theta - \tan\theta$. Replacing θ by $\frac{1}{2}x$, we get

$2\cot x = \cot\frac{1}{2}x - \tan\frac{1}{2}x$, or $\tan\frac{1}{2}x = \cot\frac{1}{2}x - 2\cot x$.

(b) From part (a), $\tan\dfrac{x}{2^n} = \cot\dfrac{x}{2^n} - 2\cot\dfrac{x}{2^{n-1}}$, so the nth partial sum of $\displaystyle\sum_{n=1}^{\infty} \dfrac{1}{2^n}\tan\dfrac{x}{2^n}$ is

$\begin{aligned}
s_n &= \dfrac{\tan(x/2)}{2} + \dfrac{\tan(x/4)}{4} + \dfrac{\tan(x/8)}{8} + \cdots + \dfrac{\tan(x/2^n)}{2^n} \\[2mm]
&= \left[\dfrac{\cot(x/2)}{2} - \cot x\right] + \left[\dfrac{\cot(x/4)}{4} - \dfrac{\cot(x/2)}{2}\right] + \left[\dfrac{\cot(x/8)}{8} - \dfrac{\cot(x/4)}{4}\right] + \cdots \\[2mm]
&\qquad + \left[\dfrac{\cot(x/2^n)}{2^n} - \dfrac{\cot(x/2^{n-1})}{2^{n-1}}\right] \\[2mm]
&= -\cot x + \dfrac{\cot(x/2^n)}{2^n} \quad \text{(telescoping sum)}
\end{aligned}$

Now $\dfrac{\cot(x/2^n)}{2^n} = \dfrac{\cos(x/2^n)}{2^n \sin(x/2^n)} = \dfrac{\cos(x/2^n)}{x} \cdot \dfrac{x/2^n}{\sin(x/2^n)} \to \dfrac{1}{x} \cdot 1 = \dfrac{1}{x}$

as $n \to \infty$ since $\dfrac{x}{2^n} \to 0$ for $x \neq 0$. Therefore, if $x \neq 0$ and $x \neq n\pi$, then

$\displaystyle\sum_{n=1}^{\infty} \dfrac{1}{2^n}\tan\dfrac{x}{2^n} = \lim_{n\to\infty}\left(-\cot x + \dfrac{1}{2^n}\cot\dfrac{x}{2^n}\right) = -\cot x + \dfrac{1}{x}$. If $x = 0$, then all terms in the

series are 0, so the sum is 0.

5. (a) At each stage, each side is replaced by four shorter sides, each of length $\frac{1}{3}$ of the side length at the preceding stage. Writing s_0 and ℓ_0 for the number of sides and the length of the side of the initial triangle, we generate the following table.

$s_0 = 3$	$\ell_0 = 1$
$s_1 = 3 \cdot 4$	$\ell_1 = \dfrac{1}{3}$
$s_2 = 3 \cdot 4^2$	$\ell_2 = \dfrac{1}{3^2}$
$s_3 = 3 \cdot 4^3$	$\ell_3 = \dfrac{1}{3^3}$
$\cdots$	$\cdots$

In general, we have $s_n = 3 \cdot 4^n$ and $\ell_n = \left(\frac{1}{3}\right)^n$, so the length of the perimeter at the nth stage of construction is $p_n = s_n \ell_n = 3 \cdot 4^n \cdot \left(\frac{1}{3}\right)^n = 3 \cdot \left(\frac{4}{3}\right)^n$.

(b) $p_n = \dfrac{4^n}{3^{n-1}} = 4\left(\dfrac{4}{3}\right)^{n-1}$. Since $\frac{4}{3} > 1$, $p_n \to \infty$ as $n \to \infty$.

(c) The area of each of the small triangles added at a given stage is one-ninth of the area of the triangle added at the preceding stage. Let a be the area of the original triangle. Then the area a_n of each of the small triangles added at stage n is $a_n = a \cdot \dfrac{1}{9^n} = \dfrac{a}{9^n}$. Since a small triangle is added to each side at every stage, it follows that the total area A_n added to the figure at the nth stage is

$$A_n = s_{n-1} \cdot a_n = 3 \cdot 4^{n-1} \cdot \frac{a}{9^n} = a \cdot \frac{4^{n-1}}{3^{2n-1}}.$$ Then the total area enclosed by the snowflake curve

is $A = a + A_1 + A_2 + A_3 + \cdots = a + a \cdot \dfrac{1}{3} + a \cdot \dfrac{4}{3^3} + a \cdot \dfrac{4^2}{3^5} + a \cdot \dfrac{4^3}{3^7} + \cdots$. After the first term,

this is a geometric series with common ratio $\frac{4}{9}$, so $A = a + \dfrac{a/3}{1 - \frac{4}{9}} = a + \dfrac{a}{3} \cdot \dfrac{9}{5} = \dfrac{8a}{5}$. But the area

of the original equilateral triangle with side 1 is $a = \frac{1}{2} \cdot 1 \cdot \sin\frac{\pi}{3} = \frac{\sqrt{3}}{4}$. So the area enclosed by the

snowflake curve is $\frac{8}{5} \cdot \frac{\sqrt{3}}{4} = \frac{2\sqrt{3}}{5}$.

7. (a) Let $a = \arctan x$ and $b = \arctan y$. Then, from Formula 14b in Appendix C,

$$\tan(a - b) = \frac{\tan a - \tan b}{1 + \tan a \tan b} = \frac{\tan(\arctan x) - \tan(\arctan y)}{1 + \tan(\arctan x)\tan(\arctan y)} \implies$$

$$\tan(a - b) = \frac{x - y}{1 + xy} \implies \arctan x - \arctan y = a - b = \arctan \frac{x - y}{1 + xy} \text{ since}$$

$-\frac{\pi}{2} < \arctan x - \arctan y < \frac{\pi}{2}$.

(b) From part (a) we have

$$\arctan \frac{120}{119} - \arctan \frac{1}{239} = \arctan \frac{\frac{120}{119} - \frac{1}{239}}{1 + \frac{120}{119} \cdot \frac{1}{239}} = \arctan \frac{\frac{28,561}{28,441}}{\frac{28,561}{28,441}} = \arctan 1 = \frac{\pi}{4}.$$

(c) Replacing y by $-y$ in the formula of part (a), we get $\arctan x + \arctan y = \arctan \dfrac{x+y}{1-xy}$. So

$$4\arctan \tfrac{1}{5} = 2\left(\arctan \tfrac{1}{5} + \arctan \tfrac{1}{5}\right) = 2\arctan \dfrac{\tfrac{1}{5}+\tfrac{1}{5}}{1-\tfrac{1}{5}\cdot\tfrac{1}{5}} = 2\arctan \tfrac{5}{12}$$

$$= \arctan \tfrac{5}{12} + \arctan \tfrac{5}{12}$$

$$= \arctan \dfrac{\tfrac{5}{12}+\tfrac{5}{12}}{1-\tfrac{5}{12}\cdot\tfrac{5}{12}} = \arctan \tfrac{120}{119}$$

Thus, from part (b), we have $4\arctan \tfrac{1}{5} - \arctan \tfrac{1}{239} = \arctan \tfrac{120}{119} - \arctan \tfrac{1}{239} = \tfrac{\pi}{4}$.

(d) From Example 7 in Section 8.6 we have $\arctan x = x - \dfrac{x^3}{3} + \dfrac{x^5}{5} - \dfrac{x^7}{7} + \dfrac{x^9}{9} - \dfrac{x^{11}}{11} + \cdots$, so

$\arctan \dfrac{1}{5} = \dfrac{1}{5} - \dfrac{1}{3\cdot 5^3} + \dfrac{1}{5\cdot 5^5} - \dfrac{1}{7\cdot 5^7} + \dfrac{1}{9\cdot 5^9} - \dfrac{1}{11\cdot 5^{11}} + \cdots$. This is an alternating series

and the size of the terms decreases to 0, so by the Alternating Series Estimation Theorem, the sum lies between s_5 and s_6, that is, $0.197395560 < \arctan \tfrac{1}{5} < 0.197395562$.

(e) From the series in part (d) we get $\arctan \dfrac{1}{239} = \dfrac{1}{239} - \dfrac{1}{3\cdot 239^3} + \dfrac{1}{5\cdot 239^5} - \cdots$. The third term is less than 2.6×10^{-13}, so by the Alternating Series Estimation Theorem, we have, to nine decimal places, $\arctan \tfrac{1}{239} \approx s_2 \approx 0.004184076$. Thus, $0.004184075 < \arctan \tfrac{1}{239} < 0.004184077$.

(f) From part (c) we have $\pi = 16\arctan \tfrac{1}{5} - 4\arctan \tfrac{1}{239}$, so from parts (d) and (e) we have

$16\,(0.197395560) - 4\,(0.004184077) < \pi < 16\,(0.197395562) - 4\,(0.004184075) \Longrightarrow$

$3.141592652 < \pi < 3.141592692$. So, to 7 decimal places, $\pi \approx 3.1415927$.

9. We start with the geometric series $\displaystyle\sum_{n=0}^{\infty} x^n = \dfrac{1}{1-x}$, $|x| < 1$, and differentiate:

$$\sum_{n=1}^{\infty} nx^{n-1} = \dfrac{d}{dx}\left(\sum_{n=0}^{\infty} x^n\right) = \dfrac{d}{dx}\left(\dfrac{1}{1-x}\right) = \dfrac{1}{(1-x)^2} \text{ for } |x| < 1 \Longrightarrow$$

$$\sum_{n=1}^{\infty} nx^n = x\sum_{n=1}^{\infty} nx^{n-1} = \dfrac{x}{(1-x)^2} \text{ for } |x| < 1. \text{ Differentiate again:}$$

$$\sum_{n=1}^{\infty} n^2 x^{n-1} = \dfrac{d}{dx}\dfrac{x}{(1-x)^2} = \dfrac{(1-x)^2 - x\cdot 2\,(1-x)\,(-1)}{(1-x)^4} = \dfrac{x+1}{(1-x)^3} \Longrightarrow$$

$$\sum_{n=1}^{\infty} n^2 x^n = \dfrac{x^2+x}{(1-x)^3} \Longrightarrow$$

$$\sum_{n=1}^{\infty} n^3 x^{n-1} = \dfrac{d}{dx}\dfrac{x^2+x}{(1-x)^3} = \dfrac{(1-x)^3\,(2x+1) - (x^2+x)\,3\,(1-x)^2\,(-1)}{(1-x)^6} = \dfrac{x^2+4x+1}{(1-x)^4} \Longrightarrow$$

$$\sum_{n=1}^{\infty} n^3 x^n = \dfrac{x^3+4x^2+x}{(1-x)^4}, \ |x| < 1. \text{ The radius of convergence is 1 because that is the radius of}$$

convergence for the geometric series we started with. If $x = \pm 1$, the series is $\sum n^3\,(\pm 1)^n$, which diverges by the Test For Divergence, so the interval of convergence is $(-1, 1)$.

11. $u = 1 + \dfrac{x^3}{3!} + \dfrac{x^6}{6!} + \dfrac{x^9}{9!} + \cdots$, $v = x + \dfrac{x^4}{4!} + \dfrac{x^7}{7!} + \dfrac{x^{10}}{10!} + \cdots$, $w = \dfrac{x^2}{2!} + \dfrac{x^5}{5!} + \dfrac{x^8}{8!} + \cdots$. The key

idea is to differentiate: $\dfrac{du}{dx} = \dfrac{3x^2}{3!} + \dfrac{6x^5}{6!} + \dfrac{9x^8}{9!} + \cdots = \dfrac{x^2}{2!} + \dfrac{x^5}{5!} + \dfrac{x^8}{8!} + \cdots = w$. Similarly,

$\dfrac{dv}{dx} = 1 + \dfrac{x^3}{3!} + \dfrac{x^6}{6!} + \dfrac{x^9}{9!} + \cdots = u$, and $\dfrac{dw}{dx} = x + \dfrac{x^4}{4!} + \dfrac{x^7}{7!} + \dfrac{x^{10}}{10!} + \cdots = v$. So $u' = w$, $v' = u$,

and $w' = v$. Now differentiate the left hand side of the desired equation:

$$\frac{d}{dx}\left(u^3 + v^3 + w^3 - 3uvw\right) = 3u^2u' + 3v^2v' + 3w^2w' - 3\left(u'vw + uv'w + uvw'\right)$$

$$= 3u^2w + 3v^2u + 3w^2v - 3\left(vw^2 + u^2w + uv^2\right) = 0 \implies$$

$u^3 + v^3 + w^3 - 3uvw = C$. To find the value of the constant C, we put $x = 0$ in the last equation and

get $1^3 + 0^3 + 0^3 - 3\left(1 \cdot 0 \cdot 0\right) = C \implies C = 1$, so $u^3 + v^3 + w^3 - 3uvw = 1$.

13. If L is the length of a side of the equilateral triangle, then the area is $A = \frac{1}{2}L \cdot \frac{\sqrt{3}}{2}L = \frac{\sqrt{3}}{4}L^2$ and so

$L^2 = \frac{4}{\sqrt{3}}A$. Let r be the radius of one of the circles. When there are n rows of circles, the figure shows

that $L = \sqrt{3}r + r + (n-2)(2r) + r + \sqrt{3}r = r\left(2n - 2 + 2\sqrt{3}\right)$, so $r = \dfrac{L}{2\left(n + \sqrt{3} - 1\right)}$. The

number of circles is $1 + 2 + \cdots + n = \dfrac{n(n+1)}{2}$ and so the total area of the circles is

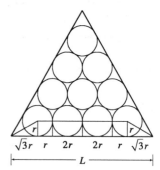

$$A_n = \frac{n(n+1)}{2}\pi r^2 = \frac{n(n+1)}{2}\pi \frac{L^2}{4\left(n + \sqrt{3} - 1\right)^2}$$

$$= \frac{n(n+1)}{2}\pi \frac{4A/\sqrt{3}}{4\left(n + \sqrt{3} - 1\right)^2} = \frac{n(n+1)}{\left(n + \sqrt{3} - 1\right)^2}\frac{\pi A}{2\sqrt{3}} \implies$$

$$\frac{A_n}{A} = \frac{n(n+1)}{\left(n + \sqrt{3} - 1\right)^2}\frac{\pi}{2\sqrt{3}} = \frac{1 + 1/n}{\left[1 + \left(\sqrt{3} - 1\right)/n\right]^2}\frac{\pi}{2\sqrt{3}}$$

$$\to \frac{\pi}{2\sqrt{3}} \text{ as } n \to \infty.$$

$\sqrt{3}r \quad r \quad 2r \qquad 2r \quad r \quad \sqrt{3}r$

$\longmapsto\!\!\!\!\!\!\!\!\!\!\!\!\!\longleftarrow L \longrightarrow\!\!\!\!\!\!\!\!\!\!\!\!\!\dashv$

15. Call the series S. We group the terms according to the number of digits in their denominators:

$$S = \underbrace{\left(1 + \frac{1}{2} + \cdots + \frac{1}{8} + \frac{1}{9}\right)}_{g_1} + \underbrace{\left(\frac{1}{11} + \cdots + \frac{1}{99}\right)}_{g_2} + \underbrace{\left(\frac{1}{111} + \cdots + \frac{1}{999}\right)}_{g_3} + \cdots$$

Now in the group g_n, there are 9^n terms, since we have 9 choices for each of the n digits in the

denominator. Furthermore, each term in g_n is less than $\dfrac{1}{10^{n-1}}$. So $g_n < 9^n \cdot \dfrac{1}{10^{n-1}} = 9\left(\dfrac{9}{10}\right)^{n-1}$.

Now $\displaystyle\sum_{n=1}^{\infty} 9\left(\frac{9}{10}\right)^{n-1}$ is a geometric series with $a = 9$ and $r = \frac{9}{10} < 1$. Therefore, by the

Comparison Test, $S = \displaystyle\sum_{n=1}^{\infty} g_n < \sum_{n=1}^{\infty} 9\left(\frac{9}{10}\right)^{n-1} = \dfrac{9}{1 - \frac{9}{10}} = 90.$

Appendixes

Appendix A Intervals, Inequalities, and Absolute Values

1. $|5 - 23| = |-18| = 18$

3. $\left|\sqrt{5} - 5\right| = -\left(\sqrt{5} - 5\right) = 5 - \sqrt{5}$ because $\sqrt{5} - 5 < 0$.

5. For $x < 2$, $x - 2 < 0$, so $|x - 2| = -(x - 2) = 2 - x$.

7. $|x + 1| = \begin{cases} x + 1 & \text{for } x + 1 \geq 0 \iff x \geq -1 \\ -(x + 1) & \text{for } x + 1 < 0 \iff x < -1 \end{cases}$

9. $\left|x^2 + 1\right| = x^2 + 1$ (since $x^2 + 1 \geq 0$ for all x).

11. $2x + 7 > 3 \iff 2x > -4 \iff x > -2$, so $x \in (-2, \infty)$.

13. $1 - x \leq 2 \iff -x \leq 1 \iff x \geq -1$, so $x \in [-1, \infty)$.

15. $0 \leq 1 - x < 1 \iff -1 \leq -x < 0 \iff 1 \geq x > 0$, so $x \in (0, 1]$.

17. $(x - 1)(x - 2) > 0$.

Case 1: $x - 1 > 0 \iff x > 1$, and $x - 2 > 0 \iff x > 2$, so $x \in (2, \infty)$.

Case 2: $x - 1 < 0 \iff x < 1$, and $x - 2 < 0 \iff x < 2$, so $x \in (-\infty, 1)$.

Thus, the solution set is $(-\infty, 1) \cup (2, \infty)$.

19. $x^2 < 3 \iff x^2 - 3 < 0 \iff \left(x - \sqrt{3}\right)\left(x + \sqrt{3}\right) < 0$.

Case 1: $x > \sqrt{3}$ and $x < -\sqrt{3}$, which is impossible.

Case 2: $x < \sqrt{3}$ and $x > -\sqrt{3}$. Thus, the solution set is $\left(-\sqrt{3}, \sqrt{3}\right)$.

Another Method: $x^2 < 3 \iff |x| < \sqrt{3} \iff -\sqrt{3} < x < \sqrt{3}$.

21. $x^3 - x^2 \leq 0 \iff x^2(x - 1) \leq 0$. Since $x^2 \geq 0$ for all x, the inequality is satisfied when $x - 1 \leq 0 \iff x \leq 1$. Thus, the solution set is $(-\infty, 1]$.

23. $x^3 > x \iff x^3 - x > 0 \iff x\left(x^2 - 1\right) > 0 \iff x\left(x - 1\right)\left(x + 1\right) > 0.$ Constructing a table:

Interval	x	$x - 1$	$x + 1$	$x\left(x - 1\right)\left(x + 1\right)$
$x < -1$	$-$	$-$	$-$	$-$
$-1 < x < 0$	$-$	$-$	$+$	$+$
$0 < x < 1$	$+$	$-$	$+$	$-$
$x > 1$	$+$	$+$	$+$	$+$

Since $x^3 > x$ when the last column is positive, the solution set is $(-1, 0) \cup (1, \infty)$.

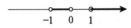

25. $1/x < 4$. This is clearly true for $x < 0$. So suppose $x > 0$. then $1/x < 4 \iff 1 < 4x \iff \frac{1}{4} < x$. Thus, the solution set is $(-\infty, 0) \cup \left(\frac{1}{4}, \infty\right)$.

27. $C = \frac{5}{9}(F - 32) \implies F = \frac{9}{5}C + 32.$ So $50 \le F \le 95 \implies 50 \le \frac{9}{5}C + 32 \le 95 \implies$ $18 \le \frac{9}{5}C \le 63 \implies 10 \le C \le 35.$ So the interval is $[10, 35]$.

29. (a) Let T represent the temperature in degrees Celsius and h the height in km. $T = 20$ when $h = 0$ and T decreases by $10°C$ for every km. Thus, $T = 20 - 10h$ when $0 \le h \le 12$.

(b) From (a), $T = 20 - 10h \implies h = 2 - T/10.$ So $0 \le h \le 5 \implies 0 \le 2 - T/10 \le 5 \implies$ $-2 \le -T/10 \le 3 \implies -20 \le -T \le 30 \implies 20 \ge T \ge -30 \implies -30 \le T \le 20.$ Thus, the range of temperatures (in °C) to be expected is $[-30, 20]$.

31. $|x + 3| = |2x + 1| \iff$ either $x + 3 = 2x + 1$ or $x + 3 = -(2x + 1).$ In the first case, $x = 2$, and in the second case, $3x = -4 \iff x = -\frac{4}{3}.$

33. By Property 5 of absolute values, $|x| < 3 \iff -3 < x < 3$, so $x \in (-3, 3)$.

35. $|x - 4| < 1 \iff -1 < x - 4 < 1 \iff 3 < x < 5$, so $x \in (3, 5)$.

37. $|x + 5| \ge 2 \iff x + 5 \ge 2$ or $x + 5 \le -2 \iff x \ge -3$ or $x \le -7$, so $x \in (-\infty, -7] \cup [-3, \infty)$.

39. $|2x - 3| \le 0.4 \iff -0.4 \le 2x - 3 \le 0.4 \iff 2.6 \le 2x \le 3.4 \iff 1.3 \le x \le 1.7$, so $x \in [1.3, 1.7]$.

41. $a(bx - c) \ge bc \iff bx - c \ge \dfrac{bc}{a} \iff bx \ge \dfrac{bc}{a} + c = \dfrac{bc + ac}{a} \iff x \ge \dfrac{bc + ac}{ab}$

43. $|ab| = \sqrt{(ab)^2} = \sqrt{a^2 b^2} = \sqrt{a^2}\sqrt{b^2} = |a|\,|b|$

Appendix B Coordinate Geometry

1. From the Distance Formula with $x_1 = 1$, $x_2 = 4$, $y_1 = 1$, $y_2 = 5$, we find the distance from $(1, 1)$ to $(4, 5)$ to be $\sqrt{(4-1)^2 + (5-1)^2} = \sqrt{3^2 + 4^2} = \sqrt{25} = 5$.

3. With $P(-3, 3)$ and $Q(-1, -6)$, the slope m of the line through P and Q is $m = \dfrac{-6-3}{-1-(-3)} = -\dfrac{9}{2}$.

5. Using $A(-2, 9)$, $B(4, 6)$, $C(1, 0)$, and $D(-5, 3)$, we have

$$|AB| = \sqrt{[4-(-2)]^2 + (6-9)^2} = \sqrt{6^2 + (-3)^2} = 3\sqrt{5},$$

$$|BC| = \sqrt{(1-4)^2 + (0-6)^2} = \sqrt{(-3)^2 + (-6)^2} = 3\sqrt{5},$$

$$|CD| = \sqrt{(-5-1)^2 + (3-0)^2} = \sqrt{(-6)^2 + 3^2} = 3\sqrt{5}, \text{ and}$$

$$|DA| = \sqrt{[-2-(-5)]^2 + (9-3)^2} = \sqrt{3^2 + 6^2} = 3\sqrt{5}. \text{ So all sides are of equal length.}$$

Moreover, $m_{AB} = \dfrac{6-9}{4-(-2)} = -\dfrac{1}{2}$, $m_{BC} = \dfrac{0-6}{1-4} = 2$, $m_{CD} = \dfrac{3-0}{-5-1} = -\dfrac{1}{2}$, and

$m_{DA} = \dfrac{9-3}{-2-(-5)} = 2$, so the sides are perpendicular. Thus, it is a square.

7. $x = 3$

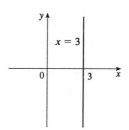

9. $xy = 0 \iff x = 0$ or $y = 0$

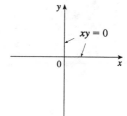

11. By the point-slope form of the equation of a line, an equation of the line through $(2, -3)$ with slope 6 is $y - (-3) = 6(x - 2)$ or $y = 6x - 15$.

13. The slope of the line through $(2, 1)$ and $(1, 6)$ is $m = \dfrac{6-1}{1-2} = -5$, so an equation of the line is $y - 1 = -5(x - 2)$ or $y = -5x + 11$.

15. By the slope-intercept form of the equation of a line, an equation of the line is $y = 3x - 2$.

17. Since the line passes through $(1, 0)$ and $(0, -3)$, its slope is $m = \dfrac{-3-0}{0-1} = 3$, so an equation is $y = 3x - 3$.

19. Since $m = 0$, $y - 5 = 0(x - 4)$ or $y = 5$.

21. Putting the line $x + 2y = 6$ into its slope-intercept form $y = -\frac{1}{2}x + 3$, we see that this line has slope $-\frac{1}{2}$. So we want the line of slope $-\frac{1}{2}$ that passes through the point $(1, -6)$: $y - (-6) = -\frac{1}{2}(x - 1) \iff y = -\frac{1}{2}x - \frac{11}{2}$.

23. $2x + 5y + 8 = 0 \iff y = -\frac{2}{5}x - \frac{8}{5}$. Since this line has slope $-\frac{2}{5}$, a line perpendicular to it would have slope $\frac{5}{2}$, so the required line is $y - (-2) = \frac{5}{2}[x - (-1)] \iff y = \frac{5}{2}x + \frac{1}{2}$.

25. $x + 3y = 0 \iff y = -\frac{1}{3}x$, so the slope is $-\frac{1}{3}$ and the y-intercept is 0.

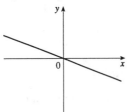

27. $3x - 4y = 12 \iff y = \frac{3}{4}x - 3$, so the slope is $\frac{3}{4}$ and the y-intercept is -3.

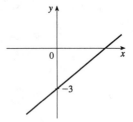

29. $\{(x, y) \mid x < 0\}$

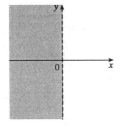

31. $\{(x, y) \mid |x| \le 2\} = \{(x, y) \mid -2 \le x \le 2\}$

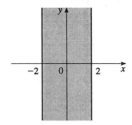

33. $\{(x, y) \mid 0 \le y \le 4, x \le 2\}$

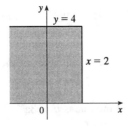

35. $\{(x, y) \mid 1 + x \le y \le 1 - 2x\}$

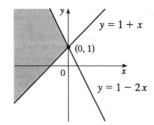

37. An equation of the circle is $(x - 3)^2 + (y + 1)^2 = 5^2 = 25$.

39. $x^2 + y^2 - 4x + 10y + 13 = 0 \iff x^2 - 4x + y^2 + 10y = -13 \iff (x^2 - 4x + 4) + (y^2 + 10y + 25) = -13 + 4 + 25 = 16 \iff (x - 2)^2 + (y + 5)^2 = 4^2$. Thus, we have a circle with center $(2, -5)$ and radius 4.

41. $2x - y = 4 \iff y = 2x - 4 \implies m_1 = 2$ and $6x - 2y = 10 \iff 2y = 6x - 10 \iff y = 3x - 5 \implies$ $m_2 = 3$. Since $m_1 \neq m_2$, the two lines are not parallel. To find the point of intersection: $2x - 4 = 3x - 5 \iff x = 1 \implies y = -2$. Thus, the point of intersection is $(1, -2)$.

43. (a) Let M be the point $\left(\dfrac{x_1 + x_2}{2}, \dfrac{y_1 + y_2}{2} \right)$. Then

$$|MP_1|^2 = \left(x_1 - \frac{x_1 + x_2}{2} \right)^2 + \left(y_1 - \frac{y_1 + y_2}{2} \right)^2 = \left(\frac{x_1 - x_2}{2} \right)^2 + \left(\frac{y_1 - y_2}{2} \right)^2 \text{ and}$$

$$|MP_2|^2 = \left(x_2 - \frac{x_1 + x_2}{2} \right)^2 + \left(y_2 - \frac{y_1 + y_2}{2} \right)^2 = \left(\frac{x_2 - x_1}{2} \right)^2 + \left(\frac{y_2 - y_1}{2} \right)^2. \text{ Hence,}$$

$$|MP_1| = |MP_2|.$$

(b) Using the midpoint formula from part (a) with $(1, 3)$ and $(7, 15)$, we get $\left(\dfrac{1 + 7}{2}, \dfrac{3 + 15}{2} \right) = (4, 9)$.

45. (a) Since the x-intercept is a, the point $(a, 0)$ is on the line, and similarly since the y-intercept is b, $(0, b)$ is on the line. Hence, the slope of the line is $m = \dfrac{b - 0}{0 - a} = -\dfrac{b}{a}$. Substituting into $y = mx + b$ gives $y = -\dfrac{b}{a} x + b \iff \dfrac{b}{a} x + y = b \iff \dfrac{x}{a} + \dfrac{y}{b} = 1$.

(b) Letting $a = 6$ and $b = -8$ gives $\dfrac{x}{6} + \dfrac{y}{-8} = 1 \iff -8x + 6y = -48 \iff 6y = 8x - 48 \iff y = \frac{4}{3} x - 8$.

47. (a)

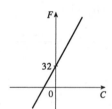

(b) Slope $= \frac{9}{5}$ represents the change in °F for every 1°C change; F-intercept $= 32$ represents the °F temperature corresponding to 0°C.

49. (a) Using the points $(N_1, T_1) = (120, 70)$ and $(N_2, T_2) = (168, 80)$, we get the slope of the line containing them to be $\dfrac{80 - 70}{168 - 120} = \dfrac{10}{48} = \dfrac{5}{24}$. A linear equation that models T as a function of N is $T - 70 = \frac{5}{24} (N - 120)$ or $T = \frac{5}{24} N + 45$.

(b) The slope of the graph is $\frac{5}{24}$, which means that an increase of 1 chirp per minute corresponds to an increase of $\left(\frac{5}{24} \right)$ °F, or, equivalently, an increase of 24 chirps per minute corresponds to an increase of 5°F. The slope represents the rate of change of temperature with respect to the chirping rate.

(c) $N = 150 \implies T = \frac{5}{24} (150) + 45 = 76.25$°F, or about 76°F.

51. (a) The water pressure increases 4.34 lb/in^2 for every 10 ft of descent, or 0.434 lb/in^2 for every foot. Since the pressure at the surface is 15 lb/in^2, a model is $P = 0.434d + 15$, where d is the depth below the ocean's surface, measured in feet.

(b) $P = 100 \implies 100 = 0.434d + 15 \implies 85 = 0.434d \implies d = \frac{85}{0.434} \approx 196$ ft.

53. If $P(x, y)$ is any point on the parabola, then the distance from P to the focus is $|PF| = \sqrt{x^2 + (y - p)^2}$ and the distance from P to the directrix is $|y + p|$. (Figure 15 in the text illustrates the case where $p > 0$.) The defining property of a parabola is that these distances are equal: $\sqrt{x^2 + (y - p)^2} = |y + p|$. We get an equivalent equation by squaring and simplifying: $x^2 + (y - p)^2 = |y + p|^2 = (y + p)^2 \iff$ $x^2 + y^2 - 2py + p^2 = y^2 + 2py + p^2 \iff x^2 = 4py$. Thus, an equation of a parabola with focus $(0, p)$ and directrix $y = -p$ is $x^2 = 4py$.

55. See Figure 21 in the text. $P(x, y)$ is a point on the ellipse when $|PF_1| + |PF_2| = 2a$, that is, $\sqrt{(x + c)^2 + y^2} + \sqrt{(x - c)^2 + y^2} = 2a$ or $\sqrt{(x - c)^2 + y^2} = 2a - \sqrt{(x + c)^2 + y^2}$. Squaring both sides, we have $x^2 - 2cx + c^2 + y^2 = 4a^2 - 4a\sqrt{(x + c)^2 + y^2} + x^2 + 2cx + c^2 + y^2$, which simplifies to $a\sqrt{(x + c)^2 + y^2} = a^2 + cx$. We square again: $a^2(x^2 + 2cx + c^2 + y^2) = a^4 + 2a^2cx + c^2x^2$, which becomes $(a^2 - c^2)x^2 + a^2y^2 = a^2(a^2 - c^2)$. From triangle $F_1 F_2 P$ in Figure 21, we see that $2c < 2a$, so $c < a$ and, therefore, $a^2 - c^2 > 0$. For convenience, let $b^2 = a^2 - c^2$. Then the equation of the ellipse becomes $b^2x^2 + a^2y^2 = a^2b^2$ or, if both sides are divided by a^2b^2, $\dfrac{x^2}{a^2} + \dfrac{y^2}{b^2} = 1$.

57. From Figure 24 in the text, $|PF_1| - |PF_2| = \pm 2a \iff \sqrt{(x + c)^2 + y^2} - \sqrt{(x - c)^2 + y^2} = \pm 2a \iff$ $\sqrt{(x + c)^2 + y^2} = \sqrt{(x - c)^2 + y^2} \pm 2a \iff (x + c)^2 + y^2 =$ $(x - c)^2 + y^2 + 4a^2 \pm 4a\sqrt{(x - c)^2 + y^2} \iff 4cx - 4a^2 = \pm 4a\sqrt{(x - c)^2 + y^2} \iff$ $c^2x^2 - 2a^2cx + a^4 = a^2(x^2 - 2cx + c^2 + y^2) \iff (c^2 - a^2)x^2 - a^2y^2 = a^2(c^2 - a^2) \iff$ $b^2x^2 - a^2y^2 = a^2b^2$ (where $b^2 = c^2 - a^2$) $\iff \dfrac{x^2}{a^2} - \dfrac{y^2}{b^2} = 1$.

59.

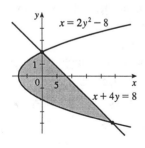

$x + 4y = 8$, $x = 2y^2 - 8$. Substitute x from the second equation into the first: $(2y^2 - 8) + 4y = 8 \iff$ $2y^2 + 4y - 16 = 0 \iff y^2 + 2y - 8 = 0 \iff$ $(y + 4)(y - 2) = 0 \iff y = -4$ or 2. So the points of intersection are $(24, -4)$ and $(0, 2)$.

61. Differentiating implicitly, $\dfrac{x^2}{a^2} + \dfrac{y^2}{b^2} = 1 \Longrightarrow \dfrac{2x}{a^2} + \dfrac{2yy'}{b^2} = 0 \Longrightarrow y' = -\dfrac{b^2 x}{a^2 y}$ $(y \neq 0)$. Thus, the slope

of the tangent line at P is $-\dfrac{b^2 x_1}{a^2 y_1}$. The slope of $F_1 P$ is $\dfrac{y_1}{x_1 + c}$ and of $F_2 P$ is $\dfrac{y_1}{x_1 - c}$. By the formula

from Focus on Problem Solving 3 (Problem 13), we have

$$
\begin{aligned}
\tan \alpha &= \frac{\dfrac{y_1}{x_1 + c} + \dfrac{b^2 x_1}{a^2 y_1}}{1 - \dfrac{b^2 x_1 y_1}{a^2 y_1 (x_1 + c)}} \\[2mm]
&= \frac{a^2 y_1^2 + b^2 x_1 (x_1 + c)}{a^2 y_1 (x_1 + c) - b^2 x_1 y_1} \\[2mm]
&= \frac{a^2 b^2 + b^2 c x_1}{c^2 x_1 y_1 + a^2 c y_1} \qquad \left[\begin{array}{c} \text{using } b^2 x_1^2 + a^2 y_1^2 = a^2 b^2 \\ \text{and } a^2 - b^2 = c^2 \end{array}\right] \\[2mm]
&= \frac{b^2 \left(c x_1 + a^2\right)}{c y_1 \left(c x_1 + a^2\right)} \\[2mm]
&= \frac{b^2}{c y_1}
\end{aligned}
$$

and

$$
\begin{aligned}
\tan \beta &= \frac{-\dfrac{b^2 x_1}{a^2 y_1} - \dfrac{y_1}{x_1 - c}}{1 - \dfrac{b^2 x_1 y_1}{a^2 y_1 (x_1 - c)}} \\[2mm]
&= \frac{-a^2 y_1^2 - b^2 x_1 (x_1 - c)}{a^2 y_1 (x_1 - c) - b^2 x_1 y_1} \\[2mm]
&= \frac{-a^2 b^2 + b^2 c x_1}{c^2 x_1 y_1 - a^2 c y_1} \\[2mm]
&= \frac{b^2 \left(c x_1 - a^2\right)}{c y_1 \left(c x_1 - a^2\right)} \\[2mm]
&= \frac{b^2}{c y_1}.
\end{aligned}
$$

Thus, $\alpha = \beta$.

Appendix C Trigonometry

1. (a) $210° = 210 \left(\frac{\pi}{180}\right) = \frac{7\pi}{6}$ rad

   (b) $9° = 9 \left(\frac{\pi}{180}\right) = \frac{\pi}{20}$ rad

3. (a) 4π rad $= 4\pi \left(\frac{180}{\pi}\right) = 720°$

   (b) $-\frac{3\pi}{8}$ rad $= -\frac{3\pi}{8} \left(\frac{180}{\pi}\right) = -67.5°$

5. Using Formula 3, $a = r\theta = 36 \cdot \frac{\pi}{12} = 3\pi$ cm.

7. Using Formula 3, $\theta = a/r = \frac{1}{1.5} = \frac{2}{3}$ rad $= \frac{2}{3} \left(\frac{180}{\pi}\right) = \left(\frac{120}{\pi}\right)° \approx 38.2°$.

9. (a) 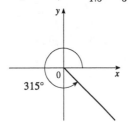   (b)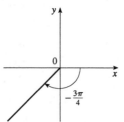

11.

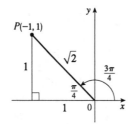

From the diagram, we see that a point on the terminal side is $P(-1, 1)$. Therefore, taking $x = -1$, $y = 1$, $r = \sqrt{2}$ in the definitions of the trigonometric ratios, we have $\sin \frac{3\pi}{4} = \frac{1}{\sqrt{2}}$, $\cos \frac{3\pi}{4} = -\frac{1}{\sqrt{2}}$, $\tan \frac{3\pi}{4} = -1$, $\csc \frac{3\pi}{4} = \sqrt{2}$, $\sec \frac{3\pi}{4} = -\sqrt{2}$, and $\cot \frac{3\pi}{4} = -1$.

13. $\sin \theta = y/r = \frac{3}{5} \Longrightarrow y = 3$, $r = 5$, and $x = \sqrt{r^2 - y^2} = 4$ (since $0 < \theta < \frac{\pi}{2}$). Therefore taking $x = 4$, $y = 3$, $r = 5$ in the definitions of the trigonometric ratios, we have $\cos \theta = \frac{4}{5}$, $\tan \theta = \frac{3}{4}$, $\csc \theta = \frac{5}{3}$, $\sec \theta = \frac{5}{4}$, and $\cot \theta = \frac{4}{3}$.

15. $\sin 35° = \dfrac{x}{10} \Longrightarrow x = 10 \sin 35° \approx 5.73576$ cm

17. $\tan \frac{2\pi}{5} = \dfrac{x}{8} \Longrightarrow x = 8 \tan \dfrac{2\pi}{5} \approx 24.62147$ cm

19.

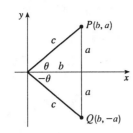

(a) From the diagram, we see that $\sin \theta = \dfrac{y}{r} = \dfrac{a}{c}$, and

$$\sin (-\theta) = \frac{-a}{c} = -\frac{a}{c} = -\sin \theta.$$

(b) Again from the diagram, we see that

$$\cos \theta = \frac{x}{r} = \frac{b}{c} = \cos (-\theta).$$

21. Using (12a), $\sin \left(\frac{\pi}{2} + x\right) = \sin \frac{\pi}{2} \cos x + \cos \frac{\pi}{2} \sin x = 1 \cdot \cos x + 0 \cdot \sin x = \cos x$.

23. Using (6), $\sin \theta \cot \theta = \sin \theta \cdot \dfrac{\cos \theta}{\sin \theta} = \cos \theta$.

25. Using (14a), we have $\tan 2\theta = \tan (\theta + \theta) = \dfrac{\tan \theta + \tan \theta}{1 - \tan \theta \tan \theta} = \dfrac{2 \tan \theta}{1 - \tan^2 \theta}$.

27. Since $\sin x = \frac{1}{3}$ we can label the opposite side as having length 1, the hypotenuse as having length 3, and use the Pythagorean Theorem to get that the adjacent side has length $\sqrt{8}$. Then, from the diagram, $\cos x = \frac{\sqrt{8}}{3}$. Similarly we have that $\sin y = \frac{3}{5}$.

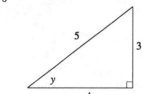

Now use (12a): $\sin (x + y) = \sin x \cos y + \cos x \sin y = \frac{1}{3} \cdot \frac{4}{5} + \frac{\sqrt{8}}{3} \cdot \frac{3}{5} = \frac{4}{15} + \frac{3\sqrt{8}}{15} = \frac{4 + 6\sqrt{2}}{15}$.

29. $2 \cos x - 1 = 0 \iff \cos x = \frac{1}{2} \implies x = \frac{\pi}{3}, \frac{5\pi}{3}$

31. Using (15a), $\sin 2x = \cos x \implies 2 \sin x \cos x - \cos x = 0 \iff \cos x (2 \sin x - 1) = 0 \iff \cos x = 0$ or $2 \sin x - 1 = 0 \implies x = \frac{\pi}{2}, \frac{3\pi}{2}$ or $\sin x = \frac{1}{2} \implies x = \frac{\pi}{6}$ or $\frac{5\pi}{6}$. Therefore, the solutions are $x = \frac{\pi}{6}, \frac{\pi}{2}, \frac{5\pi}{6}, \frac{3\pi}{2}$.

33. We know that $\sin x = \frac{1}{2}$ when $x = \frac{\pi}{6}$ or $\frac{5\pi}{6}$, and from Figure 13(a), we see that $\sin x \leq \frac{1}{2} \implies 0 \leq x \leq \frac{\pi}{6}$ or $\frac{5\pi}{6} \leq x \leq 2\pi$.

35. $\tan x = -1$ when $x = \frac{3\pi}{4}, \frac{7\pi}{4}$, and $\tan x = 1$ when $x = \frac{\pi}{4}$ or $\frac{5\pi}{4}$. From Figure 14 we see that $-1 < \tan x < 1 \implies 0 \leq x < \frac{\pi}{4}, \frac{3\pi}{4} < x < \frac{5\pi}{4}$, and $\frac{7\pi}{4} < x \leq 2\pi$.

37. $y = \cos \left(x - \frac{\pi}{3}\right)$. We start with the graph of $y = \cos x$ and shift it $\frac{\pi}{3}$ units to the right.

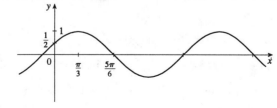

39. $y = \frac{1}{3}\tan\left(x - \frac{\pi}{2}\right)$. We start with the graph of $y = \tan x$, shift it $\frac{\pi}{2}$ units to the right and compress it to $\frac{1}{3}$ of its original vertical size.

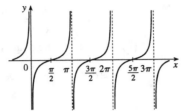

41. (a) $\sin^{-1}(0.5) = \frac{\pi}{6}$ since $\sin\frac{\pi}{6} = 0.5$ and $-\frac{\pi}{2} \leq \frac{\pi}{6} \leq \frac{\pi}{2}$.

(b) $\arctan(-1) = -\frac{\pi}{4}$ since $\tan\left(-\frac{\pi}{4}\right) = -1$ and $-\frac{\pi}{2} < -\frac{\pi}{4} < \frac{\pi}{2}$.

43. (a) $\sin\left(\sin^{-1} 0.7\right) = 0.7$ since $-1 \leq 0.7 \leq 1$.

(b) $\arcsin\left(\sin\frac{5\pi}{4}\right) = \arcsin\left(-\frac{1}{\sqrt{2}}\right) = -\frac{\pi}{4}$

45. Let $y = \sin^{-1} x$. Then $-\frac{\pi}{2} \leq y \leq \frac{\pi}{2} \implies \cos y \geq 0$, so $\cos\left(\sin^{-1} x\right) = \cos y = \sqrt{1 - \sin^2 y} = \sqrt{1 - x^2}$.

47. $g(x) = \sin^{-1}(3x + 1)$. Domain$(g) = \{x \mid -1 \leq 3x + 1 \leq 1\} = \{x \mid -\frac{2}{3} \leq x \leq 0\} = \left[-\frac{2}{3}, 0\right]$.

Range$(g) = \{y \mid -\frac{\pi}{2} \leq y \leq \frac{\pi}{2}\} = \left[-\frac{\pi}{2}, \frac{\pi}{2}\right]$.

49. From the figure we see that $x = b\cos\theta$, $y = b\sin\theta$, and from the distance formula we have that the distance c from (x, y) to $(a, 0)$ is $c = \sqrt{(x - a)^2 + (y - 0)^2} \implies$

$$c^2 = (b\cos\theta - a)^2 + (b\sin\theta)^2$$
$$= b^2\cos^2\theta - 2ab\cos\theta + a^2 + b^2\sin^2\theta$$
$$= a^2 + b^2\left(\cos^2\theta + \sin^2\theta\right) - 2ab\cos\theta$$
$$= a^2 + b^2 - 2ab\cos\theta \quad \text{[by (7)]}$$

51. Using the Law of Cosines, we have $c^2 = 1^2 + 1^2 - 2(1)(1)\cos(\alpha - \beta) = 2[1 - \cos(\alpha - \beta)]$. Now, using the distance formula, $c^2 = |AB|^2 = (\cos\alpha - \cos\beta)^2 + (\sin\alpha - \sin\beta)^2$. Equating these two expressions for c^2, we get

$2[1 - \cos(\alpha - \beta)] = \cos^2\alpha + \sin^2\alpha + \cos^2\beta + \sin^2\beta - 2\cos\alpha\cos\beta - 2\sin\alpha\sin\beta \implies$

$1 - \cos(\alpha - \beta) = 1 - \cos\alpha\cos\beta - \sin\alpha\sin\beta \implies \cos(\alpha - \beta) = \cos\alpha\cos\beta + \sin\alpha\sin\beta$.

53. In Exercise 52 we used the subtraction formula for cosine to prove the addition formula for cosine. Using that formula with $x = \frac{\pi}{2} - \alpha$, $y = \beta$, we get

$\cos\left[\left(\frac{\pi}{2} - \alpha\right) + \beta\right] = \cos\left(\frac{\pi}{2} - \alpha\right)\cos\beta - \sin\left(\frac{\pi}{2} - \alpha\right)\sin\beta \implies$

$\cos\left[\frac{\pi}{2} - (\alpha - \beta)\right] = \cos\left(\frac{\pi}{2} - \alpha\right)\cos\beta - \sin\left(\frac{\pi}{2} - \alpha\right)\sin\beta$. Now we use the identities given in the

problem, $\cos\left(\frac{\pi}{2} - \theta\right) = \sin\theta$ and $\sin\left(\frac{\pi}{2} - \theta\right) = \cos\theta$, to get $\sin(\alpha - \beta) = \sin\alpha\cos\beta - \cos\alpha\sin\beta$.

Appendix D Precise Definitions of Limits

1. On the left side of $x = 2$, we need $|x - 2| < \left|\frac{10}{7} - 2\right| = \frac{4}{7}$. On the right side, we need $|x - 2| < \left|\frac{10}{3} - 2\right| = \frac{4}{3}$. For both of these conditions to be satisfied at once, we need the more restrictive of the two to hold, that is, $|x - 2| < \frac{4}{7}$. So we can choose $\delta = \frac{4}{7}$, or any smaller positive number.

3. $\left|\sqrt{4x + 1} - 3\right| < 0.5 \iff 2.5 < \sqrt{4x + 1} < 3.5$. We plot the three parts of this inequality on the same screen and identify the x-coordinates of the points of intersection using the cursor. It appears that the inequality holds for $1.32 \le x \le 2.81$. Since $|2 - 1.32| = 0.68$ and $|2 - 2.81| = 0.81$, we choose $0 < \delta < \min\{0.68, 0.81\} = 0.68$.

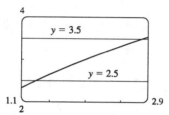

5. For $\varepsilon = 1$, the definition of a limit requires that we find δ such that $\left|(4 + x - 3x^3) - 2\right| < 1 \iff 1 < 4 + x - 3x^3 < 3$ whenever $0 < |x - 1| < \delta$. If we plot the graphs of $y = 1$, $y = 4 + x - 3x^3$ and $y = 3$ on the same screen, we see that we need $0.86 \le x \le 1.11$. So since $|1 - 0.86| = 0.14$ and $|1 - 1.11| = 0.11$, we choose $\delta = 0.11$ (or any smaller positive number). For $\varepsilon = 0.1$, we must find δ such that $\left|(4 + x - 3x^3) - 2\right| < 0.1 \iff 1.9 < 4 + x - 3x^3 < 2.1$ whenever $0 < |x - 1| < \delta$. From the graph, we see that we need $0.988 \le x \le 1.012$. So since $|1 - 0.988| = 0.012$ and $|1 - 1.012| = 0.012$, we choose $\delta = 0.012$ (or any smaller positive number) for the inequality to hold.

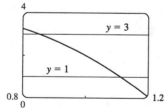 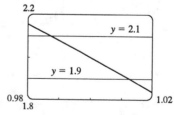

7. Given $\varepsilon > 0$, we need $\delta > 0$ such that if $|x| < \delta$ then $\left|x^3 - 0\right| < \varepsilon \iff |x|^3 < \varepsilon \iff |x| < \sqrt[3]{\varepsilon}$. Take $\delta = \sqrt[3]{\varepsilon}$. Then $|x - 0| < \delta \implies \left|x^3 - 0\right| < \delta^3 = \varepsilon$. Thus, $\lim_{x \to 0} x^3 = 0$ by the definition of a limit.

Appendixes

9. Given $\varepsilon > 0$, we need $\delta > 0$ such that if $|x - 2| < \delta$, then $|(3x - 2) - 4| < \varepsilon \iff |3x - 6| < \varepsilon \iff$
$3|x - 2| < \varepsilon \iff |x - 2| < \varepsilon/3$. So if we choose $\delta = \varepsilon/3$, then $|x - 2| < \delta \implies |(3x - 2) - 4| < \varepsilon$.

Thus, $\lim\limits_{x \to 2} (3x - 2) = 4$ by the definition of a limit.

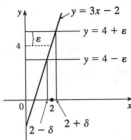

11. (a) $A = \pi r^2$ and $A = 1000 \text{ cm}^2 \implies \pi r^2 = 1000 \implies r^2 = \frac{1000}{\pi} \implies$
$r = \sqrt{\dfrac{1000}{\pi}} \quad (r > 0) \quad \approx 17.8412 \text{ cm}.$

(b) $|A - 1000| \leq 5 \implies 1000 - 5 \leq \pi r^2 \leq 1000 + 5 \implies \sqrt{\dfrac{995}{\pi}} \leq r \leq \sqrt{\dfrac{1005}{\pi}} \implies$
$17.7966 \leq r \leq 17.8858$. $\sqrt{\dfrac{1000}{\pi}} - \sqrt{\dfrac{995}{\pi}} \approx 0.04466$ and $\sqrt{\dfrac{1005}{\pi}} - \sqrt{\dfrac{1000}{\pi}} \approx 0.04455$. So if the
machinist gets the radius within 0.0445 cm of 17.8412, the area will be within 5 cm^2 of 1000.

(c) x is the radius, $f(x)$ is the area, a is the target radius given in part (a), L is the target area (1000), ε
is the tolerance in the area (5), and δ is the tolerance in the radius given in part (b).

13. $\left| \dfrac{6x^2 + 5x - 3}{2x^2 - 1} - 3 \right| < 0.2 \iff 2.8 < \dfrac{6x^2 + 5x - 3}{2x^2 - 1} < 3.2$. So we graph the three parts of this
inequality on the same screen, and find that the curve $y = \dfrac{6x^2 + 5x - 3}{2x^2 - 1}$ seems to lie between the lines
$y = 2.8$ and $y = 3.2$ whenever $x > 12.5$. So we can choose $N = 13$ (or any larger number), so that the
inequality holds whenever $x \geq N$.

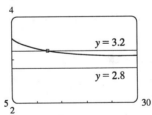

15. (a) $1/x^2 < 0.0001 \iff x^2 > 1/0.0001 = 10{,}000 \iff x > 100 \quad (x > 0)$

(b) If $\varepsilon > 0$ is given, then $1/x^2 < \varepsilon \iff x^2 > 1/\varepsilon \iff x > 1/\sqrt{\varepsilon}$. Let $N = 1/\sqrt{\varepsilon}$. Then
$x > N \implies x > \dfrac{1}{\sqrt{\varepsilon}} \implies \left| \dfrac{1}{x^2} - 0 \right| = \dfrac{1}{x^2} < \varepsilon$, so $\lim\limits_{x \to \infty} \dfrac{1}{x^2} = 0$.

17. (a)

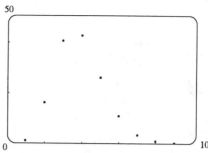

From the graph, it appears that the sequence $\left\{\dfrac{n^5}{n!}\right\}$ converges to 0, that is, $\displaystyle\lim_{n\to\infty}\dfrac{n^5}{n!} = 0$.

(b)

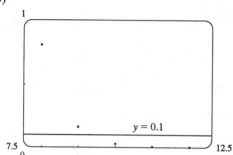

From the first graph, it seems that the smallest possible value of N corresponding to $\varepsilon = 0.1$ is 9, since $n^5/n! < 0.1$ whenever $n \geq 10$, but $9^5/9! > 0.1$. From the second graph, it seems that for $\varepsilon = 0.001$, the smallest possible value for N is 11 since $n^5/n! < 0.001$ whenever $n \geq 12$.

19. If $\displaystyle\lim_{n\to\infty}|a_n| = 0$, then $\displaystyle\lim_{n\to\infty}(-|a_n|) = 0$, and since $-|a_n| \leq a_n \leq |a_n|$, we have that $\displaystyle\lim_{n\to\infty}a_n = 0$, by the Squeeze Theorem.

Appendix F Integration of Rational Functions by Partial Fractions

1. $\dfrac{5}{2x^2 - 3x - 2} = \dfrac{5}{(2x+1)(x-2)} = \dfrac{A}{2x+1} + \dfrac{B}{x-2}$

3. $\dfrac{1}{x^4 - x^3} = \dfrac{1}{x^3(x-1)} = \dfrac{A}{x} + \dfrac{B}{x^2} + \dfrac{C}{x^3} + \dfrac{D}{x-1}$

5. $\dfrac{x^2+1}{x^2-1} = 1 + \dfrac{2}{(x-1)(x+1)} = 1 + \dfrac{A}{x-1} + \dfrac{B}{x+1}$

7. $\dfrac{x^2-2}{x(x^2+2)} = \dfrac{A}{x} + \dfrac{Bx+C}{x^2+2}$

9. $\dfrac{x^3+x^2+1}{x^4+x^3+2x^2} = \dfrac{x^3+x^2+1}{x^2(x^2+x+2)} = \dfrac{A}{x} + \dfrac{B}{x^2} + \dfrac{Cx+D}{x^2+x+2}$

11. $\displaystyle\int \dfrac{x^2}{x+1}\,dx = \int \left(x - 1 + \dfrac{1}{x+1}\right) dx = \tfrac{1}{2}x^2 - x + \ln|x+1| + C$

13. $\dfrac{4x-1}{(x-1)(x+2)} = \dfrac{A}{x-1} + \dfrac{B}{x+2} \;\Longrightarrow\; 4x - 1 = A(x+2) + B(x-1).$ Take $x = 1$ to get $3 = 3A$,

then $x = -2$ to get $-9 = -3B \;\Longrightarrow\; A = 1, B = 3.$ Now

$$\int_2^4 \dfrac{4x-1}{(x-1)(x+2)}\,dx = \int_2^4 \left(\dfrac{1}{x-1} + \dfrac{3}{x+2}\right) dx = [\ln(x-1) + 3\ln(x+2)]_2^4$$
$$= \ln 3 + 3\ln 6 - \ln 1 - 3\ln 4 = \ln(3\cdot 6^3) - \ln 4^3 = \ln\tfrac{81}{8}.$$

15. $\dfrac{2x+3}{(x+1)^2} = \dfrac{A}{x+1} + \dfrac{B}{(x+1)^2} \;\Longrightarrow\; 2x + 3 = A(x+1) + B.$ Take $x = -1$ to get $B = 1$, and equate

coefficients of x to get $A = 2.$ Now

$$\int_0^1 \dfrac{2x+3}{(x+1)^2}\,dx = \int_0^1 \left[\dfrac{2}{x+1} + \dfrac{1}{(x+1)^2}\right] dx = \left[2\ln(x+1) - \dfrac{1}{x+1}\right]_0^1$$
$$= 2\ln 2 - \tfrac{1}{2} - (2\ln 1 - 1) = 2\ln 2 + \tfrac{1}{2}.$$

17. $\dfrac{6x^2+5x-3}{x^3+2x^2-3x} = \dfrac{6x^2+5x-3}{x(x+3)(x-1)} = \dfrac{A}{x} + \dfrac{B}{x+3} + \dfrac{C}{x-1} \;\Longrightarrow$

$6x^2 + 5x - 3 = A(x+3)(x-1) + B(x)(x-1) + C(x)(x+3).$ Set $x = 0$ to get $A = 1$, then

take $x = -3$ to get $B = 3$, then set $x = 1$ to get $C = 2$:

$$\int_2^3 \dfrac{6x^2+5x-3}{x^3+2x^2-3x}\,dx = \int_2^3 \left[\dfrac{1}{x} + \dfrac{3}{x+3} + \dfrac{2}{x-1}\right] dx = [\ln x + 3\ln(x+3) + 2\ln(x-1)]_2^3$$
$$= (\ln 3 + 3\ln 6 + 2\ln 2) - (\ln 2 + 3\ln 5)$$
$$= 3\ln 6 + (\ln 3 + \ln 2) - 3\ln 5 = 4\ln 6 - 3\ln 5.$$

19. $\dfrac{5x^2 + 3x - 2}{x^3 + 2x^2} = \dfrac{5x^2 + 3x - 2}{x^2\,(x+2)} = \dfrac{A}{x} + \dfrac{B}{x^2} + \dfrac{C}{x+2}$. Multiply by $x^2\,(x+2)$ to get

$5x^2 + 3x - 2 = Ax\,(x+2) + B\,(x+2) + Cx^2$. Set $x = -2$ to get $C = 3$, and take

$x = 0$ to get $B = -1$. Equating the coefficients of x^2 gives $5 = A + C \Longrightarrow A = 2$. So

$$\int \frac{5x^2 + 3x - 2}{x^3 + 2x^2}\,dx = \int \left(\frac{2}{x} - \frac{1}{x^2} + \frac{3}{x+2}\right)\,dx = 2\ln|x| + \frac{1}{x} + 3\ln|x+2| + C.$$

21. Complete the square: $x^2 + x + 1 = \left(x + \frac{1}{2}\right)^2 + \frac{3}{4}$ and let $u = x + \frac{1}{2}$. Then

$$\int_0^1 \frac{x}{x^2 + x + 1}\,dx = \int_{1/2}^{3/2} \frac{u - 1/2}{u^2 + 3/4}\,du = \int_{1/2}^{3/2} \frac{u}{u^2 + 3/4}\,du - \frac{1}{2}\int_{1/2}^{3/2} \frac{1}{u^2 + 3/4}\,du$$

$$= \left[\frac{1}{2}\ln\left(u^2 + \frac{3}{4}\right) - \frac{1}{2}\frac{1}{\sqrt{3}/2}\tan^{-1}\left(\frac{2}{\sqrt{3}}u\right)\right]_{1/2}^{3/2}$$

$$= \frac{1}{2}\ln 3 - \frac{1}{\sqrt{3}}\left(\frac{\pi}{3} - \frac{\pi}{6}\right) = \ln\sqrt{3} - \frac{\pi}{6\sqrt{3}}.$$

23. $\dfrac{3x^2 - 4x + 5}{(x-1)\,(x^2+1)} = \dfrac{A}{x-1} + \dfrac{Bx + C}{x^2 + 1} \Longrightarrow 3x^2 - 4x + 5 = A\,(x^2 + 1) + (Bx + C)\,(x - 1).$

Take $x = 1$ to get $4 = 2A$ or $A = 2$. Now

$(Bx + C)\,(x - 1) = 3x^2 - 4x + 5 - 2\,(x^2 + 1) = x^2 - 4x + 3$. Equating coefficients of x^2 and then

comparing the constant terms, we get $B = 1$ and $C = -3$. Hence,

$$\int \frac{3x^2 - 4x + 5}{(x-1)\,(x^2+1)}\,dx = \int \left[\frac{2}{x-1} + \frac{x-3}{x^2+1}\right]\,dx = 2\ln|x-1| + \int \frac{x\,dx}{x^2+1} - 3\int \frac{dx}{x^2+1}$$

$$= 2\ln|x-1| + \frac{1}{2}\ln\left(x^2+1\right) - 3\tan^{-1}x + C$$

$$= \ln(x-1)^2 + \ln\sqrt{x^2+1} - 3\tan^{-1}x + C.$$

25. $\dfrac{1}{x^3 - 1} = \dfrac{1}{(x-1)\,(x^2+x+1)} = \dfrac{A}{x-1} + \dfrac{Bx + C}{x^2+x+1} \Longrightarrow 1 = A\,(x^2+x+1) + (Bx + C)\,(x-1).$

Take $x = 1$ to get $A = \frac{1}{3}$. Equating coefficients of x^2 and then comparing the constant terms, we get

$0 = \frac{1}{3} + B$, $1 = \frac{1}{3} - C$, so $B = -\frac{1}{3}$, $C = -\frac{2}{3} \Longrightarrow$

$$\int \frac{dx}{x^3 - 1} = \int \frac{1/3}{x-1}\,dx + \int \frac{(-1/3)x - 2/3}{x^2 + x + 1}\,dx = \frac{1}{3}\ln|x-1| - \frac{1}{3}\int \frac{x+2}{x^2+x+1}\,dx$$

$$= \frac{1}{3}\ln|x-1| - \frac{1}{3}\int \frac{x + 1/2}{x^2 + x + 1}\,dx - \frac{1}{3}\int \frac{(3/2)\,dx}{(x + 1/2)^2 + 3/4}$$

$$= \frac{1}{3}\ln|x-1| - \frac{1}{6}\ln\left(x^2 + x + 1\right) - \frac{1}{2}\left(\frac{2}{\sqrt{3}}\right)\tan^{-1}\left(\frac{x + \frac{1}{2}}{\sqrt{3}/2}\right) + K$$

$$= \frac{1}{3}\ln|x-1| - \frac{1}{6}\ln\left(x^2 + x + 1\right) - \frac{1}{\sqrt{3}}\tan^{-1}\left[\frac{1}{\sqrt{3}}\,(2x + 1)\right] + K.$$

27. $\dfrac{3x^3 - x^2 + 6x - 4}{(x^2 + 1)(x^2 + 2)} = \dfrac{Ax + B}{x^2 + 1} + \dfrac{Cx + D}{x^2 + 2} \Longrightarrow$

$3x^3 - x^2 + 6x - 4 = (Ax + B)(x^2 + 2) + (Cx + D)(x^2 + 1)$. Equating the coefficients gives

$A + C = 3, B + D = -1, 2A + C = 6$, and $2B + D = -4 \Longrightarrow A = 3, C = 0, B = -3$, and $D = 2$.

Now

$$\int \frac{3x^3 - x^2 + 6x - 4}{(x^2 + 1)(x^2 + 2)}\,dx = 3\int \frac{x - 1}{x^2 + 1}\,dx + 2\int \frac{dx}{x^2 + 2}$$

$$= 3\int \frac{x}{x^2 + 1}\,dx - 3\int \frac{1}{x^2 + 1}\,dx + 2\int \frac{dx}{x^2 + \left(\sqrt{2}\right)^2}$$

$$= \tfrac{3}{2}\ln\left(x^2 + 1\right) - 3\tan^{-1}x + \sqrt{2}\tan^{-1}\left(\tfrac{1}{\sqrt{2}}x\right) + C.$$

29.

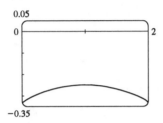

From the graph, we see that the integral will be negative, and we guess that the area is about the same as

that of a rectangle with width 2 and height 0.3, so we estimate the integral to be $-(2 \cdot 0.3) = -0.6$.

Now $\dfrac{1}{x^2 - 2x - 3} = \dfrac{1}{(x - 3)(x + 1)} = \dfrac{A}{x - 3} + \dfrac{B}{x + 1} \Longleftrightarrow 1 = (A + B)x + A - 3B$, so $A = -B$

and $A - 3B = 1 \Longleftrightarrow A = \tfrac{1}{4}$ and $B = -\tfrac{1}{4}$, so the integral becomes

$$\int_0^2 \frac{dx}{x^2 - 2x - 3} = \frac{1}{4}\int_0^2 \frac{dx}{x - 3} - \frac{1}{4}\int_0^2 \frac{dx}{x + 1} = \tfrac{1}{4}\left[\ln|x - 3| - \ln|x + 1|\right]_0^2$$

$$= \frac{1}{4}\left[\ln\left|\frac{x - 3}{x + 1}\right|\right]_0^2 = \tfrac{1}{4}\left(\ln\tfrac{1}{3} - \ln 3\right) = -\tfrac{1}{2}\ln 3 \approx -0.55.$$

31. $x^2 - 6x + 8 = (x - 3)^2 - 1$ is positive for $5 \le x \le 10$, so

$$\text{area} = \int_5^{10} \frac{dx}{(x - 3)^2 - 1} = \int_2^7 \frac{du}{u^2 - 1} \quad (\text{put } u = x - 3) \quad = \left[\tfrac{1}{2}\ln\left|\frac{u - 1}{u + 1}\right|\right]_2^7$$

$$= \tfrac{1}{2}\ln\tfrac{3}{4} - \tfrac{1}{2}\ln\tfrac{1}{3} = \tfrac{1}{2}\left(\ln 3 - 2\ln 2 + \ln 3\right) = \ln 3 - \ln 2 = \ln\tfrac{3}{2}.$$

33. (a) In Maple, we define $f(x)$, and then use `convert(f,parfrac,x);` to obtain

$$f(x) = \frac{24{,}110/4879}{5x+2} - \frac{668/323}{2x+1} - \frac{9438/80{,}155}{3x-7} + \frac{(22{,}098x + 48{,}935)/260{,}015}{x^2 + x + 5}$$

In Mathematica, we use the command `Apart`, and in Derive, we use `Expand`.

(b) $\displaystyle\int f(x)\,dx = \frac{24{,}110}{4879} \cdot \frac{1}{5}\ln|5x+2| - \frac{668}{323} \cdot \frac{1}{2}\ln|2x+1| - \frac{9438}{80{,}155} \cdot \frac{1}{3}\ln|3x-7|$

$$+\frac{1}{260{,}015}\int \frac{22{,}098\left(x+\frac{1}{2}\right) + 37{,}886}{\left(x+\frac{1}{2}\right)^2 + \frac{19}{4}}\,dx + C$$

$$= \frac{24{,}110}{4879} \cdot \frac{1}{5}\ln|5x+2| - \frac{668}{323} \cdot \frac{1}{2}\ln|2x+1| - \frac{9438}{80{,}155} \cdot \frac{1}{3}\ln|3x-7|$$

$$+\frac{1}{260{,}015}\left[22{,}098 \cdot \frac{1}{2}\ln\left(x^2 + x + 5\right) + 37{,}886 \cdot \sqrt{\frac{4}{19}}\tan^{-1}\left(\frac{1}{\sqrt{19/4}}\left(x+\frac{1}{2}\right)\right)\right] + C$$

$$= \frac{4822}{4879}\ln|5x+2| - \frac{334}{323}\ln|2x+1| - \frac{3146}{80{,}155}\ln|3x-7| + \frac{11{,}049}{260{,}015}\ln\left(x^2 + x + 5\right)$$

$$+\frac{75{,}772}{260{,}015\sqrt{19}}\tan^{-1}\left[\frac{1}{\sqrt{19}}(2x+1)\right] + C.$$

Using a CAS, we get

$$\frac{4822\ln(5x+2)}{4879} - \frac{334\ln(2x+1)}{323} - \frac{3146\ln(3x-7)}{80{,}155}$$

$$+\frac{11{,}049\ln\left(x^2 + x + 5\right)}{260{,}015} + \frac{3988\sqrt{19}}{260{,}115}\tan^{-1}\left[\frac{\sqrt{19}}{19}(2x+1)\right]$$

The main difference in this answer is that the absolute value signs and the constant of integration have been omitted. Also, the fractions have been reduced and the denominators rationalized.

35. There are only finitely many values of x where $Q(x) = 0$ (assuming that Q is not the zero polynomial). At all other values of x, $F(x)/Q(x) = G(x)/Q(x)$, so $F(x) = G(x)$. In other words, the values of F and G agree at all except perhaps finitely many values of x. By continuity of F and G, the polynomials F and G must agree at those values of x too.

More explicitly: if a is a value of x such that $Q(a) = 0$, then $Q(x) \neq 0$ for all x sufficiently close to a. Thus,

$$F(a) = \lim_{x \to a} F(x) \quad \text{(by continuity of } F\text{)} \qquad = \lim_{x \to a} G(x) \quad \text{[whenever } Q(x) \neq 0\text{]}$$

$$= G(a) \quad \text{(by continuity of } G\text{)}.$$

Appendix G   Polar Coordinates

Section G.1   Curves in Polar Coordinates

1. (a) By adding 2π to $\frac{\pi}{2}$, we obtain the point $\left(1, \frac{5\pi}{2}\right)$. The direction opposite $\frac{\pi}{2}$ is $\frac{3\pi}{2}$, so $\left(-1, \frac{3\pi}{2}\right)$ is a point that satisfies the $r < 0$ requirement.

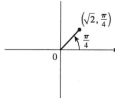

(b) $\left(-1, \frac{\pi}{5}\right)$

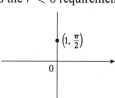

$\left(1, \frac{6\pi}{5}\right), \left(-1, \frac{11\pi}{5}\right)$

(c) $(3, 2)$

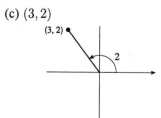

$(3, 2 + 2\pi), (-3, 2 + \pi)$

3. (a)

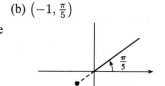

Using Equations 1 with $r = \sqrt{2}$ and $\theta = \frac{\pi}{4}$, we have
$x = r\cos\theta = \sqrt{2}\cos\frac{\pi}{4} = 1$ and
$y = r\sin\theta = \sqrt{2}\sin\frac{\pi}{4} = 1$. Thus, the point has Cartesian coordinates $(1, 1)$.

(b)

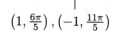

$x = 1.5\cos\frac{3\pi}{2} = 0$ and
$y = 1.5\sin\frac{3\pi}{2} = -1.5$
give us $\left(0, -\frac{3}{2}\right)$.

(c)

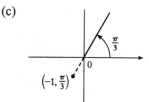

$x = -1\cos\frac{\pi}{3} = -\frac{1}{2}$ and
$y = -1\sin\frac{\pi}{3} = -\frac{\sqrt{3}}{2}$
give us $\left(-\frac{1}{2}, -\frac{\sqrt{3}}{2}\right)$.

5. (a) $(x, y) = (-1, 1)$, $r = \sqrt{(-1)^2 + 1^2} = \sqrt{2}$, $\tan\theta = y/x = -1$ and (x, y) is in quadrant II, so $\theta = \frac{3\pi}{4}$. The polar coordinates are $\left(\sqrt{2}, \frac{3\pi}{4}\right)$.

(b) $(x, y) = \left(2\sqrt{3}, -2\right)$, $r = \sqrt{12 + 4} = 4$, $\tan\theta = y/x = -\frac{1}{\sqrt{3}}$ and (x, y) is in quadrant IV, so $\theta = \frac{11\pi}{6}$. The polar coordinates are $\left(4, \frac{11\pi}{6}\right)$.

7. $r > 1$

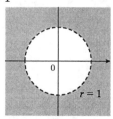

9. $0 \le r \le 2, \frac{\pi}{2} \le \theta \le \pi$

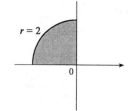

11. $3 < r < 4, -\frac{\pi}{2} \le \theta \le \pi$

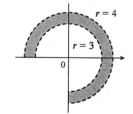

13. Since $y = r \sin \theta$, the equation $r \sin \theta = 2$ becomes $y = 2$.

15. $r = \dfrac{1}{1 - \cos \theta} \implies r - r \cos \theta = 1 \implies r = 1 + r \cos \theta \implies r^2 = (1 + r \cos \theta)^2 \implies$

$x^2 + y^2 = (1 + x)^2 = 1 + 2x + x^2 \iff y^2 = 1 + 2x$

17. $y = 5 \iff r \sin \theta = 5$

19. $x^2 + y^2 = 25 \iff r^2 = 25 \implies r = 5$

21. As in Example 4, $r = 5$
represents the circle with
center O and radius 5.

23. $r = 2 \sin \theta \implies r^2 = 2r \sin \theta \implies x^2 + y^2 = 2y \implies$
$x^2 + \left(y^2 - 2y + 1\right) = 1 \implies x^2 + (y - 1)^2 = 1$, a circle with
center $(0, 1)$ and radius 1.

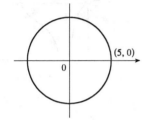

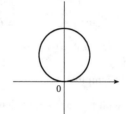

25. $r = \theta,\ \theta \geq 0$

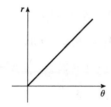

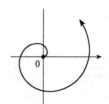

27. $r = 1 - 2 \cos \theta$

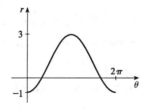

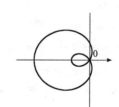

29. $r = 2 \cos 4\theta$

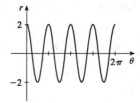

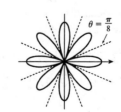

31. $r^2 = 4 \cos 2\theta$

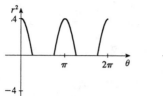 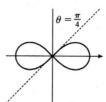

33. $x = (r) \cos \theta = (4 + 2 \sec \theta) \cos \theta = 4 \cos \theta + 2$. Now, $r \to \infty \Longrightarrow$

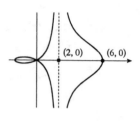

$(4 + 2 \sec \theta) \to \infty \Longrightarrow \theta \to \left(\frac{\pi}{2}\right)^-$ or $\theta \to \left(\frac{3\pi}{2}\right)^+$ (since we need only

consider $0 \le \theta < 2\pi$), so $\displaystyle\lim_{r \to \infty} x = \lim_{\theta \to \pi/2^-} (4 \cos \theta + 2) = 2$. Also,

$r \to -\infty \Longrightarrow (4 + 2 \sec \theta) \to -\infty \Longrightarrow \theta \to \left(\frac{\pi}{2}\right)^+$ or $\theta \to \left(\frac{3\pi}{2}\right)^-$, so

$\displaystyle\lim_{r \to -\infty} x = \lim_{\theta \to \pi/2^+} (4 \cos \theta + 2) = 2$. Therefore, $\displaystyle\lim_{r \to \pm\infty} x = 2 \Longrightarrow$

$x = 2$ is a vertical asymptote.

35. Using Equation 3 with $r = 3 \cos \theta$, we have

$$\frac{dy}{dx} = \frac{dy/d\theta}{dx/d\theta} = \frac{(dr/d\theta)(\sin \theta) + r \cos \theta}{(dr/d\theta)(\cos \theta) - r \sin \theta} = \frac{-3 \sin \theta \sin \theta + 3 \cos \theta \cos \theta}{-3 \sin \theta \cos \theta - 3 \cos \theta \sin \theta} = \frac{3 \left(\cos^2 \theta - \sin^2 \theta\right)}{-3 \left(2 \sin \theta \cos \theta\right)}$$

$$= -\frac{\cos 2\theta}{\sin 2\theta} = -\cot 2\theta = \frac{1}{\sqrt{3}} \text{ when } \theta = \frac{\pi}{3}.$$

Another Solution: $r = 3 \cos \theta \Longrightarrow x = r \cos \theta = 3 \cos^2 \theta, \; y = r \sin \theta = 3 \sin \theta \cos \theta \Longrightarrow$

$$\frac{dy}{dx} = \frac{dy/d\theta}{dx/d\theta} = \frac{-3 \sin^2 \theta + 3 \cos^2 \theta}{-6 \cos \theta \sin \theta} = \frac{\cos 2\theta}{-\sin 2\theta} = -\cot 2\theta = \frac{1}{\sqrt{3}} \text{ when } \theta = \frac{\pi}{3}.$$

37. $r = \theta \Longrightarrow x = r \cos \theta = \theta \cos \theta, \; y = r \sin \theta = \theta \sin \theta \Longrightarrow$

$$\frac{dy}{dx} = \frac{dy/d\theta}{dx/d\theta} = \frac{\sin \theta + \theta \cos \theta}{\cos \theta - \theta \sin \theta} = \frac{1}{-\pi/2} = -\frac{2}{\pi} \text{ when } \theta = \frac{\pi}{2}.$$

39. $r = \cos 2\theta \Longrightarrow x = r \cos \theta = \cos 2\theta \cos \theta, \; y = r \sin \theta = \cos 2\theta \sin \theta \Longrightarrow$

$$dy/d\theta = -2 \sin 2\theta \sin \theta + \cos 2\theta \cos \theta = -4 \sin^2 \theta \cos \theta + \left(\cos^3 \theta - \sin^2 \theta \cos \theta\right)$$

$$= \cos \theta \left(\cos^2 \theta - 5 \sin^2 \theta\right) = \cos \theta \left(1 - 6 \sin^2 \theta\right) = 0 \Longrightarrow$$

$\cos \theta = 0$ or $\sin \theta = \pm\frac{1}{\sqrt{6}} \Longrightarrow \theta = \frac{\pi}{2}, \frac{3\pi}{2}, \alpha, \pi - \alpha, \pi + \alpha$, or $2\pi - \alpha$ (where $\alpha = \sin^{-1} \frac{1}{\sqrt{6}}$).

So the tangent is horizontal at $\left(-1, \frac{\pi}{2}\right), \left(-1, \frac{3\pi}{2}\right), \left(\frac{2}{3}, \alpha\right), \left(\frac{2}{3}, \pi - \alpha\right), \left(\frac{2}{3}, \pi + \alpha\right)$, and $\left(\frac{2}{3}, 2\pi - \alpha\right)$.

$dx/d\theta = -2 \sin 2\theta \cos \theta - \cos 2\theta \sin \theta = -4 \sin \theta \cos^2 \theta - \left(2 \cos^2 \theta - 1\right) \sin \theta$

$$= \sin \theta \left(1 - 6 \cos^2 \theta\right) = 0 \Longrightarrow$$

$\sin \theta = 0$ or $\cos \theta = \pm\frac{1}{\sqrt{6}} \Longrightarrow \theta = 0, \pi, \frac{\pi}{2} - \alpha, \frac{\pi}{2} + \alpha, \frac{3\pi}{2} - \alpha$, or $\frac{3\pi}{2} + \alpha$ (where $\alpha = \cos^{-1} \frac{1}{\sqrt{6}}$).

So the tangent is vertical at $(1, 0), (1, \pi), \left(\frac{2}{3}, \frac{3\pi}{2} - \alpha\right), \left(\frac{2}{3}, \frac{3\pi}{2} + \alpha\right), \left(\frac{2}{3}, \frac{\pi}{2} - \alpha\right)$, and $\left(\frac{2}{3}, \frac{\pi}{2} + \alpha\right)$.

41. $r = 1 + \cos\theta \Longrightarrow x = r\cos\theta = \cos\theta\,(1 + \cos\theta),\ y = r\sin\theta = \sin\theta\,(1 + \cos\theta) \Longrightarrow$
$dy/d\theta = (1 + \cos\theta)\cos\theta - \sin^2\theta = 2\cos^2\theta + \cos\theta - 1 = (2\cos\theta - 1)(\cos\theta + 1) = 0 \Longrightarrow$
$\cos\theta = \tfrac{1}{2}$ or $-1 \Longrightarrow \theta = \tfrac{\pi}{3},\ \pi,$ or $\tfrac{5\pi}{3} \Longrightarrow$ horizontal tangents at $\left(\tfrac{3}{2}, \tfrac{\pi}{3}\right),\ (0, \pi),$ and $\left(\tfrac{3}{2}, \tfrac{5\pi}{3}\right)$.
$dx/d\theta = -(1 + \cos\theta)\sin\theta - \cos\theta\sin\theta = -\sin\theta\,(1 + 2\cos\theta) = 0 \Longrightarrow \sin\theta = 0$ or $\cos\theta = -\tfrac{1}{2} \Longrightarrow$
$\theta = 0,\ \pi,\ \tfrac{2\pi}{3},$ or $\tfrac{4\pi}{3} \Longrightarrow$ vertical tangents at $(2, 0),\ \left(\tfrac{1}{2}, \tfrac{2\pi}{3}\right),$ and $\left(\tfrac{1}{2}, \tfrac{4\pi}{3}\right)$. Note that the tangent is

horizontal, not vertical, when $\theta = \pi$, since $\displaystyle\lim_{\theta \to \pi} \frac{dy/d\theta}{dx/d\theta} = 0$ [by l'Hospital's Rule].

43. $r = a\sin\theta + b\cos\theta \Longrightarrow r^2 = ar\sin\theta + br\cos\theta \Longrightarrow x^2 + y^2 = ay + bx \Longrightarrow$
$\left(x - \tfrac{1}{2}b\right)^2 + \left(y - \tfrac{1}{2}a\right)^2 = \tfrac{1}{4}\left(a^2 + b^2\right),$ and this is a circle with center $\left(\tfrac{1}{2}b, \tfrac{1}{2}a\right)$ and radius $\tfrac{1}{2}\sqrt{a^2 + b^2}$.

45. (a) We see that the curve $r = 1 + c\sin\theta$ crosses itself at the origin, where $r = 0$ (in fact the inner
loop corresponds to negative r-values), so we solve the equation of the limaçon for $r = 0 \Longleftrightarrow$
$c\sin\theta = -1 \Longleftrightarrow \sin\theta = -1/c$. Now if $|c| < 1$, then this equation has no solution and hence there
is no inner loop. But if $c < -1$, then on the interval $(0, 2\pi)$ the equation has the two solutions
$\theta = \sin^{-1}(-1/c)$ and $\theta = \pi - \sin^{-1}(-1/c)$, and if $c > 1$, the solutions are $\theta = \pi + \sin^{-1}(1/c)$
and $\theta = 2\pi - \sin^{-1}(1/c)$. In each case, $r < 0$ for θ between the two solutions, indicating a loop.

(b) For $0 < c < 1$, the dimple (if it exists) is characterized by the fact that y has a local maximum
at $\theta = \tfrac{3\pi}{2}$. So we determine for what c-values $d^2y/d\theta^2$ is negative at $\theta = \tfrac{3\pi}{2}$, since by
the Second Derivative Test this indicates a maximum: $y = r\sin\theta = \sin\theta + c\sin^2\theta \Longrightarrow$
$dy/d\theta = \cos\theta + 2c\sin\theta\cos\theta = \cos\theta + c\sin 2\theta \Longrightarrow d^2y/d\theta^2 = -\sin\theta + 2c\cos 2\theta$. At $\theta = \tfrac{3\pi}{2}$,
this is equal to $-(-1) + 2c\,(-1) = 1 - 2c$, which is negative only for $c > \tfrac{1}{2}$. A similar argument
shows that for $-1 < c < 0$, y only has a local minimum at $\theta = \tfrac{\pi}{2}$ (indicating a dimple) for $c < -\tfrac{1}{2}$.

Note for Exercises 47 and 49: Maple is able to plot polar curves using the `polarplot`
command, or using the `coords=polar` option in a regular `plot` command. In Mathematica, use
`PolarPlot`. In Derive, change to `Polar` under `Options State`. If your graphing device cannot
plot polar equations, you must convert to parametric equations. For example, in Exercise 47,
$x = r\cos\theta = [1 + 2\sin(\theta/2)]\cos\theta,\ y = r\sin\theta = [1 + 2\sin(\theta/2)]\sin\theta$.

47. $r = 1 + 2\sin(\theta/2)$. The correct parameter   **49.** $r = \sin(9\theta/4)$. The correct parameter
interval is $[0, 4\pi]$.   interval is $[0, 8\pi]$.

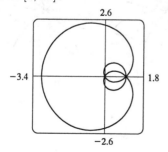

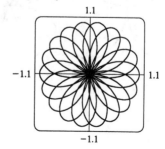

51.

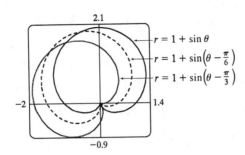

It appears that the graph of $r = 1 + \sin\left(\theta - \frac{\pi}{6}\right)$ is the same shape as the graph of $r = 1 + \sin\theta$, but rotated counterclockwise about the origin by $\frac{\pi}{6}$. Similarly, the graph of $r = 1 + \sin\left(\theta - \frac{\pi}{3}\right)$ is rotated by $\frac{\pi}{3}$. In general, the graph of $r = f(\theta - \alpha)$ is the same shape as that of $r = f(\theta)$, but rotated counterclockwise through α about the origin. That is, for any point (r_0, θ_0) on the curve $r = f(\theta)$, the point $(r_0, \theta_0 + \alpha)$ is on the curve the curve $r = f(\theta - \alpha)$, since $r_0 = f(\theta_0) = f((\theta_0 + \alpha) - \alpha)$.

53. (a) $r = \sin n\theta$. From the graphs, it seems that when n is even, the number of loops in the curve (called a rose) is $2n$, and when n is odd, the number of loops is simply n.

This is because in the case of n odd, every point on the graph is traversed twice, due to the fact that

$$r(\theta + \pi) = \sin[n(\theta + \pi)] = \sin n\theta \cos n\pi + \cos n\theta \sin n\pi = \begin{cases} \sin n\theta & \text{if } n \text{ is even} \\ -\sin n\theta & \text{if } n \text{ is odd} \end{cases}$$

$n = 2$

$n = 3$

$n = 4$

$n = 5$

(b) The graph of $r = |\sin n\theta|$ has $2n$ loops whether n is odd or even, since $r(\theta + \pi) = r(\theta)$.

$n = 2$

$n = 3$

$n = 4$

$n = 5$

55. $r = \dfrac{1 - a\cos\theta}{1 + a\cos\theta}$. We start with $a = 0$, since in this case the curve is simply the circle $r = 1$.

As a increases, the graph moves to the left, and its right side becomes flattened. As a increases through about 0.4, the right side seems to grow a dimple, which upon closer investigation (with narrower θ-ranges) seems to appear at $a \approx 0.42$ (the actual value is $\sqrt{2} - 1$). As $a \to 1$, this dimple becomes more pronounced, and the curve begins to stretch out horizontally, until at $a = 1$ the denominator vanishes at $\theta = \pi$, and the dimple becomes an actual cusp. For $a > 1$ we must choose our parameter interval carefully, since $r \to \infty$ as $1 + a\cos\theta \to 0 \Longleftrightarrow \theta \to \pm\cos^{-1}(-1/a)$. As a increases from 1, the curve splits into two parts. The left part has a loop, which grows larger as a increases, and the right part grows broader vertically, and its left tip develops a dimple when $a \approx 2.42$ (actually, $\sqrt{2} + 1$). As a increases, the dimple grows more and more pronounced.

If $a < 0$, we get the same graph as we do for the corresponding positive a-value, but with a rotation through π about the pole, as happened when c was replaced with $-c$ in Exercise 54.

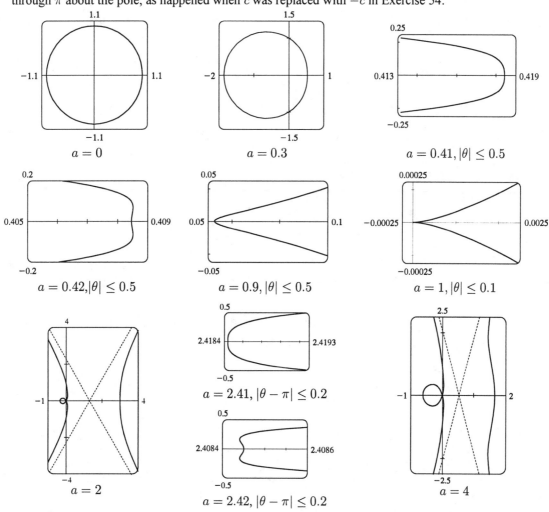

57.

$$\tan \psi = \tan(\phi - \theta) = \frac{\tan\phi - \tan\theta}{1 + \tan\phi\tan\theta} = \frac{\dfrac{dy}{dx} - \tan\theta}{1 + \dfrac{dy}{dx}\tan\theta} = \frac{\dfrac{dy/d\theta}{dx/d\theta} - \tan\theta}{1 + \dfrac{dy/d\theta}{dx/d\theta}\tan\theta}$$

$$= \frac{\dfrac{dy}{d\theta} - \dfrac{dx}{d\theta}\tan\theta}{\dfrac{dx}{d\theta} + \dfrac{dy}{d\theta}\tan\theta} = \frac{\left(\dfrac{dr}{d\theta}\sin\theta + r\cos\theta\right) - \tan\theta\left(\dfrac{dr}{d\theta}\cos\theta - r\sin\theta\right)}{\left(\dfrac{dr}{d\theta}\cos\theta - r\sin\theta\right) + \tan\theta\left(\dfrac{dr}{d\theta}\sin\theta + r\cos\theta\right)}$$

$$= \frac{r\cos\theta + r\cdot\dfrac{\sin^2\theta}{\cos\theta}}{\dfrac{dr}{d\theta}\cos\theta + \dfrac{dr}{d\theta}\cdot\dfrac{\sin^2\theta}{\cos\theta}} = \frac{r\cos^2\theta + r\sin^2\theta}{\dfrac{dr}{d\theta}\cos^2\theta + \dfrac{dr}{d\theta}\sin^2\theta} = \frac{r}{dr/d\theta}$$

Section G.2   Areas and Lengths in Polar Coordinates

1. $A = \int_0^\pi \frac{1}{2}r^2 d\theta = \int_0^\pi \frac{1}{2}\theta^2 d\theta = \left[\frac{1}{6}\theta^3\right]_0^\pi = \frac{1}{6}\pi^3$

3. $A = \int_0^{\pi/6} \frac{1}{2}(2\cos\theta)^2 d\theta = \int_0^{\pi/6} 2\cos^2\theta d\theta = \int_0^{\pi/6}(1+\cos 2\theta) d\theta$ [since $\cos^2\theta = \frac{1}{2}(1+\cos 2\theta)$]
$= \left[\theta + \frac{1}{2}\sin 2\theta\right]_0^{\pi/6} = \frac{\pi}{6} + \frac{\sqrt{3}}{4}$

5. $A = 4\int_0^{\pi/4} \frac{1}{2}r^2 d\theta = 2\int_0^{\pi/4}(4\cos 2\theta) d\theta = 8\int_0^{\pi/4}\cos 2\theta d\theta = 4\left[\sin 2\theta\right]_0^{\pi/4} = 4$

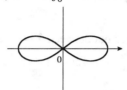

7. $A = 2\int_{-\pi/2}^{\pi/2} \frac{1}{2}(4-\sin\theta)^2 d\theta = \int_{-\pi/2}^{\pi/2}\left(16 - 8\sin\theta + \sin^2\theta\right) d\theta$
$= \int_{-\pi/2}^{\pi/2}\left(16 + \sin^2\theta\right) d\theta$ [by Theorem 5.5.6b] $= 2\int_0^{\pi/2}\left(16 + \sin^2\theta\right) d\theta$ [by Theorem 5.5.6a]
$= 2\int_0^{\pi/2}\left[16 + \frac{1}{2}(1-\cos 2\theta)\right] d\theta = 2\left[\frac{33}{2}\theta - \frac{1}{4}\sin 2\theta\right]_0^{\pi/2} = \frac{33\pi}{2}$

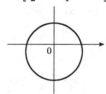

9. By symmetry, the total area is twice the area enclosed above the polar axis, so
$A = 2\int_0^\pi \frac{1}{2}r^2 d\theta = \int_0^\pi [2 + \cos 6\theta]^2 d\theta = \int_0^\pi\left(4 + 4\cos 6\theta + \cos^2 6\theta\right) d\theta$
$= \left[4\theta + 4\left(\frac{1}{6}\sin 6\theta\right) + \left(\frac{1}{24}\sin 12\theta + \frac{1}{2}\theta\right)\right]_0^\pi = 4\pi + \frac{\pi}{2} = \frac{9\pi}{2}.$

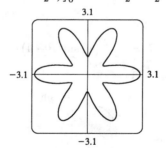

11. $A = \int_0^{\pi/5} \frac{1}{2}\sin^2 5\theta d\theta = \frac{1}{4}\int_0^{\pi/5}(1-\cos 10\theta) d\theta = \frac{1}{4}\left[\theta - \frac{1}{10}\sin 10\theta\right]_0^{\pi/5} = \frac{\pi}{20}$

13.

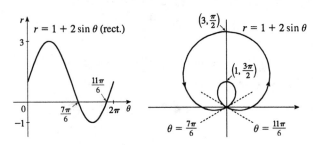

This is a limaçon, with inner loop traced out between $\theta = \frac{7\pi}{6}$ and $\frac{11\pi}{6}$ [found by solving $r = 0$].

$A = 2 \int_{7\pi/6}^{3\pi/2} \frac{1}{2} \left(1 + 2\sin\theta\right)^2 d\theta = \int_{7\pi/6}^{3\pi/2} \left(1 + 4\sin\theta + 4\sin^2\theta\right) d\theta = \left[\theta - 4\cos\theta + 2\theta - \sin 2\theta\right]_{7\pi/6}^{3\pi/2}$

$= \left(\frac{9\pi}{2}\right) - \left(\frac{7\pi}{2} + 2\sqrt{3} - \frac{\sqrt{3}}{2}\right) = \pi - \frac{3\sqrt{3}}{2}$

15. $1 - \cos\theta = \frac{3}{2} \implies \cos\theta = -\frac{1}{2} \implies \theta = \frac{2\pi}{3}$ or $\frac{4\pi}{3} \implies$

$A = 2 \int_{2\pi/3}^{\pi} \frac{1}{2} \left[\left(1 - \cos\theta\right)^2 - \left(\frac{3}{2}\right)^2 \right] d\theta = \int_{2\pi/3}^{\pi} \left(-\frac{5}{4} - 2\cos\theta + \cos^2\theta\right) d\theta$

$= \left[-\frac{5}{4}\theta - 2\sin\theta\right]_{2\pi/3}^{\pi} + \frac{1}{2} \int_{2\pi/3}^{\pi} \left(1 + \cos 2\theta\right) d\theta = -\frac{5}{12}\pi + \sqrt{3} + \frac{1}{2} \left[\theta + \frac{1}{2}\sin 2\theta\right]_{2\pi/3}^{\pi}$

$= -\frac{5}{12}\pi + \sqrt{3} + \frac{1}{6}\pi + \frac{\sqrt{3}}{8} = \frac{9\sqrt{3}}{8} - \frac{1}{4}\pi$

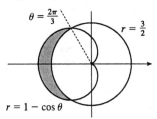

17. $4\sin\theta = 2 \iff \sin\theta = \frac{1}{2} \implies \theta = \frac{\pi}{6}$ or $\frac{5\pi}{6} \implies$

$A = 2 \int_{\pi/6}^{\pi/2} \frac{1}{2} \left[\left(4\sin\theta\right)^2 - 2^2\right] d\theta = \int_{\pi/6}^{\pi/2} \left(16\sin^2\theta - 4\right) d\theta = \int_{\pi/6}^{\pi/2} \left[8\left(1 - \cos 2\theta\right) - 4\right] d\theta$

$= \left[4\theta - 4\sin 2\theta\right]_{\pi/6}^{\pi/2} = \frac{4}{3}\pi + 2\sqrt{3}$

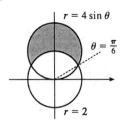

19. $A = 2 \int_0^{\pi/4} \frac{1}{2} \sin^2 \theta d\theta = \int_0^{\pi/4} \frac{1}{2} (1 - \cos 2\theta) \, d\theta = \left[\frac{1}{2}\theta - \frac{1}{4} \sin 2\theta \right]_0^{\pi/4} = \frac{1}{8}\pi - \frac{1}{4}$

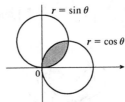

21. $\sin 2\theta = \cos 2\theta \Longrightarrow \tan 2\theta = 1 \Longrightarrow 2\theta = \frac{\pi}{4} \Longrightarrow \theta = \frac{\pi}{8} \Longrightarrow$

$A = 16 \int_0^{\pi/8} \frac{1}{2} \sin^2 2\theta d\theta = 4 \int_0^{\pi/8} (1 - \cos 4\theta) \, d\theta = 4 \left[\theta - \frac{1}{4} \sin 4\theta \right]_0^{\pi/8} = \frac{1}{2}\pi - 1$

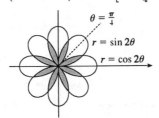

23. $A = 2 \left[\int_0^{2\pi/3} \frac{1}{2} \left(\frac{1}{2} + \cos \theta \right)^2 d\theta - \int_{2\pi/3}^{\pi} \frac{1}{2} \left(\frac{1}{2} + \cos \theta \right)^2 d\theta \right]$

$= \int_0^{2\pi/3} \left(\frac{1}{4} + \cos \theta + \cos^2 \theta \right) d\theta - \int_{2\pi/3}^{\pi} \left(\frac{1}{4} + \cos \theta + \cos^2 \theta \right) d\theta$

$= \left[\frac{\theta}{4} + \sin \theta + \frac{\theta}{2} + \frac{\sin 2\theta}{4} \right]_0^{2\pi/3} - \left[\frac{\theta}{4} + \sin \theta + \frac{\theta}{2} + \frac{\sin 2\theta}{4} \right]_{2\pi/3}^{\pi}$

$= \left(\frac{\pi}{2} + \frac{3\sqrt{3}}{8} \right) - \left(\frac{3\pi}{4} \right) + \left(\frac{\pi}{2} + \frac{3\sqrt{3}}{8} \right) = \frac{1}{4} \left(\pi + 3\sqrt{3} \right)$

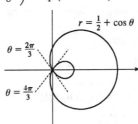

25. The curves intersect at the pole since $\left(0, \frac{\pi}{2} \right)$ satisfies $r = \cos \theta$ and $(0, 0)$ satisfies $r = 1 - \cos \theta$.

$\cos \theta = 1 - \cos \theta \Longrightarrow \cos \theta = \frac{1}{2} \Longrightarrow \theta = \frac{\pi}{3}$ or $\frac{5\pi}{3} \Longrightarrow$ the other intersection points are $\left(\frac{1}{2}, \frac{\pi}{3} \right)$ and $\left(\frac{1}{2}, \frac{5\pi}{3} \right)$.

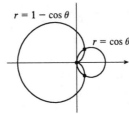

27. The pole is a point of intersection. $\sin\theta = \sin 2\theta = 2\sin\theta\cos\theta \iff \sin\theta\,(1 - 2\cos\theta) = 0 \iff$
$\sin\theta = 0$ or $\cos\theta = \frac{1}{2} \implies \theta = 0,\ \pi,\ \frac{\pi}{3},\ -\frac{\pi}{3} \implies \left(\frac{\sqrt{3}}{2}, \frac{\pi}{3}\right)$ and $\left(\frac{\sqrt{3}}{2}, \frac{2\pi}{3}\right)$ (by symmetry) are the other intersection points.

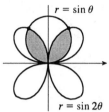

$r = \sin\theta$

$r = \sin 2\theta$

29.

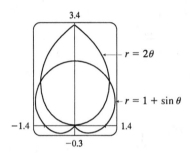

3.4

$r = 2\theta$

$r = 1 + \sin\theta$

−1.4   1.4

−0.3

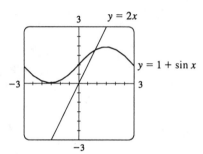

3     $y = 2x$

$y = 1 + \sin x$

−3   3

−3

From the first graph, we see that the pole is one point of intersection. By zooming in or using the cursor, we estimate the θ-values of the intersection points to be about 0.89 and $\pi - 0.89 \approx 2.25$. (The first of these values may be more easily estimated by plotting $y = 1 + \sin x$ and $y = 2x$ in rectangular coordinates; see the second graph.)

By symmetry, the total area contained is twice the area contained in the first quadrant, that is,

$$A \approx 2\int_0^{0.89} \tfrac{1}{2}(2\theta)^2\,d\theta + 2\int_{0.89}^{\pi/2} \tfrac{1}{2}(1 + \sin\theta)^2\,d\theta$$
$$= \left[\tfrac{4}{3}\theta^3\right]_0^{0.89} + \left[\theta - 2\cos\theta + \left(\tfrac{1}{2}\theta - \tfrac{1}{4}\sin 2\theta\right)\right]_{0.89}^{\pi/2} \approx 3.46$$

31. $L = \int_a^b \sqrt{r^2 + (dr/d\theta)^2}\,d\theta = \int_0^{2\pi}\sqrt{(2^\theta)^2 + [(\ln 2)\,2^\theta]^2}\,d\theta = \int_0^{2\pi} 2^\theta\sqrt{1 + \ln^2 2}\,d\theta$
$= \left[\sqrt{1 + \ln^2 2}\left(\dfrac{2^\theta}{\ln 2}\right)\right]_0^{2\pi} = \dfrac{\sqrt{1 + \ln^2 2}\,(2^{2\pi} - 1)}{\ln 2}$

33. $L = \int_0^{2\pi}\sqrt{(\theta^2)^2 + (2\theta)^2}\,d\theta = \int_0^{2\pi}\theta\sqrt{\theta^2 + 4}\,d\theta = \tfrac{1}{2}\cdot\tfrac{2}{3}\left[(\theta^2 + 4)^{3/2}\right]_0^{2\pi} = \tfrac{8}{3}\left[(\pi^2 + 1)^{3/2} - 1\right]$

35. From Figure 4 in Example 1,

$L = \int_{-\pi/4}^{\pi/4}\sqrt{r^2 + (r')^2}\,d\theta = 2\int_0^{\pi/4}\sqrt{\cos^2 2\theta + 4\sin^2 2\theta}\,d\theta \approx 2\,(1.211056) \approx 2.4221.$

Appendix H Complex Numbers

1. $(3 + 2i) + (7 - 3i) = (3 + 7) + (2 - 3)\,i = 10 - i$

3. $(3 - i)(4 + i) = 12 + 3i - 4i - (-1) = 13 - i$

5. $\overline{12 + 7i} = 12 - 7i$

7. $\dfrac{2 + 3i}{1 - 5i} = \dfrac{2 + 3i}{1 - 5i} \cdot \dfrac{1 + 5i}{1 + 5i} = \dfrac{2 + 10i + 3i + 15\,(-1)}{1 - 25\,(-1)} = \dfrac{-13 + 13i}{26} = -\tfrac{1}{2} + \tfrac{1}{2}i$

9. $\dfrac{1}{1 + i} = \dfrac{1}{1 + i} \cdot \dfrac{1 - i}{1 - i} = \dfrac{1 - i}{1 - (-1)} = \dfrac{1 - i}{2} = \tfrac{1}{2} - \tfrac{1}{2}i$

11. $i^3 = i^2 \cdot i = (-1)\,i = -i$

13. $\sqrt{-25} = \sqrt{25}\,i = 5i$

15. $\overline{3 + 4i} = 3 - 4i,\ |3 + 4i| = \sqrt{3^2 + 4^2} = \sqrt{25} = 5$

17. $\overline{-4i} = \overline{0 - 4i} = 0 + 4i = 4i,\ |-4i| = \sqrt{0^2 + (-4)^2} = 4$

19. $4x^2 + 9 = 0 \iff 4x^2 = -9 \iff x^2 = -\tfrac{9}{4} \iff x = \pm\sqrt{-\tfrac{9}{4}} = \pm\sqrt{\tfrac{9}{4}}\,i = \pm\tfrac{3}{2}i.$

21. By the quadratic formula, $x^2 - 8x + 17 = 0 \iff$
$$x = \frac{8 \pm \sqrt{(-8)^2 - 4\,(1)\,(17)}}{2\,(1)} = \frac{8 \pm \sqrt{-4}}{2} = \frac{8 \pm 2i}{2} = 4 \pm i.$$

23. By the quadratic formula, $z^2 + z + 2 = 0 \iff z = \dfrac{-1 \pm \sqrt{1^2 - 4\,(1)\,(2)}}{2\,(1)} = \dfrac{-1 \pm \sqrt{-7}}{2} = -\tfrac{1}{2} \pm \tfrac{\sqrt{7}}{2}i.$

25. For $z = -3 + 3i$, $r = \sqrt{(-3)^2 + 3^2} = 3\sqrt{2}$ and $\tan\theta = \tfrac{3}{-3} = -1 \implies \theta = \tfrac{3}{4}\pi$ (since z lies in the second quadrant). Therefore, $-3 + 3i = 3\sqrt{2}\left(\cos\tfrac{3\pi}{4} + i\sin\tfrac{3\pi}{4}\right).$

27. For $z = 3 + 4i$, $r = \sqrt{3^2 + 4^2} = 5$ and $\tan\theta = \tfrac{4}{3} \implies \theta = \tan^{-1}\tfrac{4}{3}$ (since z lies in the second quadrant). Therefore, $3 + 4i = 5\left[\cos\left(\tan^{-1}\tfrac{4}{3}\right) + i\sin\left(\tan^{-1}\tfrac{4}{3}\right)\right].$

29. For $z = \sqrt{3} + i$, $r = \sqrt{\left(\sqrt{3}\right)^2 + 1^2} = 2$ and $\tan\theta = \tfrac{1}{\sqrt{3}} \implies \theta = \tfrac{\pi}{6} \implies$
$z = 2\left(\cos\tfrac{\pi}{6} + i\sin\tfrac{\pi}{6}\right).$ For $w = 1 + \sqrt{3}i$, $r = 2$ and $\tan\theta = \sqrt{3} \implies \theta = \tfrac{\pi}{3} \implies$
$w = 2\left(\cos\tfrac{\pi}{3} + i\sin\tfrac{\pi}{3}\right).$ Therefore, $zw = 2 \cdot 2\left[\cos\left(\tfrac{\pi}{6} + \tfrac{\pi}{3}\right) + i\sin\left(\tfrac{\pi}{6} + \tfrac{\pi}{3}\right)\right] = 4\left(\cos\tfrac{\pi}{2} + i\sin\tfrac{\pi}{2}\right),$
$z/w = \tfrac{2}{2}\left[\cos\left(\tfrac{\pi}{6} - \tfrac{\pi}{3}\right) + i\sin\left(\tfrac{\pi}{6} - \tfrac{\pi}{3}\right)\right] = \cos\left(-\tfrac{\pi}{6}\right) + i\sin\left(-\tfrac{\pi}{6}\right),$ and $1 = 1 + 0i =$
$1\,(\cos 0 + i\sin 0) \implies 1/z = \tfrac{1}{2}\left[\cos\left(0 - \tfrac{\pi}{6}\right) + i\sin\left(0 - \tfrac{\pi}{6}\right)\right] = \tfrac{1}{2}\left[\cos\left(-\tfrac{\pi}{6}\right) + i\sin\left(-\tfrac{\pi}{6}\right)\right].$ For
$1/z$, we could also use the formula that precedes Example 5 to obtain $1/z = \tfrac{1}{8}\left(\cos\tfrac{\pi}{6} - i\sin\tfrac{\pi}{6}\right).$

Appendixes

31. For $z = 2\sqrt{3} - 2i$, $r = \sqrt{\left(2\sqrt{3}\right)^2 + (-2)^2} = 4$ and $\tan\theta = \frac{-2}{2\sqrt{3}} = -\frac{1}{\sqrt{3}} \implies$

$\theta = -\frac{\pi}{6} \implies z = 4\left[\cos\left(-\frac{\pi}{6}\right) + i\sin\left(-\frac{\pi}{6}\right)\right]$. For $w = -1 + i$, $r = \sqrt{2}$,

$\tan\theta = \frac{1}{-1} = -1 \implies \theta = \frac{3\pi}{4} \implies z = \sqrt{2}\left(\cos\frac{3\pi}{4} + i\sin\frac{3\pi}{4}\right)$. Therefore,

$zw = 4\sqrt{2}\left[\cos\left(-\frac{\pi}{6} + \frac{3\pi}{4}\right) + i\sin\left(-\frac{\pi}{6} + \frac{3\pi}{4}\right)\right] = 4\sqrt{2}\left(\cos\frac{7\pi}{12} + i\sin\frac{7\pi}{12}\right)$,

$z/w = \frac{4}{\sqrt{2}}\left[\cos\left(-\frac{\pi}{6} - \frac{3\pi}{4}\right) + i\sin\left(-\frac{\pi}{6} - \frac{3\pi}{4}\right)\right] = \frac{4}{\sqrt{2}}\left[\cos\left(-\frac{11\pi}{12}\right) + i\sin\left(-\frac{11\pi}{12}\right)\right]$

$= 2\sqrt{2}\left(\cos\frac{13\pi}{12} + i\sin\frac{13\pi}{12}\right)$, and

$1/z = \frac{1}{4}\left[\cos\left(-\frac{\pi}{6}\right) - i\sin\left(-\frac{\pi}{6}\right)\right] = \frac{1}{4}\left(\cos\frac{\pi}{6} + i\sin\frac{\pi}{6}\right)$.

33. For $z = 1 + i$, $r = \sqrt{2}$ and $\tan\theta = \frac{1}{1} = 1 \implies \theta = \frac{\pi}{4} \implies z = \sqrt{2}\left(\cos\frac{\pi}{4} + i\sin\frac{\pi}{4}\right)$. So by De Moivre's Theorem,

$(1 + i)^{20} = \left[\sqrt{2}\left(\cos\frac{\pi}{4} + i\sin\frac{\pi}{4}\right)\right]^{20} = \left(2^{1/2}\right)^{20}\left(\cos\frac{20\cdot\pi}{4} + i\sin\frac{20\cdot\pi}{4}\right)$

$= 2^{10}\left(\cos 5\pi + i\sin 5\pi\right) = 2^{10}\left[-1 + i\left(0\right)\right] = -2^{10} = -1024$.

35. For $z = 2\sqrt{3} + 2i$, $r = 4$ and $\tan\theta = \frac{2}{2\sqrt{3}} = \frac{1}{\sqrt{3}} \implies \theta = \frac{\pi}{6} \implies z = 4\left(\cos\frac{\pi}{6} + i\sin\frac{\pi}{6}\right)$. So by De Moivre's Theorem,

$\left(2\sqrt{3} + 2i\right)^5 = \left[4\left(\cos\frac{\pi}{6} + i\sin\frac{\pi}{6}\right)\right]^5 = 4^5\left(\cos\frac{5\pi}{6} + i\sin\frac{5\pi}{6}\right) = 1024\left[-\frac{\sqrt{3}}{2} + \frac{1}{2}i\right]$

$= -512\sqrt{3} + 512i$.

37. $1 = 1 + 0i = 1\left(\cos 0 + i\sin 0\right)$. Using Equation 3 with $r = 1$, $n = 8$, and $\theta = 0$, we have

$w_k = 1^{1/8}\left[\cos\left(\frac{0 + 2k\pi}{8}\right) + i\sin\left(\frac{0 + 2k\pi}{8}\right)\right] = \cos\frac{k\pi}{4} + i\sin\frac{k\pi}{4}$, where $k = 0, 1, 2, \ldots, 7$.

$w_0 = 1\left(\cos 0 + i\sin 0\right) = 1$ $w_4 = 1\left(\cos\pi + i\sin\pi\right) = -1$

$w_1 = 1\left(\cos\frac{\pi}{4} + i\sin\frac{\pi}{4}\right) = \frac{1}{\sqrt{2}} + \frac{1}{\sqrt{2}}i$ $w_5 = 1\left(\cos\frac{5\pi}{4} + i\sin\frac{5\pi}{4}\right) = -\frac{1}{\sqrt{2}} - \frac{1}{\sqrt{2}}i$

$w_2 = 1\left(\cos\frac{\pi}{2} + i\sin\frac{\pi}{2}\right) = i$ $w_6 = 1\left(\cos\frac{3\pi}{2} + i\sin\frac{3\pi}{2}\right) = -i$

$w_3 = 1\left(\cos\frac{3\pi}{4} + i\sin\frac{3\pi}{4}\right) = -\frac{1}{\sqrt{2}} + \frac{1}{\sqrt{2}}i$ $w_7 = 1\left(\cos\frac{7\pi}{4} + i\sin\frac{7\pi}{4}\right) = \frac{1}{\sqrt{2}} - \frac{1}{\sqrt{2}}i$

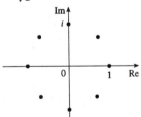

39. $0 = 0 + i = 1\left(\cos\frac{\pi}{2} + i\sin\frac{\pi}{2}\right)$. Using Equation 3 with $r = 1$, $n = 3$, and $\theta = \frac{\pi}{2}$, we have

$$w_k = 1^{1/3}\left[\cos\left(\frac{\frac{\pi}{2} + 2k\pi}{3}\right) + i\sin\left(\frac{\frac{\pi}{2} + 2k\pi}{3}\right)\right], \text{ where } k = 0, 1, 2.$$

$$w_0 = \left(\cos\frac{\pi}{6} + i\sin\frac{\pi}{6}\right) = \frac{\sqrt{3}}{2} + \frac{1}{2}i$$

$$w_1 = \left(\cos\frac{5\pi}{6} + i\sin\frac{5\pi}{6}\right) = -\frac{\sqrt{3}}{2} + \frac{1}{2}i$$

$$w_2 = \left(\cos\frac{9\pi}{6} + i\sin\frac{9\pi}{6}\right) = -i$$

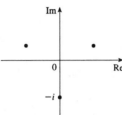

41. Using Euler's formula (6) with $y = \frac{\pi}{2}$, we have $e^{i\pi/2} = \cos\frac{\pi}{2} + i\sin\frac{\pi}{2} = 0 + 1i = i$.

43. Using Euler's formula with $y = \frac{3\pi}{4}$, we have $e^{i3\pi/4} = \cos\frac{3\pi}{4} + i\sin\frac{3\pi}{4} = -\frac{1}{\sqrt{2}} + \frac{1}{\sqrt{2}}i$.

45. Using Equation 7 with $x = 2$ and $y = \pi$, we have
$$e^{2+i\pi} = e^2 e^{i\pi} = e^2\left(\cos\pi + i\sin\pi\right) = e^2\left(-1 + 0\right) = -e^2.$$

47. Take $r = 1$ and $n = 3$ in De Moivre's Theorem to get $\left[1\left(\cos\theta + i\sin\theta\right)\right]^3 = 1^3\left(\cos 3\theta + i\sin 3\theta\right) \implies$
$\left(\cos\theta + i\sin\theta\right)^3 = \cos 3\theta + i\sin 3\theta \implies$
$\cos^3\theta + 3\left(\cos^2\theta\right)\left(i\sin\theta\right) + 3\left(\cos\theta\right)\left(i\sin\theta\right)^2 + \left(i\sin\theta\right)^3 = \cos 3\theta + i\sin 3\theta \implies$
$\left(\cos^3\theta - 3\sin^2\theta\cos\theta\right) + \left(3\sin\theta\cos^2\theta - \sin^3\theta\right)i = \cos 3\theta + i\sin 3\theta$. Equating real and imaginary
parts gives $\cos 3\theta = \cos^3\theta - 3\sin^2\theta\cos\theta$ and $\sin 3\theta = 3\sin\theta\cos^2\theta - \sin^3\theta$.

49. $F(x) = e^{rx} = e^{(a+bi)x} = e^{ax+bxi} = e^{ax}\left(\cos bx + i\sin bx\right) = e^{ax}\cos bx + i\left(e^{ax}\sin bx\right) \implies$
$$F'(x) = \left(e^{ax}\cos bx\right)' + i\left(e^{ax}\sin bx\right)'$$
$$= \left(ae^{ax}\cos bx - be^{ax}\sin bx\right) + i\left(ae^{ax}\sin bx + be^{ax}\cos bx\right)$$
$$= a\left[e^{ax}\left(\cos bx + i\sin bx\right)\right] + b\left[e^{ax}\left(-\sin bx + i\cos bx\right)\right]$$
$$= ae^{rx} + b\left[e^{ax}\left(i^2\sin bx + i\cos bx\right)\right]$$
$$= ae^{rx} + bi\left[e^{ax}\left(\cos bx + i\sin bx\right)\right]$$
$$= ae^{rx} + bie^{rx}$$
$$= (a + bi)e^{rx}$$
$$= re^{rx}$$